MODERN GENERAL TOPOLOGY

North-Holland Mathematical Library

VOLUME 33

NORTH-HOLLAND
AMSTERDAM · NEW YORK · OXFORD

MODERN GENERAL TOPOLOGY

JUN-ITI NAGATA

University of Amsterdam

Second revised edition

1985

NORTH-HOLLAND
AMSTERDAM · NEW YORK · OXFORD

First edition: 1968
Revised edition: 1974
Second revised edition: 1985

(The first edition and the revised edition were published as
Volume VII in the series Bibliotheca Mathematica)

ISBN: 0444 876553

Published by:
ELSEVIER SCIENCE PUBLISHERS B.V.
P.O. Box 1991
1000 BZ Amsterdam
The Netherlands

Sole distributors for the U.S.A. and Canada:
ELSEVIER SCIENCE PUBLISHING COMPANY, INC.
52, Vanderbilt Avenue
New York, NY 10017
U.S.A.

Library of Congress Cataloging in Publication Data

Nagata, Jun-iti, 1925–
    Modern general topology.

    (North-Holland mathematical library ; v. 33)
    Bibliography: p.
    Includes index.
    1. Topology. I. Title. II. Series.
  QA611.N25 1985   514′.322   85-4415
  ISBN 0-444-87655-3

PREFACE

This book is intended to be a text as well as an advanced reference book. To achieve these somewhat inconsistent objectives and to avoid boring the beginner before he reaches the interesting part of the theory, much abstract description as well as too much attention to the more abstract spaces has been avoided. For the same reason, ease of understanding was the primary consideration in choosing the basic topological tools. For example, the concept of filter rather than that of net was emphasized in the discussion of convergence; and as far as possible it was avoided to involve the product space $X \times X$ into the discussion of coverings or uniform structures of X. Thus the reader will find the methods used here are rather more popular than novel.

The first consideration in choosing the topics was to include as many aspects as possible from the vast amount of interesting results which have come about in the great development of modern general topology during the past twenty years. Nevertheless, an encyclopedic exposition is not the aim of this book, and the topics were selected from only the most significant results which may serve as the foundations for further development of the theory. The author regrets that he had to exclude many interesting topics. However, the bibliography is fairly extensive.

Thus the reader will find in this book not only the routine material of ordinary texts but also some advanced discussions in the theory of general topology. For example, advanced topics are treated in the discussions of paracompact spaces, metric spaces and extensions of mappings.

Since no knowledge of topology or of set theory is assumed and since the methods used here are popular, even the beginning student will feel no difficulty in reading the book while he can expect to obtain a bird's-eye view of modern general topology. However, in the hope that the book will suit different levels of readers, some theories were consciously divided into two parts – elementary discussions in an earlier chapter and advanced ones in a later chapter. Thus we can suggest to the reader who wants to study general topology only as a background for other branches of mathematics to skip, for example, Sections 3, 4 and 6 of Chapter IV, the last three sections of Chapter V, Sections 4 and 6–8 of Chapter VI, the last six sections of Chapter VII, and Chapter VIII; and to the undergraduate level reader to read only the first three chapters. (But we wish that he comes back to the book later!)

V

Finally the author wishes to express his heartfelt thanks to those who helped him in so many respects and especially to Professor Johannes de Groot who encouraged and helped him to devote this book to this fundamental area of mathematics, to Dr. and Mrs. George E. Strecker who corrected the manuscript in its English expression and to Dr. Jos v. d. Slot and Mr. Albert Verbeek who kindly read the manuscript. The author also wishes to acknowledge that most of this book was written during his stay at the Institute for Advanced Study, under support of the National Science Foundation and he wishes to express his gratitude to them.

JUN-ITI NAGATA

PREFACE TO THE REVISED EDITION

There have been remarkable developments in various aspects of general topology since the first edition of this book was published in 1968. Accordingly, a substantial amount of descriptions on important new results, new references as well as new exercise problems have been added to make the book up to date, while errors in the old edition have been corrected. However, it has not been intended to drastically revise the main discussions because the author believes that those theories discussed in the original edition still form the foundation for the whole system of general topology, and besides this is not an appropriate time for a major revision because no one can tell with confidence what the shape of general topology will be after this unusually active, fast moving, and somewhat chaotic era of this science. The author likes to express his thanks to Prof. J. E. Vaughan and other mathematicians who kindly pointed out errors in the old edition and advised him in various respects.

JUN-ITI NAGATA

PREFACE TO THE SECOND REVISED EDITION

The main part of this edition was written during 1975–1981 when the author worked as professor of pure mathematics at the University of Amsterdam. Then he gave various lectures and seminars and was inspired by meetings and discussions with mathematicians and students there. Thus the present edition is greatly affected by his experience in Amsterdam, and in fact many of the new contents were selected directly from his notebook prepared for lectures at the University of Amsterdam.

In contrast to the old (revised) edition of 1974, which inherited the style of the initial edition with relatively small revisions, this 2nd revised edition is the product of a major innovation. On the whole, the introductory part, Chapters I–III, came from the old edition with numerous but minor corrections as well as some additional exercise questions. On the other hand, Chapters IV–VII were wholly innovated both in the topics and descriptions, and the new Chapter VIII was added to fit the great developments of general topology that occurred since 1968 when the initial edition of this book appeared. The reader will find many new topics in Chapters IV–VIII, e.g., theory of Wallman–Shanin's compactification, realcompact space, various generalizations of paracompactness, generalized metric spaces, Dugundji type extension theory, linearly ordered topological space, theory of cardinal functions, dyadic space, etc. that were, according to the author's opinion, mostly special or isolated topics some twenty years ago but now settle down in the mainstream of general topology.

The author wishes to conclude this preface with his heartfelt thanks to many people who gave him various suggestions and inspirations for preparing this innovated edition of 'Modern General Topology'.

JUN-ITI NAGATA

CONTENTS

CHAPTER I

INTRODUCTION

The purpose of this chapter is to prepare the reader for the theory of general topology, i.e., it presents set theory and point set theory in the Euclidean plane. The latter is a special and concrete model of topology that will help the reader understand abstract concepts of general topology. This is the reason why the final section of this chapter is devoted to the elementary theory of topology in the Euclidean plane, although in Chapter II the reader will meet again the same terminologies and discussions in their generality.

1. Set

We assume that the concept of *set* is obvious. Let A be a set; then we mean by an *element* of A any particle which belongs to the set A and we denote by $a \in A$ the fact that a is an element of A and by $b \notin A$ the fact that b is not an element of A. We mean by an *empty* (or *vacant*) *set* a set which contains no element and we denote it by $\emptyset$. If A contains infinitely (finitely) many elements, then we call it an *infinite* (*finite*) *set*. Let A and B be two sets. If every element of A is also an element of B, then we call A a *subset* of B and denote the relation by $A \subset B$ and its negation by $A \not\subset B$. A itself and $\emptyset$ are also considered to be subsets of A. If $A \subset B$ and $B \subset A$, then A and B coincide. We denote the coincidence by $A = B$ and its negation by $A \neq B$. If A is a subset of B and they do not coincide, then we call A a *proper subset* of B and denote the relation by $A \subset B$.

Example I.1. The set A of all natural numbers, the set B of all real numbers, the set C of all real-valued continuous functions, are examples of (infinite) sets. Between the sets A and B cited above there are the relations, $A \subset B$, $A \subsetneqq B$, $B \not\subset A$.

Suppose that to every element a of A there corresponds one and only one element, $f(a)$, of B; then we call this correspondence f a *mapping* or a *map* or more precisely a *unique mapping* of (from) the set A into the set B and we call $f(a)$ the *image* of a in B. If an element b of B is the image of an element a of A by the mapping f, then we call a an *inverse image* of b. Of course an inverse image of b may not be uniquely determined by the element b, but the inverse images of b form a subset of A denoted by $f^{-1}(b)$. Let A' be a subset of A; then we denote by $f(A')$ the subset of B consisting of the images of the elements of A' under the mapping f, i.e.

$$f(A') = \{f(a) \mid a \in A'\},^1$$

and call it the image of A'. Similarly, let B' be a subset of B; then we define the *inverse image* $f^{-1}(B')$ of B' by

$$f^{-1}(B') = \{a \mid a \in A, f(a) \in B'\}.$$

If $f(A) = B$, then we call f a mapping of A *onto* B or a *surjection*. If f is a mapping of A into B and if the inverse image $f^{-1}(b)$ of each element b of $f(A)$ consists of one element, i.e. if for every pair of distinct elements a and a' of A, $f(a) \neq f(a')$ holds, then we call f a *one-to-one mapping* of A into B or an *injection*. If f is a one-to-one mapping from A onto B, then it is called a *bijection*. If f is a bijection from A onto itself such that $f(a) = a$ for all $a \in A$, then f is called an *identity mapping*. In this case the mapping which maps $b \in B$ to $f^{-1}(b)$ is also a one-to-one mapping of B onto A called the *inverse mapping of f*, and is denoted by f^{-1}. If there exists a one-to-one mapping of A onto B, then A and B are said to be *equivalent* and we denote this by $A \sim B$. It is easily seen that:

A) *The relation $\sim$ satisfies the following three conditions*:
    (i) $A \sim A$,
    (ii) *if $A \sim B$, then $B \sim A$*,
    (iii) *if $A \sim B$ and $B \sim C$, then $A \sim C$*.

Example I.2. A real-valued function of a real variable is a mapping of a subset of the set R of all real numbers into R. For example, a function

---

[1] In this book we often use such a notation to show how the set is constructed, e.g. $\{f(a) \mid a \in A'\}$ denotes the set consisting of all $f(a)$ such that $a \in A'$, and $\{a \mid a \in A, f(a) \in B'\}$ denotes the set of all a such that $a \in A$ and $f(a) \in B'$.

$f(n) = 2n$, $n = 0$, ± 1, $\pm 2, \ldots$, is a mapping of the set R' of all integers into R', or more precisely a one-to-one mapping of R' onto the set R'' of all even numbers, which shows $R' \sim R''$.

Still assuming that A and B are two sets, we now define their *sum* (or *union*) $A \cup B$, their *intersection* $A \cap B$ and their *difference* $A - B$ by

$$A \cup B = \{a \mid a \in A \text{ or } a \in B\},$$

$$A \cap B = \{a \mid a \in A \text{ and } a \in B\}$$

and

$$A - B = \{a \mid a \in A \text{ and } a \notin B\}.$$

If $A \cap B = \emptyset$, then we call A and B *disjoint*. Note that for $A - B$ to be defined, it is not necessary that B be a subset of A; as a matter of fact $A - B = A - (A \cap B)$. We often call $A - B$ the complement of B with respect to A or merely the *complement* of B if there is little likelihood of confusion. We also use notations like 2^A and B^A to denote the set of all subsets of A and the set of all mappings of A into B, respectively. We often consider a *collection* of sets such as

$$\mathcal{A} = \{A_\gamma \mid \gamma \in \Gamma\}.$$

This denotes that $\mathcal{A}$ is the collection of the sets A_γ indexed by the elements γ of an index set Γ. (Γ may be an infinite set.) Since a collection is essentially a set whose elements are sets, we may use such terminologies as finite collection, subcollection, etc. without establishing a specific definition for them. We may call an element of a collection a *member* of the collection. Now we define the *sum* $\cup \{A_\gamma \mid \gamma \in \Gamma\}$ and the *intersection* $\cap \{A_\gamma \mid \gamma \in \Gamma\}$ of all the sets belonging to $\mathcal{A}$ by

$$\cup \{A_\gamma \mid \gamma \in \Gamma\} = \{a \mid a \in A_\gamma \text{ for some element } \gamma \text{ of } \Gamma\}$$

and

$$\cap \{A_\gamma \mid \gamma \in \Gamma\} = \{a \mid a \in A_\gamma \text{ for every element } \gamma \text{ of } \Gamma\}.$$

We often replace the notations $\cup \{A_\gamma \mid \gamma \in \Gamma\}$ and $\cap \{A_\gamma \mid \gamma \in \Gamma\}$ with $\cup A_\gamma$ and $\cap A_\gamma$, respectively, in cases where confusion is not likely. If the index set Γ is the set of all natural numbers $\{1, 2, 3, \ldots\}$, then we prefer to denote the sum and intersection by $\cup_{i=1}^{\infty} A_i$ and $\cap_{i=1}^{\infty} A_i$, respectively. We use similar designations in the case that Γ is finite.

Example I.3. Consider a set M and a relation $a \sim b$ defined for some of the pairs (a, b) of elements of M. Then for every element a of M we define a subset $M_a = \{b \mid b \in M, a \sim b\}$ of M. If the relation $\sim$ satisfies the following conditions:

   (i) $a \sim a$,

   (ii) if $a \sim b$, then $b \sim a$,

   (iii) if $a \sim b$ and $b \sim c$, then $a \sim c$,

then for each pair (a, b), either $M_a \cap M_b = \emptyset$ or $M_a = M_b$ is true. For, if $M_a \cap M_b \neq \emptyset$, then take $c \in M_a \cap M_b$. For a given $b' \in M_b$ we obtain $a \sim c$, $b \sim c$ and $b \sim b'$ which imply $a \sim c$, $c \sim b$, and $b \sim b'$. Hence $a \sim b'$, i.e. $b' \in M_a$. Therefore $M_b \subset M_a$. In the same way we can prove that $M_a \subset M_b$ and thus $M_a = M_b$. Therefore the original set M is decomposed into the sum of the disjoint sets M_a. For example, let M denote the set of all integers and let $a \sim b$ for $a, b \in M$ denote that $a - b$ is a multiple of 2 (this is usually denoted by $a \equiv b \pmod 2$); then this relation $\sim$ satisfies (i), (ii), (iii), and M is decomposed into two sets, the set M_0 of the even integers and the set M_1 of the odd integers.

In the calculation of sets the following rules are fundamental:

B)
$$A \cup B = B \cup A, \qquad A \cap B = B \cap A,$$
$$A \cup (B \cup C) = (A \cup B) \cup C, \qquad A \cap (B \cap C) = (A \cap B) \cap C,$$
$$A \cap [\cup \{A_\gamma \mid \gamma \in \Gamma\}] = \cup \{A \cap A_\gamma \mid \gamma \in \Gamma\},$$
$$A \cup [\cap \{A_\gamma \mid \gamma \in \Gamma\}] = \cap \{A \cup A_\gamma \mid \gamma \in \Gamma\};$$

de Morgan's rule:

$$E - \cup \{A_\gamma \mid \gamma \in \Gamma\} = \cap \{E - A_\gamma \mid \gamma \in \Gamma\},$$
$$E - \cap \{A_\gamma \mid \gamma \in \Gamma\} = \cup \{E - A_\gamma \mid \gamma \in \Gamma\}.$$

Proof. Let us give only the proof of the last formula as an example of a set theoretical proof. First, it follows from the obvious relation

$$\cap \{A_\gamma \mid \gamma \in \Gamma\} \subset A_\gamma \quad \text{for every } \gamma \in \Gamma$$

that

$$E - \cap \{A_\gamma \mid \gamma \in \Gamma\} \supset E - A_\gamma \quad \text{for every } \gamma \in \Gamma.$$

Hence

$$E - \cap \{A_\gamma \mid \gamma \in \Gamma\} \supset \cup \{E - A_\gamma \mid \gamma \in \Gamma\}. \tag{1}$$

To show the inverse inclusion relation, we suppose that p is a given point of $E - \cap \{A_\gamma \mid \gamma \in \Gamma\}$. Then $p \in E$ and $p \notin \cap \{A_\gamma \mid \gamma \in \Gamma\}$, which means $p \in E$ and $p \notin A_\gamma$ for some $\gamma \in \Gamma$. Therefore $p \in E - A_\gamma$ for some $\gamma \in \Gamma$ and hence $p \in \cup \{E - A_\gamma \mid \gamma \in \Gamma\}$. Thus the relation

$$E - \cap \{A_\gamma \mid \gamma \in \Gamma\} \subset \cup \{E - A_\gamma \mid \gamma \in \Gamma\}$$

is established because p is a given element of $E - \cap \{A_\gamma \mid \gamma \in \Gamma\}$. This combined with (1) implies

$$E - \cap \{A_\gamma \mid \gamma \in \Gamma\} = \cup \{E - A_\gamma \mid \gamma \in \Gamma\}.$$

Let A and B be two sets. Then we mean by the *cartesian product* of A and B the set of all pairs (a, b) where a is an element of A and b is an element of B, in this order. We denote the cartesian product of A and B by $A \times B$, i.e.

$$A \times B = \{(a, b) \mid a \in A, b \in B\}.$$

Furthermore, we may consider a (not necessarily finite) collection $\{A_\gamma \mid \gamma \in \Gamma\}$ of disjoint sets and define the cartesian product $\prod \{A_\gamma \mid \gamma \in \Gamma\}$ of its members. We choose one element a_γ from each A_γ and make a class $\{a_\gamma \mid \gamma \in \Gamma\}$ composed of these elements. Then the cartesian product $\prod \{A_\gamma \mid \gamma \in \Gamma\}$ is the set consisting of all those classes, i.e.

$$\prod \{A_\gamma \mid \gamma \in \Gamma\} = \{\{a_\gamma \mid \gamma \in \Gamma\} \mid a_\gamma \in A_\gamma \text{ for every } \gamma \in \Gamma\}.$$

Each a_γ is called the γ-*coordinate* of the element $a = \{a_\gamma \mid \gamma \in \Gamma\}$ of the cartesian product. By convention a_γ, p_γ, etc. denote the γ-coordinates of a, p, etc. respectively.

Example I.4. Let E^1 be the set of all points on the real line $(-\infty, +\infty)$. Then the cartesian product $E^1 \times E^1$ is the set E^2 of all points on the Euclidean plane. The cartesian product $E^1 \times E^2$ is the set E^3 of all points in the 3-dimensional Euclidean space.

2. Cardinal numbers

We call a set a *countable set* if it is equivalent with the set $\{1, 2, 3, \ldots\}$

of the natural numbers. Thus a countable set A is a set in which all elements are numbered, i.e. A can be expressed as $A = \{a_1, a_2, a_3, \ldots\} = \{a_i \mid i = 1, 2, 3, \ldots\}$. As is easily seen, the set of the integers, the set of the rational numbers, etc. are countable sets. The sum and the cartesian product of two countable sets are also countable. Every infinite subset of a countable set is also countable. We often call a set *at most countable* if it is either countable or finite. On the contrary, the set of the real numbers and the set of the irrational numbers are not countable; roughly speaking, they contain many more elements than a countable set does. Thus with reason we assign a cardinal number to each class of equivalent sets, i.e., we say that two sets have the same *cardinal number* (or *power*) if and only if they are equivalent. We denote by $\mathfrak{a}$ or $\aleph_0$ the cardinal number of countable sets and by $\mathfrak{c}$ the cardinal number of the set of all real numbers.

Example I.5. Let Q be the set of all rational numbers; then every element of Q can be expressed as n/m where n is an integer and m is a positive integer. We can number all rational numbers in turn:

$$0/1; \quad 1/1, -1/1; \quad 2/1, -2/1, 1/2, -1/2; \quad 3/1, -3/1, 1/3, -1/3;$$

$$4/1, -4/1, 3/2, -3/2, \ldots,$$

where the kth group consists of all rational numbers of the form n/m with $|n| + m = k$. Thus Q is countable.

Let $A = \{a_1, a_2, a_3, \ldots\}$ and $B = \{b_1, b_2, b_3, \ldots\}$ be countable sets. Then their cartesian product $A \times B$ consists of the elements

$$(a_1, b_1) \longrightarrow (a_1, b_2) \quad (a_1, b_3) \longrightarrow (a_1, b_4) \quad (a_1, b_5) \longrightarrow \cdots$$

$$(a_2, b_1) \quad (a_2, b_2) \quad (a_2, b_3) \quad (a_2, b_4) \quad \cdots$$

$$(a_3, b_1) \quad (a_3, b_2) \quad (a_3, b_3) \quad (a_3, b_4) \quad \cdots$$

$$(a_4, b_1) \quad (a_4, b_2) \quad (a_4, b_3) \quad (a_4, b_4) \quad \cdots$$

$$(a_5, b_1) \quad \cdots$$

Thus we can number them as indicated in the above, starting at (a_1, b_1) and hence $A \times B$ is also countable. This fact can be also interpreted as that the sum of countably many countable sets is countable.

We can uniquely express every real number x, where $0 < x \leq 1$, as an infinite decimal of the form $0.n_1 n_2 n_3 \ldots$; for example, 0.5 can be expressed as $0.4999 \ldots$. (Note that such an expression is uniquely determined if we require $n_i \neq 0$ for infinitely many i.) To show that the set M of those real numbers is uncountable, we assume to the contrary that we have numbered all elements of M as

$$a_1 = 0.n_{11} n_{12} n_{13} \ldots$$

$$a_2 = 0.n_{21} n_{22} n_{23} \ldots$$

$$a_3 = 0.n_{31} n_{32} n_{33} \ldots$$

For each i, we choose an integer n'_{ii} with $1 \leq n'_{ii} \leq 9$, $n'_{ii} \neq n_{ii}$. Then $0.n'_{11} n'_{22} n'_{33} \ldots$ is an element of M but $a_i \neq 0.n'_{11} n'_{22} n'_{33} \ldots$ for every i because $n'_{ii} \neq n_{ii}$. This contradiction shows that M is uncountable. Thus the set R of all real numbers is also uncountable. The set Q' of the irrational numbers is also uncountable. Because if it were countable, then the sum R of Q' and the set of the rational numbers would be countable, contradicting the conclusion above.

Generally, we denote by $|A|$ the cardinal number of a set A; for example, $|N| = $ for the set N of the natural numbers. We define order and calculation between cardinal numbers as follows. Let A and B be two disjoint sets; then

$$|A| \leq |B| \quad \text{if } A \text{ is equivalent with a subset of } B,$$

$$|A| < |B| \quad \text{if } |A| \leq |B| \text{ and } |A| \neq |B|,$$

$$|A| + |B| = |A \cup B|,$$

$$|A| \times |B| = |A \times B|,$$

$$|A|^{|B|} = |A^B|.$$

We can also define infinite sum and product of cardinal numbers. Let A_γ, $\gamma \in \Gamma$ be given disjoint sets; then

$$\sum_{\gamma \in \Gamma} |A_\gamma| = |\cup \{A_\gamma \mid \gamma \in \Gamma\}|, \qquad \prod_{\gamma \in \Gamma} |A_\gamma| = |\prod \{A_\gamma \mid \gamma \in \Gamma\}|.$$

Let μ denote an infinite cardinal number. Then we can assert the following inequality which implies that there are infinitely many infinite cardinal numbers.

A) $2^\mu > \mu$.

Proof. Let $|M| = \mu$. Then by the definition, $2^\mu = |2^M|$. It is clear that M is equivalent with the subset of 2^M consisting of all one element sets. Therefore $2^\mu \geqslant \mu$. Thus it suffices to show that 2^M is equivalent with no subset of M. To do so, we assume the contrary. Then to each element P of 2^M, there corresponds an element $f(P)$ of M by a one-to-one mapping f. We should note that P is an element of 2^M and a subset of M as well. Putting

$$P_0 = \{f(P) \mid P \in 2^M, f(P) \notin P\},$$

we obtain a subset P_0 of M which is an element of 2^M at the same time. But now we have reached a contradiction, because P_0 corresponds to no element of M by the one-to-one mapping f. To explain it more precisely, we assume that $f(P_0)$ exists. Then it must be true that either $f(P_0) \in P_0$ or $f(P_0) \notin P_0$, but each of them obviously contradicts the definition of P_0.

Example I.6. Let $\{\mu_\gamma \mid \gamma \in \Gamma\}$ be a set of cardinal numbers. Then $\mu = \sum_{\gamma \in \Gamma} \mu_\gamma$ is obviously a cardinal number satisfying $\mu \geqslant \mu_\gamma$ for every $\gamma \in \Gamma$. Therefore by A) 2^μ is a cardinal number which is greater than every μ_γ. If we considered the set M of all cardinal numbers, then we should obtain a cardinal number ν greater than every cardinal number in M, i.e. $\nu \notin M$. This contradicts that M contains all cardinal numbers. This example shows us that we should not regard M as a set in the usual sense. In the same way we should not consider such a set as the set of all sets. It is the purpose of *axiomatic set theory* to re-establish set theory on the foundation of a more exact concept of 'set' to avoid antinomies like these. We call the following statement *continuum hypothesis* (*CH*): There is no cardinal number μ such that $\mathfrak{a} < \mu < \mathfrak{c}$. A significant merit of axiomatic set theory is that it showed the independence of CH from a group of more basic axioms of set theory.[1]

Now we can assert the following important proposition which will be needed later to prove that two given cardinal numbers are comparable with each other, just as are finite cardinal numbers.

---

[1] See, for example, A. Fraenkel [1], K. Gödel [1].

B) (Bernstein's theorem). *Let μ and ν be two cardinal numbers. If $\mu \leqslant \nu$ and $\nu \leqslant \mu$, then $\mu = \nu$.*

Proof. To prove this theorem, it suffices to show that if a set A is equivalent with a subset D of a set B by a one-to-one mapping f and if B is equivalent with a subset C of A by a one-to-one mapping g, then A and B are equivalent.

Now $f(A) = D$, $g(B) = C$. We put

$$A - C = A_1,$$

$$f(A_1) = B_1, \qquad g(B_1) = A_2, \qquad f(A_2) = B_2, \qquad g(B_2) = A_3, \ldots.$$

Furthermore, we put

$$A - \bigcup_{n=1}^{\infty} A_n = P, \qquad B - \bigcup_{n=1}^{\infty} B_n = Q.$$

Then $A_n \sim B_n$, $n = 1, 2, \ldots$, because $f(A_n) = B_n$.

Since f and g are one-to-one, we can easily see that $A_1, A_2, \ldots$ as well as $B_1, B_2, \ldots$ are disjoint sets. Therefore

$$\bigcup_{n=1}^{\infty} A_n \sim \bigcup_{n=1}^{\infty} B_n.$$

On the other hand, it is easily seen that P and Q are equivalent by the mapping g, because it follows from the definitions of $A_1, A_2, \ldots, B_1$, $B_2, \ldots$ that $q \in Q$ implies $g(q) = p \in P$ and conversely $p \in P$ implies $g^{-1}(p) = q \in Q$. Thus we obtain $A \sim B$, because each of A and B is decomposed into two disjoint sets as

$$A = P \cup \left(\bigcup_{n=1}^{\infty} A_n \right), \qquad B = Q \cup \left(\bigcup_{n=1}^{\infty} B_n \right),$$

where $P \sim Q$ and $\bigcup_{n=1}^{\infty} A_n \sim \bigcup_{n=1}^{\infty} B_n$.

Example I.7. Let R be the set of all real numbers and M its subset consisting of all numbers in the closed interval $[-1, 1] = \{x \mid -1 \leqslant x \leqslant 1\}$. Since M is a subset of R, $|M| \leqslant |R|$. On the other hand, it is obvious that $f(x) = \tan(\pi/2)x$ gives a one-to-one mapping of the subset $M' = (-1, 1) = \{x \mid -1 < x < 1\}$ of M onto R, and hence $|R| \leqslant |M|$. Therefore it follows from B) that $|M| = |R| = \mathfrak{c}$.

3. Ordinal numbers

Let D be a set with order $<$ defined between some pairs (a, b) of its elements such that

(i) $a < b$ and $b < c$ implies $a < c$,

(ii) for every two elements a, b of D, there is an element c satisfying $a < c$ and $b < c$.

Then D is called a *directed set*.

Let A be a set with order $\leq$ defined for some pairs (a, b) of its elements such that

(i) $a \leq a$,

(ii) $a \leq b$ and $b \leq a$ imply $a = b$,

(iii) $a \leq b$ and $b \leq c$ imply $a \leq c$.

Then A is called a *partially ordered set*. In a partially ordered set A we denote the fact that $a \leq b$ and $a \neq b$ by $a < b$. Suppose A is a partially ordered set with B a subset of A. If an element a of A satisfies $a \geq b$ $(a \leq b)$ for every element b of B, then we call a an *upper (lower) bound* of B. If a is an upper (lower) bound of B such that $a \leq a'$ $(a \geq a')$ for every upper (lower) bound a' of B, then a is called the *supremum* (*infimum*) or the *lowest* (or *least*) *upper bound* (*greatest lower bound*) of B and denoted by $a = \sup B$ ($\inf B$). The supremum (infimum) of a subset B is uniquely determined if it exists. The supremum of a two element set $\{a, b\}$ is denoted by $a \vee b$ and its infimum is denoted by $a \wedge b$. If every two element set $\{a, b\}$ has the supremum $a \vee b$ and the infimum $a \wedge b$, then the partially ordered set A is called a *lattice*. Every lattice is a directed set if we regard its order $\leq$ as the order $<$ in the definition of directed set.

Example I.8. The set of all natural numbers is a directed set with its usual order, and it is also a lattice. Let D be the set of all rectangles in the Euclidean plane with their centers at the origin. For two rectangles a, $b \in D$, we define that $a < b$ if and only if the rectangle b is contained in the rectangle a. Then D turns out to be a directed set.

Let M be a set. For two elements M', M'' of 2^M, we define that $M' \leq M''$ if and only if $M' \subset M''$. Then we easily see that 2^M with this order is a lattice. The sum $M' \cup M''$ and intersection $M' \cap M''$ of M' and M'' are the supremum and infimum in the lattice, respectively. We shall obtain another lattice if we define that $M' \leq M''$ means $M' \supset M''$. In a similar way we can regard every subcollection of 2^M as a partially ordered set.

Let C denote the set of all real-valued continuous functions defined over the whole real line $(-\infty, \infty)$; then C is a lattice with its usual order: $f \leq g$ if and only if $f(x) \leq g(x)$ for every x.

Let A be a partially ordered set; then note that some two element subset may have no order between its members. Generally, we call two elements a, b of a partially ordered set *comparable* if either $a \leq b$ or $a \geq b$. If every two elements of A are comparable, then we call A a *totally-ordered set* or a *linearly ordered set*. If A is a totally ordered set such that every non-empty subset B of A has infimum, which belongs to B, then A is called a *well-ordered set*. In this case the infimum is often called the *smallest (or first) element of B*.

Let A and B be two partially ordered sets and f a mapping of A into B such that $a \leq a'$ for a, $a' \in A$ implies that $f(a) \leq f(a')$ in B. Then f is called a *homomorphic mapping* (or *homomorphism*) of A into B. If f is a one-to-one homomorphic mapping of A onto B and the inverse mapping f^{-1} is also homomorphic, then f is called an *isomorphic mapping* (or an *isomorphism*) of A onto B, and A and B are said to be *isomorphic*. In the case that A and B are lattices, a homomorphic mapping f of A into B does not necessarily preserve the operations $a \vee b$ or $a \wedge b$. If $f(a \vee b) = f(a) \vee f(b)$ and $f(a \wedge b) = f(a) \wedge f(b)$, then f is called a *lattice homomorphism*. Two totally ordered sets A and B are said to have the same *order type* if they are isomorphic. The order type of a well-ordered set is called the *ordinal number* of the set.

We define order between two ordinal numbers α of A and β of B as follows: $\alpha < \beta$ if and only if A is isomorphic with a subset $B(b)$ of B such that $B(b) = \{b' \mid b' \in B, b' < b\}$ for an element b of B. We call such a subset $B(b)$ of B a *cut* of B. Now we can assert that every two ordinal numbers are comparable.[1]

Example I.9. The set N of the positive integers $1, 2, 3, \ldots$ and the set N' of $1, 2, 3, \ldots; 1^*, 2^*, 3^*, \ldots (n^* > m)$ are well-ordered sets. On the other hand, for example, the set A of the real numbers x with $0 \leq x \leq 1$ is not well-ordered though it is totally ordered, because the subset $(0, 1] = \{x \mid 0 < x \leq 1\}$ of A has no first element. The ordinal numbers of N and N' are denoted by ω and 2ω, respectively. By the above definition, $\omega < 2\omega$ holds. In the following discussion we shall see that for every

---

[1] Readers who are not familiar with set-theoretical arguments are recommended to skip over A), B), C), D) and the proofs of Theorem I.1 and Corollaries.

ordinal number α, there is a 'next' ordinal number α', i.e. $\alpha < \alpha'$ and there is no ordinal number between α and α'. We denote the ordinal number α' by $\alpha + 1$.

A) *Let A be a well-ordered set and f an isomorphic mapping of A onto a subset B of A. Then $a \leq f(a)$ for every element a of A.*

Proof. Assume the contrary, i.e. $a > f(a)$ for some $a \in A$. Then putting $A' = \{a \mid a > f(a)\}$ we obtain a non-empty subset A' of A. Since A is well-ordered, there is the first element a_0 of A'. Since $a_0 \in A'$,

$$a_0 > f(a_0) \tag{1}$$

which implies

$$f(a_0) \notin A'$$

because of the definition of a_0. Therefore by the definition of A' we obtain

$$f(a_0) \leq f(f(a_0)). \tag{2}$$

On the other hand, since f is an isomorphic mapping, from (1) it follows that

$$f(a_0) > f(f(a_0)).$$

This contradicts (2), and hence the validity of the proposition is established.

B) *A well-ordered set A is isomorphic with no cut of A.*

Proof. Assume, contrary to the assertion, that f is an isomorphic mapping of A onto a cut $A(b) = \{a \mid a \in A, a < b\}$ of A. Then by A) we obtain $b \leq f(b)$, but, on the other hand, $f(b)$ must belong to $A(b)$, i.e. $f(b) < b$. Thus we have reached a contradiction, which means that the proposition is true.

Proposition B)[1] implies that $\alpha = \beta$ and $\alpha < \beta$ does not happen simultaneously, but we can prove more, as seen in the following.

---

[1] Generally we mean by B) Proposition B) of the same section and chapter, by 1.B) Proposition B) of Section 1 of the same chapter, and by I.1.B) Proposition B) of Section 1 of Chapter I.

C) *Let α and β be two ordinal numbers. Then no two of the following three cases occur at the same time:*

$$\alpha < \beta, \qquad \alpha = \beta, \qquad \alpha > \beta.$$

Proof. Let A and B be well-ordered sets with the ordinal numbers α and β, respectively. If we assume that $\alpha < \beta$ and $\beta < \alpha$, then A is isomorphic with a cut of B, and B is isomorphic with a cut of A. As is easily seen, this means that A is isomorphic with a cut of A itself, but this is impossible by B).

D) *Let τ be an ordinal number and $W(\tau)$ the set of the ordinal numbers less than τ and not less than 0. For two elements α and β of $W(\tau)$, we denote by $\alpha \leq \beta$ the relation that $\alpha < \beta$ or $\alpha = \beta$. Then with respect to this order, $\leq$, $W(\tau)$ turns out to be a well-ordered set with the ordinal number τ.*

Proof. If $\tau = 0$, then $W(\tau) = \emptyset$, and hence the assertion is true. Therefore we shall prove that $W(\tau)$ for $\tau > 0$ is a well-ordered set. In view of C) it is clear that $W(\tau)$ is a totally ordered set. Let τ be the ordinal number of a well-ordered set T. Then every element α of $W(\tau)$ is the ordinal number of a cut $T(a) = \{t \mid t \in T, t < a\}$ of T, because $\alpha < \tau$. Put $f(\alpha) = a$, then we easily see that f is an isomorphic mapping of $W(\tau)$ onto the well-ordered set T with the ordinal number τ. Therefore $W(\tau)$ is also a well-ordered set with the ordinal number τ.

Theorem I.1. *Let α and β be two ordinal numbers. Then one and only one of the following three cases occurs:*

$$\alpha < \beta, \qquad \alpha = \beta, \qquad \alpha > \beta.$$

Proof. In C) we have shown that at most one of the three cases occurs, so we shall now prove that at least one of them occurs.

Using the same notation as in D), we put

$$W(\alpha) \cap W(\beta) = C.$$

Since C is a subset of a well-ordered set $W(\alpha)$, it is also well-ordered. Let us denote by γ the ordinal number of C; then we claim that $\gamma \leq \alpha$ and $\gamma \leq \beta$.

In the case that $C = W(\alpha)$, we obtain $\gamma = \alpha$ since by D) the well-ordered set $W(\alpha)$ has the ordinal number α. In the case that $C \subsetneqq W(\alpha)$, we denote by δ the first element (ordinal number) of $W(\alpha) - C$. Suppose η is a given ordinal number with $\eta < \delta$; then $\eta \in W(\alpha)$ since $\eta < \delta < \alpha$. Furthermore, from the definition of δ it follows that $\eta \in C$. Conversely, suppose that $\eta \in C$; then $\eta < \delta$ or $\eta = \delta$ or $\eta > \delta$ because $W(\alpha)$ is well-ordered by D). But $\eta \geq \delta$ combined with $\eta \in C \subset W(\beta)$ implies $\beta > \eta \geq \delta$, and hence $\delta \in W(\beta)$, and accordingly $\delta \in C$ which contradicts the definition of δ. Therefore we obtain $\eta < \delta$. Hence we conclude that $C = W(\delta)$ and hence C is a cut of $W(\alpha)$. Therefore $\gamma < \alpha$. Thus we have proved $\gamma \leq \alpha$ for the ordinal number γ of C. The assertion $\gamma \leq \beta$ can be proved quite analogously.

Now, we can consider the following four cases:

$\gamma < \alpha$ and $\gamma < \beta$,

$\gamma = \alpha$ and $\gamma = \beta$ which implies $\alpha = \beta$,

$\gamma = \alpha$ and $\gamma < \beta$ which implies $\alpha < \beta$,

$\gamma < \alpha$ and $\gamma = \beta$ which implies $\alpha > \beta$.

In the first case we obtain $\gamma \in C = W(\alpha) \cap W(\beta)$, which implies that $W(\gamma)$ is a cut of C. But, since the ordinal numbers of C and $W(\gamma)$ are both γ, this implies $\gamma < \gamma$ contradicting C). Thus the first case cannot happen, and hence the proof of the theorem is complete.

Corollary 1. *Every set A of ordinal numbers is a well-ordered set.*

Proof. Since by the theorem, A is a totally-ordered set, it suffices to show only that every non-empty subset, A', of A has the first element. Take an element α' of A'. If α' is not the first element of A', then we denote by $W(\alpha')$ the set of the ordinal numbers less than α'. Put $W(\alpha') \cap A' = A''$; then, since by D) $W(\alpha')$ is well-ordered, A'' is also well-ordered. Therefore A'' contains the first number α''. Let β be a given element of A'. Then by the theorem, either $\alpha'' \leq \beta$ or $\beta < \alpha''$ holds. But $\beta < \alpha''$ implies $\beta < \alpha'' < \alpha'$, and hence $\beta \in W(\alpha')$, i.e. $\beta \in A''$. This contradicts the fact that α'' is the first element of A''. Therefore $\alpha'' \leq \beta$ which means that α'' is is the first element of A'.

Corollary 2. *Let α and β be the ordinal numbers of a well-ordered set A and its subset B respectively. Then $\alpha \geq \beta$.*

Proof. If we assume the contrary, $\alpha < \beta$, then A is isomorphic with a cut $B(a) = \{b \mid b \in B,\ b < a\}$ of B. We denote by f this isomorphic mapping of A onto $B(a)$. Since $B(a) \subset B \subset A$, f is an isomorphic mapping of A onto its subset. Therefore it follows from A) that $a \leq f(a)$. On the other hand, since $f(a) \in B(a)$, we obtain $a > f(a)$ which is a contradiction. Hence we conclude that $\alpha \geq \beta$.

Corollary 3. *Every set of cardinal numbers is well-ordered.*

Proof. Let μ be a cardinal number and ω_μ the least ordinal number with cardinality μ. Then it is obvious that $\mu < \mu'$ if and only if $\omega_\mu < \omega_{\mu'}$. Thus this corollary follows from Corollary 1.

Example I.10. Let α_1, α_2, α_3, ... be ordinal numbers of well-ordered sets A_1, A_2, A_3, ... respectively. Then we order $A = \bigcup_{i=1}^{\infty} A_i$ as follows:

If $a, b \in A_i$, then the order between a, b in A is the same as that in A_i.

If $a \in A_i$, $b \in A_j$, $i < j$, then $a < b$.

We can easily see that A is a well-ordered set. Let us denote by α its ordinal number; then by Corollary 2 of Theorem I.1, $\alpha \geq \alpha_i$, $i = 1, 2, \ldots$. Moreover, if $|A_i| \leq \aleph_0$, $i = 1, 2, \ldots$, then $|A| \leq \aleph_0$ by Example I.5. Thus there is an ordinal number with cardinal number $\leq \aleph_0$, which is greater than each of countably many given ordinal numbers with cardinals $\leq \aleph_0$. Another remark: In view of Corollary 3 we often denote each infinite cardinal number by $\aleph_\alpha$ for an ordinal number α, implying that $\aleph_\alpha < \aleph_\beta$ if and only if $\alpha < \beta$. Thus $\aleph_1$ denotes the least uncountable cardinal number.

We shall often use the following method of proof in succeeding chapters.

Theorem I.2 (Transfinite Induction). *Suppose $P(\alpha)$ denotes a proposition with respect to the ordinal number α. If the following statements* (i) *and* (ii) *are verified, then $P(\alpha)$ is true for every ordinal number α.*

(i) *$P(0)$ is true,*

(ii) *supposing α is a given ordinal number with $\alpha > 0$, if $P(\beta)$ is true for every $\beta < \alpha$, then $P(\alpha)$ is also true.*

Proof. Contrary to the assertion, we assume that there is an ordinal number α for which $P(\alpha)$ is not true. Then by Corollary 1 of Theorem I.1, there exists the first number α_0 among those α. It follows from (i) that $\alpha_0 > 0$. By the definition of α_0, $P(\beta)$ is true for every $\beta < \alpha_0$, and hence by (ii) $P(\alpha_0)$ must be true, but this is a contradiction. Thus $P(\alpha)$ is true for every α.

4. Zermelo's theorem and Zorn's lemma[1]

In this section we shall first prove that every set can be well-ordered with a proper order. For this purpose we need the following axiom which we have implicitly used in the preceding sections, too.

Zermelo's axiom of choice. Let $\mathscr{A}$ be a collection of non-empty sets. Then we can define a mapping φ of $\mathscr{A}$ into $\bigcup \{A \mid A \in \mathscr{A}\}$ such that $\varphi(A) \in A$ for every $A \in \mathscr{A}$.[2]

A) *Let A be a non-empty set. By Zermelo's axiom of choice we choose an element $\varphi(B)$ from every non-empty subset B of A and put $B' = B - \varphi(B)$, $\emptyset' = \emptyset$. Then $B \supset C \supset B'$ for subsets B and C of A implies that $B = C$ or $C = B'$.*

Proof. Since B' is either smaller than B by one element $\varphi(B)$, or is equal to B, this assertion is obvious.

B) *Let us consider the following three conditions on a collection $\mathscr{P}$ of subsets of A:*
  (i) *$A \in \mathscr{P}$,*
  (ii) *$B_\gamma \in \mathscr{P}$, $\gamma \in \Gamma$ [3] implies $\bigcap \{B_\gamma \mid \gamma \in \Gamma\} \in \mathscr{P}$,*
  (iii) *$B \in \mathscr{P}$ implies $B' \in \mathscr{P}$.*
Denote by $\mathscr{P}_0$ the intersection of all collections satisfying (i)–(iii). Then $\mathscr{P}_0$ also satisfies the same conditions; namely, $\mathscr{P}_0$ is the smallest collection which satisfies (i)–(iii).

Proof. The easy proof of this assertion is left to the reader.

C) *We introduce order $\leqslant$ into the collection $\mathscr{P}_0$ of B), defining $B \leqslant C$ for B, $C \in \mathscr{P}_0$ if and only if $B \supset C$. Then $\mathscr{P}_0$ is a totally-ordered set.*

Proof. We denote by $\mathscr{P}_0'$ the set of the elements of $\mathscr{P}_0$ which are comparable with every element of $\mathscr{P}_0$. To begin with, we shall verify the following assertion:
  (1) For every $C \in \mathscr{P}_0'$ and $B \in \mathscr{P}_0$, either $C \subset B$ or $C' \supset B$, where $C' = C - \varphi(C)$ as defined in A).

---

[1] Readers who are not familiar with set-theoretical arguments are advised to skip A), B), C), D) and the proofs of Theorem I.3 and Theorem I.4.

[2] In other words, we can simultaneously choose a representative $\varphi(A)$ from every set $A \in \mathscr{A}$.

[3] For brevity, we use the convention $B_\gamma \in \mathscr{P}$, $\gamma \in \Gamma$, to mean that $B_\gamma \in \mathscr{P}$ for *every* $\gamma \in \Gamma$.

To prove this, we denote by $\mathscr{P}_0''$ the set of all elements $B \in \mathscr{P}_0$ which satisfy the condition (1). To prove $\mathscr{P}_0'' = \mathscr{P}_0$, it suffices to show that $\mathscr{P}_0''$ satisfies (i)–(iii) of B), because $\mathscr{P}_0$ is the smallest collection satisfying those three conditions. First (i) is clearly satisfied by $\mathscr{P}_0''$ because $C \subset A$ for every $C \in \mathscr{P}_0'$. To see (ii) we suppose $B_\gamma \in \mathscr{P}_0''$, $\gamma \in \Gamma$ and $C \in \mathscr{P}_0'$. Then each B_γ satisfies either $C \subset B_\gamma$ or $C' \supset B_\gamma$. If $C \subset B_\gamma$ for every γ, then

$$C \subset \cap \{B_\gamma \mid \gamma \in \Gamma\}.$$

If $C' \supset B_\gamma$ for some γ, then

$$C' \supset \cap \{B_\gamma \mid \gamma \in \Gamma\}.$$

Therefore we can conclude

$$\cap \{B_\gamma \mid \gamma \in \Gamma\} \in \mathscr{P}_0''.$$

Finally, to see (iii), suppose $B \in \mathscr{P}_0''$ and $C \in \mathscr{P}_0'$. If $C' \supset B$ or $B = C$, then $C' \supset B'$ because $B \supset B'$. If $B \not\supseteq C$, then we recall that C is comparable with every element of $\mathscr{P}_0$ since $C \in \mathscr{P}_0'$. Hence either $C \supset B'$ or $B' \supset C$ holds. In the first case it follows from A) that $C = B'$. Thus in either case we obtain $C \subset B'$. Therefore B' satisfies the condition of B in (1), i.e. $B' \in \mathscr{P}_0''$. Thus $\mathscr{P}_0''$ satisfies (i)–(iii) and hence $\mathscr{P}_0'' = \mathscr{P}$, which assures us of the validity of (1) for every $B \in \mathscr{P}_0$ and $C \in \mathscr{P}_0'$.

Now, let us turn to the proof that $\mathscr{P}_0$ is totally-ordered. For that purpose, it suffices to show that $\mathscr{P}_0' = \mathscr{P}_0$. The procedure of the proof is somewhat similar to the previous proof that $\mathscr{P}_0'' = \mathscr{P}_0$. Namely we shall show that $\mathscr{P}_0'$ satisfies (i)–(iii). For, if we could do so, then $\mathscr{P}_0' = \mathscr{P}_0$ would follow from the fact that $\mathscr{P}_0' \subset \mathscr{P}_0$ while $\mathscr{P}_0$ is the smallest collection satisfying (i)–(iii).

(i) is clearly satisfied by $\mathscr{P}_0'$. It is also easy to show (ii) by an argument analogous with that for $\mathscr{P}_0''$ in the preceding part of the proof. Thus the only problem is to show the validity of (iii) for $\mathscr{P}_0'$. Let $C \in \mathscr{P}_0'$, $B \in \mathscr{P}_0$. Then by (1) either $C \subset B$ or $C' \supset B$ holds. In the former case we obtain $C' \subset B$ because $C' \subset C$. Therefore in either case C' and B are comparable. Thus we obtain $C' \in \mathscr{P}_0'$, proving (iii). Therefore $\mathscr{P}_0' = \mathscr{P}_0$ which completes the proof of C).

D) $\mathscr{P}_0$ *is a well-ordered set.*

Proof. Let $\mathscr{P}_1$ be a given non-empty subcollection of $\mathscr{P}_0$. If $A \in \mathscr{P}_1$, then A is the first element of $\mathscr{P}_1$ (see C)). If $A \notin \mathscr{P}_1$, then we put

$$\mathscr{P}_2 = \{B \mid B \in \mathscr{P}_0 - \mathscr{P}_1, \ B \text{ is a lower bound of } \mathscr{P}_1\}.$$

Since $A \in \mathscr{P}_2$, $\mathscr{P}_2$ is not empty. Put

$$B_0 = \cap\{B \mid B \in \mathscr{P}_2\};$$

then B_0 is obviously a lower bound for $\mathscr{P}_1$. B_0 belongs to either $\mathscr{P}_1$ or $\mathscr{P}_2$. If $B_0 \in \mathscr{P}_1$, then by the definition of B_0, B_0 is the first element of $\mathscr{P}_1$. If $B_0 \in \mathscr{P}_2$, then we claim that B_0' is the first element of $\mathscr{P}_1$. For every element B of $\mathscr{P}_1$, $B_0 \not\supseteq B$ is satisfied. Since B and B_0' are comparable, either $B_0' \supset B$ or $B_0' \subset B$. In the latter case we obtain from A) that $B_0' = B$. Thus in any case $B_0' \supset B$, i.e. $B_0' \leqslant B$. Therefore B_0' is a lower bound of $\mathscr{P}_1$. Since $B_0' \in \mathscr{P}_1$ is clear by the definition of B_0 (because otherwise $B_0' \in \mathscr{P}_2$, and hence $B_0 \subset B_0'$, i.e. $B_0 = B_0'$, a contradiction), B_0' is the first element of $\mathscr{P}_1$. Thus we have proved that $\mathscr{P}_1$ has a first element, and hence $\mathscr{P}_0$ is a well-ordered set.

Theorem I.3 (Zermelo's theorem). *We can introduce into every set A such an appropriate order that A becomes a well-ordered set.*

Proof. We denote by $\mathscr{P}_0$ the collection of subsets of A defined in B). As proved in D), $\mathscr{P}_0$ is a well-ordered set with the order defined in C). Let us define a mapping f of A into $\mathscr{P}_0$ by

$$f(a) = \cap\{B \mid a \in B, B \in \mathscr{P}_0\}, \quad a \in A.$$

By virtue of the conditions (i) and (ii) of B), f actually maps every element a of A into $\mathscr{P}_0$. We note that

$$a \in f(a) = B_a,$$

where we put $f(a) = B_a$ to emphasize that $f(a)$ is a set belonging to $\mathscr{P}_0$. On the other hand, since $B_a \neq \emptyset$, we obtain $B_a' \subsetneqq B_a$. Since $B_a' \in \mathscr{P}_0$ by (iii) of B), it follows from the definition of B_a that $a \notin B_a'$. In view of the fact that $B_a - B_a' = \varphi(B_a)$, we have $a = \varphi(B_a)$. Hence if $a \neq a'$, then $\varphi(B_a) \neq \varphi(B_{a'})$, which implies $B_a \neq B_{a'}$, i.e. $f(a) \neq f(a')$. Thus the mapping f is a one-to-one mapping of A onto a subset $f(A)$ of $\mathscr{P}_0$. Now we define order in A as follows:

$$a < a' \text{ in } A \text{ if and only if } f(a) < f(a') \text{ in } \mathscr{P}_0.$$

Then A is a partially ordered set isomorphic with $f(A)$, because $f(A)$ as a subset of $\mathscr{P}_0$ is partially ordered. Since by D) $\mathscr{P}_0$ is well-ordered, $f(A)$ is also well-ordered. Therefore A is also a well-ordered set with respect to the order introduced above.

Combining this theorem with 3.D) we obtain the following convenient corollary which is a generalization of the fact that every countable set M can be expressed as $M = \{x_i \mid i = 1, 2, 3, \ldots\}$.

Corollary 1. *Every set A can be expressed as*

$$A = \{x_\alpha \mid 0 \leqslant \alpha < \tau\}$$

for an ordinal number τ.

Corollary 2. *If μ and ν are two cardinal numbers, then one and only one of the following three cases occurs*:

$$\mu < \nu, \qquad \mu = \nu, \qquad \mu > \nu.$$

Proof. 2.B) implies that no two cases simultaneously occur. On the other hand, let A and B be well-ordered sets such that $|A| = \mu$, $|B| = \nu$; then we can well order A and B by use of the theorem. Suppose the well-ordered sets A and B have ordinal numbers α and β respectively. Then, by Theorem I.1, either $\alpha < \beta$ or $\alpha = \beta$ or $\alpha > \beta$ holds. If $\alpha < \beta$, then $\mu \leqslant \nu$. If $\alpha = \beta$, then $\mu = \nu$. If $\alpha > \beta$, then $\mu \geqslant \nu$. Thus at least one of the three relations must hold.

Corollary 3. *For an arbitrary infinite cardinal number μ, $\mathfrak{a}\mu = \mu$.*

Proof. Take a set M with $|M| = \mu$. By use of the theorem we well-order M denoting its ordinal number with τ. Namely,

$$M = \{x_\alpha \mid 0 \leqslant \alpha < \tau\}.$$

If for an ordinal number α, $\{\beta \mid \beta < \alpha\}$ has no last number, then we call α a *limit number*. Then we can easily verify that M is decomposed as

$$M = \bigcup \{M_\alpha \mid \alpha \text{ is a limit number with } 0 \leqslant \alpha < \tau\},$$

where

$$M_\alpha = \{x_\alpha, x_{\alpha+1}, x_{\alpha+2}, \ldots\}.$$

Since $M_\alpha \cap M_{\alpha'} = \emptyset$ for different ordinal numbers α, α', M is decomposed into disjoint countable subsets. Therefore $\mu = \mathfrak{a}\nu$, where ν is a cardinal number, and hence $\mathfrak{a}\mu = \mathfrak{a}^2\nu = \mathfrak{a}\nu = \mu$. (See Exercise I.)

Let B be a given set; then we consider a property P on subsets of B. If P is satisfied by a subset B' of B if and only if it is satisfied by every finite subset of B', then we call P a *finite property* on subsets of B. Now, we suppose a is an element of a partially ordered set A. If a satisfies $a = a'$ for every element a' of A such that $a \leqslant a'$ ($a \geqslant a'$), then we call a a *maximal* (*minimal*) *element* of A. In other words, a maximal element is an element a such that there is no element a' with $a' > a$. As the reader may notice, a partially ordered set can have more than one maximal or minimal elements.

Example I.11. Let A be a partially ordered set; then 'totally-ordered' is a finite property on subsets of A, because a subset A' of A is totally-ordered if and only if every two point subset of A' is totally-ordered.

Let us consider the partially ordered set 2^M defined in Example I.8. We say that a subset $\mathcal{M}$ of 2^M satisfies P if and only if every finite number of elements $M_1, \ldots, M_k$ of $\mathcal{M}$ satisfies $\bigcap_{i=1}^{k} M_i \neq \emptyset$. Then P is a finite property. $\emptyset$ is a minimal and also the smallest element of 2^M. If we consider the partially ordered set $2^M - \{\emptyset\}$, then each one point set of M is a minimal element.

In the forthcoming discussions we shall often use the following convenient lemmas which are equivalent with Zermelo's axiom of choice.

Zorn's lemma. *Let A be a non-empty partially ordered set. If every totally-ordered subset of A has a supremum, then A has a maximal element.*

Hausdorff's lemma. *Let P be a finite property on subsets of a set A. Then there exists a maximal set among the subsets of A which satisfy P.*

Theorem I.4. *Zermelo's axiom of choice, Zorn's lemma and Hausdorff's lemma are equivalent.*

Proof. *Zermelo's axiom $\Rightarrow$ Zorn's lemma.*

To begin with, we define a mapping f of A, a given partially ordered set satisfying the condition of Zorn's lemma, into itself as follows: To each element a of A we assign a subset A_a defined by

$$A_a = \begin{cases} \{a\} & \text{if } a \text{ is a maximal element of } A, \\ \{b \mid b \in A, b > a\} & \text{if } a \text{ is no maximal element of } A. \end{cases}$$

Then by use of Zermelo's axiom of choice, we can define a mapping f such that

$$f(a) \in A_a, \quad a \in A.$$

It is obvious that $f(a)$ is a mapping of A into itself such that
 (1) $a \leqslant f(a)$,
 (2) $a = f(a)$ if and only if a is a maximal element of A.
Since A is non-empty, we choose a fixed element a_0 of A. Then we can construct a totally-ordered subset A_0 of A such that
  (i) $a_0 \in A_0$,
  (ii) $f(A_0) \subset A_0$, and
  (iii) $\sup A_0 \in A_0$.
To construct A_0, we consider subsets A' of A satisfying (i), (ii) and
  (iii)' for every non-empty totally-ordered subset B of A',

$$\sup B \in A'.$$

For example, A itself satisfies those conditions. We denote by A_0 the intersection of all subsets A' satisfying (i), (ii) and (iii)'. Then, observe the following fact that will be needed later. Since the set $\{a \mid a \in A, a \geqslant a_0\}$ obviously satisfies (i), (ii) and (iii)',

$$a \geqslant a_0 \quad \text{for every } a \in A_0. \tag{3}$$

It is also clear that A_0 satisfies (i), (ii) and (iii)'. Therefore all we have to prove is that A_0 is totally-ordered. For this purpose, we put

$$A_0' = \{a' \mid a' \in A_0; f(a) \leqslant a' \text{ or } a' \leqslant a \text{ for every } a \in A_0\}. \tag{4}$$

Now, let us show that $A_0' = A_0$. To do so, it suffices to show that A_0' satisfies (i), (ii) and (iii)', because A_0 is the smallest set satisfying the three conditions while $A_0' \subset A_0$.
  We shall first show that for $a \in A_0$ and $a' \in A_0'$
  (5) either $a \leqslant a'$ or $f(a') \leqslant a$.
Denote by A_0'' the set of all elements $a \in A_0$ satisfying (5) for every $a' \in A_0'$. Then we can assert that A_0'' satisfies (i), (ii) and (iii)'. It follows from (3) that $a_0 \leqslant a'$ for every $a' \in A_0'$, i.e. $a_0 \in A_0''$, proving (i) for A_0''. To

deal with (ii), we suppose $a \in A_0''$ and $a' \in A_0'$. Then by the definition of A_0'', either $a < a'$ or $a = a'$ or $f(a') \le a$. If $a < a'$, then by (4) $f(a) \le a'$. If $a = a'$, then $f(a) = f(a')$. If $f(a') \le a$, then from (1) we get $f(a') \le f(a)$. Therefore $f(a)$ satisfies the condition of a in (5), and hence $f(a) \in A_0''$, i.e. (ii) is satisfied by A_0''. As for (iii)', let B be a non-empty totally-ordered subset of A_0'' and a' an element of A_0'. Note that by the hypothesis of Zorn's lemma sup B exists and belongs to A_0 (recall that A_0 satisfies (iii)'). If $a \le a'$ for every element a of B, then sup $B \le a'$. If $f(a') \le a$ for some $a \in B$, then $f(a') \le$ sup B (see (5)). Thus sup B satisfies the condition of a in (5) for every $a' \in A_0'$, and hence sup $B \in A_0''$, i.e. (iii)' is also satisfied by A_0''. Since $A_0'' \subset A_0$ and A_0 is the smallest set satisfying (i), (ii) and (iii)', we have $A_0'' = A_0$. Thus (5) is established for every $a \in A_0$.

Now, let us turn to the proof that A_0' satisfies (i), (ii) and (iii)'. It clearly follows from (3) and (4) that (i) is satisfied by A_0'. To see (ii), let $a' \in A_0'$, $a \in A_0$. Then by (5) either $a < a'$ or $a = a'$ or $f(a') \le a$. In the first case we obtain from (4) that $f(a) \le a'$, which combined with (1) implies that $f(a) \le f(a')$. In the second case we obtain $f(a) = f(a')$. Thus $f(a')$ satisfies either $f(a) \le f(a')$ or $f(a') \le a$ for each $a \in A_0$, and hence we conclude, in view of (4), $f(a') \in A_0'$, proving (ii) for A_0'. Finally, referring to (iii)', we suppose B is a totally-ordered subset of A_0' and a is a given element of A_0. If $a \ge a'$ for every $a' \in B$, then $a \ge$ sup B. If $a \not\ge a'$ for some $a' \in B$, then it follows from (4) that $f(a) \le a' \le$ sup B. Therefore in view of (4), we obtain sup $B \in A_0'$. Thus A_0' is a subset of A_0 satisfying (i), (ii) and (iii)', and hence $A_0' = A_0$. Therefore from the definition (4) of A_0' as well as (1), we can conclude that any two elements of A_0 are comparable, i.e. A_0 is totally-ordered. Hence A_0 is a non-empty totally-ordered set satisfying (ii), (iii)' and accordingly (iii), as well. Now we put $a_1 =$ sup A_0; then by (iii), $a_1 \in A_0$, and hence by (ii), $f(a_1) \in A_0$. Since a_1 is the supremum of A_0 and $a_1 \le f(a_1)$ (by (1)), we obtain $a_1 = f(a_1)$. Thus, in view of (2), we have reached the conclusion that a_1 is a maximal element of A_0, which proves Zorn's lemma.

Zorn's lemma $\Rightarrow$ Hausdorff's lemma.

We denote by $\mathscr{A}$ the collection of the subsets of A with the finite property P. We define order between two elements of $\mathscr{A}$ by the ordinary inclusion relation, regarding $\mathscr{A}$ as a partially ordered set.

Let $\mathscr{A}'$ be a totally-ordered subcollection of $\mathscr{A}$; then we easily see that

$$A' = \bigcup \{B \mid B \in \mathscr{A}'\}$$

has P, because P is a finite property. Therefore $A' =$ sup $\mathscr{A}'$ in $\mathscr{A}$. Thus we can apply Zorn's lemma on $\mathscr{A}$ to get a maximal element A_0, which is the maximal set having the property P.

Hausdorff's lemma $\Rightarrow$ Zermelo's axiom.

Let $\mathscr{A}$ be a collection of non-empty sets. We denote by $\mathscr{L}$ the set of the pairs (A, a) where A is an element of $\mathscr{A}$ and a is an element of A. A subset $\mathscr{L}'$ of $\mathscr{L}$ is said to have property P if and only if every two elements of $\mathscr{L}'$ have distinct first coordinates, i.e., for every $(A, a) \in \mathscr{L}'$ and $(A', a') \in \mathscr{L}'$, $A = A'$ implies $a = a'$. Then P is clearly a finite property. Therefore by use of Hausdorff's lemma we can find a maximal subset $\mathscr{L}_0$ of $\mathscr{L}$ having the property P. Then every element A of $\mathscr{A}$ appears in one and only one element of $\mathscr{L}_0$. For, if we assume $A \in \mathscr{A}$ does not appear in any element of $\mathscr{L}_0$, then we take $a \in A$ and construct

$$\mathscr{L}_1 = \mathscr{L}_0 \cup \{(A, a)\}.$$

Then $\mathscr{L}_1 \supsetneq \mathscr{L}_0$ and $\mathscr{L}_1$ has P, contradicting the fact that $\mathscr{L}_0$ is a maximal set having the property. Therefore for every $A \in \mathscr{A}$, an element $a \in A$ is determined by the condition $(A, a) \in \mathscr{L}_0$. Now, we can define a mapping f over $\mathscr{A}$ by

$$f(A) = a, \quad (A, a) \in \mathscr{L}_0.$$

Since $a \in A$, Zermelo's axiom is established.

5. Topology of Euclidean plane

In the present section we shall deal with point sets in the Euclidean plane E^2 to help the reader to understand the concept of topological space. As a matter of fact, the theory of point sets in Euclidean spaces gives the simplest example of general topology, and historically the investigation of the former theory by G. Cantor in the late 19th century led to the establishment of the concept of topological space by F. Hausdorff, M. Fréchet, C. Kuratowski and the other mathematicians in the early 20th century, while the foundation of further development of general topology was established by A. Tychonoff, P. Urysohn, P. Alexandroff[1] and others.

[1] The paper [3] of P. Alexandroff is highly recommended to the reader as a historical survey of the modern development in general topology. See his paper [5] too, for recent developments.

As is well known, the concept of convergence of a point sequence is very significant in E^2, especially in analysis of E^2. We may say it is fundamental in the study of E^2. This concept is closely related with such other concepts as neighborhood, closure, open set, etc. as seen in the following.

We denote by $\rho(p, q)$ the distance between two points p, q of E^2 and by $S_\varepsilon(p)$ for a positive ε the domain in the circle with center p and radius ε, i.e.

$$S_\varepsilon(p) = \{q \mid \rho(p, q) < \varepsilon\}.$$

$S_\varepsilon(p)$ is often called the ε-*neighborhood* of p.

Moreover, for a point p and a point set A of E^2 (i.e. a subset of E^2), we define the distance between p and A by

$$\rho(p, A) = \inf\{\rho(p, q) \mid q \in A\}.$$

Let p be a point of E^2 and U a point set of E^2, which contains p. If U contains an ε-neighborhood of p for some $\varepsilon > 0$, then we call U a *neighborhood* (or *nbd*[1] for brevity) of p.

Example I.12. We consider an x-y coordinate system in E^2 with the origin $p_0 = (0, 0)$. Then for every $\varepsilon > 0$ and $\delta > 0$, the following sets are all examples of nbds of p_0: $S_\varepsilon(p_0)$, $U = \{(x, y) \mid x^2/\varepsilon^2 + y^2/\delta^2 < 1\}$, $V = \{(x, y) \mid |x| < \varepsilon$ and $|y| < \delta\}$, $W = \{(x, y) \mid (x - \varepsilon/2)^2 + y^2 \leqslant \varepsilon^2\}$. On the other hand $P = \{(x, y) \mid (x - \varepsilon)^2 + (y - \delta)^2 \leqslant \varepsilon^2 + \delta^2\}$ and $Q = \{(x, 0) \mid |x| < \varepsilon\}$ are not nbds of p_0 although they contain p_0.

We often adopt notations like $U(p)$, $V(p)$, etc. to denote nbds of p. We can easily see that nbds have the following properties.

A) (i) *Let $U(p)$ and $V(p)$ be nbds of a point p; then $U(p) \cap V(p)$ is also a nbd of p,*

(ii) *if a set U contains a nbd of p, then U itself is also a nbd of p,*

(iii) *every nbd $U(p)$ of a point p contains a subset V containing p such that V is a nbd of every point belonging to V.*

---

[1] We pronounce it 'neighborhood'.

Proof. Since the proofs of (i) and (ii) are easy, they are left to the reader. Referring to (iii), $U(p)$ contains an ε-nbd, $S_\varepsilon(p)$, which can be easily seen to satisfy the required condition for V.

The convergence[1] of a point sequence is characterized by use of the concept of nbds, and vice versa, as seen in the following assertions whose easy proofs will be left to the reader.

B) *Let* $\{p_n \mid n = 1, 2, \ldots\}$ *be a point sequence of* E^2. *Then it converges to a point p of* E^2 *if and only if for every nbd* $U(p)$ *of p and for some number n,*

$$p_i \in U(p) \quad \text{for every } i \geqslant n.$$

C) *A set* U *is a nbd of* p *if and only if for every point sequence* $\{p_n \mid n = 1, 2, \ldots\}$ *converging to p, we can choose a number n such that*

$$p_i \in U \quad \text{for every } i \geqslant n.$$

Let A be a point set of E^2 and p a point of A. If A is a nbd of p, then p is called an *inner point* of A. If q is an inner point of $E^2 - A$, then it is called an *exterior point* of A. If a point p of E^2 is neither an inner point nor an exterior point of A, then it is called a *boundary point* of A. We call the set of all boundary points of A the *boundary* of A. We can define open sets and closed sets in terms of the concept of nbd as follows. A point set U of E^2 is called an *open set* if every point p of U is an inner point of U, or we may express this definition as follows: U is an open set if for every point $p \in U$, we can choose a positive ε such that $S_\varepsilon(p) \subset U$.

A point set F of E^2 is called a *closed set* if and only if $E^2 - F$ is an open set.

We obtain the following easily shown but important properties of open sets whose proofs will be left to the reader.

D) (i) E^2 *and* $\emptyset$ *are open sets,*
   (ii) *the intersection of finitely many open sets is open,*
   (iii) *the union of (not necessarily finitely many) open sets is open.*

---

[1] As is well known, a point sequence $\{p_n \mid n = 1, 2, \ldots\}$ is said to converge to a point p if for every $\varepsilon > 0$ there is n such that $\rho(p, p_i) < \varepsilon$ whenever $i \geqslant n$.

By use of de Morgan's rule, we can deduce from D) the following properties of closed sets.

E) (i) E^2 and $\emptyset$ are closed sets,

(ii) the union of finitely many closed sets is closed,

(iii) the intersection of (not necessarily finitely many) closed sets is closed.

In view of the condition (iii) of A), we can characterize nbds of a point by use of open sets as follows:

F) A subset U of E^2 is a nbd of p if and only if there is an open subset V such that $p \in V \subset U$.

Example I.13. Let us consider the sets cited in Example I.12. Every point of $S_\varepsilon(p_0)$ is easily seen to be an inner point of it, and accordingly $S_\varepsilon(p_0)$ is an open set. In the same way, U, V are also open sets, but W, P, Q are not open. The inner points of W are the points which satisfy $(x - \varepsilon/2)^2 + y^2 < \varepsilon^2$, while Q has no inner point. The boundary of $S_\varepsilon(p_0)$ is the set of the points on the circle, i.e. $\{(x, y) \mid x^2 + y^2 = \varepsilon^2\}$, and the boundaries of U, V, W, P are the closed curves which surround the respective domains, while all points of Q plus $(\varepsilon, 0)$, $(-\varepsilon, 0)$ form the boundary of Q.

W and P are examples of closed sets. Although $S_\varepsilon(p_0)$ itself is not closed, the union of $S_\varepsilon(p_0)$ and its boundary forms a closed set. E^2 and $\emptyset$ are the only subsets of E^2 which are open and closed at the same time. Roughly speaking, closed sets are sets fringed with the boundary, and open sets are sets stripped of the boundary. Note that 'non-closed' does not necessarily mean 'open' in topology, as in an ordinary conversation; for example, the set Q in Example I.12 is neither closed nor open. Also note that condition (ii) of D) (E)) fails to be true for infinitely many open (closed) sets. The intersection of the $1/n$-nbds $S_{1/n}(p)$, $n = 1, 2, \ldots$ is the one point set $\{p\}$ which is closed but not open.

In the above we have learned that the concept of nbds is equivalent with that of convergence, in the sense that the former is characterized by the latter, and vice versa. Likewise, the concept of open sets, and also that of closed sets, is equivalent to that of nbds and therefore to convergence, too. Now, let us turn to another important concept, closure of set, which is also closely related to the concept of convergence; we may even say that they are equivalent. Let A be a subset of E^2. We denote by $\bar{A}$ the set of the points p with $\rho(p, A) = 0$ and call it the *closure* of A.

Example I.14. Consider the open set $S_\varepsilon(p_0)$ in Example I.13; then

$$\overline{S_\varepsilon(p_0)} = S_\varepsilon(p_0) \cup B(S_\varepsilon(p_0)),$$

where $B(S_\varepsilon(p_0))$ denotes the boundary of $S_\varepsilon(p_0)$. As a matter of fact, we can assert that $\bar{A} = A \cup B(A)$ for every subset A of E^2. Thus we may say that the closure of a set is the original set with its boundary added. We should note that the concept of boundary in its exact definition may not always agree with our intuition. For example, consider the set R of all rational points in E^2, i.e. the points whose x-y coordinates are both rational. Then the boundary of R coincides with the whole of E^2 and accordingly $\bar{R} = E^2$. Generally such a set D that satisfies $\bar{D} = E^2$ is said to be dense in E^2. The set S of all irrational points is also dense in E^2.

We obtain the following properties of closure:

G) (i) $\bar{\emptyset} = \emptyset$,
   (ii) $\bar{A} \supset A$,
   (iii) $\overline{A \cup B} = \bar{A} \cup \bar{B}$,
   (iv) $\bar{\bar{A}} = \bar{A}$.

Proof. (i) and (ii) are obvious. In view of (ii), $\bar{\bar{A}} \supset \bar{A}$ and $\overline{A \cup B} \supset \bar{A} \cup \bar{B}$ are also obvious. Therefore we shall show

$$\overline{A \cup B} \subset \bar{A} \cup \bar{B} \quad \text{and} \quad \bar{\bar{A}} \subset \bar{A}.$$

To show the first relation, suppose p is a given point of $\overline{A \cup B}$. Then $\rho(p, A \cup B) = 0$ which means that there exist points $p_n \in A \cup B$, $n = 1$, $2, \ldots$, such that $\rho(p, p_n) < 1/n$. Since $p_n \in A \cup B$, either A or B contains an infinite subsequence $p_{n_1}, p_{n_2}, \ldots$ of $\{p_n \mid n = 1, 2, \ldots\}$. For example, suppose A does so; then we can easily see that $\rho(p, A) = 0$ which means $p \in \bar{A}$. Thus we have verified

$$\overline{A \cup B} \subset \bar{A} \cup \bar{B}.$$

To verify $\bar{\bar{A}} \subset \bar{A}$, we suppose p is a given point of $\bar{\bar{A}}$. Then $\rho(p, \bar{A}) = 0$ which means that there exist points $p_n \in \bar{A}$, $n = 1, 2, \ldots$, such that

$\rho(p, p_n) < 1/n$. Since $p_n \in \bar{A}$, there exist points $q_n \in A$, $n = 1, 2, \ldots$, such that $\rho(p_n, q_n) < 1/n$. Thus

$$\rho(p, q_n) \leq \rho(p, p_n) + \rho(p_n, q_n) < 2/n$$

for $q_n \in A$, $n = 1, 2, \ldots$. Therefore $\rho(p, A) = 0$, which means $p \in \bar{A}$, and thus the assertion is proved.

The relationship between closure and convergence is described in the following assertions.

H) $\bar{A} = \{p \mid$ there exists a point sequence $\{p_n \mid n = 1, 2, \ldots\}$ converging to p such that $p_n \in A$, $n = 1, 2, \ldots\}$.

I) A point sequence $\{p_n \mid n = 1, 2, \ldots\}$ converges to a point p if and only if for every subsequence $\{p_{n_i} \mid i = 1, 2, \ldots\}$ of $\{p_n \mid n = 1, 2, \ldots\}$, $p \in \overline{\{p_{n_i}\}}$ holds.

Proof. We shall prove only I). If $\{p_n\}$ converges to p, then its subsequence $\{p_{n_i}\}$ also converges to p. Therefore for every $\varepsilon > 0$ and for some i, $\rho(p, p_{n_i}) < \varepsilon$, which means $\rho(p, A) = 0$, where $A = \{p_{n_i}\}$. Hence $p \in \bar{A}$.

Conversely, if $\{p_n\}$ does not converge to p, then for some $\varepsilon > 0$, there exists an arbitrarily large number n such that $\rho(p, p_n) \geq \varepsilon$. Therefore we can choose a sequence $n_1 < n_2 < \cdots$ such that

$$\rho(p, p_{n_i}) \geq \varepsilon, \quad i = 1, 2, \ldots.$$

Thus $\rho(p, A) \geq \varepsilon$ for $A = \{p_{n_i}\}$, and hence $p \notin \bar{A}$.

We can also establish a direct relationship between closure and nbds as the reader may have already realized. But now, we do not like to be involved in such detailed discussions because we shall handle them in more generality in the following chapters.

Example I.15. Concepts such as limit point, continuous functions, etc. which are defined in terms of convergence, can be also defined by use of one of the equivalent terminologies, nbd, open set, closed set or closure. As is well known, a point p is called a *limit point* of a set A if for every $\varepsilon > 0$, there are infinitely many points q of A for which $\rho(p, q) < \varepsilon$. To define this terminology by use of closure, for example, we can say that p is a limit point of A if and only if

$$p \in \overline{A - \{p\}}.$$

Now, we define *convergence of set sequence* which will help the reader to understand the concept of filter in the following chapter. We mean by a set sequence a sequence $A_1, A_2, \ldots$ of non-empty subsets of E^2 such that

$$A_1 \supset A_2 \supset \cdots.$$

The set sequence $\{A_n \mid n = 1, 2, \ldots\}$ is said to converge to a point p if for every $\varepsilon > 0$ there exists an n for which $A_n \subset S_\varepsilon(p)$. We take a point $p_n \in A_n$ from each member, A_n, of the set sequence. Then we obtain a point sequence $\{p_n\}$ which is called a point sequence derived from $\{A_n\}$. We can easily show that the set sequence $\{A_n\}$ converges to p if and only if every point sequence derived from $\{A_n\}$ converges to p. Conversely, we consider a given point sequence $\{p_n\}$. Putting $A_n = \{p_n, p_{n+1}, \ldots\}$ we obtain a set sequence $\{A_n\}$. Then we call $\{A_n\}$ the set sequence derived from $\{p_n\}$. A point sequence $\{p_n\}$ converges to p if and only if the set sequence derived from $\{p_n\}$ converges to p. Thus we can regard the two concepts, convergence of point sequence and convergence of set sequence, as equivalent.

Example I.16. The concept of set sequence often appears in an introductory part of calculus where, for example, the following proposition plays a significant role: Every set sequence $\{A_n\}$ converges to a point p if $\lim_{n \to \infty}$ diameter of $A_n = 0$; if moreover each A_n is a closed set, then $\bigcap_{n=1}^{\infty} A_n = \{p\}$.

We should take note that all the discussions (except some examples) in this section are also valid for three-dimensional Euclidean space E^3 and more generally for every metric space X, where we call a set X a *metric space* if to every two elements p, q, of X, a non-negative real number $\rho(p, q)$ called the distance between p and q is assigned, such that
   (i) $\rho(p, q) = 0$ if and only if $p = q$,
   (ii) $\rho(p, q) = \rho(q, p)$,
   (iii) $\rho(p, q) \leqslant \rho(p, r) + \rho(r, q)$, for every $r \in X$.

Exercise I[1]

1. Prove

$$A \cap [\cup \{A_\gamma \mid \gamma \in \Gamma\}] = \cup \{A \cap A_\gamma \mid \gamma \in \Gamma\},$$

$$E - \cup \{A_\gamma \mid \gamma \in \Gamma\} = \cap \{E - A_\gamma \mid \gamma \in \Gamma\},$$

$$[\cup \{A_\gamma \mid \gamma \in \Gamma\}] \cap [\cup \{B_\delta \mid \delta \in \Delta\}] = \cup \{A_\gamma \cap B_\delta \mid (\gamma, \delta) \in \Gamma \times \Delta\},$$

where A, A_γ, etc. denote given sets.

2. Prove that the following sets are countable sets: The set of the solutions of all algebraic equations with rational coefficients. The set of disjoint open intervals on the real line $(-\infty, \infty)$. The set of the points at which a given monotone increasing (decreasing) real-valued function of one real variable is not continuous.

3. Let λ, μ, ν be given cardinal numbers. Then prove:

$$(\lambda\mu)\nu = \lambda(\mu\nu), \qquad \lambda(\mu + \nu) = \lambda\mu + \lambda\nu,$$

$$\lambda^\mu \lambda^\nu = \lambda^{\mu+\nu}, \qquad (\lambda^\mu)^\nu = \lambda^{\mu\nu}.$$

4. Prove: A totally-ordered set A is well-ordered if and only if A does not contain a sequence $\{a_i \mid i = 1, 2, \ldots\}$ of elements such that $a_1 > a_2 > a_3 > \cdots$.

5. We consider a decomposition of a given totally-ordered set A such that $A = B \cup C$, $B \cap C = \emptyset$, and $b < c$ whenever $b \in B$ and $c \in C$.

Prove that A is well-ordered if and only if C has the first element for any such decomposition.

6. Prove: For a given partially ordered set A, there is a totally-ordered subset A' of A such that every upper bound of A' belongs to A'.

---

[1] As a rule the exercises in this book do not contain new topics which result from rather extensive studies or which could make additional sections of the text. Such supplementary topics will rather be found in footnotes, in examples, or in descriptions without proof. The exercises do not contain many special types of questions which require artificial techniques of proof, but consist mostly of questions that are rather easy to prove and are often related to handy small theorems or examples. They are primarily aimed to aid understanding, while at the same time serving as part of the main discussion. Proofs of the propositions which were left to the reader (e.g. 5.B)) may be also regarded as exercise problems though usually they are not restated in the exercise section.

7. Let A be a subset of E^2. Then prove that

$$\bar{A} = A \cup B(A) = A \cup A^*,$$

where $B(A)$ and A^* denote the boundary and the set of the limit points of A, respectively. Furthermore, prove that A is a closed set if and only if $\bar{A} = A$.

CHAPTER II

BASIC CONCEPTS IN TOPOLOGICAL SPACES

At the end of the last section of Chapter I, we noted that the whole discussion there, essentially based on the concept of convergence in E^2, could be extended to metric spaces. However, even the existence of distance is no prerequisite for the concept of convergence. In fact, we can discuss convergence in a topological space, which is even more general than a metric space. We may say a topological space is a set endowed with the concept of convergence. From a practical point of view, however, defining a topological space by means of convergence itself is not the best way. Rather, we prefer to define a topological space first with the concept of open sets, nbds or closure, which are essentially equivalent with convergence. Then we define convergence of the space. We shall find open sets, nbds and closure are often more convenient than convergence not only for defining a topological space but also for making a study of the space. In this chapter we shall begin with the concept of open sets and derive the other concepts from it.

1. Topological space

Definition II.1. Let X be a set and $\mathcal{O}$ be a collection of subsets of X which satisfies:
  (i) $\emptyset \in \mathcal{O}$, $X \in \mathcal{O}$,
  (ii) if $U_i \in \mathcal{O}$, $i = 1, \ldots, k$, then $\bigcap_{i=1}^{k} U_i \in \mathcal{O}$,
  (iii) if $U_\gamma \in \mathcal{O}$, $\gamma \in \Gamma$, then $\bigcup \{U_\gamma \mid \gamma \in \Gamma\} \in \mathcal{O}$,
where the index set Γ is not necessarily finite.

Then we call every set belonging to $\mathcal{O}$ an *open set* and X a *topological space* or a *T-space*. The collection $\mathcal{O}$ is called the *topology of X*.[1]

---

[1] To be precise, we should call the pair $\langle X, \mathcal{O} \rangle$ a topological space. But for brevity we usually say 'X is a topological space'. The same applies to metric spaces and other spaces with various structures that will appear later in this book.

Definition II.2. Consider two topologies $\mathcal{O}$ and $\mathcal{O}'$ for a given set X. If $\mathcal{O} \subset \mathcal{O}'$, then the topology $\mathcal{O}$ is called *weaker* than $\mathcal{O}'$ ($\mathcal{O}'$ is *stronger* than $\mathcal{O}$).

Example II.1. The Euclidean plane E^2 is one of the most popular examples of a topological space. In fact, we have seen in I.5.D) that the collection of the open sets of E^2 satisfies (i), (ii) and (iii) of Definition II.1.

Generally every metric space X is a topological space with the topology $\mathcal{O} = \{U \mid U \subset X$, and for every $p \in U$ and for some $\varepsilon > 0$, $S_\varepsilon(p) \subset U\}$, where $S_\varepsilon(p) = \{q \mid q \in X, \rho(p, q) < \varepsilon\}$ ($\mathcal{O}$ is called the metric topology of X). Among examples of metric spaces are n-dimensional Euclidean space $E^n = \{(x_1, \ldots, x_n) \mid x_i, i = 1, \ldots, n$, are real numbers$\}$ with distance

$$\rho((x_1, \ldots, x_n), (y_1, \ldots, y_n)) = \left(\sum_{i=1}^{n} (x_i - y_i)^2\right)^{1/2},$$

and *Hilbert space* $H = \{(x_1, x_2, \ldots) \mid x_i, i = 1, 2, \ldots$, are real numbers, $\sum_{i=1}^{\infty} x_i^2 < +\infty\}$ with distance

$$\rho((x_1, x_2, \ldots), (y_1, y_2, \ldots)) = \left(\sum_{i=1}^{\infty} (x_i - y_i)^2\right)^{1/2}.$$

To give an example of a non-metric topological space, we consider the set R_5 of all ordinal numbers of the countable well-ordered sets. Namely,

$$R_5 = \{\alpha \mid 0 \leqslant \alpha < \omega_1\},$$

where ω_1 denotes the smallest non-countable ordinal number. We define a collection $\mathcal{O}$ of subsets of R_5 as follows:

$$\mathcal{O} = \{U \mid U \subset R_5, \text{ for every } \alpha \in U \text{ with } \alpha > 0$$
$$\text{and for some } \beta < \alpha, (\beta, \alpha] \subset U\} \cup \{0\},$$

where

$$(\beta, \alpha] = \{\gamma \mid \beta < \gamma \leqslant \alpha\}.$$

Then we can easily see that $\mathcal{O}$ satisfies (i)–(iii) of Definition II.1. Namely, R_5 is a topological space. We can introduce a topology into any set of ordinal numbers in a similar way. Thus defined topology is called the *order topology*.

Suppose X is a given set. We denote by $\mathcal{O}_1$ the collection consisting of $\emptyset$ and X only, and by $\mathcal{O}_2$ the collection of all subsets of X. $\mathcal{O}_1$ and $\mathcal{O}_2$ are easily verified to be topologies of X. In fact $\mathcal{O}_1$ and $\mathcal{O}_2$ are the weakest and the strongest topologies of X, respectively. Generally speaking, interesting topologies are those between $\mathcal{O}_1$ and $\mathcal{O}_2$. A topological space with the strongest (weakest) topology is called a *discrete space* (*indiscrete space*).

Definition II.3. Let X be a topological space and p a point of X. If a subset U of X contains an open set V which contains p, then U is called a *neighborhood* (often abbreviated as *nbd*) of p.

This definition implies that every open set containing p is a nbd of p. We call such a nbd an *open nbd* of p.

Example II.2. Let us deal with the examples E^2 and R_5 in Example II.1. In E^2 every set containing $S_\varepsilon(p)$ for some $\varepsilon > 0$ is a nbd of p, while in R_5 every set containing $(\beta, \alpha]$ for some $\beta < \alpha$ is a nbd of α if $\alpha \neq 0$, and every set containing 0 is a nbd of 0.

A) *With respect to Definition* II.3, *we denote by* $\mathcal{U}(p)$ *the collection of all nbds of* p. *Then it satisfies*:

 (i) $X \in \mathcal{U}(p)$,
 (ii) *if* $U \in \mathcal{U}(p)$, *then* $p \in U$,
 (iii) *if* $U \in \mathcal{U}(p)$, $V \supset U$, *then* $V \in \mathcal{U}(p)$,
 (iv) *if* $U, V \in \mathcal{U}(p)$, *then* $U \cap V \in \mathcal{U}(p)$,
 (v) *if* $U \in \mathcal{U}(p)$, *then there exists a set* V *such that* $p \in V \subset U$ *and such that* $V \in \mathcal{U}(q)$ *for every point* $q \in V$.

Proof. The conditions (i), (ii) and (iii) are direct consequences of Definition II.3. The condition (iv) is derived from the definition combined with the condition (ii) of open sets in Definition II.1. The condition (v) is directly derived from the assertion that if $U \in \mathcal{U}(p)$, then there is an open set V such that $p \in V \subset U$, and this assertion is implied by Definition II.3 itself.

B) *A subset* U *of a topological space* X *is an open set if and only if* U *is a nbd of each point* p *of* U.

Proof. Suppose U is an open set of X and $p \in U$. Then it follows directly from Definition II.3 that U is a nbd of p.

Conversely, suppose U is a subset of X satisfying the condition of B). Then, by Definition II.3, for each point p of U we can select an open set $V(p)$ such that $p \in V(p) \subset U$. Therefore

$$U = \bigcup \{V(p) \mid p \in U\},$$

which means that U is the union of open sets, and hence from (iii) of Definition II.1 it follows that U is an open set.

Though we have defined a topological space initially by means of open sets in Definition II.1, we can start with nbds to define a topological space if we want to do so.

C) *Let X be a set in which to each element p a collection $\mathcal{U}(p)$ of subsets of X is assigned. Suppose $\mathcal{U}(p)$ satisfies (i)–(v) of A). Then we call each element of $\mathcal{U}(p)$ a nbd of p. Now, we define that a subset U of X is an open set if and only if it satisfies the condition B). Then thus defined open sets satisfy the conditions (i)–(iii) of Definition II.1, i.e. X is a topological space.*

Proof. The conditions (i), (ii) and (iii) of Definition II.1 are direct consequences of (i), (iv) and (iii) of A), respectively.

Proposition C) shows that to make a topological space from a set we may first define the totality of nbds instead of defining open sets. But, now, there arises a question. Let X be a topological space defined by use of Definition II.1. Using Definition II.3, we can define the nbds in X. They, of course, satisfy (i)–(v) of A), and hence we can again define open sets by use of the nbds as shown in C). Do those newly defined open sets coincide with the original open sets? If not, we might get confused about the concept of open sets in a topological space. Fortunately, there is no chance for such confusion to occur. Namely we can assert the following:

D) *We suppose X is a topological space defined by Definition II.1, denoting its topology by $\mathcal{O}$, i.e. the collection of the open sets of X. We define the nbds of X by use of $\mathcal{O}$ and Definition II.3. By use of those nbds and C) we again define open sets and denote by $\mathcal{O}'$ the collection of the newly defined open sets. Then $\mathcal{O} = \mathcal{O}'$.*

Proof. It is directly derived from B) which implies that $\mathcal{O}$ has the same relation to the nbds as $\mathcal{O}'$ does.

The relationship between the two concepts, open sets and nbds, are reciprocal. The following proposition shows that we shall likewise have no trouble even if we begin with nbds to define a topological space.

E) *Let X be a topological space defined with nbds satisfying* (i)–(v) *of* A). *We denote by $\mathcal{U}(p)$ the collection of the nbds of each point p of X. By use of those nbds and* C) *we define open sets of X. Then by use of the open sets and Definition* II.3 *we define nbds for each point p of X denoting by $\mathcal{U}'(p)$ the collection of those newly defined nbds of p. Then $\mathcal{U}(p) = \mathcal{U}'(p)$ for each point p of X.*

Proof. Let $U \in \mathcal{U}(p)$; then by (v) of A) and C) there is an open set V such that $p \in V \subset U$. Hence by Definition II.3, $U \in \mathcal{U}'(p)$.

Conversely, let $U \in \mathcal{U}'(p)$; then by Definition II.3, there is an open set V such that $p \in V \subset U$. Since V is an open set containing p, it follows from C) that $V \in \mathcal{U}(p)$. Therefore by (iii) of A) $U \in \mathcal{U}(p)$. Thus $\mathcal{U}(p) = \mathcal{U}'(p)$ holds for each point p of X.

In view of propositions A), C), D) and E) we can conclude that there is no difference between the results whether we adopt open sets or nbds to define a topological space. Now, let us proceed to extend the other terminologies on point sets of E^2 to topological spaces.

Definition II.4. Let X be a topological space. A subset F of X is called a *closed set* if $X - F$ is an open set.[1]

F) *The collection $\mathcal{C}$ of the closed sets of X satisfies the following conditions:*
    (i) *$\emptyset \in \mathcal{C}$, $X \in \mathcal{C}$,*
    (ii) *if $F_i \in \mathcal{C}$, $i = 1, \ldots, k$, then $\bigcup_{i=1}^{k} F_i \in \mathcal{C}$,*
    (iii) *if $F_\gamma \in \mathcal{C}$, $\gamma \in \Gamma$, then $\bigcap\{F_\gamma \mid \gamma \in \Gamma\} \in \mathcal{C}$,*
where the index set Γ is not necessarily finite.

Proof. We shall prove only the condition (iii). By use of de Morgan's rule,

$$X - \bigcap\{F_\gamma \mid \gamma \in \Gamma\} = \bigcup\{X - F_\gamma \mid \gamma \in \Gamma\}.$$

Since F_γ is closed, $X - F_\gamma$ is open, and hence it follows from (iii) of Definition II.1 that $\bigcup\{R - F_\gamma \mid \gamma \in \Gamma\}$ is open. Therefore its complement, $\bigcap\{F_\gamma \mid \gamma \in \Gamma\}$, is a closed set.

---

[1] Generally, we may denote by F^c the complement $X - F$ of a subset F of X if there is no fear of confusion.

Since the definitions of open set and closed set are reciprocal, we can define a topological space by means of closed sets as well, but the detailed discussion will be left to the reader.

A set A of a topological space X is called a G_δ-set (an F_σ-set) if it is the intersection of countably many open sets (the union of countably many closed sets, respectively).

Let A be a set of a topological space X and p a point of X. If A is a nbd of p, or in other words if there is a nbd $U(p)$ of p such that $U(p) \subset A$, then p is called an *inner point* of A. We denote by $A°$ (or Int A) the set of the inner points of A and call it the *interior* or *inner part* of A. If p is an inner point of $X - A$, then it is called an *exterior point* of A. If p is neither an inner nor an exterior point of A, i.e. if every nbd U of p intersects both $X - A$ and A, then p is called a *boundary point* of A. We denote by $B(A)$ (or Bdr A) the set of the boundary points of A and call it the *boundary* of A.

G) $A°$ *is an open set.* $B(A)$ *is a closed set.*

Proof. The easy proof of this proposition will be left to the reader.

Example II.3. Let R_5 be the topological space of ordinal numbers given in Examples II.1 and II.2. Then $F = [\omega, 2\omega] = \{\alpha \mid \omega \leq \alpha \leq 2\omega\}$ is a closed set for which $F° = U$ and $B(F) = \{\omega\}$, where $U = (\omega, 2\omega]$. U is an open set for which $U° = U$, $B(U) = \emptyset$. Note that U is a closed set at the same time. As a matter of fact, R_5 contains, besides R_5 and $\emptyset$, infinitely many sets which are open and closed at the same time. (Such a set is called a *clopen* set.)

Example II.4. Let $\mathcal{B}$ be a collection of subsets of a topological space X. $\mathcal{B}$ is called a *Borel field* if it satisfies:

   (i) $A_i \in \mathcal{B}$, $i = 1, 2, \ldots$, imply $\bigcap_{i=1}^{\infty} A_i \in \mathcal{B}$,

   (ii) $A \in \mathcal{B}$ implies $X - A \in \mathcal{B}$.

For a given collection $\mathcal{A}$ of subsets of X, we denote by $\mathcal{B}(\mathcal{A})$ the intersection of all Borel fields containing $\mathcal{A}$ as a subcollection. Then $\mathcal{B}(\mathcal{A})$ is the smallest Borel field containing $\mathcal{A}$. Let us denote by $\mathcal{C}$ the collection of all closed sets of X. Then every set $B \in \mathcal{B}(\mathcal{C})$ is called a *Borel set*. F_σ-sets and G_δ-sets are Borel sets, but the Borel sets are not exhausted by those two types of sets.[1]

---

[1] For the theory of Borel sets, see C. Kuratowski [3].

2. Open basis and neighborhood basis

Reviewing Section 5 of Chapter I, we notice that in the definition of nbds of E^2, the ε-nbd, $S_\varepsilon(p)$, plays an important and specific role. In fact, we often define a topological space by first defining some specific nbds and then, in terms of these, all nbds, rather than defining all the nbds at once. The same circumstances occur when defining a topological space by means of open sets.

Definition II.5. Let X be a topological space, $\mathscr{U}$ a collection of open sets of X. If every open set of X can be expressed as the union of sets belonging to $\mathscr{U}$, then $\mathscr{U}$ is called an *open basis* of X (or *base* for X). Let $\mathscr{V}$ be a collection of open sets in X. If the collection $\Delta\mathscr{V}$ of all finite intersections of sets belonging to $\mathscr{V}$ is an open basis of X, then $\mathscr{V}$ is called an *open subbasis* of X (or *subbase* for X).

Definition II.6. Let $\mathscr{G}$ be a collection of closed sets of X. If every closed set of X can be expressed as the intersection of sets belonging to $\mathscr{G}$, then $\mathscr{G}$ is called a *closed basis* (*closed base*) of X. A collection $\mathscr{H}$ of closed subsets of X is called a *closed subbasis* (*closed subbase*) if the collection of all finite sums of elements of $\mathscr{H}$ is a closed basis of X.

A) *A collection $\mathscr{U}$ of open sets of a topological space X is an open basis of X if and only if for each open set V of X and each point p of V, there is some $U \in \mathscr{U}$ such that $p \in U \subset V$.*

Proof. The easy proof is left to the reader.

B) *Let $\mathscr{U}$ be an open basis of a topological space X; then it satisfies*
(i) $\emptyset \in \mathscr{U}$,
(ii) *if U_1, $U_2 \in \mathscr{U}$ and $p \in U_1 \cap U_2$, then there is $U_3 \in \mathscr{U}$ such that $p \in U_3 \subset U_1 \cap U_2$,*
(iii) $\bigcup\{U \mid U \in \mathscr{U}\} = X$.
Conversely, let X be a set and $\mathscr{U}$ a collection of subsets of X which satisfies (i)–(iii). Then, if we denote by $\mathscr{O}$ the collection of the unions of sets belonging to $\mathscr{U}$, then $\mathscr{O}$ satisfies (i)–(iii) of Definition II.1. Thus X is a topological space with $\mathscr{U}$ as an open basis.

Proof. The first half of the assertion is almost obvious in view of the property of open sets and A). Therefore we shall prove only the latter half, i.e. that $\mathscr{O}$ derived from $\mathscr{U}$ satisfies (i)–(iii) of Definition II.1.

The condition (i) of Definition II.1 follows from (i) and (iii) of B). The condition (iii) of Definition II.1 is obvious from the definition of $\mathcal{O}$. Thus it suffices to verify (ii) of Definition II.1. To do so, we note that by repeated use of (ii) of B) the following assertion can be verified:

(1) If $U_1, \ldots, U_k \in \mathcal{U}$, $p \in U_1 \cap \cdots \cap U_k$, then there is some $U \in \mathcal{U}$ such that $p \in U \subset \cap_{i=1}^{k} U_i$.

Now, assume that $V_1, \ldots, V_k \in \mathcal{O}$ and $p \in \cap_{i=1}^{k} V_i$; then $p \in V_i$, $i = 1, \ldots, k$. Since each V_i is the union of sets belonging to $\mathcal{U}$, for each i there is some $U_i \in \mathcal{U}$ such that

$$p \in U_i \subset V_i.$$

The statement (1) assures us that there is some $U \in \mathcal{U}$ such that

$$p \in U \subset \bigcap_{i=1}^{k} U_i.$$

Therefore

$$p \in U \subset \bigcap_{i=1}^{k} V_i,$$

where $U \in \mathcal{U}$. This means that $\cap_{i=1}^{k} V_i$ can be expressed as the union of sets belonging to $\mathcal{U}$ and hence $\cap_{i=1}^{k} V_i \in \mathcal{O}$ proving (iii) of Definition II.1. Finally it is obvious that $\mathcal{U}$ is an open basis of the topological space X with the topology $\mathcal{O}$.

C) *Let $\mathcal{G}$ be a closed basis of a topological space X; then it satisfies*:
  (i) $X \in \mathcal{G}$,
  (ii) *if $G_1, G_2 \in \mathcal{G}$ and $p \notin G_1 \cup G_2$, then there is some $G_3 \in \mathcal{G}$ such that $p \notin G_3 \supset G_1 \cup G_2$,*
  (iii) $\cap \{F \mid F \in \mathcal{G}\} = \emptyset$.
  Conversely, let X be a set and $\mathcal{G}$ a collection of subsets of X which satisfies (i)–(iii). Put $\mathcal{U} = \{X - G \mid G \in \mathcal{G}\}$, then $\mathcal{U}$ satisfies (i)–(iii) of B). Thus X is a topological space with $\mathcal{G}$ as a closed basis.

Proof. The easy proof is left to the reader.

Definition II.7. Let p be a point of a topological space X. A collection $\mathcal{V}(p)$ of nbds of p is called a *nbd basis* of p if for every nbd U of p, there exists $V \in \mathcal{V}(p)$ such that $p \in V \subset U$.

D) *Let $\mathcal{V}(p)$ be a nbd basis of a point p of X. Then it satisfies the following conditions:*

(i) $\mathcal{V}(p) \neq \emptyset$,

(ii) *if $U \in \mathcal{V}(p)$, then $p \in U$,*

(iii) *if $U \in \mathcal{V}(p)$ and $V \in \mathcal{V}(p)$, then there exists $W \in \mathcal{V}(p)$ such that $W \subset U \cap V$,*

(iv) *if $U \in \mathcal{V}(p)$, then there exists a set V such that $p \in V \subset U$ and such that for every point $q \in V$, there is some $W \in \mathcal{V}(q)$ satisfying $W \subset V$.*

Proof. The easy proof is left to the reader.

To define a topological space, we may give only a nbd basis of each point instead of the totality of nbds as seen in the following.

E) *Let X be a set in which to each element p of X a collection $\mathcal{V}(p)$ of subsets of X is assigned. Suppose $\mathcal{V}(p)$ satisfies (i)–(iv) of D). If we define a collection $\mathcal{U}(p)$ for each $p \in X$ by*

$$\mathcal{U}(p) = \{U \mid U \supset V \text{ for some } V \in \mathcal{V}(p)\},$$

then $\mathcal{U}(p)$ satisfies the conditions (i)–(v) of 1.A). Thus X is a topological space with $\mathcal{U}(p)$ as the collection of the nbds of point p while $\mathcal{V}(p)$ forms a nbd basis of p.

Proof. The easy proof is left to the reader.

Example II.5. In view of Definition II.3, we know that in every topological space X the collection of the open nbds of a point p form a nbd basis of p. In fact, we can get along with only open nbds in most discussions related to nbds. In E^2, or more generally in every metric space X, the collection $\mathcal{V}(p)$ of the ε-nbds, $S_\varepsilon(p)$, for $\varepsilon > 0$, forms a nbd basis of p. The subcollection $\mathcal{V}'(p) = \{S_{1/n}(p) \mid n = 1, 2, \ldots\}$ of $\mathcal{V}(p)$ also forms a nbd basis of p. It is also easily seen that $\cup \{\mathcal{V}(p) \mid p \in X\}$ and $\cup \{\mathcal{V}'(p) \mid p \in X\}$ are open bases of X. Generally two nbd bases (open bases) which generate the same topology are called *equivalent*. Thus $\mathcal{V}(p)$ and $\mathcal{V}'(p)$ are equivalent.

As for R_5 in Example II.1, the collection $\mathcal{U}(\alpha) = \{(\beta, \alpha] \mid \beta < \alpha\}$ for $\alpha \neq 0$ and $\mathcal{U}(\alpha) = \{\{0\}\}$ for $\alpha = 0$ forms a nbd basis of α, and $\{\{0\}, (\beta, \alpha] \mid 0 \leq \beta < \alpha < \omega_1\}$ is an open basis. Let us give two more examples of topological spaces defined by means of nbd bases.

Let R_2 be the set of all real numbers. For each point p of R_2, we define a collection $\mathcal{U}(p)$ of subsets by

$$\mathcal{U}(p) = \{(p - \varepsilon, p + \varepsilon) \mid \varepsilon > 0\}^1 \qquad \text{for } p \neq 0,$$

$$\mathcal{U}(p) = \{(-\varepsilon, \varepsilon) - \{1, \tfrac{1}{2}, \tfrac{1}{3}, \ldots\} \mid \varepsilon > 0\} \quad \text{for } p = 0.$$

Then we can easily show that $\mathcal{U}(p)$ satisfies the conditions (i)–(iv) of D). Thus R_2 turns out to be a topological space with a different topology from the usual metric topology.

Let

$$R_4 = \{(x, y) \mid y \geq 0, -\infty < x < +\infty\},$$

i.e. R_4 is a half plane of E^2. Then

$$\mathcal{U}((x, y)) = \{S_\varepsilon((x, y)) \cap R_4 \mid \varepsilon > 0\} \qquad \text{for } y \neq 0,$$

$$\mathcal{U}((x, y)) = \{S_\varepsilon((x, \varepsilon)) \cup \{(x, 0)\} \mid \varepsilon > 0\} \quad \text{for } y = 0,$$

also satisfies the conditions of D); hence R_4 is a topological space. For the real line E^1, the collection $\mathcal{V} = \{(-\infty, a), (b, +\infty) \mid a, b \in E^1\}$ of half-infinite intervals is a subbase.

3. Closure

Definition II.8. Let A be a subset of a topological space X. We denote by $\bar{A}$ the intersection of all closed sets containing A as a subset. $\bar{A}$ is called the *closure* of A. Each point of $\bar{A}$ is called a *contact point* of A.

The following proposition is practically another expression of definition II.8, because the intersection of closed sets is closed.

A) *The closure $\bar{A}$ of A is the smallest closed set which contains A.*

---

[1] We denote by (a, b), $[a, b]$, $[a, b)$ and $(a, b]$ for real numbers a, b with $a < b$ the intervals $\{x \mid a < x < b\}$, $\{x \mid a \leq x \leq b\}$, $\{x \mid a \leq x < b\}$ and $\{x \mid a < x \leq b\}$, respectively. We may also use the same notation (a, b) to mean a point with coordinates a, b but there will be little likelihood of confusion.

B) *Closure satisfies the following conditions*:
   (i) $\bar{\emptyset} = \emptyset$,
  (ii) $\bar{A} \supset A$,
 (iii) $\overline{A \cup B} = \bar{A} \cup \bar{B}$,
 (iv) $\bar{\bar{A}} = \bar{A}$.

Proof. (i) follows from A) and the fact that $\emptyset$ is a closed set. (ii) also follows from A). To prove (iii) we note that Definition II.8 implies the following assertion: if $C \supset D$, then $\bar{C} \supset \bar{D}$. Thus
  (1) $\overline{A \cup B} \supset \bar{A} \cup \bar{B}$.
On the other hand it follows from (ii) that $\bar{A} \cup \bar{B}$ is a closed set containing $A \cup B$. Thus it follows from A) that

$$\overline{A \cup B} \subset \bar{A} \cup \bar{B}.$$

Combining this with the assertion (1) we obtain (iii).

To prove (iv) we note that by A) $\bar{A}$ is the smallest closed set containing $\bar{A}$. However, $\bar{A}$ itself is closed, and hence $\bar{\bar{A}} = \bar{A}$ proving (iv).

C) *A subset F of a topological space X is closed if and only if $\bar{F} = F$. Consequently a set U is open if and only if $\overline{X - U} = X - U$.*

Proof. This is a direct consequence of A).

D) *Let A and p be a set and a point, respectively, of a topological space X. Then $p \in \bar{A}$ if and only if every nbd of p intersects A.*

Proof. Suppose there is a nbd U of p which does not intersect A. We may assume U is open. Then $X - U$ is a closed set containing A. Thus by A) we get $X - U \supset \bar{A}$ which combined with $p \in U$ implies $p \notin \bar{A}$.

Conversely, if $p \notin \bar{A}$, then $X - \bar{A}$ is a nbd of p which does not intersect A.

E) *Let U and p be a set and a point of a topological space X. Then U is a nbd of p if and only if $p \notin \overline{X - U}$.*

Proof. If U is a nbd of p, then since $U \cap (X - U) = \emptyset$, from D) it follows that $p \notin \overline{X - U}$.

Conversely, if $p \notin \overline{X - U}$, then again by D), there is a nbd V of p such that

$$V \cap (X - U) = \emptyset .$$

This implies that $V \subset U$, and hence U is also a nbd of p.

Example II.6. Proposition D) may help us to determine the closure of a given subset. From the same proposition we can also derive that

$$\bar{A} = A \cup B(A) .$$

For example, in the topological space R_5 given in Example II.1, we consider a set $A = (\omega, \ 2\omega) = \{\alpha \mid \omega < \alpha < 2\omega\}$. Then $\bar{A} = (\omega, \ 2\omega] = \{\alpha \mid \omega < \alpha \leq 2\omega\}$. In a discrete space X, $\bar{A} = A$ holds for every subset A of X (see Example II.1). On the other hand, if X has the weakest topology, then $\bar{A} = X$ if $A \neq \emptyset$ and $\bar{A} = \emptyset$ if $A = \emptyset$.

F) *Let X be a set. To every subset A of X, we assign a subset $\bar{A}$ satisfying* (i)–(iv) *of* B). *Now we define that a subset U of X is open if and only if it satisfies the condition of proposition* C):

$$\overline{X - U} = X - U .$$

Then the open sets thus defined satisfy (i)–(iii) *of Definition* II.1. *Thus X is a topological space.*

Proof. It follows from (ii) of B) that

$$\overline{X - \emptyset} = \bar{X} = X = X - \emptyset ,$$

which means that $\emptyset$ is open. It follows from (i) of B) that

$$\overline{X - X} = \bar{\emptyset} = \emptyset = X - X ,$$

which means that X is open. Suppose U_i, $i = 1, \ldots, k$ are open sets; then

$$\overline{X - U_i} = X - U_i , \quad i = 1, \ldots, k .$$

Therefore by a repeated use of (iii) of B) we get

$$X - \overline{\bigcap_{i=1}^{k} U_i} = \overline{\bigcup_{i=1}^{k}(X - U_i)} = \bigcup_{i=1}^{k}\overline{(X - U_i)}$$

$$= \bigcup_{i=1}^{k}(X - U_i) = X - \bigcap_{i=1}^{k} U_i ,$$

which means that $\bigcap_{i=1}^{k} U_i$ is open, proving (ii) of Definition II.1. Suppose U_γ, $\gamma \in \Gamma$ are open sets; then

$$\overline{X - U_\gamma} = X - U_\gamma, \quad \gamma \in \Gamma.$$

On the other hand, we note that (iii) of B) implies that

$$\bar{A} \supset \bar{B} \quad \text{if } A \supset B. \tag{1}$$

Therefore,

$$\overline{X - \cup\{U_\gamma \mid \gamma \in \Gamma\}} = \overline{\cap\{X - U_\gamma \mid \gamma \in \Gamma\}} \subset \cap\{\overline{X - U_\gamma} \mid \gamma \in \Gamma\}$$

$$= \cap\{X - U_\gamma \mid \gamma \in \Gamma\} = X - \cup\{U_\gamma \mid \gamma \in \Gamma\}$$

because

$$\cap\{X - U_\gamma \mid \gamma \in \Gamma\} \subset X - U_\gamma \quad \text{for each } \gamma \in \Gamma.$$

Thus $\cup\{U_\gamma \mid \gamma \in \Gamma\}$ is an open set, proving (iii) of Definition II.1.

Example II.7. We shall give an example of a topological space defined by closure. Let us consider the set

$$R_1 = E^1 \cup \{p\},$$

where E^1 denotes 1-dimensional Euclidean space (or the real line) and p a point which does not belong to E^1. For every subset A of R_1, we define $\bar{A}$ by

$$\bar{A} = \begin{cases} \overline{A - \{p\}}^{E^1} \cup \{p\} & \text{if } A \text{ is an infinite set}, \\ A & \text{if } A \text{ is a finite set}, \end{cases}$$

where $\overline{M}^{E^1}$ denotes the closure of M in E^1 with respect to the ordinary (metric) topology of E^1. Then we can easily verify that $\bar{A}$ satisfies (i)–(iv) of B) and accordingly R_1 is a topological space.

G) *Starting with a topological space X defined with topology $\mathcal{O}$ satisfying (i)–(iii) of Definition II.1, we define closure by use of $\mathcal{O}$ and Definition II.8 and then new open sets (topology) by use of the closure and F). We denote by $\mathcal{O}'$ the newly defined topology. Then $\mathcal{O} = \mathcal{O}'$.*

Proof. It is a direct consequence of C) and F).

H) *Starting with a topological space X defined with closure* $\bar{A}$ *satisfying* (i)–(iv) *of* B), *we define open sets by use of the closure and* F) *and then new closure* $\tilde{A}$ *by use of the open sets and Definition* II.8. *Then* $\tilde{A} = \bar{A}$ *for each subset A of X.*

Proof. By definition

$$\tilde{A} = \cap\{F \mid X - F \text{ is an open set such that } F \supset A\}$$
$$= \cap\{F \mid \bar{F} = F, F \supset A\}.$$

Since, as seen in (1) of the proof of F), $F \supset A$ implies $\bar{F} \supset \bar{A}$, the above equality turns into

$$\tilde{A} = \cap\{F \mid \bar{F} = F, F \supset \bar{A}\}. \tag{1}$$

Therefore

$$\tilde{A} \supset \bar{A}.$$

On the other hand, it follows from (iv) of B) that

$$\bar{\bar{A}} = \bar{A}, \qquad \bar{A} \supset \bar{A};$$

hence $\bar{A}$ satisfies the condition of F appearing in (1), i.e. $\tilde{A} \subset \bar{A}$. This combined with the previous inequality implies $\tilde{A} = \bar{A}$.

In view of G) and H), we can assert that there is no difference in the results whether we adopt the concept of open set or of closure to define a topological space. After all we can define a topological space initially by any one of the following means: open sets (open basis), nbds (nbd basis), or closure.

Though there are several other methods for defining topological spaces, those mentioned here are the most popular and practical ones. When we deal with a given abstract topological space X, we may assume that all of the three terminologies: open sets, nbds and closure (which satisfy Definition II.1, 1.A) and 3.B), respectively) are defined in X, while the relationships between them are given by Definition II.3, 1.B), Definition II.8, C), D) and E).

Let A be a set and p a point of a topological space X. If $p \in \overline{A - \{p\}}$, then p is called a *cluster point* (or *accumulation point*) of A. (Note that p

does not necessarily belong to A.) If a point p of A is not a cluster point of A, then p is called an *isolated point* of A.[1] Every cluster point of A is a contact point of A, but the converse is not true. As easily seen, $\bar{A}$ consists of the cluster points and the isolated points of A. Let A and B be subsets of X such that $B \subset \bar{A}$. Then A is said to be *dense* in B. If A is dense in X, then it is called a *dense subset* of X.

Example II.8. A subset A of a topological space X is called a *border set* if $X - A$ is a dense set of X. A subset A whose closure $\bar{A}$ is a border set is called a *nowhere dense set*. If A is a sum of countably many nowhere dense sets, then it is called a set of the *first category*. If A is not of the first category, then it is called a set of the *second category*. In the real line E^1 the set Q of the rational numbers as well as the set $E^1 - Q$ of the irrational numbers are dense, border sets. We can easily see that Q is of the first category because it is a countable set, and a one point set is nowhere dense in E^1. To show that $E^1 - Q$ is of the second category, we assume the contrary. Then $E^1 = (E^1 - Q) \cup Q$ is also of the first category, and hence

$$E^1 = \bigcup_{i=1}^{\infty} A_i$$

for nowhere dense sets A_i, $i = 1, 2, \ldots$. Note that generally a set A is nowhere dense if and only if for every open set U there is an open set V such that $V \subset U$ and $V \cap A = \emptyset$. Now applying this observation to A_1 and the open segment $I_1 = (0, 1)$, we get an open set V_1 such that

$$V_1 \subset I_1 \quad \text{and} \quad V_1 \cap A_1 = \emptyset.$$

Since every segment contained in V_1 satisfies the same condition, we may assume without loss of generality that V_1 is an open segment with

$$\bar{V}_1 \cap A_1 = \emptyset.$$

Also observe that the length of $V_1 \leq 1$. Repeating this process, we obtain open segments $V_2, V_3, V_4, \ldots$ such that

$$V_1 \supset V_2 \supset V_3 \supset \cdots,$$

$$\bar{V}_i \cap A_i = \emptyset, \quad i = 1, 2, \ldots,$$

the length of $V_i \leq 1/i$.

Since $\bar{V}_1 \supset \bar{V}_2 \supset \bar{V}_3 \supset \cdots$ is a sequence of closed segments whose lengths

---

[1] Some people use the notation A' to denote the set of all cluster points of A, but in this book we give no such specific meaning to the notation A'.

converge to zero, we get a point $p = \bigcap_{i=1}^{\infty} \bar{V}_i$. But it follows from $\bar{V}_i \cap A_i = \emptyset$ that $p \notin A_i$ for every i, contradicting the assumption $E^1 = \bigcup_{i=1}^{\infty} A_i$. Thus $E^1 - Q$ must be of the second category.

4. Convergence

Although we have suggested in the prologue of this chapter that topology is practically equivalent with the concept of convergence, and although we can actually define a topological space by means of convergence if we wish, in this book we rather prefer, from a practical point of view, to derive the concept of convergence from the other fundamental concept, nbd. To discuss convergence in general topological spaces, we need to extend the concepts of point sequence and set sequence to those of net and filter respectively. It seems to the author that both of them have their own merit; net is more intuitive while filter is often easier to handle and is conveniently related to another significant concept, covering, as we shall see later.

Definition II.9. Let $\mathcal{F}$ be a non-empty collection of subsets of a topological space X satisfying:
   (i) $\emptyset \notin \mathcal{F}$,
  (ii) if $A \in \mathcal{F}$ and $X \supset B \supset A$, then $B \in \mathcal{F}$,
 (iii) if $A \in \mathcal{F}$ and $B \in \mathcal{F}$, then $A \cap B \in \mathcal{F}$.
Then $\mathcal{F}$ is called a *filter*.
   If $\mathcal{F}$ is a maximal collection satisfying (i)–(iii), i.e., if there is no collection which contains $\mathcal{F}$ as a proper subcollection and satisfies (i)–(iii), then $\mathcal{F}$ is called a *maximal filter* (or *ultra-filter*).[1]

Definition II.10. Let $\mathcal{F}$ be a filter of a topological space X and p a point of X. If every nbd U of p belongs to $\mathcal{F}$, then $\mathcal{F}$ is said to *converge* to p. We denote this by $\mathcal{F} \to p$. In view of the condition (ii) of Definition II.9, we observe that $\mathcal{F}$ converges to p if and only if for every nbd U of p, there is $A \in \mathcal{F}$ such that $A \subset U$. If $p \in \bar{A}$ for every set $A \in \mathcal{F}$, then we call p a *cluster point* of $\mathcal{F}$.

A) *Let $\mathcal{F}'$ be a collection of sets satisfying the* finite intersection property (f.i.p.), *i.e., for every finitely many members $F_1, \ldots, F_k$ of $\mathcal{F}'$, $\bigcap_{i=1}^{k} F_i \neq \emptyset$;*

---

[1] Note that the concept of filter is actually a set-theoretic one, and X may be assumed just to be a set in this definition.

then there is a maximal filter $\mathcal{F}$ containing $\mathcal{F}'$ as a subcollection. (Note that every filter has f.i.p.)

Proof. Let $\Phi = \{\mathcal{F} \mid \mathcal{F}$ is a collection of sets with f.i.p. such that $\mathcal{F} \supset \mathcal{F}'\}$; then we define order between two elements $\mathcal{F}_1$, $\mathcal{F}_2$ of Φ by the usual inclusion relation, $\mathcal{F}_1 \subset \mathcal{F}_2$. It is easily seen that Φ is a partially ordered set whose every totally ordered subset has a supremum. Therefore by Zorn's lemma there is a maximal element $\mathcal{F}$ of Φ. Let us show that $\mathcal{F}$ is a filter. It is clear that $\mathcal{F}$ satisfies (i) of Definition II.9. If $A, B \in \mathcal{F}$, then

$$\mathcal{G} = \{A \cap B\} \cup \mathcal{F} \in \Phi$$

is obvious, and hence $A \cap B \in \mathcal{F}$, because $\mathcal{F}$ is maximal. To prove (ii) for $\mathcal{F}$, we assume $B \supset A \in \mathcal{F}$. Then

$$\mathcal{G}' = \{B\} \cup \mathcal{F} \in \Phi$$

is obvious, and hence $B \in \mathcal{F}$. Thus $\mathcal{F}$ is the desired maximal filter.

B) *A filter $\mathcal{F}$ is maximal if and only if every set A intersecting every member of $\mathcal{F}$ belongs to $\mathcal{F}$.*

Proof. If $\mathcal{F}$ is a maximal filter and A intersects every member of $\mathcal{F}$, then

$$\mathcal{F}' = \{C \mid C \supset A \cap B \text{ for some } B \in \mathcal{F}\}$$

is easily verified to be a filter containing $\mathcal{F}$ as a subcollection and A as a member. Since $\mathcal{F}$ is maximal, $\mathcal{F} = \mathcal{F}'$ which implies $A \in \mathcal{F}$.

Conversely, let $\mathcal{F}$ be a filter satisfying the condition and $\mathcal{F}'$ a filter such that $\mathcal{F}' \supset \mathcal{F}$. Let A be a given member of $\mathcal{F}'$; then by (i) and (iii) of Definition II.9, A intersects every member of $\mathcal{F}'$ and especially every member of $\mathcal{F}$. Therefore $A \in \mathcal{F}$, which means $\mathcal{F}' = \mathcal{F}$. Thus $\mathcal{F}$ is a maximal filter.

C) *Let $\mathcal{F}$ be a filter and p a point. If $\mathcal{F} \to p$, then p is a cluster point of $\mathcal{F}$.*

Conversely, if $\mathcal{F}$ is a maximal filter and p is a cluster point of $\mathcal{F}$, then $\mathcal{F} \to p$.

Proof. Let $\mathcal{F} \to p$ and $A \in \mathcal{F}$. Then every nbd U of p belongs to $\mathcal{F}$ by Definition II.10, and hence $A \cap U \neq 0$. This implies $p \in \bar{A}$, and hence p is a cluster point of $\mathcal{F}$.

Conversely, suppose p is a cluster point of a maximal filter $\mathcal{F}$. Then by Definition II.10, each nbd U of p intersects every member of $\mathcal{F}$. Hence from B) it follows that $U \in \mathcal{F}$. Therefore $\mathcal{F} \to p$.

It is often convenient to loosen the conditions in the definition of filter. Let $\mathcal{F}'$ be a subcollection of a filter $\mathcal{F}$ such that for every $F \in \mathcal{F}$, there is $F' \in \mathcal{F}'$ with $F' \subset F$. We call such an $\mathcal{F}'$ a *filter basis* of $\mathcal{F}$. Then $\mathcal{F}'$ satisfies:

(i) $\emptyset \notin \mathcal{F}'$,
(ii) for every $F_1, F_2 \in \mathcal{F}'$ there exists $F_3 \in \mathcal{F}'$ with $F_3 \subset F_1 \cap F_2$.

Conversely, given a collection $\mathcal{F}'$ satisfying (i), (ii), then putting

$$\mathcal{F} = \{F \mid F \supset F' \text{ for some } F' \in \mathcal{F}'\}$$

we get a filter $\mathcal{F}$ which contains $\mathcal{F}'$ as a filter basis. $\mathcal{F}'$ is said to *generate* $\mathcal{F}$. Let $\mathcal{F}'$ be a filter basis and p a point. If for every nbd U of p, there is $F' \in \mathcal{F}'$ with $F' \subset U$, then we say that $\mathcal{F}'$ *converges* to p (denoted by $\mathcal{F}' \to p$). Suppose $\mathcal{F}'$ is a filter basis of $\mathcal{F}$; then it is easily seen that $\mathcal{F}' \to p$ if and only if $\mathcal{F} \to p$. *A cluster point* of a filter basis is defined in the same way as in Definition II.10.

We consider a collection of closed sets satisfying (i), (ii) (for a closed set B) and (iii) of Definition II.9 and call it a *closed filter*. The terminologies for filters are mostly valid for closed filters, too. To define the convergence of a closed filter we take the observation in Definition II.10 as the definition.

Example II.9. Let p be a point of a topological space X. Then the collection $\mathcal{U}(p)$ of the nbds of p satisfies (i)–(iii) of Definition II.9, and hence it is a filter. Moreover, it is clear that $\mathcal{U}(p)$ converges to p.

The collection $\mathcal{F}(p)$ of all sets containing p also forms a filter converging to p. We can see that $\mathcal{F}(p)$ is a maximal filter. For if $\mathcal{G}$ is a filter containing $\mathcal{F}(p)$ as a subcollection, then every member B of $\mathcal{G}$ satisfies $B \ni p$, because $\{p\}$ and B are both members of the filter $\mathcal{G}$. Therefore $B \in \mathcal{F}(p)$, which means $\mathcal{G} = \mathcal{F}(p)$.

The sequence $\mathcal{H} = \{H_i \mid i = 1, 2, \ldots\}$ of half lines $H_i = (i, \infty)$, $i = 1, 2, \ldots$, on E^1 satisfies (i) and (iii) but not (ii) of Definition II.9. However, putting

$$\mathcal{F} = \{A \mid E^1 \supset A \supset H_i \text{ for some } i\}$$

we get a filter $\mathcal{F}$ of E^1. Thus $\mathcal{H}$ is a filter basis of $\mathcal{F}$. As is easily seen, neither $\mathcal{F}$ nor $\mathcal{H}$ converges. Generally we can say every set sequence (defined at the end of Section 5 of Chapter I) is a filter basis.

Let p, q be different points of E^2, then the collection $\mathcal{F}$ of the subsets of E^2 containing both of p, q is a filter. It is easily seen that p, q are cluster points of $\mathcal{F}$, but $\mathcal{F}$ converges to no point.[1]

Definition II.11. Let Δ be a non-empty directed set and X a topological space. A mapping φ of Δ into X is called a *net* or more precisely a net on Δ. We may regard a net as a set $\{\varphi(\delta) \mid \delta \in \Delta\}$ of points of X indexed by the elements of Δ. For convenience, we use notation $\varphi(\Delta \mid >)$ to denote a net on Δ and $\varphi(\delta)$ to denote the individual point of the net.

Let $\varphi(\Delta \mid >)$ be a net of a topological space X and A a subset of X. If there is $\delta_0 \in \Delta$ such that $\varphi(\delta) \in A$ for every $\delta > \delta_0$, then $\varphi(\Delta \mid >)$ is said to be *residual* in A (or *eventually* in A). If for every $\delta_0 \in \Delta$, there is $\delta \in \Delta$ such that $\delta > \delta_0$, $\varphi(\delta) \in A$, then $\varphi(\Delta \mid >)$ is said to be *cofinal* in A. If for every subset A of X, $\varphi(\Delta \mid >)$ is residual either in A or in $X - A$, then it is called a *maximal net* (or *ultra-net*).

Definition II.12. If a net $\varphi(\Delta \mid >)$ is residual in every nbd U of a point p, then it is said to *converge* to p. We denote this by $\varphi(\Delta \mid >) \to p$.

If $\varphi(\Delta \mid >)$ is cofinal in every nbd U of p, then p is called a *cluster point* of $\varphi(\Delta \mid >)$.

Let $\mathcal{F} = \{A_\delta \mid \delta \in \Delta\}$ be a filter; then we can regard Δ as a directed set, defining that $\delta > \delta'$ if and only if $A_\delta \subset A_{\delta'}$. (We assume that $A_\delta \neq A_{\delta'}$ if $\delta \neq \delta'$). A net $\varphi(\Delta \mid >)$ on the directed set Δ satisfying $\varphi(\delta) \in A_\delta$ for every $\delta \in \Delta$ is called a *net derived* from the filter $\mathcal{F}$.

Conversely, let $\varphi(\Delta \mid >)$ be a given net. Putting

$$\mathcal{F} = \{A \mid \varphi(\Delta \mid >) \text{ is residual in } A\},$$

we obtain a filter $\mathcal{F}$ which is called the *filter derived* from the net $\varphi(\Delta \mid >)$.

As seen in the following, we can discuss convergence of filters and that of nets in parallel.

D) *A filter $\mathcal{F}$ converges to a point p if and only if every net derived from $\mathcal{F}$ converges to p.*

[1] Extensive studies on filters have been done by J. Schmidt and other mathematicians. See J. Schmidt [1], [2], G. Bruno–J. Schmidt [1], G. Grimeisen [1].

Proof. Suppose $\mathscr{F} \to p$ and $\mathscr{F} = \{A_\delta \mid \delta \in \Delta\}$. Let $\varphi(\Delta \mid >)$ be a net derived from $\mathscr{F}$, i.e. $\varphi(\delta) \in A_\delta$, $\delta \in \Delta$. Given a nbd U of p, then $U = A_{\delta_0}$ for some $\delta_0 \in \Delta$. If $\delta > \delta_0$, then $A_\delta \subset A_{\delta_0} = U$ which implies $\varphi(\delta) \in U$. Therefore $\varphi(\Delta \mid >) \to p$.

Conversely, suppose $\mathscr{F} \not\to p$, i.e. $\mathscr{F}$ does not converge to p. Then for some nbd U of p, $A_\delta \not\subset U$ for every $\delta \in \Delta$. Hence, we can choose a point $\varphi(\delta) \in A_\delta - U$ for each $\delta \in \Delta$. Then $\varphi(\Delta \mid >)$ is a net derived from $\mathscr{F}$, but does not converge to p.

E) *A net $\varphi(\Delta \mid >)$ converges to a point p if and only if the filter derived from $\varphi(\Delta \mid >)$ converges to p.*

Proof. Let $\mathscr{F}$ be the filter derived from $\varphi(\Delta \mid >)$, i.e.

$$\mathscr{F} = \{A \mid \varphi(\Delta \mid >) \text{ is residual in } A\}.$$

If $\varphi(\Delta \mid >) \to p$, then for a given nbd U of p, $\varphi(\Delta \mid >)$ is residual in U. Hence $U \in \mathscr{F}$. Therefore $\mathscr{F} \to p$.

Conversely, let $\mathscr{F} \to p$. Then every nbd U of p satisfies $U \in \mathscr{F}$, and hence $\varphi(\Delta \mid >)$ is residual in U. Therefore $\varphi(\Delta \mid >) \to p$.

Example II.10. The set N of the natural numbers is a directed set with respect to the natural order. A net $\varphi(N \mid >)$ on N is a point sequence. In this respect, the concept of net is an extension of that of point sequence. However, the property of net is sometimes tricky. Suppose $\Delta = \{(n, m) \mid n, m = 1, 2, \ldots\}$. Define that $(n, m) > (n', m')$ if $n \geqslant n'$, $m \geqslant m'$; then Δ turns out to be a directed set. Let us define a net $\varphi(\delta \mid >)$ of E^2 on Δ by

$$\varphi((n, m)) = (1/n, 1/m).$$

Then $\varphi(\delta \mid >) \to (0, 0)$ is obvious. Though $\Delta' = \{(n, 1) \mid n = 1, 2, \ldots\}$ is an infinite directed subset of N, $(0, 0)$ is neither a convergence point nor cluster point of the subnet $\varphi(\Delta' \mid >)$.

Generally, a subset Δ' of a directed set Δ is called *cofinal* in Δ if for each $\delta \in \Delta$, there is $\delta' \in \Delta'$ with $\delta' > \delta$. Suppose Δ' is cofinal in Δ. Then we can assert that $\varphi(\Delta \mid >) \to p$ implies $\varphi(\Delta' \mid >) \to p$. But still $\varphi(\Delta' \mid >)$ is somewhat different, in its property, from a subsequence of a point sequence. Let Δ be the directed set of all countable ordinal numbers with the usual order. For each $\delta \in \Delta$, we can uniquely find a sequence

$$\delta_1 < \delta_2 < \cdots < \delta_n = \delta$$

of ordinal numbers such that δ_1 is a limit ordinal number and such that each δ_{i+1} is the immediate successor of δ_i. Then we define a net $\varphi(\Delta \mid >)$ of E^1 on Δ by $\varphi(\delta) = 1/n$ for the natural number n defined as in the above. Then we can easily see that 0 is a cluster point of $\varphi(\Delta \mid >)$, but there is no cofinal subset Δ' of Δ for which $\varphi(\Delta' \mid >) \to 0$. (Recall that every point sequence in E^1 clustering to 0 contains a subsequence which converges to 0.)

5. Covering

Let X be a topological space and $\mathcal{U}$ a collection of subsets of X. We shall often call a collection of subsets merely a *collection*. If the elements of $\mathcal{U}$ cover X, i.e., if

$$\bigcup \{U \mid U \in \mathcal{U}\} = X,$$

then we call $\mathcal{U}$ a *covering* (*cover*) of X. A covering consisting of finitely (countably, infinitely) many sets is called a *finite* (*countable*, *infinite*) *covering*. If a covering consists of closed (open) sets only, then it is called a *closed* (*open*) *covering*.

Let $\mathcal{U}$ be a covering of X. If every point p of X has a nbd U which intersects at most finitely (countably) many sets belonging to $\mathcal{U}$, then $\mathcal{U}$ is called a *locally finite* (*locally countable*) covering. We often index a given locally finite covering $\mathcal{U}$ as

$$\mathcal{U} = \{U_\alpha \mid \alpha \in A\}.$$

Generally speaking, despite the local finiteness of $\mathcal{U}$, every nbd of a point p of X may intersect U_α for infinitely many indexes α, because U_α might be equal to $U_{\alpha'}$, even if $\alpha \neq \alpha'$. However, in this book *we make it convention to index $\mathcal{U}$ so that each point p of X has a nbd which intersects U_α for only finitely many $\alpha \in A$.* Similar conventions apply to similar coverings such as point-finite coverings which are defined below and discrete coverings which will be defined in Section 1, Chapter V.

A covering $\mathcal{U}$ is called a *point-finite* (*point-countable*) *covering* if every point p of X is contained in at most finitely many (countably many) members of $\mathcal{U}$. A covering $\mathcal{U}$ is called a *disjoint covering* (or *decomposition*) if each two members of $\mathcal{U}$ are disjoint.

Let $\mathcal{U}$ and $\mathcal{V}$ be coverings of X. If $\mathcal{U} \subset \mathcal{V}$, then $\mathcal{U}$ is called a *subcovering* of $\mathcal{V}$. On the other hand, if for each set U belonging to $\mathcal{U}$, there is a set V belonging to $\mathcal{V}$ such that $U \subset V$, then we call $\mathcal{U}$ a *refinement* of $\mathcal{V}$ and we denote this relation by $\mathcal{U} < \mathcal{V}$. It is clear that $\mathcal{U} \subset \mathcal{V}$ implies $\mathcal{U} < \mathcal{V}$, but the converse is not true. We should note that $\mathcal{U} < \mathcal{V}$ and $\mathcal{V} < \mathcal{U}$ does not necessarily imply $\mathcal{U} = \mathcal{V}$. Moreover, if $\mathcal{U}$ and $\mathcal{V}$ are two coverings, we define their *union* and *intersection* by

$$\mathcal{U} \vee \mathcal{V} = \mathcal{U} \cup \mathcal{V} = \{U \mid U \in \mathcal{U} \text{ or } U \in \mathcal{V}\},$$

$$\mathcal{U} \wedge \mathcal{V} = \{U \cap V \mid U \in \mathcal{U}, V \in \mathcal{V}\}.$$

If $\mathcal{U}$ and $\mathcal{V}$ are open coverings, then $\mathcal{U} \vee \mathcal{V}$ and $\mathcal{U} \wedge \mathcal{V}$ are also open coverings. The following property is also obvious.

A) $\mathcal{U} \wedge \mathcal{V} < \mathcal{U} < \mathcal{U} \vee \mathcal{V}$ *for every pair of coverings* $\mathcal{U}$, $\mathcal{V}$.

We can extend the definition to arbitrarily many coverings $\mathcal{U}_\alpha$, $\alpha \in A$, as

$$\vee \{\mathcal{U}_\alpha \mid \alpha \in A\} = \cup \{\mathcal{U}_\alpha \mid \alpha \in A\},$$

$$\wedge \{\mathcal{U}_\alpha \mid \alpha \in A\} = \{\cap \{U_\alpha \mid \alpha \in A\} \mid U_\alpha \in \mathcal{U}_\alpha, \alpha \in A\}.$$

Given a point p, a set A, and a collection $\mathcal{U}$ in X, we often use the following notations:

$$S(p, \mathcal{U}) = \cup \{U \mid p \in U \in \mathcal{U}\},$$

$$S(A, \mathcal{U}) = \cup \{U \mid U \in \mathcal{U}, U \cap A \neq \emptyset\},$$

$$S^1(p, \mathcal{U}) = S(p, \mathcal{U}),$$

$$S^n(p, \mathcal{U}) = S(S^{n-1}(p, \mathcal{U}), \mathcal{U}), \quad n = 2, 3, \ldots,$$

$$S^1(A, \mathcal{U}) = S(A, \mathcal{U}),$$

$$S^n(A, \mathcal{U}) = S(S^{n-1}(A, \mathcal{U}), \mathcal{U}), \quad n = 2, 3, \ldots,$$

$$\bar{\mathcal{U}} = \{\bar{U} \mid U \in \mathcal{U}\},$$

$$\mathcal{U}^\Delta = \{S(p, \mathcal{U}) \mid p \in X\},$$

$$\mathcal{U}^* = \{S(U, \mathcal{U}) \mid U \in \mathcal{U}\},$$

$$\mathcal{U}^{\Delta\Delta} = (\mathcal{U}^\Delta)^\Delta, \qquad \mathcal{U}^{**} = (\mathcal{U}^*)^*, \text{ etc.}$$

$S(p, \mathcal{U})$ $(S(A, \mathcal{U}))$ is called the *star* of p (of A) with respect to $\mathcal{U}$. If $\mathcal{U}$ is an open covering, then $S^n(p, \mathcal{U})$ and $S^n(A, \mathcal{U})$ are open sets, and therefore $\mathcal{U}^\Delta$ and $\mathcal{U}^*$ are also open coverings. We often consider coverings $\mathcal{U}, \mathcal{V}$ such that

$$\mathcal{U}^\Delta < \mathcal{V} \quad \text{or} \quad \mathcal{U}^* < \mathcal{V}.$$

In the former case $\mathcal{U}$ is called a *delta-refinement* of $\mathcal{V}$, and in the latter case a *star-refinement*. A covering $\mathcal{U}$ is called a *normal covering* if there is a sequence $\mathcal{U}_1, \mathcal{U}_2, \ldots$ of open coverings such that

$$\mathcal{U} > \mathcal{U}_1^* > \mathcal{U}_1 > \mathcal{U}_2^* > \mathcal{U}_2 > \cdots .^1$$

We should note that all the above notions and notations are generally valid for collections (of subsets), too. For example, we may use terminologies like open collection, locally finite collection etc. without further comment. But remember that *we shall use in this book the terminologies refinement, delta-refinement and star-refinement only for coverings*. Also remember that the convention established for indexing locally finite coverings will be applied to locally finite collections and other similar collections.

B) $\mathcal{U} < \mathcal{U}^\Delta < \mathcal{U}^* < \mathcal{U}^{\Delta\Delta}$ *for every covering* $\mathcal{U}$.

Proof. Since $\mathcal{U} < \mathcal{U}^\Delta < \mathcal{U}^*$ is clear from the definitions of $\mathcal{U}^\Delta$ and $\mathcal{U}^*$, we consider a given member $S(U, \mathcal{U})$ of $\mathcal{U}^*$ to show that $\mathcal{U}^* < \mathcal{U}^{\Delta\Delta}$. Choose a fixed point $p \in U$. Suppose U' is an arbitrary member of $\mathcal{U}$ intersecting U, then there is some $p' \in U' \cap U$. Hence we obtain

$$U' \subset S(p', \mathcal{U}). \tag{1}$$

On the other hand,

$$p \in U \subset S(p', \mathcal{U}),$$

and thus $S(p', \mathcal{U})$ is a member of $\mathcal{U}^\Delta$ containing p. Therefore

---

[1] As will be seen from the following Proposition B), this condition is equivalent with the existence of a sequence $\mathcal{V}_1, \mathcal{V}_2, \ldots$ of open coverings such that $\mathcal{U} > \mathcal{V}_1^\Delta > \mathcal{V}_1 > \mathcal{V}_2^\Delta > \mathcal{V}_2 > \cdots$. Such a sequence $\{\mathcal{U}_i \mid i = 1, 2, \ldots\}$ or $\{\mathcal{V}_i \mid i = 1, 2, \ldots\}$ is called a *normal sequence*.

$$S(p', \mathcal{U}) \subset S(p, \mathcal{U}^{\Delta}).$$

Combining this with (1), we obtain $U' \subset S(p, \mathcal{U}^{\Delta})$ for every $U' \in \mathcal{U}$ intersecting U, i.e.

$$S(U, \mathcal{U}) \subset S(p, \mathcal{U}^{\Delta}) \in \mathcal{U}^{\Delta\Delta}.$$

Therefore

$$\mathcal{U}^* < \mathcal{U}^{\Delta\Delta}.$$

C) *Let $\mathcal{U}$ be a locally finite collection. Then*

$$\bigcup \{\bar{U} \mid U \in \mathcal{U}\} = \overline{\bigcup \{U \mid U \in \mathcal{U}\}}.$$

Proof. Let $p \notin \bigcup \{\bar{U} \mid U \in \mathcal{U}\}$; then by the local finiteness of $\mathcal{U}$, there is a nbd V of p which intersects at most finitely many members of $\mathcal{U}$, say $U_1, \ldots, U_n$. Since $p \notin \bar{U}_i$, each $X - \bar{U}_i$ is an open nbd of p. Therefore

$$W = V \cap [\cap \{X - \bar{U}_i \mid i = 1, \ldots, n\}]$$

is a nbd of p. For $i = 1, \ldots, n$ we get

$$W \cap U_i \subset (X - \bar{U}_i) \cap \bar{U}_i = \emptyset.$$

For any $U \in \mathcal{U}$ which is different from each U_i, we get

$$W \cap U \subset V \cap U = \emptyset.$$

Thus W is a nbd of p intersecting no member of U. Therefore

$$p \notin \overline{\bigcup \{U \mid U \in \mathcal{U}\}},$$

which means

$$\bigcup \{\bar{U} \mid U \in \mathcal{U}\} \supset \overline{\bigcup \{U \mid U \in \mathcal{U}\}}.$$

Since

$$\bigcup \{\bar{U} \mid U \in \mathcal{U}\} \subset \overline{\bigcup \{U \mid U \in \mathcal{U}\}}$$

follows from the fact that

$$\bar{U} \subset \overline{\bigcup\{U \mid U \in \mathcal{U}\}} \quad \text{for all } U \in \mathcal{U},$$

we can conclude the validity of the equality.

The following is a direct consequence of C).

D) *Let $\mathcal{F}$ be a locally finite closed collection; then $F = \bigcup\{F' \mid F' \in \mathcal{F}\}$ is a closed set.*

Example II.11. Let $\mathcal{U} = \{(2n-1, \quad 2n+1), \quad (2n, \quad 2n+2) \mid n = 0, \quad \pm 1,$ $\pm 2, \ldots\}$, where $(2n-1, 2n+1)$ and $(2n, 2n+2)$ denote open intervals on E^1. Then $\mathcal{U}$ is a locally finite (but infinite) open covering of E^1.

Let X be a metric space; then

$$\mathcal{S}_\varepsilon = \{S_\varepsilon(p) \mid p \in X\},$$

where $S_\varepsilon(p)$ denotes the ε-nbd of p, is an open covering of X. We can easily show that $\mathcal{S}_{\varepsilon/3} = \{S_{\varepsilon/3}(p) \mid p \in X\}$ is a star refinement of $\mathcal{S}_\varepsilon$, because

$$S(S_{\varepsilon/3}(p), \mathcal{S}_{\varepsilon/3}) \subset S_\varepsilon(p).$$

Therefore $\mathcal{S}_\varepsilon$ is a normal covering.

6. Mapping

Definition II.13. Let X and Y be topological spaces and f a mapping of X into Y. Suppose $p \in X$, $q \in Y$ and $f(p) = q$. If for every nbd V of q, $f^{-1}(V)$ is a nbd of p, then we call f *continuous* at the point p. In view of (iii) of 1.A) we may express the definition as follows: If for every nbd V of q, there is a nbd U of p such that $f(U) \subset V$, then f is said to be continuous at p.

If f is continuous at every point of X, then it is called a *continuous mapping*. In this book we mean by *a continuous function* on (or over) a topological space X a continuous mapping of X into E^1, i.e. a real-valued continuous function (unless the contrary is explicitly stated).[1]

A) *Let X, Y and Z be topological spaces. If f and g are continuous mappings of X into Y and of Y into Z, respectively, then $g \circ f$ is a continuous mapping of X into Z, where we denote by $g \circ f$ the composite mapping which maps $p \in X$ to $g(f(p)) \in Z$.*

Proof. It is a direct consequence of Definition II.13.

B) *Let f be a mapping of a topological space X into a topological space Y. Then f is a continuous mapping if and only if it satisfies one of the following conditions*:

(i) *for every open set V of Y, $f^{-1}(V)$ is an open set of X,*

(ii) *for every closed set G of Y, $f^{-1}(G)$ is a closed set of X,*

(iii) *for every subset A of X, $f(\bar{A}) \subset \overline{f(A)}$,*

(iv) *if a filter basis $\mathscr{F}$ of X converges to a point p in X, then the filter basis $f(\mathscr{F}) = \{f(A) \mid A \in \mathscr{F}\}$ converges to $f(p)$ in Y.*

Proof. The proof will proceed as follows: Definition II.13 $\Rightarrow$ (i) $\Rightarrow$ (ii) $\Rightarrow$ (iii) $\Rightarrow$ (iv) $\Rightarrow$ Definition II.13.

Definition II.13 $\Rightarrow$ (i).

Let f be a continuous mapping of X into Y, and V a given open set of Y. Suppose $p \in f^{-1}(V)$; then $f(p) \in V$. Since V is an open set, it is a nbd of $f(p)$. Therefore by Definition II.13, $f^{-1}(V)$ is a nbd of p. Since p is a given point of $f^{-1}(V)$, $f^{-1}(V)$ is an open set. (Recall 1.B).)

(i) $\Rightarrow$ (ii).

Let f be a mapping of X into Y, satisfying (i), and G a closed set of Y. Then $Y - G$ is an open set of Y, and hence $f^{-1}(Y - G)$ is an open set of X by (i). Since

$$f^{-1}(G) = X - f^{-1}(Y - G),$$

$f^{-1}(G)$ is a closed set of X.

(ii) $\Rightarrow$ (iii).

---

[1] For a function f over a topological space X, we may use notations like $f(A) = 1$ to mean that $f(p) = 1$ for every point p of a subset A of X, or $0 \le f \le 1$ to mean that $0 \le f(p) \le 1$ for every point p of X. Thus, to be precise, $f(A) = 1$ means that $f(A) \subset \{1\}$.

Let f be a mapping of X into Y satisfying (ii), and A a subset of X. By condition (ii), $f^{-1}(\overline{f(A)})$ is a closed set of X and satisfies

$$f^{-1}(\overline{f(A)}) \supset f^{-1}(f(A)) \supset A .$$

Therefore

$$f^{-1}(\overline{f(A)}) = \overline{f^{-1}(\overline{f(A)})} \supset \bar{A} .$$

This implies that

$$\overline{f(A)} \supset f(f^{-1}(\overline{f(A)})) \supset f(\bar{A})$$

proving (iii).

(iii) $\Rightarrow$ (iv).

Let f be a mapping of X into Y satisfying (iii), and $\mathscr{F}$ a filter basis converging to $p \in X$. It is obvious that $f(\mathscr{F})$ is a filter basis. Then f satisfies the condition of Definition II.13, i.e., if V is a nbd of $f(p)$, then $f^{-1}(V)$ is a nbd of p. For, suppose $f^{-1}(V)$ is no nbd of p; then by 3.E)

$$p \in \overline{X - f^{-1}(V)} .$$

This combined with (iii) implies

$$f(p) \in f(\overline{X - f^{-1}(V)}) \subset \overline{f(X - f^{-1}(V))} \subset \overline{Y - V}$$

contradicting the fact that V is a nbd of $f(p)$. Now, let us turn to the proof that $f(\mathscr{F}) \rightarrow f(p)$. Suppose V is a given nbd of $f(p)$; then $f^{-1}(V)$ is a nbd of $f(p)$ as shown in the above. Hence $F \subset f^{-1}(V)$ for some $F \in \mathscr{F}$, which implies $f(F) \subset V$. Thus $f(\mathscr{F}) \rightarrow f(p)$.

(iv) $\Rightarrow$ Definition II.13.

Let f be a mapping of X into Y satisfying (iv). Suppose p is a point of X and V is a nbd of $f(p)$. We consider the filter $\mathscr{U}(p)$ of all nbds of p. Then since $\mathscr{U}(p) \rightarrow p$, by (iv) we obtain

$$f(\mathscr{U}(p)) \rightarrow f(p) .$$

Hence

$$F \subset V \quad \text{for some } F \in f(\mathscr{U}(p)),$$

which means that $f(U) \subset V$ for some nbd U of p. This implies that

$$U \subset f^{-1}(f(U)) \subset f^{-1}(V),$$

and hence $f^{-1}(V)$ is a nbd of p. Thus f is a continuous mapping.

C) *Suppose X and X' are topological spaces with the same points but different topologies. Then the topology of X is stronger than that of X' if and only if the identity mapping $f(p) = p$ is a continuous mapping of X onto X'.*

Proof. The easy proof is left to the reader.

Definition II.14. Let f be a continuous mapping of a topological space X onto a topological space Y. If f is a one-to-one mapping and the inverse mapping f^{-1} of f is also a continuous mapping of Y onto X, then we call f a *homeomorphism* or *topological mapping* of X onto Y. If there is a homeomorphism of X onto Y, then we call X and Y *homeomorphic*.

Example II.12. The concept of continuous mapping is a generalization of that of real-valued continuous function. In fact a real-valued continuous function with n real variables is a continuous mapping of E^n into E^1 if it is defined for every value of the variables. Every monotone continuous function with a real variable gives an example of a topological mapping. For example, the mapping f in Example I.7 is a topological mapping of the open interval $(-1, 1)$ onto E^1. Thus $(-1, 1)$ and E^1 are homeomorphic.

Let us consider an Euclidean plane $E^2 = \{(x, y, z) \mid z = 0\}$ and a sphere $S^2 = \{(x, y, z) \mid x^2 + y^2 + (z - 1)^2 = 1\}$ in E^3. To each point $p \in S^2$ we assign the intersection $f(p)$ of E^2 and the straight line connecting p and $p_0 = (0, 0, 2)$. Then it is easily seen that f is a topological mapping of $S - \{p_0\}$ onto E^2. Thus they are homeomorphic. On the other hand, E^2 and E^1 are not homeomorphic, because E^1 minus one point can be decomposed into two disjoint, non-empty, open sets while this is not the case for E^2. *Two homeomorphic topological spaces have the same topological properties and hence in topology, they are often regarded as the same space.*

Among various types of mappings appearing in topology, continuous mappings including topological mappings are the most important ones, but there are many other interesting conditions for mappings. Here we shall give two of them, leaving some others to the later chapters. Let f be a mapping of a topological space X onto a topological space Y. If every closed set of X is mapped by f onto a closed set of Y, i.e., if the image $f(F)$ of each closed set F of X is a closed set of Y, then we call f a *closed mapping*. If every open set of X is mapped by f onto an open set of Y, then f is called an *open mapping*. In view of (ii) (or (i)) of B), we can say that a one-to-one mapping f of X onto Y is a homeomorphism if and only if f is a closed (or open) continuous mapping.

Example II.13. The mapping f defined by $f((x, y)) = x$ is a continuous mapping of E^2 onto E^1, but it is not a closed mapping. Because

$$F = \{(x, y) \mid y = \tan \tfrac{1}{2}\pi x, |x| < 1\}$$

is a closed set of E^2 while

$$f(F) = \{x \mid |x| < 1\}$$

is not closed in E^1. The mapping g defined by

$$g(x) = \begin{cases} \dfrac{1}{x} - 1, & 0 < x < 1, \\[2mm] \dfrac{1}{x} + 1, & -1 < x < 0, \\[2mm] 0, & \text{otherwise}, \end{cases}$$

is a closed, but non-continuous mapping of E^1 onto itself.

7. Subspace, product space, quotient space and inverse limit space

In this section we shall describe several methods to construct new topological spaces from one or more given topological spaces.

A) *Let X be a given topological space with topology $\mathcal{O}$ (the collection of the open sets) and X' a subset of X. Put $\mathcal{O}' = \{X' \cap U \mid U \in \mathcal{O}\}$; then $\mathcal{O}'$ satisfies the conditions (i)–(iii) of Definition II.1 with respect to X', i.e. X' is a topological space with topology $\mathcal{O}'$.*

Proof. It is left to the reader.

Definition II.15. The topological space X' defined in A) is called a *subspace of X*.

The following assertions in which X' denotes a subspace of a topological space X are direct consequences of the definition of subspace. Their proofs are left to the reader.

B) *Let $F' \subset X'$; then F' is a closed set of the subspace X' if and only if $F' = X' \cap F$ for a closed set F of X.*

C) *Let A be a subset of X' and denote by $\bar{A}'$ the closure of A in the subspace X'. Then $\bar{A}' = X' \cap \bar{A}$, where $\bar{A}$ denotes the closure of A in X.*

D) *Let p be a point of X'. Then a subset U' of X' is a nbd of p in the subspace X' if and only if $U' = X' \cap U$ for some nbd U of p in X.*

Example II.14. Let f be a continuous mapping of a subspace X' of a topological space X into a topological space Y. If there is a continuous mapping g of X into Y such that $f(p) = g(p)$ for all $p \in X'$, then we call g a (*continuous*) *extension* of f over X and f (*continuously*) *extendable over X*. On the other hand we call f the *restriction* of g to X'. For example, let $X = [0, 1]$ with the ordinary metric topology, $X' = (0, 1]$. Then $f(x) = 1/x$, $x \in X'$, is not extendable over X. On the other hand $f'(x) = x \sin 1/x$, $x \in X'$, can be extended over X by putting $g(0) = 0$ and $g(x) = x \sin 1/x$, $x \in X'$.

Now, let us turn to the definition of product space. Let X_i, $i = 1, \ldots, n$, be given topological spaces and $p = \{p_i \mid i = 1, \ldots, n\}$ a given point of the cartesian product of X_i, $i = 1, \ldots, n$. For each i, we choose a nbd U_i of p_i in X_i to construct a subset $\prod_{i=1}^{n} U_i$ of the cartesian product $\prod_{i=1}^{n} X_i$. Then the collection

$$\mathcal{V}(p) = \left\{ \prod_{i=1}^{n} U_i \mid U_i \text{ is a nbd of } p_i \text{ for each } i \right\}$$

of those subsets satisfies the conditions (i)–(iv) of 2.D). The topological space $\prod_{i=1}^{n} X_i$ defined with this nbd basis, $\mathcal{V}(p)$, is called the *product space* of $X_1, \ldots, X_n$. To define the product space of infinitely many topological spaces, we usually proceed as follows.

E) *Let* X_γ, $\gamma \in \Gamma$, *be topological spaces, and* $p = \{p_\gamma \mid \gamma \in \Gamma\}$ *a point of their cartesian product* $\prod\{X_\gamma \mid \gamma \in \Gamma\}$. *We choose an arbitrary nbd* U_γ *of* p_γ *in* X_γ *for each* γ *such that* $U_\gamma \neq X_\gamma$ *for at most finitely many* γ. *Then we construct a subset* $\prod\{U_\gamma \mid \gamma \in \Gamma\}$ *of the cartesian product. Denote by* $\mathcal{V}(p)$ *the collection of those subsets, i.e.* $\mathcal{V}(p) = \{\prod\{U_\gamma \mid \gamma \in \Gamma\} \mid$ *for each* $\gamma \in \Gamma$, U_γ *is a nbd of* p_γ *in* X_γ; $U_\gamma = X_\gamma$ *except for at most finitely many* $\gamma\}$. *Then* $\mathcal{V}(p)$ *satisfies the conditions* (i)–(iv) *of* 2.D), *i.e., the cartesian product* $\prod\{X_\gamma \mid \gamma \in \Gamma\}$ *turns out to be a topological space with nbd basis* $\mathcal{V}(p)$.

Proof. The easy proof of this assertion is left to the reader.

Definition II.16. The topological space $\prod\{X_\gamma \mid \gamma \in \Gamma\}$ defined in E) is called the *product space* of the topological spaces X_γ, $\gamma \in \Gamma$.

Example II.15. In E) and accordingly in Definition II.16, too, we may replace a nbd U_γ of p_γ with a nbd belonging to a nbd basis of p_γ. More precisely, we consider a nbd basis $\mathcal{V}(p_\gamma)$ of every point p_γ of X_γ and for each $\gamma \in \Gamma$. (We assume $X_\gamma \in \mathcal{V}(p)$, $\gamma \in \Gamma$.) Then we define

$$\mathcal{V}'(p) = \left\{ \prod\{V_\gamma \mid \gamma \in \Gamma\} \mid V_\gamma \in \mathcal{V}(p_\gamma),\ \gamma \in \Gamma; \right.$$
$$\left. V_\gamma = X_\gamma \text{ except for at most finitely many } \gamma \right\}.$$

This we take as a nbd basis of $p = \{p_\gamma \mid \gamma \in \Gamma\} \in \prod\{X_\gamma \mid \gamma \in \Gamma\}$. $\mathcal{V}'(p)$ may be different from $\mathcal{V}(p)$ given in E), but they are easily seen to be equivalent.

Viewing this, we can see that E^2 is the product space of E^1 with itself and more generally E^n is the product space of n copies of E^1. We often deal with the subspace

$$I^\omega = \{(x_1, x_2, \ldots) \mid 0 \leqslant x_i \leqslant 1/i,\ i = 1, 2, \ldots\}$$

of Hilbert space. It is easily seen that I^ω is homeomorphic with the product space of countably many closed segments $[0, 1/i]$, $i = 1, 2, \ldots$. I^ω is called the *Hilbert cube*. Generally, if topological spaces X_γ and Y_γ are homeomorphic for each $\gamma \in \Gamma$, then we can easily show that their product spaces $\prod\{X_\gamma \mid \gamma \in \Gamma\}$ and $\prod\{Y_\gamma \mid \gamma \in \Gamma\}$ are homeomorphic. Thus the Hilbert cube I^ω is homeomorphic with the product space of countably many copies of the unit segment $[0, 1]$ because $[0, 1/i]$ and $[0, 1]$ are homeomorphic.

Let $\{X_\gamma \mid \gamma \in \Gamma\}$ be a collection of topological spaces and $X = \prod X_\gamma$ their cartesian product. We denote by π_γ the *projection* from X onto X_γ, i.e. $\pi_\gamma(\{x_\gamma \mid \gamma \in \Gamma\}) = x_\gamma$. (Every projection is obviously continuous.) Thus we can define the topology of the product space as the one derived from the subbase $\{\pi_\gamma^{-1}(U) \mid \gamma \in \Gamma, \ U$ is an open set of $X_\gamma\}$.

F) *Let X be a topological space and $\mathcal{D}$ a decomposition of X, i.e. a covering of X whose any two distinct elements are disjoint. Moreover, we assume that $\mathcal{D}$ contains no empty set as an element.*

Then we denote by φ the mapping of X onto $\mathcal{D}$ which maps every point p of X to the member of $\mathcal{D}$ containing p. Now, we call a subset $\mathcal{D}'$ of $\mathcal{D}$ an open set if and only if $\varphi^{-1}(\mathcal{D}')$ is an open set of X. Then the collection $\mathcal{O}$ of open sets thus defined satisfies the conditions (i)–(iii) of Definition II.1, i.e. $\mathcal{D}$ turns out to be a topological space. We denote this topological space by $X(\mathcal{D})$.

Proof. The easy proof is left to the reader.

Definition II.17. The topological space $X(\mathcal{D})$ defined in F) is called the *decomposition space* (or *quotient space*) of X with respect to $\mathcal{D}$. The mapping φ of X onto $X(\mathcal{D})$ is called the *natural mapping* or a *quotient mapping*. In other words, given a mapping φ from a space X onto a space Y such that V is an open set of Y if and only if $\varphi^{-1}(V)$ is open in X, then Y is called a quotient space of X, and φ a quotient mapping.

Example II.16. We may intuitively consider that the quotient space $X(\mathcal{D})$ is obtained from X by identifying the points belonging to the same member of $\mathcal{D}$.[1] We also note that a set of $X(\mathcal{D})$ is closed if and only if its inverse image by the natural mapping is a closed set of X.

Let E be an equivalence relation between points of a topological space X satisfying (i)–(iii) of Example I.3. Then by use of the equivalence, we can decompose X into disjoint sets which we consider form a decomposition $\mathcal{D}$ of X. We may use notation X/E, $X/\sim$ or $X/\mathcal{D}$ instead of $X(\mathcal{D})$ to denote the decomposition space.

Conversely, we can regard every decomposition space $X(\mathcal{D})$ as having

---

[1] A variation of this concept is adjunction space. Let X and Y be topological spaces and f a continuous mapping from a closed set A of X into Y. Then the discrete sum of X and Y is the sum $X \cup Y$ of the two sets X and Y with the topology $\{U \mid U \subset X \cup Y, \ U \cap X$ is open in X, and $U \cap Y$ is open in $Y\}$. Then the quotient space $X \cup_f Y$ obtained from the discrete sum by identifying each $x \in A$ with $f(x)$ is called the *adjunction space*.

been induced by an equivalence relation $\sim$ which is defined by: $a \sim b$ provided that $a, b \in D$ for some $D \in \mathcal{D}$. Let E^1 be the real line and E the relation between two points p, q of E^1, defined by: $p \sim q$ provided that $p - q =$ an integer. As is easily seen, E satisfies (i)–(iii) in Example I.3. The decomposition space E^1/E is homeomorphic with a circle S^1, which is a subspace of E^2. Further elementary examples: A cylinder $\{(x, y, z) \in E^3 \,|\, x^2 + y^2 = 1\}$ is a quotient space of E^2 and also the product space of E^1 and a circle. A torus is a quotient space of a cylinder and accordingly of E^2. It is also the product space of two circles.

G) *Let $\mathcal{D}$ be a decomposition of a topological space X and φ the natural mapping of X onto $\mathcal{D}$. Then the topology of $X(\mathcal{D})$ is the strongest topology of $\mathcal{D}$ for which φ is continuous.*

Proof. It directly follows from the definition of decomposition space.

Suppose f is a continuous mapping of a topological space X onto a topological space Y. Define a decomposition $\mathcal{D}$ of X by

$$\mathcal{D} = \{f^{-1}(q) \,|\, q \in Y\}.$$

Then we can establish a one-to-one mapping between Y and $X(\mathcal{D})$ by letting $q \in S$ correspond to $f^{-1}(q) \in X(\mathcal{D})$. Generally speaking, this mapping is not a homeomorphism, though it is an open mapping.

H) *Let f be a closed or open continuous mapping of a topological space X onto a topological space Y. Then Y is homeomorphic with $X(\mathcal{D})$, where $\mathcal{D} = \{f^{-1}(q) \,|\, q \in Y\}$.*

Proof. It is left to the reader.

A decomposition $\mathcal{D}$ of a topological space X is called *upper semi-continuous* if for each $D \in \mathcal{D}$ and each open set U containing D, there is an open set V such that $D \subset V \subset U$, and V is a union of members of $\mathcal{D}$.

I) *The natural mapping φ of X onto $X(\mathcal{D})$ is closed if and only if the decomposition $\mathcal{D}$ is upper semi-continuous.*

Proof. Let $\mathcal{D}$ be upper semi-continuous. Suppose F is a closed set of X. Put

$$U = X - S(F, \mathscr{D}) = \bigcup\{D \mid D \in \mathscr{D}, D \cap F = \emptyset\}.$$

Since

$$\varphi(F) = X(\mathscr{D}) - \varphi(U),$$

we shall prove that $\varphi(U)$ is open to show that $\varphi(F)$ is closed. Assume $q \in \varphi(U)$. Then $\varphi^{-1}(q)$ is a member of $\mathscr{D}$ contained in $X - F$. Since $\mathscr{D}$ is upper semi-continuous, there is an open set V which is the union of members of $\mathscr{D}$ satisfying $\varphi^{-1}(q) \subset V \subset X - F$. Thus $\varphi(V)$ is an open set of $X(\mathscr{D})$ satisfying

$$q \in \varphi(V) \subset \varphi(U).$$

Therefore $\varphi(U)$ is a nbd of q, which means that $\varphi(U)$ is an open set. Thus φ is a closed mapping.

Conversely, suppose φ is a closed mapping. Assume U is an open set of X containing $D \in \mathscr{D}$. Then, since $X - U$ is closed, $\varphi(X - U)$ is a closed set of $X(\mathscr{D})$. Therefore $W = X(\mathscr{D}) - \varphi(X - U)$ is an open nbd of $\varphi(D)$. Hence $V = \varphi^{-1}(W)$ is an open set and is the union of sets belonging to $\mathscr{D}$ satisfying

$$D \subset V \subset U.$$

This proves that $\mathscr{D}$ is upper semi-continuous.

Finally, let us define the concept of inverse limit space.

Definition II.18. Let A be a directed set and $\{X_\alpha \mid \alpha \in A\}$ a collection of topological spaces. We suppose that to every pair α, β of elements of A with $\alpha > \beta$, there is assigned a continuous mapping π_β^α (called a *bonding mapping*) of X_α into X_β such that

$$\pi_\gamma^\beta \circ \pi_\beta^\alpha = \pi_\gamma^\alpha \quad \text{for } \alpha > \beta > \gamma.$$

Then we call the system $\{X_\alpha, \pi_\beta^\alpha \mid \alpha, \beta \in A, \alpha > \beta\}$ an *inverse system* (*inverse spectrum*) of topological spaces. If A is the set of all natural numbers with the natural order, then we call the inverse system an *inverse sequence*.

Now, for a given inverse system $\{X_\alpha, \pi_\beta^\alpha\}$, we define a subspace X of the product space $\prod\{X_\alpha \mid \alpha \in A\}$ as follows: $X = \{p \mid p = \{p_\alpha \mid \alpha \in A\}$ and

$\pi_\beta^\alpha(p_\alpha) = p_\beta$ for every α, $\beta \in A$ with $\alpha > \beta$}. Then we call X the *limit space* of the inverse system and denote it by

$$X = \lim_{\leftarrow} \{X_\alpha, \pi_\beta^\alpha \mid \alpha, \beta \in A, \alpha > \beta\}.$$

Let us define the concept of direct limit space (a less used twin of the above defined concept of inverse limit space).

Generally, let $\{X_\alpha \mid \alpha \in A\}$ be a collection of disjoint topological spaces; then we introduce a topology into the sum $X = \cup \{X_\alpha \mid \alpha \in A\}$ by defining that a subset U of X is an open set if and only if $U \cap X_\alpha$ is an open set of X_α for every $\alpha \in A$. Then X with this topology is called the *discrete sum* or *topological sum* of X_α, $\alpha \in A$.

Now, assume that $\{X_\alpha \mid \alpha \in A\}$ is a given collection of topological spaces and A a directed set. Furthermore, we assume that for every pair α, β of elements of A with $\alpha < \beta$, there is assigned a continuous mapping φ_β^α from X_α into X_β such that

$$\varphi_\gamma^\alpha = \varphi_\gamma^\beta \circ \varphi_\beta^\alpha \quad \text{for } \alpha < \beta < \gamma.$$

Then we call the system $\{X_\alpha, \varphi_\beta^\alpha \mid \alpha, \beta \in A, \alpha < \beta\}$ a *directed (inductive) system* of topological spaces. We may use the term 'spectrum' instead of 'system' as we did for an inverse system. Let X denote the discrete sum of X_α, $\alpha \in A$. Two points p_α and p_β of X are called equivalent if $p_\alpha \in X_\alpha$, $p_\beta \in X_\beta$ and $\varphi_\gamma^\alpha(p_\alpha) = \varphi_\gamma^\beta(p_\beta)$ for some γ. Now the quotient space X^* of X with respect to this equivalence is called the *limit space* of the directed system $\{X_\alpha, \varphi_\beta^\alpha\}$ and denoted by

$$X^* = \lim_{\rightarrow} \{X_\alpha, \varphi_\beta^\alpha \mid \alpha, \beta \in A, \alpha < \beta\}.$$

Example II.17. Let $I^n = \{(x_1, \ldots, x_n) \mid 0 \le x_i \le 1, \ i = 1, \ldots, n\}$, $n = 1$, $2, \ldots$, be n-dimensional cubes. For natural numbers, n, m with $n > m$, we define a mapping π_m^n of I^n onto I^m by

$$\pi_m^n((x_1, \ldots, x_n)) = (x_1, \ldots, x_m).$$

Then we can easily see that the limit space

$$I^\infty = \lim_{\leftarrow} \{I^n, \pi_m^n \mid n, m = 1, 2, \ldots; n > m\}$$

of the inverse sequence $\{I^n, \pi_m^n\}$ is homeomorphic to Hilbert cube I^ω by the mapping

$$f(p) = (x_1, \tfrac{1}{2}x_2, \tfrac{1}{3}x_3, \ldots) \in I^\omega,$$

where

$$p = \{(x_1), (x_1, x_2), (x_1, x_2, x_3), \ldots\} \in I^\infty.$$

Example II.18. Let A' be a cofinal subset of a directed set A. Then

$$X = \lim_{\leftarrow} \{X_\alpha, \pi_\beta^\alpha \mid \alpha, \beta \in A, \alpha > \beta\}$$

and

$$X' = \lim_{\leftarrow} \{X_{\alpha'}, \pi_{\beta'}^{\alpha'} \mid \alpha', \beta' \in A', \alpha' > \beta'\}$$

are homeomorphic with each other. To see it, we define a mapping f of X onto X' by

$$f(\{p_\alpha \mid \alpha \in A\}) = \{p_{\alpha'} \mid \alpha' \in A'\}.$$

Then f is one-to-one. For, if $\{p_\alpha \mid \alpha \in A\} \neq \{q_\alpha \mid \alpha \in A\}$, then $p_\alpha \neq q_\alpha$ for some $\alpha \in A$. Since A' is cofinal in A, there is $\alpha' \in A'$ with $\alpha' > \alpha$. It follows from

$$\pi_\alpha^{\alpha'}(p_{\alpha'}) = p_\alpha, \qquad \pi_\alpha^{\alpha'}(q_{\alpha'}) = q_\alpha$$

that $p_{\alpha'} \neq q_{\alpha'}$, i.e.

$$\{p_{\alpha'} \mid \alpha' \in A'\} \neq \{q_{\alpha'} \mid \alpha' \in A'\}.$$

It is clear that f is continuous. To see that f^{-1} is continuous, we consider a nbd $U(p_{\alpha_0})$ of p_{α_0} in X_{α_0} with $\alpha_0 \notin A'$. Then

$$U = X \cap [U(p_{\alpha_0}) \times \prod \{X_\alpha \mid \alpha \in A, \alpha \neq \alpha_0\}]$$

is a nbd of $p = \{p_\alpha \mid \alpha \in A\}$. Choose $\alpha_1 \in A'$ with $\alpha_1 > \alpha_0$; then there is a nbd $V(p_{\alpha_1})$ of p_{α_1} in X_{α_1} such that

$$\pi_{\alpha_0}^{\alpha_1}(V(p_{\alpha_1})) \subset U(p_{\alpha_0}),$$

because $\pi_{\alpha_0}^{\alpha_1}$ is a continuous mapping with $\pi_{\alpha_0}^{\alpha_1}(p_{\alpha_1}) = p_{\alpha_0}$. Now, putting

$$V = X' \cap [V(p_{\alpha_1}) \times \prod \{X_{\alpha'} \mid \alpha' \in A', \alpha' \neq \alpha_1\}],$$

we obtain a nbd V of $p' = \{p_{\alpha'} \mid \alpha' \in A'\}$ such that $f^{-1}(V) \subset U$. Of course, U is a special type of nbd of p, but in view of the above argument, we can easily see that $f^{-1}(V) \subset U$ holds for every nbd U of p and for some nbd V of p'. Thus f is a homeomorphism of X onto X'.

If all X_α are the same topological space X, and each π_β^α is the identity mapping, then $\varprojlim\{X_\alpha, \pi_\beta^\alpha \mid \alpha, \beta \in A, \alpha > \beta\}$ is homeomorphic to X. If A contains a cofinal subset A' such that $X_{\alpha'} = X$ for all $\alpha' \in A'$ and $\pi_{\beta'}^{\alpha'}(p) = p$, $\alpha', \beta' \in A'$, we obtain the same conclusion.

8. Connectedness

Definition II.19. A topological space X is called *connected* if it cannot be decomposed into the sum of two non-empty, disjoint closed sets. A subset X' of X is called connected if the subspace X' is a connected space. A topological space X is called *locally connected* if every point p of X has a nbd basis consisting of connected sets.

A) *Let A_γ, $\gamma \in \Gamma$, be connected subsets of a topological space X. If $\cap\{A_\gamma \mid \gamma \in \Gamma\} \neq \emptyset$, then $\cup\{A_\gamma \mid \gamma \in \Gamma\}$ is a connected set.*

Proof. Assume the contrary; then

$$\cup A_\gamma = F \cup G, \qquad F \cap G = \emptyset$$

for some non-empty closed sets F, G of the subspace $\cup A_\gamma$. Choose a point $p \in \cap A_\gamma$; then either F or G contains p. Suppose, for example, $p \in F$. Since $G \neq \emptyset$, $G \cap A_\gamma \neq \emptyset$ for some γ. Now putting

$$F \cap A_\gamma = F', \qquad G \cap A_\gamma = G',$$

we get non-empty closed sets F' and G' of the subspace A_γ. Moreover $A_\gamma = F' \cup G'$ and $F' \cap G' = \emptyset$ are obvious, and hence A_γ is not connected, contradicting the initial hypothesis. Thus $\cup A_\gamma$ must be connected.

B) *Let A be a connected subset of a topological space X. If $A \subset B \subset \bar{A}$, then B is connected. Thus $\bar{A}$ is also connected.*

Proof. Assume the contrary, i.e. let

$$B = F \cup G, \qquad F \cap G = \emptyset,$$

where F and G are non-empty closed sets of the subspace B. Note that F and G are also open sets of the subspace B. Choose a point $p \in F$. Then, since $p \in F$, $p \in \bar{A}$, and F is a nbd of p in the subspace B, we obtain

$$F \cap A \neq \emptyset.$$

In the same way we obtain $G \cap A \neq \emptyset$. Therefore putting

$$F' = F \cap A, \qquad G' = G \cap A,$$

we get non-empty closed sets F' and G' of the subspace A. Moreover $A = F' \cup G'$ and $F' \cap G' = \emptyset$ are obvious, and hence A is not connected, contradicting the initial hypothesis. Thus B must be connected.

C) *Let p be a point of a topological space X. If we denote by $K(P)$ the sum of the connected subsets of X which contain p, then $K(P)$ is a connected closed set.*

Proof. By A), $K(P)$ is a connected set. Therefore by B), $\overline{K(P)}$ is also a connected set which contains $K(P)$. Hence it follows from the definition of $K(P)$ that $\overline{K(P)} \subset K(P)$. This means that $K(P)$ is a closed set.

Definition II.20. The connected closed set $K(P)$ given in C) is called a *component* of X. If every component of a topological space X contains only one point, then X is called a *totally disconnected* (or *hereditarily disconnected*) *space*.

It is easily seen that if two components intersect, then they coincide. Hence we can assert the following:

D) *Every topological space X is decomposed into the sum of disjoint components.*

E) *Let X_α, $\alpha \in A$, be connected topological spaces; then their product space $X = \prod \{X_\alpha \mid \alpha \in A\}$ is also connected.*

Proof. Assume that X is decomposed into the sum of two disjoint, closed sets F and G. Let $F \neq \emptyset$; then we can select a point $p = \{p_\alpha \mid \alpha \in A\} \in F$. Now, we can assert that every point $q = \{q_\alpha \mid \alpha \in A\}$ which has at most finitely many coordinates different from those of p, belongs to F. To prove this, it suffices to show that if $q = \{q_\alpha \mid \alpha \in A\}$ differs from p by only one coordinate, i.e. $q_{\alpha_0} \neq p_{\alpha_0}$, $q_\alpha = p_\alpha$ for $\alpha \neq \alpha_0$, then $q \in F$. The above is sufficient because the general case can be proved by repeated use of this assertion. Put

$$P = \{\{p'_\alpha \mid \alpha \in A\} \mid p'_\alpha = p_\alpha \text{ for } \alpha \neq \alpha_0\} ;$$

then we can easily see that P is a subset of X which is homeomorphic to X_{α_0}. Therefore P is connected. On the other hand, P is decomposed into two disjoint closed sets $P \cap F$ and $P \cap G$. Since $p \in P \cap F \neq \emptyset$, it must be that $P \cap G = \emptyset$, i.e. $P \cap F = P$. Thus we have proved that

$$q \in P \subset F.$$

Now, we put

$$Q = \{\{q_\alpha \mid \alpha \in A\} \mid q_\alpha = p_\alpha \text{ except for at most finitely many } \alpha\} .$$

Then, as shown in the above, $Q \subset F$, and hence $\bar{Q} \subset \bar{F}$. On the other hand, by the definition of product space, Q is obviously a dense subset of X. Therefore $\bar{Q} = X$ which implies $\bar{F} = X$. Since F is closed, we obtain $F = X$, $G = \emptyset$. Thus X is a connected space.

Example II.19. Among the most popular examples of connected spaces is E^n. To see the connectedness of E^1, we suppose $E^1 = F \cup G$ for disjoint closed sets F, G. If both F and G are non-empty, then there is an interval $I = [a, b]$ such that $I \cap F \neq \emptyset$. $I \cap G \neq \emptyset$. Assume, for example, $b \in I \cap G$. Let $c = \sup (I \cap F)$. Since $I \cap F$ is closed, $c \in I \cap F$. Hence $c < b$ and $J = (c, b] \subset I \cap G$. Since $I \cap G$ is closed, $c \in \bar{J} \subset I \cap G$ which is a contradiction. Thus E^1 is connected. Therefore by E), E^n is connected for every n. On the other hand, the subspace consisting of the rational points of E^n is a totally disconnected space.

Let X be a topological space and $\mathscr{D}$ the collection of the components of X. Then the decomposition space $X(\mathscr{D})$ consists of a point if X is connected. Generally we can assert that $X(\mathscr{D})$ is a totally disconnected space. For, let K be a component of $X(\mathscr{D})$; then the inverse image $\varphi^{-1}(K)$ of K by the natural mapping φ is a closed set of X and contains at least one component of X. To show that $\varphi^{-1}(K)$ is connected, assume the contrary. Then $\varphi^{-1}(K) = F \cup G$ for non-empty disjoint, closed sets F and G of $\varphi^{-1}(K)$. Every component contained in $\varphi^{-1}(K)$ is contained either in F or in G, for the component cannot meet both of F and G because of its connectedness. Now, let $\varphi(F) = L$, $\varphi(G) = M$; then by the above remark, L and M are non-empty, disjoint sets whose union is K. On the other hand it is clear that $\varphi^{-1}(L) = F$, $\varphi^{-1}(M) = G$, and hence L and M are closed sets of K by the condition of decomposition space. This contradicts the fact that K is connected. Therefore $\varphi^{-1}(K)$ is connected. Consequently it coincides with a component P of X. This implies that $K = \varphi(P)$ contains only one point proving that $X(\mathscr{D})$ is totally disconnected.

Example II.20. Suppose every two points p, q of a topological space X can be joined by a curve, i.e. there is a continuous mapping f of the unit segment $[0, 1]$ into X such that $f(0) = p$, $f(1) = q$. Then we call X *arcwise connected* or *path-connected*. Every arcwise connected space is connected, but the converse is not true. For example, the subspace $\{(x, y) \mid y = \sin 1/x,\ x \neq 0\} \cup \{(0, y) \mid -1 \leqslant y \leqslant 1\}$ of E^2 is connected but non-arcwise connected.

Definition II.21. Let p be a point of a topological space X. Then we define a *quasi-component* $Q(p)$ by

$$Q(p) = \bigcap \{A \mid A \text{ is a clopen set of } X \text{ such that } p \in A\}.$$

F) *Generally* $K(p) \subset Q(p)$.

Proof. The easy proof is left to the reader.

G) *Every topological space X is decomposed into the sum of mutually disjoint quasi-components.*

Proof. Observe that for every p, $q \in X$, either $K(p) \subset Q(q)$ or $K(p) \cap Q(q) = \emptyset$ holds. Because otherwise there is a clopen set C such that

$$K(p) \cap C \neq \emptyset, \qquad K(p) \cap (X - C) \neq \emptyset,$$

contradicting the connectedness of $K(p)$. Thus the present proposition follows from D).

Example II.21. $Q(p)$ and $K(p)$ can differ. Let

$$F_n = \{(x, y) \in E^2 \mid y = 1/n\}, \quad n = 1, 2, \ldots .$$

Then consider the subspace

$$Y = \bigcup_{n=1}^{\infty} F_n \cup \{(0, 0), (1, 0)\} \quad \text{of } E^2.$$

In Y,

$$Q((0, 0)) = Q((1, 0)) = \{(0, 0), (1, 0)\},$$

while

$$K((0, 0)) = \{(0, 0)\}, \qquad K((1, 0)) = \{(1, 0)\}.^{[1]}$$

Exercise II

1. Let X be a given set. Then $\mathcal{O} = \{U \mid U \subset X, X - U \text{ is finite or } U = \emptyset\}$ is a topology of X. (Such a topological space is called a *cofinite space*.)

2. For a set X and a collection $\mathcal{U}(p)$, $p \in X$, of subsets of X, the conditions (i)–(v) of 2.D) are equivalent with (i)–(iv) plus (v)' if $U \in \mathcal{U}(p)$, then there exists a set $V' \in \mathcal{U}(p)$ such that for every point $q \in V'$ there is $W \in \mathcal{U}(q)$ with $W \subset U$. (To derive (v) from (i)–(iv) plus (v)', put $V = \{q \mid W \subset U \text{ for some } W \in \mathcal{U}(q)\}$.)

3. $A°$ is the largest open set contained in A. Consequently A is an open set if and only if $A° = A$.

4. A is a closed set if and only if $A \supset B(A)$.

5. An open set U is called a *regular open set* if $\bar{U}° = U$. Then $\bar{A}°$ is a regular open set for every subset A of a topological space X. (A closed set F is called a *regular closed set* if $\overline{F°} = F$.)

---

[1] As for further discussions on connectedness, see, e.g., books like C. Kuratowski [4], C. T. Whyburn [1], R. Engelking [2], where one can find rather extensive information on this aspect; the last gives detailed arguments on disconnectedness and 0-dimensionality. As for articles on various topics, see, e.g., W. Sierpiński [1], [2] for interesting classical theorems, B. Banaschewski [1], G. T. Whyburn [2], J. de Groot–R. H. McDowell [2], M. Henriksen–J. Isbell [1] for local connectedness, A. P. Kombarov [1], E. Pol–R. Pol [1] for totally disconnectedness, E. Michael [10], J. T. Goodykoontz, Jr. [1] for connectedness of spaces of subsets and of functions and also T. Tanaka [1], B. Knaster–A. Lelek [1], J. Mycielski [1], A. Lelek [1], R. Duda [1] for various interesting results.

6. Give the relationship between the two concepts, nbd basis and open basis.

7. Let $\mathscr{G}$ be a closed basis of a topological space X. Then $\bar{A} = \cap\{G \mid A \subset G \in \mathscr{G}\}$ for every subset A of X. A mapping f from Z into X is continuous if $f^{-1}(G)$ is closed in Z for every $G \in \mathscr{G}$.

8. $\bar{A} = X - (X - A)^{\circ}$.

9. Prove $\overline{A \cap B} \subset \bar{A} \cap \bar{B}$. Give an example to show the equality does not necessarily hold.

10. A is a closed set if and only if every cluster point of A belongs to A.

11. For a closed set F, $F = F^{\circ} \cup B(F)$, $F^{\circ} = \overline{F^{\circ}}^{\circ}$.

12. A subset A of a topological space X is nowhere dense if and only if for every non-empty open set U of X, there is a non-empty open set V such that $V \subset U$ and $V \cap A = \emptyset$.

13. A filter $\mathscr{F}$ converges to p if and only if p is a cluster point of every filter which contains $\mathscr{F}$ as a subcollection.

14. A filter $\mathscr{F}$ of a topological space X is maximal if and only if for every subset A of X, either A or $X - A$ belongs to $\mathscr{F}$.

15. A net derived from a maximal filter is maximal.

16. The filter derived from a maximal net is maximal.

17. Discuss the relationship between cluster points of nets and those of filters in view of 4.D), E). (Revise Definition II.12 of derived net slightly to allow A_δ and $A_{\delta'}$ coincide for different δ and δ'.)

18. A subset U of a topological space X is a nbd of $p \in X$ if and only if every filter converging to p contains U as a member.

19. Let A be a subset of a topological space X and put $B = \{p \mid$ there is a filter $\mathscr{F}$ such that $A \in \mathscr{F} \to p\}$. If a filter $\mathscr{F}'$ satisfies $B \in \mathscr{F}' \to q$, then $q \in B$.

20. Let X be a set. Suppose convergence of filters in X is defined so that:
    (i) if for a filter $\mathscr{F}$ and a point p of X, $\{p\} \in \mathscr{F}$, then $\mathscr{F} \to p$,
    (ii) if $\mathscr{F} \to p$ and $\mathscr{G}$ is a filter with $\mathscr{G} \supset \mathscr{F}$, then $\mathscr{G} \to p$,
    (iii) the condition of Exercise 19 is satisfied.
Now, we define that a subset U of X is a nbd of a point p of X if and only if U and p satisfy the condition of Exercise 18. Then the thus defined nbds satisfy (i)–(v) of 1.A), and therefore X is a topological space.

21. A subset U of a topological space X is a nbd of p if and only if for every net $\varphi(\Delta \,|\,>)$ converging to p, $U \cap \varphi(\Delta) \neq \emptyset$.

22. Let $\varphi(\Delta \,|\,>)$ be a net; then prove:

(i) if $\varphi(\Delta \,|\,>) \to p$, then for every cofinal subset Δ' of Δ, $\varphi(\Delta' \,|\,>) \to p$,

(ii) if $\varphi(\Delta \,|\,>) \to p$, $\varphi_\delta(\Gamma_\delta \,|\,>) \to \varphi(\delta)$, then there is a net $\psi(\Gamma \,|\,>) \to p$ such that $\psi(\Gamma) \subset \bigcup \{\varphi_\delta(\Gamma_\delta) \,|\, \delta \in \Delta\}$.

23. Let X be a set. If convergence of nets of X is defined so that they satisfy (i), (ii) of Exercise 22 and

(iii) if $\varphi(\delta) = p$, $\delta \in \Delta$, then $\varphi(\Delta \,|\,>) \to p$,

then defining nbd by the condition of Exercise 21 we see (i)–(v) of 1.A) is satisfied, i.e. X is a topological space.

24. If a covering $\mathcal{U}$ of a topological space is locally finite, then $\bar{\mathcal{U}}$ is also locally finite.

25. If $\mathcal{U}$ is an open covering of a topological space X, then for each subset A of X, $S^\infty(A, \mathcal{U}) = \bigcup_{n=1}^{\infty} S^n(A, \mathcal{U})$ is a clopen set of X.

26. If f is a mapping of a topological space X onto a topological space Y and $\mathcal{F}$ is a filter of X, then $f(\mathcal{F}) = \{f(F) \,|\, F \in \mathcal{F}\}$ is a filter of S. If f is simply an into mapping, then $\{G \,|\, G \subset Y, f^{-1}(G) \in \mathcal{F}\}$ is a filter of Y.

27. Let f be a mapping of a topological space X into a topological space Y. Then f is continuous if and only if for every net $\varphi(\delta \,|\,>)$ of X with $\varphi(\Delta \,|\,>) \to p$, $f \circ \varphi(\Delta \,|\,>) \to f(p)$ holds in Y.

28. A continuous mapping of a topological space X into a topological space Y is a continuous mapping of X onto the subspace $f(X)$ of Y.

29. Let f be a mapping from a topological space X onto a topological space Y. Then f is closed and continuous iff $f(\bar{A}) = \overline{f(A)}$ for every $A \subset X$.

30. Let $X \supset X' \supset A$, where X is a topological space. We denote by $A^{o\prime}$ and $B'(A)$ the interior and boundary of A in the subspace X'. Then prove

$$A^\circ \cap X' \subset A^{o\prime}, \qquad B(A) \cap X' \supset B'(A).$$

Give examples to show that the equality does not necessarily hold in either of the cases.

31. Let A be a dense subset of a topological space X and f, g continuous maps from X into a topological space Y whose every distinct points have mutually disjoint nbds. If $f = g$ on A, then $f = g$ on X.

32. Let f be a continuous mapping of a topological space X onto a topological space Y. Suppose that for any topological space Z and any mapping g of Y into Z, the continuity of $g \circ f$ implies the continuity of g. Then Y is homeomorphic to the decomposition space $X(\mathscr{D})$, where $\mathscr{D} = \{f^{-1}(q) \mid q \in S\}$, i.e. f is a quotient mapping.

33. (i) Let $\{X_\gamma \mid \gamma \in \Gamma\}$ be a collection of topological spaces. Assume $U_\gamma \subset X_\gamma$, $\gamma \in \Gamma$; then $\overline{\prod \{U_\gamma \mid \gamma \in \Gamma\}} = \prod \{\bar{U}_\gamma \mid \gamma \in \Gamma\}$ in the product space $\prod X_\gamma$.

(ii) Let $A \subset X$, $B \subset Y$ for topological spaces X and Y. Then $B(A \times B) = (B(A) \times \bar{B}) \cup (\bar{A} \times B(B))$ in $X \times Y$, where boundary and closure of A and B are those in X and Y, respectively.

34. Let X_γ, $\gamma \in \Gamma$, be topological spaces. For each point $p = \{p_\gamma \mid \gamma \in \Gamma\}$ of the cartesian product $X = \prod \{X_\gamma \mid \gamma \in \Gamma\}$, we define a collection $\mathscr{W}(p)$ of subsets of X as follows: $\mathscr{W}(p) = \{\prod \{U_\gamma \mid \gamma \in \Gamma\} \mid U_\gamma$ is a nbd of p_γ in X_γ for each $\gamma \in \Gamma\}$ where all U_γ may differ from X_γ. Then $\mathscr{W}(p)$ satisfies (i)–(iv) of 2.D), i.e. X turns out to be a topological space called the product space with the *strong topology* or *box topology*. The product space defined in Definition II.16. is sometimes said to have the *weak topology*.

35. Let f be a continuous mapping of a topological space X onto a topological space Y. If X is connected, then Y is also connected.

36. Let F and G be closed sets of a topological space X such that $F \cap G$ and $F \cup G$ are connected. Then F and G are connected sets.

37. Give an example of a topological space which is connected but not locally connected and of a topological space which is locally connected but not connected.

38. A topological space X is arcwise connected if and only if X is connected and each point of X has an arcwise connected nbd.

39. Let f be a one-to-one continuous mapping from the real line onto itself. Then f is a topological mapping.[1]

40. If X is locally connected, then each component of X is an open set. The converse is not true.

41. Construct a closed subset C of $[0, 1]$ as follows: Let $J_0 = [0, 1]$, $J_1 = [0, 1/3] \cup [2/3, 1]$, $J_2 = [0, 1/9] \cup [2/9, 3/9] \cup [6/9, 7/9] \cup [8/9, 1], \ldots$ Generally

---

[1] It is an interesting problem to find out more general conditions of spaces under which one-to-one continuous mappings are topological mappings. See E. Duda [1].

J_n is obtained from J_{n-1} by subtracting the middle one third open interval from each closed interval constituting J_{n-1}. Then $C = \bigcap_{i=0}^{\infty} J_i$ is called *Cantor discontinuum*.

(i) C is the set of all real numbers of the form $\sum_{n=1}^{\infty} a_n/3^n$ for $a_n = 0$ or 2,

(ii) C is homeomorphic to the product space of countably many copies of the two point discrete space $\{0, 1\}$,

(iii) C is a totally disconnected space with no isolated point.

CHAPTER III

VARIOUS TOPOLOGICAL SPACES

1. T_1, T_2, regular and completely regular spaces

Definition III.1. A topological space X is called a T_1-*space* if for every two distinct points p, q of X, there are nbds U of p and V of q such that $q \notin U$ and $p \notin V$. X is called a T_2-*space* or a *Hausdorff space* if for every two distinct points p, q of X, there are nbds U of p and V of q such that $U \cap V = \emptyset$.

A T_1-space is characterized as a topological space in which every point forms a closed set. The following proposition characterizes T_2-spaces.

A) *A topological space X is a T_2-space if and only if every filter converges to at most one point.*

Proof. Let X be a T_2-space and $\mathscr{F}$ a filter of X. Suppose $\mathscr{F} \to p$ and $p \neq q$. By Definition III.1, there are disjoint nbds U of p and V of q. Since $\mathscr{F} \to p$, $U \in \mathscr{F}$. If $V \in \mathscr{F}$, then $U \cap V = \emptyset$ for two members of the filter, contradicting the definition of filter. Therefore $V \notin \mathscr{F}$, and hence $\mathscr{F}$ does not converge to q.

Conversely, let X be a non T_2-space. Then there are two distinct points p, q, of X such that every pair of nbds U of p and V of q intersect. Thus $\mathscr{F} = \{M \mid M \supset U \cap V \text{ for some nbds } U \text{ of } p \text{ and } V \text{ of } q\}$ is a filter. Since every nbd of p as well as every nbd of q belongs to $\mathscr{F}$, we get $\mathscr{F} \to p$, $\mathscr{F} \to q$. Thus the assertion is proved.

Example III.1. Every indiscrete space is not a T_1-space if it contains at least two points. The topological space X_1 in Example II.7 is a T_1-space but not T_2.

Definition III.2. A T_1-space X is called a *regular space*[1] if for every point p of X and every closed set F of X satisfying $p \notin F$, there are open sets U and V such that $p \in U$, $F \subset V$ and $U \cap V = \emptyset$.

A T_1-space X is called a *completely regular space*[1] (or *Tychonoff space*) if for every point p of X and every closed set F of X satisfying $p \notin F$, there is a continuous function f over X such that

$$f(p) = 0, \qquad f(F) = 1 \quad \text{and} \quad 0 \leqslant f \leqslant 1 .$$

B) *A T_1-space X is regular if and only if for every open nbd W of each point p of X, there is an open set U such that*

$$p \in U \subset \bar{U} \subset W .$$

Proof. Let X be a T_1-space satisfying the condition. Given a point p and a closed set F such that $p \notin F$, then $W = X - F$ is an open nbd of p. Therefore we can choose an open set U such that

$$p \in U \subset \bar{U} \subset W .$$

Then U and $V = X - \bar{U}$ are open sets satisfying the condition of Definition III.2, and hence X is regular.

Conversely, let X be regular and W an open nbd of a point p of X. Then there are open sets U and V such that

$$p \in U, \qquad X - W \subset V, \qquad U \cap V = \emptyset .$$

Hence $U \subset X - V$, which implies

$$\bar{U} \subset \overline{X - V} = X - V \subset W$$

since $X - V$ is closed. Then U satisfies the desired condition.

Example III.2. The space R_2 in Example II.5 is T_2 but not regular.

To construct a regular space which is not completely regular,[2] we consider a subset $E = \{(x, y) \mid y \geqslant 0\}$ of E^2 and add one point p to E to define $R_3' = E \cup \{p\}$. We define a special topology for R_3' as follows.

---

[1] Some people mean by a regular (completely regular) space merely a T-space satisfying this condition.

[2] We owe this simple example to A. Mysior [2].

Every point $(x, y) \in E$ with $y > 0$ is defined as an isolated point of R_3'. To each point $(x, 0) \in E$ we assign two segments I_x and J_x, defined by

$$I_x = \{(x, y') \in E \mid 0 \le y' \le 2\},$$
$$J_x = \{(x', y') \in E \mid x' = y' + x, 0 \le y' \le 2\}.$$

Then $\{I_x \cup J_x - P \mid P$ is a finite set which does not contain $(x, 0)\}$ is defined as a nbd base of $(x, 0)$ in R_3'. On the other hand $\{H_n \mid n = 1, 2, \ldots\}$, where

$$H_n = \{p\} \cup \{(x, y) \in E \mid x \ge n\},$$

is defined as a nbd base of p in R_3'. Then it is easy to see that R_3' is a regular space.

To prove that R_3' is not completely regular, let

$$F_n = \{(x, 0) \mid n - 1 \le x \le n\}, \quad n = 1, 2, \ldots.$$

Then F_1 is a closed set such that $p \notin F_1$. Assume that f is a real-valued continuous function on R_3' satisfying

$$f(F_1) = 1, \quad 0 \le f \le 1.$$

We can prove that $f(p) = 1$ (and accordingly $f(p) \neq 0$) as follows. For that purpose, we shall show that $f^{-1}(1) \cap F_n$ is infinite for every n, by use of induction on n.

The assertion is obviously true for $n = 1$. Assume that $f^{-1}(1) \cap F_n$ is infinite. Select a countable subset A of $f^{-1}(1) \cap F_n$. For each $(x, 0) \in A$,

$$J_x - f^{-1}(1) = \bigcup_{n=1}^{\infty} G_n,$$

where
$$G_n = \{z \in J_x \mid 0 \le f(z) \le 1 - 1/n\}.$$

Observe that each G_n is a closed set of R_3'. Since $(x, 0) \notin G_n$, $(x, 0) \notin \bar{G}_n$, and hence G_n is a finite set. Thus $J_x - f^{-1}(1)$ is at most countable, and so is

$$A' = \bigcup \{J_x - f^{-1}(1) \mid (x, 0) \in A\}.$$

Now, denote by π the projection of E onto the x-axis, i.e. $\pi((x, y)) = x$.

Then $\pi(A') \cap F_{n+1}$ is at most countable. Hence $F_{n+1} - \pi(A')$ is infinite. Let $(x', 0) \in F_{n+1} - \pi(A')$; then

$$I_{x'} \cap J_x \cap f^{-1}(1) \neq \emptyset \quad \text{for all } x \in A.$$

Thus, for every nbd U of $(x', 0)$,

$$U \cap J_x \cap f^{-1}(1) \neq \emptyset \quad \text{for some } x \in A.$$

This implies that

$$f((x', 0)) = 1.$$
Hence
$$F_{n+1} - \pi(A') \subset f^{-1}(1),$$

i.e. $f^{-1}(1) \cap F_{n+1}$ is infinite, proving our assertion. Since $H_n \supset F_{n+1}$, it follows that

$$H_n \cap f^{-1}(1) \neq \emptyset \quad \text{for every } n.$$

Therefore $f(p) = 1$, i.e., R'_3 is not completely regular.

C) *Every subspace of a T_1 (T_2, regular, completely regular) space is T_1 (T_2, regular, completely regular).*
    The product space of T_1 (T_2, regular, completely regular) spaces is T_1 (T_2, regular, completely regular).

Proof. The easy proof is left to the reader.

Completely regular spaces are characterized by the following important theorem which is very interesting, since it gives a concrete expression to the (abstract) completely regular spaces.

Theorem III.1 (Tychonoff's imbedding theorem). *A topological space X is completely regular if and only if it is homeomorphic with a subspace of the product space of copies of the unit segment $[0, 1]$.*

Proof. The 'if' part is a direct consequence of C), because $[0, 1]$ is obviously completely regular.
    To prove the 'only if' part, we suppose X is a completely regular space.

Let $\{f_\alpha \mid \alpha \in A\}$ denote the totality of continuous functions over X such that $0 \leq f_\alpha \leq 1$. For the index set A, we construct the product space

$$P = \prod \{I_\alpha \mid \alpha \in A\}$$

of the unit segments $I_\alpha = [0, 1]$, $\alpha \in A$. To every point p of X, we assign a point

$$f(p) = \{f_\alpha(p) \mid \alpha \in A\}$$

of P. Thus we get a mapping f of X into P.

Now, all we have to do is to show that f is a topological mapping. If p, q are distinct points of X, then since X is completely regular, and therefore T_1, $\{q\}$ is a closed set. Hence by Definition III.2, there is a continuous function f_α such that

$$f_\alpha(p) = 0, \qquad f_\alpha(q) = 1 \quad \text{and} \quad 0 \leq f_\alpha \leq 1.$$

Hence $f(p)$ and $f(q)$ are different in their α-coordinates, i.e. $f(p) \neq f(q)$. Thus f is a one-to-one mapping. To see the continuity of f, we consider a given point $p \in X$ and a given nbd V of $f(p)$ in P. By the definition of product space, there are nbds V_{α_i} of $f_{\alpha_i}(p)$ in I_{α_i}, $i = 1, \ldots, k$, such that

$$V' = \prod_{i=1}^{k} V_{\alpha_i} \times \prod \{I_\alpha \mid \alpha \neq \alpha_i, i = 1, \ldots, k\} \subset V.$$

Since $f_{\alpha_i}(p)$, $i = 1, \ldots, k$, are continuous functions, there is a nbd U of p such that

$$f_{\alpha_i}(q) \in V_{\alpha_i}, \quad i = 1, \ldots, k,$$

for every point $q \in U$. Hence $q \in U$ implies

$$f(q) = \{f_\alpha(q) \mid \alpha \in A\} \in V' \subset V,$$

i.e. $f(U) \subset V$. Thus by Definition II.13, f is a continuous mapping.

Finally, to prove the continuity of the inverse mapping f^{-1} of f, we suppose U is a given nbd of a given point p of X. Since X is completely regular, there is a continuous function f_α such that

$$f_\alpha(p) = 0, \qquad f_\alpha(X - U) = 1 \quad \text{and} \quad 0 \le f_\alpha \le 1 .$$

Though $X - U$ is not necessarily closed, we can construct such a function f_α by considering an open nbd U' of p which is contained in U. Now

$$V = U_\alpha \times \prod \{I_{\alpha'} \mid \alpha' \ne \alpha\}$$

for $U_\alpha = [0, 1)$, is a nbd of $f(p)$ in P. If $p' \not\in U$, then $f_\alpha(p') = 1$, which implies $f(p') \not\in V$. Therefore $f(p') \in V$ implies $p' \in U$. In other words, $q' \in V \cap f(X)$ implies $f^{-1}(q') \in U$. Thus f^{-1} is also a continuous mapping, and hence f is a topological mapping of X onto a subspace $f(X)$ of P.

Generally a topological mapping f from X into Y is called a *topological imbedding*.

D) *Let f be a mapping from a topological space X into the product space $\prod \{X_\alpha \mid \alpha \in A\}$ of topological spaces X_α, $\alpha \in A$. Then f is continuous if and only if $\pi_\alpha \circ f$ is continuous for every $\alpha \in A$.*

Proof. It is practically proved and used in the proof of Theorem III.1.

2. Normal space and fully normal space

Definition III.3. A T_1-space X is called a *normal space* if for every pair of disjoint closed sets F, G of X, there are open sets U, V such that $U \supset F$, $V \supset G$ and $U \cap V = \emptyset$.

A T_1-space X is called a *fully normal space* if for every open covering $\mathcal{U}$ of X, there is an open covering $\mathcal{V}$ such that $\mathcal{V}^* < \mathcal{U}$.[1]

A) *A T_1-space X is normal if and only if for every closed set F and every open set W with $F \subset W$, there is an open set U such that*

$$F \subset U \subset \bar{U} \subset W .$$

---

[1] Some people mean by a normal (fully normal) space a T-space satisfying this condition. The six conditions given in Definitions III.1, 2, 3, are called *separation axioms*. There are some other separation axioms which are generally not so important. We shall deal with another separation axiom, T_0, in Chapter VII.

Proof. The proof of this assertion is quite analogous to that of 1.B) and is left to the reader.

B) *A T_1-space X is fully normal if and only if for every open covering $\mathscr{U}$ of X, there is an open covering $\mathscr{V}$ such that $\mathscr{V}^\Delta < \mathscr{U}$.*

Proof. This is a direct consequence of II.5.B).

Example III.3. The topological space R_4 in Example II.5 is completely regular, but not normal. It is clear that R_4 is T_1. To see that R_4 is completely regular, let $p \in R_4$ and F a closed set of R_4 such that $p \notin F$. If p is not on the x-axis, then it is clear that there exists a continuous function f satisfying the condition of Definition III.2. Therefore we assume that p is a point on the x-axis, i.e. $p = (x, 0)$. Then we can choose a nbd $U_\varepsilon(p) = S'_\varepsilon(p) \cup \{p\}$ such that $U_\varepsilon(p) \cap F = \emptyset$, where we denote by $S'_\varepsilon(p)$ the open set bounded by the circle with radius ε touching the x-axis at p, i.e. $S'_\varepsilon(p) = S_\varepsilon((x, \varepsilon))$. For every point $q \in S'_\varepsilon(p)$, we consider the circle passing through q and touching the x-axis at p and denote by $\eta(q)$ its radius. Then we define a continuous function g over R_4 by

$$g(p) = 0\,,$$

$$g(q) = \begin{cases} \eta(q) & \text{if } q \in S'_\varepsilon(p)\,, \\ \varepsilon & \text{if } q \notin U_\varepsilon(p)\,. \end{cases}$$

Putting $f = (1/\varepsilon)g$, we obtain a continuous function f over R_4 satisfying

$$f(p) = 0, \qquad f(F) = 1 \quad \text{and} \quad 0 \leqslant f \leqslant 1\,.$$

Therefore R_4 is completely regular.

To show that R_4 is not normal, we consider two sets

$$G = \{(x, 0) \mid x \text{ is rational}\}\,,$$

$$H = \{(x, 0) \mid x \text{ is irrational}\}\,,$$

on the x-axis. Then G and H are obviously disjoint closed sets of R_4. Now, we can assert that if U and V are open sets such that $U \supset G$ and $V \supset H$, then $U \cap V \neq \emptyset$. For, if we assume the contrary, then $U \cap V = \emptyset$. Put

$$H_n = \{q \mid q \in H,\ S'_{1/n}(q) \subset V\},$$

then

$$H = \bigcup_{n=1}^{\infty} H_n.$$

Every point $p = (x, 0) \in G$ has a nbd $U_\varepsilon(p) = S'_\varepsilon(p) \cup \{p\}$ with $\varepsilon < 2/n$ which is contained in U and accordingly does not meet V. Hence it is easily seen that the open interval $(x - \varepsilon, x + \varepsilon)$ on the x-axis does not meet H_n. From now on, let us regard the x-axis as E^1 with the usual topology. Then it follows from the above argument that $x \notin \bar{H}_n$ for every $x \in G$. This implies

$$G \subset E^1 - \bar{H}_n$$

in E^1. Hence

$$E^1 = \bar{G} = \overline{E^1 - \bar{H}_n},$$

and hence H_n is a nowhere dense set. Therefore H is a set of the first category, but this is a contradiction since H is of the second category, as shown in Example II.8. Thus we obtain $U \cap V \neq \emptyset$, i.e. R_4 is not normal. R_4 is often called *Niemytzki space*.

The topological space R_5 in Example II.1 is normal, but not fully normal. It is clear that R_5 is T_1. To see that R_5 is normal, let F and G be disjoint closed sets of R_5. For every point $\alpha \in F \cup G$, we choose, by transfinite induction on the ordinal number α, an open nbd $U(\alpha)$ of the form $(\beta, \alpha]$ such that $U(\alpha) \cap U(\beta) = \emptyset$ if $\alpha \in F$, $\beta \in G$, $U(0) = \{0\}$ if $0 \in F \cup G$.

Assume that we have defined $U(\beta)$ for every $\beta \in F \cup G$ with $\beta < \alpha$ and that $\alpha \in F \cup G$. For example, let $\alpha \in F$; then there is $\gamma < \alpha$ such that $(\gamma, \alpha] \cap G = \emptyset$ because G is a closed set which does not contain α (see Example II.5). Putting $U(\alpha) = (\gamma, \alpha]$, we get a nbd of α which does not meet $U(\beta)$ for every $\beta \in G$ with $\beta < \alpha$. Thus we can define desired open nbds $U(\alpha)$ for every $\alpha \in F \cup G$. Then $U = \bigcup \{U(\alpha) \mid \alpha \in F\}$ and $V = \bigcup \{U(\beta) \mid \beta \in G\}$ are disjoint open sets containing F and G respectively. Thus R_5 is normal.

Now observe that every increasing point sequence $\alpha_1 < \alpha_2 < \alpha_3 < \cdots$ of R_5 converges to its supremum

$$\sup\{\alpha_i \mid i = 1, 2, \ldots\} = \min\{\beta \mid \beta \in R_5,\ \beta \geq \alpha_i \text{ for } i = 1, 2, \ldots\}.$$

(Note that there is a countable ordinal number β such that $\beta \geqslant \alpha_i$, $i = 1, 2, \ldots$. Because, if not, then $R_5 = \{\beta \mid \beta$ is an ordinal number such that $\beta < \alpha_i$ for some $i\}$, which is impossible because R_5 is uncountable while the set on the right of the above equality is countable.)

Assume that R_5 is fully normal. Then for the open covering $\mathcal{U} = \{[0, \alpha] \mid \alpha \in R_5\}$ we can select an open covering $\mathcal{V}$ such that $\mathcal{V}^{\Delta} < \mathcal{U}$. Select an arbitrary point α_1 of R_5. Then

$$S(\alpha_1, \mathcal{V}) \subset [0, \beta_1] \quad \text{for some } \beta_1 \in R_5.$$

Select $\alpha_2 > \beta_1$. Then

$$S(\alpha_2, \mathcal{V}) \subset [0, \beta_2] \quad \text{for some } \beta_2 \in R_5.$$

Repeating the same process we get a point sequence

$$\alpha_i \leqslant \beta_1 < \alpha_2 \leqslant \beta_2 < \alpha_3 \leqslant \beta_3 < \cdots$$

such that

$$S(\alpha_i, \mathcal{V}) \subset [0, \beta_i], \quad i = 1, 2, \ldots.$$

Then, as observed in the above, $\{\alpha_i\}$ converges to $\alpha = \sup\{\alpha_i\}$. Then $\alpha > \beta_i$, $i = 1, 2, \ldots$, and hence

$$\alpha \notin \bigcup_{i=1}^{\infty} S(\alpha_i, \mathcal{V}).$$

On the other hand, since $S(\alpha, \mathcal{V})$ is a nbd of α,

$$\alpha_i \in S(\alpha, \mathcal{V}) \quad \text{for some } i.$$

This implies $\alpha \in S(\alpha_i, \mathcal{V})$, a contradiction. Hence R_5 is not fully normal.

A property which distinguishes normal and fully normal spaces from spaces in the former section is that a subspace of a normal space is not necessarily normal. In fact, as shown later in Example IV.3, even a subspace of a fully normal space may fail to be normal. Furthermore, the product space of two fully normal spaces is not necessarily normal, as shown in Example III.5. We shall show later in Example V.2 that even the product space of a fully normal space and a metric space need not be normal.

Theorem III.2 (Urysohn's lemma). *A T_1-space X is normal if and only if for every disjoint closed sets F, G of X, there is a continuous function f defined over X such that*

$$f(F) = 0, \qquad f(G) = 1 \quad and \quad 0 \le f \le 1.$$

Proof. Sufficiency. Let F and G be disjoint closed sets and f a function satisfying the condition. Then putting

$$U = \{p \mid f(p) < \tfrac{1}{2}\}, \qquad V = \{p \mid f(p) > \tfrac{1}{2}\},$$

we obtain open sets U, V which satisfy

$$U \supset F, \qquad V \supset G, \qquad U \cap V = \emptyset.$$

Therefore X is a normal space.

Necessity. Let F and G be disjoint closed sets of a normal space X. Then by A) we can construct an open set U such that

$$F \subset U \subset \bar{U} \subset X - G.$$

We put $U = U(1/2)$. By use of a similar process, we can construct open sets $U(1/2^2)$, $U(3/2^2)$ such that

$$F \subset U\left(\frac{1}{2^2}\right) \subset \overline{U\left(\frac{1}{2^2}\right)} \subset U\left(\frac{1}{2}\right) \subset \overline{U\left(\frac{1}{2}\right)} \subset U\left(\frac{3}{2^2}\right) \subset \overline{U\left(\frac{3}{2^2}\right)} \subset X - G.$$

By repeating this process, we define open sets $U(m/2^n)$, $m = 1, 2, \ldots$, $2^n - 1$, for every positive integer n. To give a precise definition by induction on the number n, we assume that we have defined $U(m/2^{n-1})$, $m = 1, \ldots, 2^{n-1} - 1$, such that

$$F \subset U\left(\frac{1}{2^{n-1}}\right) \subset \overline{U\left(\frac{1}{2^{n-1}}\right)} \subset \cdots \subset U\left(\frac{m-1}{2^{n-1}}\right)$$

$$\subset U\left(\frac{m}{2^{n-1}}\right) \subset \overline{U\left(\frac{m}{2^{n-1}}\right)} \subset \cdots \subset U\left(\frac{2^{n-1}-1}{2^{n-1}}\right) \subset X - G.$$

Since $U(2m/2^n)$ $(= U(m/2^{n-1}))$, $m = 1, \ldots, 2^{n-1} - 1$, are already defined, we construct, by use of A), open sets $U((2m - 1)/2^n)$, $m = 1, \ldots, 2^{n-1}$, such that

$$\overline{U\left(\frac{m-1}{2^{n-1}}\right)} \subset U\left(\frac{2m-1}{2^n}\right) \subset \overline{U\left(\frac{2m-1}{2^n}\right)} \subset U\left(\frac{m}{2^{n-1}}\right).$$

Thus we can construct open sets $U(m/2^n)$, $m = 1, \ldots, 2^n - 1$, for every positive integer n such that

$$F \subset U\left(\frac{1}{2^n}\right) \subset \cdots \subset \overline{U\left(\frac{m-1}{2^n}\right)} \subset U\left(\frac{m}{2^n}\right)$$

$$\subset \cdots \subset \overline{U\left(\frac{2^n - 1}{2^n}\right)} \subset X - G. \tag{1}$$

Adding to those $U(m/2^n)$ we define

$$U(1) = X \quad \text{and} \quad U(0) = \emptyset.$$

It follows from (1) that

$$\overline{U(r)} \subset U(r')$$

for any r, r' with $r < r'$, and $r, r' \in \{m/2^n \mid m = 0, \ldots, 2^n; n = 1, 2, \ldots\}$. For, if $r < r'$, then we can find a common denominator 2^n to express r and r' as $r = m/2^n$, $r' = m'/2^n$ with $m < m'$.

Now, we define a real-valued function f over X by

$$f(p) = \inf\{r \mid p \in U(r)\}, \quad p \in X,$$

where we note that r is a rational number of the form $m/2^n$ with $0 \leq m \leq 2^n$. This function f obviously satisfies $0 \leq f \leq 1$. Since $F \subset U(1/2^n)$, $n = 1, 2, \ldots, f(F) = 0$ is clear. On the other hand, since each point of G belongs to $U(r)$ only for $r = 1$, $f(G) = 1$ is also clear. Thus all we have to show is the continuity of f. Let p be a given point of X with $0 < f(p) < 1$. Then for a given $\varepsilon > 0$, we select a positive integer n with $1/2^{n-1} < \varepsilon$. Furthermore, we take the positive integer m defined by

$$\frac{m-1}{2^n} < f(p) \leq \frac{m}{2^n}. \tag{2}$$

Then by the definition of $f(p)$, we obtain

$$p \notin \overline{U\left(\frac{m-1}{2^n}\right)}.$$

This is true because, if we assume $p \in \overline{U((m-1)/2^n)}$, then by (1), $p \in U(r)$ for every $r > (m-1)/2^n$ which implies $f(p) \leq (m-1)/2^n$ contradicting (2). On the other hand, we also obtain

$$p \in U\left(\frac{m+1}{2^n}\right).^1$$

This follows since, by the definition of $f(p)$, $p \notin U((m+1)/2^n)$ implies $f(p) \geq (m+1)/2^n$, contradicting (2). Thus we can conclude that

$$p \in U\left(\frac{m+1}{2^n}\right) - \overline{U\left(\frac{m-1}{2^n}\right)} = V,$$

i.e., V is an open nbd of p. Suppose q is a given point of V; then it is obvious that

$$\frac{m-1}{2^n} \leq f(q) \leq \frac{m+1}{2^n}.$$

This combined with (2) implies that

$$|f(p) - f(q)| < \frac{2}{2^n} < \varepsilon.$$

Thus

$$f(V) \subset S_\varepsilon(f(p)).$$

We have assumed $0 < f(p) < 1$, but in the special case, $f(p) = 0$ or $f(p) = 1$, we can find, in a similar way, a nbd V of p satisfying the same condition. Therefore f is continuous at every point p and hence it is continuous over X. This proves the theorem.

---

[1] We can assume $m + 1 \leq 2^n$ without loss of generality, because we may choose as large n as we like.

Theorem III.3 (Tietze's extension theorem). *Let X be a T_1-space. Then X is normal if and only if for every closed set F of X and every (real-valued) continuous function f over F, there exists a continuous extension of f over X.*

Proof. Sufficiency. Let F and G be disjoint closed sets. Then we define a continuous function f over the closed set $F \cup G$ by

$$f(F) = 0, \qquad f(G) = 1 .$$

Since F and G are disjoint, f is certainly a continuous function over the subspace $F \cup G$. Hence f can be extended to a continuous function φ over X. Put

$$U = \{p \mid \varphi(p) < \tfrac{1}{2}\}, \qquad V = \{p \mid \varphi(p) > \tfrac{1}{2}\} .$$

Then by the continuity of φ, U and V are open sets satisfying

$$U \supset F, \qquad V \supset G \quad \text{and} \quad U \cap V = \emptyset .$$

Therefore X is a normal space.

Necessity. Let f be a real-valued continuous function defined over a closed set F of a normal space X. First we assume that f is bounded, i.e.,

$$|f(p)| \leq \alpha, \quad p \in F .$$

Putting

$$G = \{p \mid f(p) \leq -\tfrac{1}{3}\alpha\}, \qquad H = \{p \mid f(p) \geq \tfrac{1}{3}\alpha\} ,$$

we obtain disjoint closed sets G, H. Since X is normal, by Urysohn's lemma, there is a continuous function g over X such that

$$g(G) = 0, \qquad g(H) = 1 \quad \text{and} \quad 0 \leq g \leq 1 .$$

Putting

$$\varphi_1(p) = \tfrac{2}{3}\alpha(g(p) - \tfrac{1}{2}), \quad p \in X ,$$

we get a continuous function φ_1 over X satisfying

$$|\varphi_1(p)| \leq \tfrac{1}{3}\alpha, \quad p \in X ,$$

and

$$|f(p) - \varphi_1(p)| \leq \tfrac{2}{3}\alpha, \quad p \in F .$$

By applying an argument analogous with that for f, but this time for the function

$$\psi_1 = f - \varphi_1,$$

we obtain a continuous function φ_2 satisfying

$$|\varphi_2(p)| \leq \frac{2}{3^2}\alpha, \quad p \in X,$$

and

$$|\psi_1(p) - \varphi_2(p)| \leq (2/3)^2\alpha, \quad p \in F.$$

By repeating this process we obtain continuous functions φ_n over X and ψ_n over F such that

$$|\varphi_n(p)| \leq \frac{2^{n-1}}{3^n}\alpha, \quad p \in X, \tag{1}$$

$$|\psi_n(p)| \leq (2/3)^n\alpha, \quad p \in F,$$

and

$$\psi_{n+1}(p) = \psi_n(p) - \varphi_{n+1}(p), \quad p \in F.$$

It follows from

$$\psi_1 = f - \varphi_1,$$
$$\psi_2 = \psi_1 - \varphi_2,$$
$$\psi_3 = \psi_2 - \varphi_3,$$
$$\vdots$$
$$\psi_n = \psi_{n-1} - \varphi_n,$$

that

$$\psi_n = f - \sum_{i=1}^{n} \varphi_i,$$

and hence

$$|\psi_n(p)| = \left| f(p) - \sum_{i=1}^{n} \varphi_i(p) \right| \leq (2/3)^n\alpha, \quad p \in F. \tag{2}$$

It follows from (1) that $\sum_{i=1}^{n} \varphi_i(p)$ uniformly converges over X.[1] Therefor $\varphi(p) = \sum_{i=1}^{\infty} \varphi_i(p)$ is a continuous function over X. In view of (2) we know that

$$\varphi(p) = f(p), \quad p \in F,$$

i.e., φ is a continuous extension of f over X.

We should note that by (1) φ satisfies the same condition, $|\varphi| \leqslant \alpha$, over X as f does over F.

Now, let us deal with a non-bounded continuous function f over F. Then $\tan^{-1} f$ is a continuous function over F satisfying

$$|\tan^{-1} f| < \pi/2.$$

Therefore, using the preceding result, we can construct a continuous extension Φ of $\tan^{-1} f$ over X such that $|\Phi| \leqslant \pi/2$. Put

$$G = \{p \mid \Phi(p) = \pi/2 \text{ or } \Phi(p) = -\pi/2\};$$

then G is a closed set of X which does not intersect F. By use of Urysohn's lemma we define a continuous function g over X such that

$$g(F) = 1, \qquad g(G) = 0 \quad \text{and} \quad 0 \leqslant g \leqslant 1.$$

Putting $\Phi' = g\Phi$, we obtain a continuous function Φ' over X such that

$$|\Phi'| < \pi/2,$$

and

$$\Phi'(p) = \tan^{-1} f(p), \quad p \in F.$$

Thus $\varphi = \tan \Phi'$ is a continuous extension of f over X.

Corollary. *We denote by S^n the n-dimensional sphere, i.e. the subspace*

---

[1] A sequence $\{f_n \mid n = 1, 2, \ldots\}$ of functions over a topological space X is said to *uniformly converge* to a function f if for every $\varepsilon > 0$, there is n_0 such that $|f_n(p) - f(p)| < \varepsilon$ for every $n \geqslant n_0$ and $p \in X$. As well known in elementary analysis, $\{f_n \mid n = 1, 2, \ldots\}$ uniformly converges if and only if for every $\varepsilon > 0$, there is an n_0 such that $|f_n(p) - f_m(p)| < \varepsilon$ for every n, $m \geqslant n_0$ and $p \in X$. In this case, if f_n, $n = 1, 2, \ldots$, are continuous functions, then their limit function f is also continuous.

$$S^n = \left\{ (x_1, \ldots, x_{n+1}) \,\middle|\, \sum_{i=1}^{n+1} x_i^2 = 1 \right\}$$

of E^{n+1}. Let f be a continuous mapping of a closed set F of a normal space X into S^n. Then we can continuously extend f over an open set of X which contains F.

Proof. Let

$$f(p) = (f_1(p), \ldots, f_{n+1}(p)), \quad p \in F.$$

Then $f_1, \ldots, f_{n+1}$ are continuous functions over F satisfying

$$\sum_{i=1}^{n+1} f_i^2(p) = 1 .$$

Therefore by the theorem we can extend f_i to a continuous function φ_i over X. Putting

$$U = \left\{ p \,\middle|\, p \in X, \sum_{i=1}^{n+1} \varphi_i^2(p) > 0 \right\},$$

we get an open set U of X containing F. Now $(p \in U)$

$$\varphi(p) = \left(\varphi_1(p) \middle/ \left(\sum_{i=1}^{n+1} \varphi_i^2(p) \right)^{1/2}, \ldots, \varphi_{n+1}(p) \middle/ \left(\sum_{i=1}^{n+1} \varphi_i^2(p) \right)^{1/2} \right)$$

is a continuous mapping of U into S^n satisfying

$$\varphi(p) = f(p), \quad p \in F.$$

Example III.4. The condition that F is closed is crucial in Tietze's extension theorem. If we drop it, then the theorem does not hold even for E^1. Let

$$f(x) = \sin 1/x, \quad x \in F = (0, \infty) .$$

Then f is continuous over F, but we cannot continuously extend it over E^1, because $\lim_{x \to +0} f(x)$ is indeterminate.

Example III.5. In Example III.3 we showed that the Niemytzki space R_4 is not normal. Here we shall prove the same by use of Tietze's theorem. Generally we denote by $C(X)$ the set of all real-valued continuous functions defined on a topological space X. Now, assume that R_4 is normal and put

$$F = \{(x, y) \in R_4 \mid y = 0\}.$$

Then F is a closed set of R_4 and it is a discrete space when regarded as a subspace of R_4. Hence every real-valued function defined on F is continuous and accordingly can be continuously extended over R_4 by use of Tietze's theorem. Thus we obtain

$$|C(R_4)| \geq |C(F)| = \mathfrak{c}^{\mathfrak{c}}, \tag{1}$$

where $\mathfrak{c}$ denotes the cardinal number of the set of all real numbers. On the other hand,

$$A = \{(x, y) \in R_4 \mid x \text{ and } y \text{ are rationals}\}$$

is dense in R_4. Observe that every continuous function on R_4 is decided by the values on A. Namely, if $f, g \in C(R_4)$ satisfy $f(p) = g(p)$ for all $p \in A$, then $f = g$. (See Exercise II.31.) Thus

$$|C(R_4)| = |C(A)| \leq \mathfrak{c}^{a} = \mathfrak{c},$$

because A is countable. But this contradicts (1). Hence R_4 is not normal.

Next, we shall show, by an example, that the product of two normal spaces is not necessarily normal. Let S be the set of all real numbers. We define the topology of S by use of the open base

$$\{(x - \varepsilon, x] \mid x \in S, \varepsilon > 0\}.$$

Then S turns out to be a topological space called *Sorgenfrey line*. S is normal, because if F and G are disjoint closed sets of S, then for each $x \in F$ and $y \in G$ find $\varepsilon(x) > 0$ and $\delta(y) > 0$ such that

$$(x - \varepsilon(x), x] \cap G = \emptyset, \qquad (y - \delta(y), y] \cap F = \emptyset.$$

Put

$$U = \bigcup \{(x - \varepsilon(x), x] \mid x \in F\},$$
$$V = \bigcup \{(y - \delta(y), y] \mid y \in G\};$$

then it is easy to see that $U \cap V = \emptyset$, which proves the normality of S.

To see that $S \times S$ is not normal, assume the contrary. Then note that

$$D = \{(x, y) \in S \times S \mid x + y = 0\}$$

is a closed set of S and a discrete subspace. Also observe that

$$A = \{(x, y) \in S \times S \mid x \text{ and } y \text{ are rationals}\}$$

is a countable dense set of $S \times S$. The rest of the proof goes on like the above proof on R_4. Namely, by use of Tietze's theorem,

$$|C(S \times S)| \geqslant |C(D)| = \mathfrak{c}^{\mathfrak{c}} .$$

On the other hand,

$$|C(S \times S)| = |C(A)| \leqslant \mathfrak{c}^{a} = \mathfrak{c} ,$$

which is a contradiction. Thus $S \times S$ is not normal. We shall prove later in Example III.8 that S is fully normal. Also observe that S is totally disconnected.

Theorem III.4. *In the following sequence of classes of topological spaces, each class is included by its precursor*: T_1, T_2, *regular, completely regular, normal, fully normal.*
    This relation is non-reversible.

Proof. By use of Urysohn's lemma, we can see that every normal space is completely regular. To show that every fully normal space is normal, we consider two given disjoint closed sets F, G of a fully normal space X. Since $\mathscr{U} = \{X - F, X - G\}$ is an open covering of X, there is an open covering $\mathscr{V}$ of X such that $\mathscr{V}^{\Delta} < \mathscr{U}$. Putting

$$U = S(F, \mathscr{V}) \quad \text{and} \quad V = S(G, \mathscr{V}) ,$$

we obtain open sets U, V containing F and G respectively. If we assume that $U \cap V \neq \emptyset$, then there are V_1, $V_2 \in \mathscr{V}$ such that

$$V_1 \cap F \neq \emptyset, \qquad V_2 \cap G \neq \emptyset \quad \text{and} \quad V_1 \cap V_2 \neq \emptyset .$$

Hence, for a point $p \in V_1 \cap V_2$,

$$S(p, \mathcal{V}) \supset V_1 \cup V_2,$$

which implies

$$S(p, \mathcal{V}) \cap F \neq \emptyset \quad \text{and} \quad S(p, \mathcal{V}) \cap G \neq \emptyset,$$

i.e., $S(p, \mathcal{V})$ is contained in no member of $\mathcal{U}$. This contradicts $\mathcal{V}^\Delta < \mathcal{U}$, and hence

$$U \cap V = \emptyset.$$

Therefore X is normal. The rest of the proof is easy and so it is left to the reader. Remember that counter examples showing that this relation is non-reversible were given in Examples III.1, 2, 3.

The following assertion will be used later.

C) *Let $\mathcal{U} = \{U_\alpha \mid \alpha \in A\}$ be a point-finite open covering of a normal space X. Then there is an open covering $\mathcal{V} = \{V_\alpha \mid \alpha \in A\}$ such that*

$$\bar{V}_\alpha \subset U_\alpha, \quad \alpha \in A.$$

In this case we say we can *shrink* $\mathcal{U}$ to $\bar{\mathcal{V}}$.

Proof. By use of Corollary 1 to Theorem I.3 we well-order the elements of $\mathcal{U}$ so that

$$\mathcal{U} = \{U_\alpha \mid 0 \leq \alpha < \tau\},$$

where τ denotes a definite ordinal number. Now, we shall define an open set V_α for every α such that

$$X - [\bigcup \{V_\beta \mid \beta < \alpha\} \cup \bigcup \{U_\gamma \mid \gamma > \alpha\}] \subset V_\alpha \subset \bar{V}_\alpha \subset U_\alpha. \tag{1}_\alpha$$

For that purpose we shall use transfinite induction on the ordinal number α. First, let us define V_0. Since $\mathcal{U}$ is an open covering, $X - \bigcup \{U_\gamma \mid \gamma > 0\}$ is a closed set contained in U_0. Therefore by A) we can construct an open set V_0 such that

$$X - \bigcup \{U_\gamma \mid \gamma > 0\} \subset V_0 \subset \bar{V}_0 \subset U_0.$$

Assume that we have defined V_β for every $\beta < \alpha$. Then note that

$$\mathcal{U}' = \{V_\beta \mid \beta < \alpha\} \cup \{U_\gamma \mid \gamma \geqslant \alpha\}$$

forms a covering of X. For, if p is a given point of X which is not contained in U_γ for every $\gamma \geqslant \alpha$, then, since $\mathcal{U}$ is point-finite, there is the last ordinal number β such that $\beta < \alpha$ and $p \in U_\beta$. If $p \notin V_{\beta'}$ for every $\beta' < \beta$, then

$$p \in X - [\cup\{V_{\beta'} \mid \beta' < \beta\} \cup \cup\{U_{\beta''} \mid \beta'' > \beta\}]$$

which is contained in V_β by the induction hypothesis $(1)_\beta$. Hence $p \in V_\beta$, which proves that $\mathcal{U}'$ is a covering.

Thus $X - [\cup\{V_\beta \mid \beta < \alpha\} \cup \cup \{U_\gamma \mid \gamma > \alpha\}]$ is a closed set contained in U_α, and hence by use of A) we can construct an open set V_α such that

$$X - [\cup\{V_\beta \mid \beta < \alpha\} \cup \cup\{U_\gamma \mid \gamma > \alpha\}] \subset V_\alpha \subset \bar{V}_\alpha \subset U_\alpha .$$

Thus we can construct V_α satisfying $(1)_\alpha$ for every α with $\alpha < \tau$. Since each V_α is an open set satisfying $\bar{V}_\alpha \subset U_\alpha$, we must only prove that $\{V_\alpha \mid \alpha < \tau\}$ is a covering of X, but the proof is quite analogous to the proof that $\mathcal{U}'$ is a covering, so it is left to the reader.

3. Compact space and paracompact space

In the preceding sections we have learned a sequence of conditions which belong to the category of separation axioms. Now, here is another important group of conditions for a topological space, i.e. compactness and related conditions, which are rather different from the separation axioms in their nature.

Definition III.4. A topological space X is called a *compact space* (or *bicompact space*) if every open covering $\mathcal{U}$ of X contains a finite subcovering $\mathcal{V}$, i.e., a finite open covering $\mathcal{V}$ such that $\mathcal{V} \subset \mathcal{U}$. A topological space X is called a *paracompact space* if for every open covering $\mathcal{U}$ of X, there is a locally finite open covering $\mathcal{V}$ such that $\mathcal{V} < \mathcal{U}$.[1]

---

[1] A subset of a topological space is called a *compact* (*paracompact*) *set* if it is compact (paracompact) as a subspace. A paracompact T_2-space is often called a paracompact space, but in this book we do not assume T_2 when speaking of a general paracompact space.

The following is a direct consequence of Definition III.4.

A) *Every compact space is paracompact.*

B) *Let f be a continuous mapping of a compact space X onto a topological space Y. Then Y is also compact.*

Proof. Let $\mathcal{V}$ be an open covering of Y. Then

$$f^{-1}(\mathcal{V}) = \{f^{-1}(V) \mid V \in \mathcal{V}\}$$

is an open covering of X. Since X is compact, there is a finite subcovering $\{f^{-1}(V_1), \ldots, f^{-1}(V_k)\}$ of $f^{-1}(\mathcal{V})$. Then $\{V_1, \ldots, V_k\}$ is a finite subcovering of $\mathcal{V}$, proving the compactness of Y.

C) *Every closed set of a compact space is compact. Every compact set of a T_2-space is closed.*

Proof. Let us prove only the latter half. Let A be a non-closed set of a T_2-space X. Then there is a point $p \in \bar{A} - A$. For each point $q \in A$, select an open nbd $U(q)$ of q and a nbd $V_q(p)$ of p such that $U(q) \cap V_q(p) = \emptyset$. Then

$$\mathcal{U} = \{A \cap U(q) \mid q \in A\}$$

is an open covering of A. But for every k,

$$[U(q_1) \cup \cdots \cup U(q_k)] \cap [V_{q_1}(p) \cap \cdots \cap V_{q_k}(p)] = \emptyset,$$

which implies

$$U(q_1) \cup \cdots \cup U(q_k) \not\supset A$$

because

$$V_{q_1}(p) \cap \cdots \cap V_{q_n}(p) \cap A \neq \emptyset.$$

Thus $\mathcal{U}$ contains no finite subcovering, proving that A is non-compact.

Later, in Theorem IV.1, we shall see that the product space of compact spaces is also compact.

D) *Let X be a compact space and Y a T_2-space. Then every continuous one-to-one mapping of X onto Y is a topological mapping.*

Proof. All we have to prove is that f is closed. Suppose F is a closed set of X; then it is compact by C). Hence by B) $f(F)$ is compact, and hence by C) it is closed in Y, proving that f is a closed and, consequently, a topological mapping.

Theorem III.5. *A topological space X is compact if and only if it satisfies one of the following conditions*:

    (i) *every closed collection with finite intersection property has a non-empty intersection,*

    (ii) *every filter of X has a cluster point,*

    (iii) *every maximal filter of X converges.*

Proof. Compactness $\Rightarrow$ (i). Let X be a compact space and $\mathscr{G}$ a closed collection with f.i.p. Suppose $\mathscr{G}$ has empty intersection. Then $\{X - G \mid G \in \mathscr{G}\}$ is an open covering of X. Hence there is a finite subcovering $\{X - G_1, \ldots, X - G_n\}$. This implies $\bigcap_{i=1}^{n} G_i = \emptyset$ contradicting the f.i.p. of $\mathscr{G}$. Therefore $\mathscr{G}$ has non-empty intersection.

(i) $\Rightarrow$ (ii). Suppose $\mathscr{F}$ is a filter of a topological space X satisfying (i). Then

$$\bar{\bar{\mathscr{F}}} = \{\bar{F} \mid F \in \mathscr{F}\}$$

is a closed collection with f.i.p. Therefore by (i), its intersection contains at least one point p. This means that $p \in \bar{F}$ for every $F \in \mathscr{F}$, i.e. p is a cluster point of $\mathscr{F}$.

(ii) $\Rightarrow$ (iii). If X is a topological space satisfying (ii), then every maximal filter of X has a cluster point which is, by II.4.C), simultaneously a convergence point of the filter.

(iii) $\Rightarrow$ compactness. Finally, we suppose $\mathscr{U}$ is a given open covering of a topological space X satisfying (iii). Assume that $\mathscr{U}$ has no finite subcovering. Then

$$\mathscr{F}' = \{X - U \mid U \in \mathscr{U}\}$$

has f.i.p. Therefore, by II.4.A), we can construct a maximal filter $\mathscr{F}$ which contains $\mathscr{F}'$ as a subcollection. By (iii) we have $\mathscr{F} \to p$ for some point p of X. Since by II.4.C) p is a cluster point of $\mathscr{F}$, $p \in \bar{F}$ for every member F of $\mathscr{F}$ and especially

$$p \in \overline{X - U} = X - U$$

for every $U \in \mathcal{U}$. But this contradicts the fact that $\mathcal{U}$ is a covering of X. Thus $\mathcal{U}$ must have a finite subcovering, and hence X is compact.

Corollary. *A topological space X is compact if and only if it satisfies one of the following conditions*:
  (i) *every net of X has a cluster point,*
  (ii) *every maximal net of X converges.*[1]

Proof. Compactness $\Rightarrow$ (i). Let $\varphi(\Delta \mid >)$ be a net of compact space X. Then by (ii) of the theorem, the filter derived from φ has a cluster point which is, as easily seen, a cluster point of φ, too.

(i) $\Rightarrow$ (ii). If X satisfies (i), then every maximal net φ of X has a cluster point p. Then for each nbd U of p, φ is residual either in U or in $X - U$. But, since φ is cofinal in U, it cannot be residual in $X - U$. Thus φ is residual in U, proving $\varphi(\Delta \mid >) \rightarrow p$.

(ii) $\Rightarrow$ compactness. Suppose $\mathcal{F}$ is a maximal filter of a topological space X satisfying (ii) and $\varphi(\Delta \mid >)$ is a net derived from $\mathcal{F}$. Then φ is maximal (Exercise II.15), and hence $\varphi(\Delta \mid >) \rightarrow p$ for some $p \in X$. Now p is easily seen to be a cluster point of $\mathcal{F}$, but since $\mathcal{F}$ is maximal, by II.4.C), $\mathcal{F} \rightarrow p$. Thus by the theorem X is compact.

E) *Every paracompact T_2-space X is fully normal.*

Proof. First let us prove that X is regular. Suppose F is a closed set of X and p is a point of X which is not in F. Since X is T_2, for every point q of F, there is an open nbd $V(q)$ of q and a nbd $U_q(p)$ of p such that

$$V(q) \cap U_q(p) = \emptyset.$$

Hence $p \notin \overline{V(q)}$. We assign to each point q of F such an open nbd $V(q)$ and let

$$\mathcal{V} = \{X - F, V(q) \mid q \in F\}.$$

Then $\mathcal{V}$ is an open covering of X, and hence by the paracompactness of X there is a locally finite open covering $\mathcal{W}$ of X such that $\mathcal{W} < \mathcal{V}$. Putting

---

[1] G. Helmberg [1] gave another interesting necessary and sufficient condition for compactness with respect to nets of continuous functions.

$$W = S(F, \mathcal{W}),$$

we obtain an open set W which contains F. Then we can assert that $p \notin \bar{W}$. For, if

$$W' \in \mathcal{W}, \qquad W' \cap F \neq \emptyset,$$

then since $\mathcal{W} < \mathcal{V}$, $W' \subset V(q)$ for some $q \in F$. This combined with $p \notin \overline{V(q)}$ implies that $p \notin \bar{W}'$. Therefore, by use of II.5.C), we obtain

$$p \notin \cup \{\bar{W}' \mid W' \cap F \neq \emptyset, W' \in \mathcal{W}\}$$

$$= \overline{\cup \{W' \mid W' \cap F \neq \emptyset, W' \in \mathcal{W}\}} = \bar{W}.$$

This means that p has an open nbd U satisfying $W \cap U = \emptyset$, proving that X is regular.

Now, to prove the normality of X, we suppose F and G are given disjoint closed sets of X. To every point p of F, we assign an open nbd $V(p)$ such that

$$\overline{V(p)} \cap G = \emptyset;$$

this is possible by virtue of the regularity of X (see 1.B)). Putting

$$\mathcal{V} = \{X - F, V(p) \mid p \in F\},$$

we get an open covering $\mathcal{V}$ of X. Since X is paracompact, there is a locally finite open covering $\mathcal{U}$ such that

$$\mathcal{U} < \mathcal{V}.$$

Put

$$U = S(F, \mathcal{U});$$

then U is an open covering containing F. Furthermore, U satisfies

$$\bar{U} = \overline{\cup \{U' \mid U' \in \mathcal{U}, U' \cap F \neq \emptyset\}}$$

$$= \cup \{\bar{U}' \mid U' \in \mathcal{U}, U' \cap F \neq \emptyset\} \subset X - G.$$

This is true because it follows from $\mathcal{U} < \mathcal{V}$ that every $U' \in \mathcal{U}$ with $U' \cap F \neq \emptyset$ must be contained in $V(p)$ for some $p \in F$, and hence

$$\bar{U}' \subset \overline{V(p)} \subset X - G.$$

Thus from 2.A) we can conclude that X is normal.

Finally, let us prove that X is fully normal. To do this we suppose $\mathcal{U}$ is a given open covering of X. Since X is paracompact, there is a locally finite open covering $\mathcal{V}$ with

$$\mathcal{V} < \mathcal{U}.$$

Suppose

$$\mathcal{V} = \{ V_\alpha \mid \alpha \in A \} ;$$

then by 2.C) we can construct an open covering

$$\mathcal{W} = \{ W_\alpha \mid \alpha \in A \}$$

such that

$$\bar{W}_\alpha \subset V_\alpha, \quad \alpha \in A.$$

Let A' be a subset of the index set A; then we define a subset $P(A')$ of X by

$$P(A') = [\cap \{ V_\alpha \mid \alpha \in A' \}] \cap [\cap \{ X - \bar{W}_\alpha \mid \alpha \in A - A' \}].$$

Let us prove that $P(A')$ is an open set. Since $\mathcal{V}$ is locally finite, $\bar{\mathcal{W}}$ is also locally finite, and hence

$$\cap \{ X - \bar{W}_\alpha \mid \alpha \in A - A' \} = X - \cup \{ \bar{W}_\alpha \mid \alpha \in A - A' \}$$

is an open set by II.5.D). On the other hand, since $\mathcal{V}$ is locally finite,

$$\cap \{ V_\alpha \mid \alpha \in A' \} = \emptyset$$

if A' is infinite, and it is open if A' is finite. Therefore $P(A')$ is an open set for each subset A' of A. It is easily seen that each point p of X is contained in $P(A')$ where

$$A' = \{ \alpha \mid p \in \bar{W}_\alpha \}.$$

Thus

$$\mathcal{P} = \{ P(A') \mid A' \subset A \}$$

is an open covering of X. To prove that $\mathscr{P}^\Delta < \mathscr{V}$, we consider a given point p of X. Since $\mathscr{W}$ is a covering, $p \in W_{\alpha_0}$ for some $\alpha_0 \in A$. Let $p \in P(A')$; then in view of the definition of $P(A')$ we know that $\alpha_0 \in A'$, because if $\alpha_0 \in A - A'$, then

$$P(A') \subset X - \bar{W}_{\alpha_0} \not\ni p$$

contradicting $p \in P(A')$. Therefore

$$P(A') \subset V_{\alpha_0}.$$

Since this is valid for every $P(A')$ containing p, we obtain

$$S(p, \mathscr{P}) \subset V_{\alpha_0} \in \mathscr{V}.$$

This proves that $\mathscr{P}^\Delta < \mathscr{V}$ and therefore

$$\mathscr{P}^\Delta < \mathscr{U}.$$

Thus X is fully normal.

Combining E) with A), we get

F) *Every compact T_2-space is fully normal.*

A topological space X is called *locally compact* if every point p of X has a nbd U which is compact. Generally, for a given property (P) of space, we can define a *locally (P)-space* to be a space each point of which has a nbd with the property (P).[1] For example, we can consider terminologies like locally regular, locally normal, locally fully normal, etc.

We also consider hereditary properties of spaces. A space X is called a *hereditarily (P)-space* if every subspace of X (including X itself) has (P). For example, a topological space X is called a hereditarily normal space if every subspace of X is normal. As we have seen, T_1, T_2, regular and completely regular are hereditary conditions, i.e., every T_1 (T_2, regular, completely regular) space is a hereditarily T_1 (T_2, regular, completely regular) space at the same time. But this is not the case for properties like

---

[1] Note that there are a few exceptions like the definition of locally connected space.

normal and fully normal. Thus terminologies like hereditarily normal and hereditarily fully normal have their own meaning.

G) *Every locally compact T_2-space X is completely regular.*

Proof. Let F be a closed set of X and $p \notin F$. Suppose G is a compact nbd of p. Then by F) G is fully normal and accordingly completely regular. Hence there is a continuous function f from G into $[0, 1]$ such that $f(p) = 0$, $f((F \cap G) \cup B(G)) = 1$. Define a function g from X into $[0, 1]$ by $g(x) = f(x)$ if $x \in G$, $g(x) = 1$ if $x \in X - G$. Then g is obviously continuous. Thus X is completely regular.

Example III.6. As taught in an introductory part of calculus, every bounded closed set of E^2 (or more generally E^n) is compact. Conversely, it is easily seen that every compact set of E^2 is bounded and closed. As a matter of fact, we may say compactness is a generalization of the bounded closedness in E^n. Another simple example of a compact space is a point sequence $\{p_i\}$ converging to p in a topological space X. Then the subspace $\{p, p_1, p_2, \ldots\}$ of X is compact. In view of the fundamental role of the above theorem on bounded closed sets (Heine–Borel theorem) in calculus, we can imagine the significance of compact spaces not only in topology but also in the field of analysis. Note that E^n itself is non-compact but locally compact. We shall see later that E^n is also paracompact. E^n is the sum of countably many compact sets (such a space is called *σ-compact*).

To give another example of a compact space, we consider R_5 in Example II.1 and put

$$R_6 = R_5 \cup \{\omega_1\}.$$

We define a nbd basis for each point of R_5 as in Example II.5 and

$$\mathcal{U}(\omega_1) = \{(\beta, \omega_1] \mid \beta < \omega_1\}$$

for the newly added point ω_1. Suppose $\mathcal{U}$ is an open covering of R_6. To each $\alpha \in R_6$ ($\alpha \neq 0$) we assign a nbd $V(\alpha) = (\beta, \alpha]$ such that $V(\alpha) \subset U$ for some $U \in \mathcal{U}$ and $V(0) = \{0\}$. To prove that $\mathcal{V} = \{V(\alpha) \mid \alpha \in R_6\}$ contains a finite subcovering, we assume the contrary. Let

$$V(\omega_1) = (\beta_1, \omega_1], \qquad V(\beta_1) = (\beta_2, \beta_1], \qquad V(\beta_2) = (\beta_3, \beta_2], \ldots.$$

Then since $\beta_1 > \beta_2 > \beta_3 > \cdots$, the subset $\{\beta_1, \beta_2, \ldots\}$ of R_6 has no first number, which is impossible. Thus $\mathscr{V}$ has a finite subcovering, and so does $\mathscr{U}$. Hence R_6 is compact. Since R_6 is obviously T_2, it is fully normal by F). On the other hand, as shown in Example III.3, R_5 is not fully normal and thus neither paracompact because of E). Hence a subspace of a compact space is not necessarily paracompact, and a subspace of a fully normal space is not necessarily fully normal.

Turning to an example of a paracompact space, every discrete space with infinitely many points is easily seen to be a non-compact, para-compact space.

4. Axioms of countability

Definition III.5. Let X be a topological space. Then the minimal infinite cardinal number of an open basis of X is called the *weight* of X and denoted by $w(X)$, i.e., $w(x) = \min\{|\mathscr{U}| \mid \mathscr{U}$ is an open base of $X\} + \aleph_0$. If the weight of X is $\aleph_0$, i.e., if there is an at most countable open basis, then X is said to satisfy the *second axiom of countability*. Similarly we define the *character* $\chi(X)$, *density* $d(X)$ and *Lindelöf number* $l(X)$ by $\chi(X) = \min\{\mu| $ every point x of X has a nbd base $\mathscr{U}(x)$ with $|\mathscr{U}(x)| \leq \mu\} + \aleph_0$, $d(X) = \min\{|D| \mid D$ is a dense subset of $X\} + \aleph_0$, $l(X) = \min\{\mu|$ every open covering $\mathscr{U}$ of X has a subcover $\mathscr{V}$ with $|\mathscr{V}| \leq \mu\} + \aleph_0$. If $\chi(X) = \aleph_0$, i.e. if every point of X has a nbd basis consisting of at most countably many nbds, then X is said to satisfy the *first axiom of countability*.[1] If $l(X) = \aleph_0$, i.e. if every open covering of X contains an at most countable subcovering, then X is said to be a *Lindelöf space*. If $d(X) = \aleph_0$, i.e. if X has a dense subset consisting of at most countably many points, then it is called a *separable space*.

A) *Every subspace of a space satisfying the 1st (2nd) axiom of countability also satisfies the 1st (2nd) axiom of countability. The product space of at most countably many spaces satisfying the 1st (2nd) axiom of countability also satisfies the 1st (2nd) axiom of countability. The product of at most countably many separable spaces is separable.*[2]

---

[1] A topological space satisfying 1st axiom of countability (2nd axiom of countability) is often called a *1st countable* (*2nd countable*) space.

[2] It is known that the product of at most c number of separable spaces is separable. See Corollary 1 to Theorem VIII.9.

Proof. The easy proof is left to the reader.

B) *If a topological space X satisfies the 2nd axiom of countability, then it satisfies the 1st axiom of countability and is Lindelöf and separable, too.*

Proof. Let $\mathcal{U}$ be a countable open basis of X and p a given point of X. Then

$$\mathcal{U}(p) = \{U \mid p \in U \in \mathcal{U}\}$$

forms a nbd basis of p with $|\mathcal{U}(p)| \leqslant \aleph_0$. This proves that X satisfies the 1st axiom of countability.

Let $\mathcal{V}$ be a given open covering of X. Since $\mathcal{U}$ is an open basis, for each point p of X there is an open set $U(p) \in \mathcal{U}$ such that

$$p \in U(p) \subset V$$

for some $V \in \mathcal{V}$. Now

$$\mathcal{U}' = \{U(p) \mid p \in X\}$$

is a subcollection of $\mathcal{U}$, and hence it is an at most countable open covering of X. For every $U(p)$, we choose an open set $V(p) \in \mathcal{V}$ such that $U(p) \subset V(p)$, where $V(p) = V(p')$ if $U(p) = U(p')$. Then

$$\mathcal{V}' = \{V(p) \mid p \in X\}$$

is an at most countable subcovering of $\mathcal{V}$. This proves that X is Lindelöf.

Finally, if we choose a point $p(U)$ from each member U of $\mathcal{U}$, then $D = \{p(U) \mid U \in \mathcal{U}\}$ forms an at most countable dense subset of X. Therefore X is separable.

C) *Every Lindelöf, regular space is paracompact and fully normal.*

Proof. Suppose $\mathcal{W}$ is a given open covering of X. Since X is regular, for every point p of X, we can choose an open nbd $U(p)$ such that $\overline{U(p)} \subset W(p)$ for some $W(p) \in \mathcal{W}$. Since X is Lindelöf, we choose a countable subcovering $\{U(p_i) \mid i = 1, 2, \ldots\}$ from the open covering $\{U(p) \mid p \in X\}$. Putting

$$W_i = W(p_i) - \bigcup_{j=1}^{i-1} \overline{U(p_j)}, \quad i = 1, 2, \ldots,$$

we get open sets W_i, $i = 1, 2, \ldots$, satisfying $W_i \subset W(p_i)$. Hence

$$\mathcal{W}' = \{W_i \mid i = 1, 2, \ldots\}$$

is an open collection such that $\mathcal{W}' < \mathcal{W}$. To show that $\mathcal{W}'$ is a locally finite covering, we consider a given point p of X. Suppose $W(p_i)$ is the first member of $\{W(p_i) \mid i = 1, 2, \ldots\}$ which contains p. Then it is clear that $p \in W_i$. Hence $\mathcal{W}'$ covers X. Since $\{U(p_i) \mid i = 1, 2, \ldots\}$ is a covering, $p \in U(p_{i_0})$ for some i_0. On the other hand, by the definition of W_i,

$$U(p_{i_0}) \cap W_i = \emptyset$$

for all i such that $i > i_0$. Therefore $\mathcal{W}'$ is locally finite. This proves that X is paracompact. By use of 3.E), we also conclude that X is fully normal.

Example III.7. The real line E^1 satisfies the second axiom of countability. For $\mathcal{U} = \{S_{1/n}(p) \mid p$ is a rational number, $n = 1, 2, \ldots\}$ is easily seen to be a countable open basis of E^1. Therefore, by A), every Euclidean space E^n and the Hilbert cube I^ω, too, satisfy the second axiom of countability. Furthermore, every subset of I^ω also satisfies the same axiom. It is obvious that every compact space is Lindelöf. For example, R_6 given in Example III.6 is compact and hence Lindelöf, too. On the other hand, since the supremum of a countable set of $R_6 - \{\omega_1\}$ belongs to $R_6 - \{\omega_1\}$, we can easily show that ω_1 has no countable nbd base. Therefore R_6 does not satisfy the first axiom of countability. To the contrary, R_5 in Example II.1 satisfies the first axiom of countability, but is not Lindelöf. For, if we choose for every $\alpha \in R_5$ different from 0 an ordinal number $\beta(\alpha)$ with $\beta(\alpha) < \alpha$, then

$$\mathcal{V} = \{\{0\}, (\beta(\alpha), \alpha] \mid 0 < \alpha < \omega_1\}$$

is an open covering of R_5 which has no countable subcovering.

It is also easy to see that neither R_5 nor R_6 is separable. R_1 in Example II.7 is separable, because the rational numbers on E^1 form a dense subset of R_1, but p has no countable nbd basis, i.e., R_1 does not satisfy the first axiom of countability. We can also easily see that R_4 in Example II.5 is separable, but not Lindelöf. R_4 also shows us that a subspace of a separable space does not need to be separable, because the x-axis is a nonseparable subspace of R_4.

Example III.8. Let us prove that the Sorgenfrey line S defined in Example III.5 is Lindelöf. Let $\mathcal{U}$ be a given open covering of S. It suffices to prove that every interval $[a, b]$ in S is covered by countably many elements of $\mathcal{U}$, because S is a countable sum of such intervals. Let

$$L = \{x \in [a, b] \mid [x, b] \text{ can be covered by countably many elements of } \mathcal{U}\}.$$

Since $b \in L$, $L \neq \emptyset$. Put

$$c = \inf L.$$

It is obvious that $c < b$. Assume $a < c$. Choose $x_1, x_2, \ldots \in [a, b]$ such that $c < \cdots < x_2 < x_1 \leqslant b$. Then obviously $x_i \in L$, $i = 1, 2, \ldots$, and hence $c \in L$, because

$$[c, b] = \{c\} \cup \left(\bigcup_{i=1}^{\infty} [x_i, b] \right).$$

Now, for some $\varepsilon > 0$ such that $a < c - \varepsilon < c$, $(c - \varepsilon, c]$ is contained in some element of $\mathcal{U}$. Thus $c - \varepsilon/2 \in L$, which contradicts the definition of c. This proves that $a = c$, and therefore $a \in L$, i.e., $[a, b]$ is covered by countably many elements of $\mathcal{U}$. Hence S is Lindelöf.

By C) S is paracompact and fully normal, too. On the other hand, as we saw in Example III.5, $S \times S$ is not normal. Hence it is neither paracompact nor Lindelöf. This example shows that the product space of two paracompact (Lindelöf, fully normal) spaces is not necessarily paracompact (Lindelöf, fully normal). Also observe that S is separable and 1st countable but not 2nd countable.

In Section 2 we have characterized all completely regular spaces as subspaces of the product space of unit segments. Now, we can give a concreter materialization to regular spaces satisfying the second axiom of countability.

Theorem III.6 (Urysohn's imbedding theorem). *A regular space satisfies the second axiom of countability if and only if it is homeomorphic with a subset of the Hilbert cube I^ω.*

Proof. Since the 'if' part is a direct consequence of A) (see Example III.7), we shall concern ourselves only with the 'only if' part. Let $\mathcal{U}$ be a countable open basis of a regular space X satisfying the second axiom of countability. We consider a pair $P = (U, V)$ of members of $\mathscr{U}$ satisfying $\bar{U} \subset V$ and denote by $\mathscr{P} = \{P_i \mid i = 1, 2, \ldots\}$ the totality of such pairs. Note that $\mathscr{P}$ is a countable collection since $\mathscr{U}$ is countable. Let $P_i = (U_i, V_i)$ be a member of $\mathscr{P}$, where $\bar{U}_i \subset V_i$. Then, since X is normal, by B) and C), we can construct, by use of Urysohn's lemma, a continuous function f_i over X such that

$$f_i(U_i) = 0, \qquad f_i(X - V_i) = 1 \quad \text{and} \quad 0 \leqslant f_i \leqslant 1 .$$

Now we define by

$$f(p) = (f_1(p), \tfrac{1}{2}f_2(p), \tfrac{1}{3}f_3(p), \ldots), \quad p \in X,$$

a continuous mapping of X into I^ω.

Let W be a nbd of a point p of X; then, since $\mathscr{U}$ is an open basis of the regular space X, we can choose $U, V \in \mathscr{U}$ such that

$$p \in U \subset \bar{U} \subset V \subset W.$$

Thus (U, V) is a member of $\mathscr{P}$; hence we may suppose, say $P_i = (U, V)$, i.e. $U = U_i$, $V = V_i$. Therefore

$$\frac{1}{i}f_i(p) = 0, \qquad \frac{1}{i}f_i(q) = \frac{1}{i} \quad \text{for } q \notin V,$$

which implies that

$$\rho(f(p), f(q)) \geqslant 1/i \quad \text{for every } q \notin W.$$

It is easily seen that the last statement implies that f is one-to-one, and f^{-1} is continuous. Thus f is a homeomorphism of X onto a subset of I^ω.

The following corollary, which is a modification of Tychonoff's imbedding theorem, follows by use of the technique of the proof of Theorem III.6.

Corollary. *A topological space X is a Tychonoff space with weight $\leqslant \mu$ if and only if it is homeomorphic to a subspace of I^μ, where μ denotes an infinite cardinal number and I^μ the product of μ-copies of the segment $[0, 1]$.*

5. Metric space

Definition III.6. Let X be a set. We call X a *metric space* if for every pair p, q of elements of X, the *distance* (or *metric*) $\rho(p, q)$ between them is defined satisfying the following conditions:

   (i) $\rho(p, q)$ is a non-negative real number,
  (ii) $\rho(p, q) = 0$ if and only if $p = q$,
 (iii) $\rho(p, q) = \rho(q, p)$,
 (iv) $\rho(p, q) \leqslant \rho(p, r) + \rho(r, q)$ for every $r \in X$.

A) *Let X be a metric space. If we define by $\mathcal{U}(p) = \{S_{1/n}(p) \mid n = 1, 2, \ldots\}$ the nbd basis of each point p of X, then X is a fully normal topological space satisfying the first axiom of countability.*[1]

Proof. It is clear that X is a topological space satisfying the first axiom of countability. To prove that X is fully normal, we consider a given open covering $\mathcal{U}$ of X. To every point p of X, we assign a spherical nbd $S_{\varepsilon(p)}(p)$ such that $\varepsilon(p) \leqslant 1$ and $S_{\varepsilon(p)}(p) \subset U$ for some $U \in \mathcal{U}$. Then $\mathcal{S} = \{S_{\varepsilon(p)}(p) \mid p \in X\}$ is an open covering of X with $\mathcal{S} < \mathcal{U}$.

Now, we can show that the open covering $\mathcal{S}' = \{S_{\varepsilon(p)/4}(p) \mid p \in X\}$ is a delta-refinement of $\mathcal{S}$. For, let p be a given point of X; then we put

$$\eta = \sup\{\varepsilon(p') \mid p \in S_{\varepsilon(p')/4}(p')\}.$$

It follows from the definition of η that $p \in S_{\varepsilon(p_0)/4}(p_0)$ for some p_0 with

$$\varepsilon(p_0) > \tfrac{3}{4}\eta. \tag{1}$$

Let $S_{\varepsilon(q)/4}(q)$ be a given member of $\mathcal{S}'$ which contains p. Then for each point $r \in S_{\varepsilon(q)/4}(q)$, we obtain, in view of (1),

$$\rho(p_0, r) \leqslant \rho(p_0, p) + \rho(p, q) + \rho(q, r)$$
$$< \tfrac{1}{4}\varepsilon(p_0) + \tfrac{1}{4}\varepsilon(q) + \tfrac{1}{4}\varepsilon(q) \leqslant \tfrac{1}{4}\varepsilon(p_0) + \tfrac{1}{2}\eta < \varepsilon(p_0),$$

and hence $r \in S_{\varepsilon(p_0)}(p_0)$, i.e.,

$$S_{\varepsilon(q)/4}(q) \subset S_{\varepsilon(p_0)}(p_0).$$

---

[1] Remember that $S_\varepsilon(p) = \{q \mid \rho(p, q) < \varepsilon\}$. $S_\varepsilon(p)$ is called the ε-*nbd* or a *spherical nbd* of p.

This means that

$$S(p, \mathcal{S}') \subset S_{\varepsilon(p_0)}(p_0) \in \mathcal{S}.$$

Therefore $\mathcal{S}'$ is a delta-refinement of $\mathcal{S}$ and accordingly of $\mathcal{U}$, too. Thus X is fully normal.

B) *For metric spaces, the three conditions, 2nd axiom of countability, Lindelöf and separability are equivalent.*

Proof. In 4.B), we have proved that the second axiom of countability implies the other two conditions. To prove the converse, suppose X is a separable metric space; then there is a countable dense set $D = \{p_1, p_2, \ldots\}$ of X. Put

$$\mathcal{S}_n = \{S_{1/n}(p_i) \mid i = 1, 2, \ldots\}, \qquad \mathcal{S} = \bigcup_{n=1}^{\infty} \mathcal{S}_n ;$$

then $\mathcal{S}$ is a countable open basis of X. To see it, let p be a point of an open set U of X. Then $S_\varepsilon(p) \subset U$ for some $\varepsilon > 0$. Choose a positive integer n with $2/n < \varepsilon$; then, since D is dense in X, there is $p_i \in D$ with $\rho(p_i, p) < 1/n$. Now

$$p \in S_{1/n}(p_i) \subset S_\varepsilon(p) \subset U,$$

and hence $\mathcal{S}$ is an open basis of X. In a similar way we can also prove that Lindelöf implies the 2nd axiom of countability.

Example III.9. Euclidean space E^n and Hilbert space H are the most popular examples of separable metric spaces. In H the subset $A = \{(x_1, x_2, \ldots) \mid$ for some n, $x_1, \ldots, x_n$ are rational, and $x_j = 0$ for all $j \geq n + 1\}$ forms a dense countable set.

Let B be a set of real-valued bounded functions over a topological space X. Defining the distance between $f, g \in B$ by

$$\rho(f, g) = \sup\{|f(p) - g(p)| \mid p \in X\},$$

we obtain a metric space called a *function space*. Function spaces are not generally separable, but we can construct a simpler example of a non-separable metric space by considering the discrete sum of uncountably many non-empty metric spaces. Let I_α, $\alpha \in A$, be copies of the unit segment $[0, 1]$. In their union $\cup \{I_\alpha \mid \alpha \in A\}$ we identify all zeros to get a star-shaped set $S(A)$. Then we define the distance between two points of $S(A)$ by

$$\rho(p, q) = \begin{cases} |p - q| & \text{if } p, q \in I_\alpha, \\ p + q & \text{if } p \in I_\alpha, q \in I_\beta, \text{ where } \alpha \neq \beta. \end{cases}$$

Now, we can easily verify that $S(A)$ is a metric space, which is called a (*metric*) *star-space* with the index set A. A star space is non-separable if its index set A is uncountable. $S(A)$ may be called sometimes a (*metric*) *hedgehog*.

Here is another interesting example of non-separable metric space. Let A be a set and

$$P(A) = \prod \{E_\alpha \mid \alpha \in A\}$$

be the cartesian product of the copies E_α of the real line $E^1 = (-\infty, \infty)$. Put

$$H(A) = \left\{ p \mid p = \{p_\alpha \mid \alpha \in A\} \in P(A), \sum_\alpha p_\alpha^2 < +\infty \right\}.$$

Then for each pair $p = \{p_\alpha \mid \alpha \in A\}$ and $q = \{q_\alpha \mid \alpha \in A\}$ of elements of $H(A)$ we define the distance by

$$\rho(p, q) = \sqrt{\sum_\alpha (p_\alpha - q_\alpha)^2}.$$

Then we can easily show that $H(A)$ is a metric space. This metric space is called a *generalized Hilbert space* (with the index set A). If A is a countable set, then $H(A)$ is the separable Hilbert space. If A is uncountable, then $H(A)$ is non-separable.

Baire's zero-dimensional space is also an interesting metric space. Suppose A is a given set. We denote by $N(A)$ the set of all sequences $(\alpha_1, \alpha_2, \ldots)$ of elements α_i of A. The distance between two points $\alpha = (\alpha_1, \alpha_2, \ldots)$ and $\beta = (\beta_1, \beta_2, \ldots)$ of $N(A)$ is defined by

$$\rho(\alpha, \beta) = 1/\min\{k \mid \alpha_k \neq \beta_k\} \quad \text{if } \alpha \neq \beta, \qquad \rho(\alpha, \alpha) = 0.$$

Then $N(A)$ turns out to be a metric space called *Baire's zero-dimensional space* (or *generalized Baire's zero-dimensional space*) with respect to A. If A is the set of all natural numbers, then by mapping $\alpha = (\alpha_1, \alpha_2, \ldots) \in N(A)$ to the continued fraction

$$\alpha_1 + \cfrac{1}{\alpha_2 + \cfrac{1}{\alpha_3 + \cdots}}$$

we know that $N(A)$ is homeomorphic with the subset $P = \{x \in E^1 \mid x$ is irrational and $>1\}$ of E^1.

Finally, let us give a method to introduce a metric into the collection of the closed subsets of a metric space. Let X be a metric space with a bounded metric ρ and 2^X the collection of all non-empty closed sets of X. For $\varepsilon > 0$ and $C \in 2^X$, let $S_\varepsilon(C) = \{p \mid \rho(p, C) < \varepsilon\}$.[1] For $C, K \in 2^X$, we define

$$\rho'(C, K) = \inf\{\varepsilon \mid C \subset S_\varepsilon(K) \text{ and } K \subset S_\varepsilon(C)\}.$$

Then ρ' satisfies the conditions of metric and is called the *Hausdorff metric* of 2^X.[2]

Example III.10. Every subspace of a metric space X is also a metric space with the same distance as in X. Let X and Y be metric spaces with metrics ρ_1 and ρ_2 respectively; then their cartesian product $X \times Y$ turns out to be a metric space with the metric function

$$\rho((p, q), (p', q')) = \sqrt{(\rho_1(p, p'))^2 + (\rho_2(q, q'))^2},$$

which induces the topology of the product space.

---

[1] Generally for a subset A of a metric space, we define its *diameter* by $\delta(A) = \sup\{\rho(p, q) \mid p, q \in A\}$, and the *distance* between two sets A and B by $\rho(A, B) = \inf\{\rho(x, y) \mid x \in A, y \in B\}$.

[2] There is another, interesting method to introduce a metric into the space of compact subsets of a metric space. See K. Borsuk [3], J. de Groot [2]. See also J. Flachsmeyer [3]. Let X be a regular space and 2^X the collection of all non-empty closed subsets of X. For any finite collection $\{U_1, \ldots, U_n\}$ of open sets of X, we define a subcollection of 2^X by

$$\langle U_1, \ldots, U_n \rangle = \left\{ F \mid F \in 2^X, F \subset \bigcup_{i=1}^{n} U_i, F \cap U_i \neq \emptyset \text{ for } i = 1, \ldots, n \right\}.$$

Let all such collections $\langle U_1, \ldots, U_n \rangle$ form an open basis of 2^X. Then 2^X turns out to be a T_2-space. This topology is called the *finite topology* of 2^X. An interesting result on this topology is due to J. Keesling [1], who proved that continuum hypothesis implies that 2^X is normal if and only if X is compact. Because 2^X is compact for compact X, this result means that 2^X is normal if and only if it is compact.

Now, let X_i, $i = 1, 2, \ldots$, be metric spaces with metric functions ρ_i, $i = 1, 2, \ldots$ respectively. Then

$$\rho'_i(p, q) = \min\{1/i, \rho_i(p, q)\}$$

is a metric function such that $\rho'_i \leqslant 1/i$. We denote by X'_i the metric space with metric ρ'_i. X_i and X'_i are easily seen to have the same topology. Now

$$\rho'(p, q) = \left(\sum_{i=1}^{\infty} \rho'_i(p_i, q_i)^2\right)^{1/2}, \quad p = (p_1, p_2, \ldots), \quad q = (q_1, q_2, \ldots),$$

gives a metric of the cartesian product $\prod_{i=1}^{\infty} X'_i$. The metric ρ' induces the topology of the product space $\prod_{i=1}^{\infty} X_i$. In other words, the metric space $\prod_{i=1}^{\infty} X'_i$ with metric ρ' is homeomorphic with the product space $\prod_{i=1}^{\infty} X_i$ by the identity mapping.

Definition III.7. A mapping of a metric space X into a metric space Y is called a *uniformly continuous mapping*, if for every $\varepsilon > 0$, there is $\delta > 0$ such that $\rho(p, p') < \delta$ for $p, p' \in X$ implies $\rho'(f(p), f(p')) < \varepsilon$ in Y, where ρ and ρ' denote the metrics in X and Y, respectively.

This definition is a generalization of the well-known definition of a uniformly continuous function on a Euclidean space. The following proposition is also a generalization of the theorem in calculus: A continuous function defined over a bounded closed set A of E^n is uniformly continuous.

C) *Every continuous mapping f of a compact metric space X into a metric space Y is uniformly continuous.*[1]

Proof. Suppose ε is a given positive number. Then for every point p of X, there is $\delta(p) > 0$ such that

$$f(S_{\delta(p)}(p)) \subset S_{\varepsilon/2}(f(p)).$$

Since X is compact, we can cover X with finitely many of $S_{\delta(p)/2}(p)$, say, $S_{\delta(p_i)/2}(p_i)$, $i = 1, \ldots, n$. Put

---

[1] J. de Groot [1] proved another interesting property for such a mapping f; namely f is topological on a G_δ-set X' of X such that $f(X')$ is dense G_δ in $f(X)$.

$$\delta = \min\{\tfrac{1}{2}\delta(p_i) \mid i = 1, \ldots, n\}.$$

Then, if $\rho(p, p') < \delta$ for $p, p' \in X$, then we choose p_i such that $p \in S_{\delta(p_i)/2}(p_i)$. Since $\delta \leqslant \tfrac{1}{2}\delta(p_i)$, $p' \in S_{\delta(p_i)}(p_i)$. Since

$$f(S_{\delta(p_i)}(p_i)) \subset S_{\varepsilon/2}(f(p_i)),$$

we obtain

$$\rho'(f(p), f(p')) \leqslant \rho'(f(p), f(p_i)) + \rho'(f(p_i), f(p')) < \varepsilon.$$

Therefore f is a uniformly continuous mapping.

Example III.11. Proposition C) does not hold if we drop the compactness of X. For example, $f(x) = \tan x$, $|x| < \pi/2$ is a continuous mapping of $(-\pi/2, \pi/2)$ into E^1, but it is not uniformly continuous.

Definition III.8. Let $\mathscr{F}$ be a filter of a metric space X. If for every $\varepsilon > 0$, there exists a point $p(\varepsilon) \in X$ and a member $F(\varepsilon)$ of $\mathscr{F}$ such that

$$F(\varepsilon) \subset S_\varepsilon(p(\varepsilon)),$$

then we call $\mathscr{F}$ a *Cauchy filter*.

Let $\varphi(\Delta \mid >)$ be a net of a metric space X. If for every $\varepsilon > 0$, there exists a point $p(\varepsilon) \in X$ such that $\varphi(\Delta \mid >)$ is residual in $S_\varepsilon(p(\varepsilon))$, then we call φ a *Cauchy net*.

If every Cauchy filter of X converges, then we call X a *complete metric space*.

D) *If a Cauchy filter of a metric space has a cluster point, then it converges to the cluster point.*

Proof. The easy proof is left to the reader.

E) *A metric space X is complete if and only if every Cauchy net of X converges.*

Proof. Note that every net derived from a Cauchy filter is Cauchy and the filter derived from a Cauchy net is Cauchy. The proof in detail is left to the reader.

We can describe the condition of completeness in terms of ordinary point sequences as follows:

F) *A metric space X is complete if and only if every Cauchy sequence $\{p_i \mid i = 1, 2, \ldots\}$ of points converges.*

Proof. In view of E), the necessity of the condition is clear because a Cauchy point sequence is a Cauchy net on the directed set N of the positive integers.

To see the sufficiency, we suppose $\mathscr{F}$ is a given Cauchy filter of X. For $n = 1, 2, \ldots$, we choose points $p(1/n) \in X$ and $F(1/n) \in \mathscr{F}$ such that

$$F(1/n) \subset S_{1/n}(p(1/n)), \quad n = 1, 2, \ldots.$$

Then $\{S_{1/n}(p(1/n)) \mid n = 1, 2, \ldots\}$ has f.i.p. because $\mathscr{F}$ does so. This implies that $\{p(1/n) \mid n = 1, 2, \ldots\}$ is a Cauchy sequence of points. For, let ε be a given positive number; then choose n_0 such that $2/n_0 < \varepsilon$. Then it is easy to show $p(1/n) \in S_\varepsilon(p(1/n_0))$ for every $n \geq n_0$ by use of the fact that

$$S_{1/n}(p(1/n)) \cap S_{1/n_0}(p(1/n_0)) \neq \emptyset.$$

This means that $\{p(1/n) \mid n = 1, 2, \ldots\}$ is Cauchy. Therefore by the hypothesis, $\{p(1/n)\}$ converges to a point p of X. Now we can easily see that $\mathscr{F}$ also converges to p proving that X is complete.

The following theorem is often used to prove topological theorems and shows the significance of complete metric spaces.

Theorem III.7 (Baire's theorem). *Let U_n, $n = 1, 2, \ldots$, be open dense subsets of a complete metric space X. Then $\bigcap_{n=1}^{\infty} U_n$ is dense in X.*

Proof. Let q be a given point of X and $V(q)$ a given open nbd of q. Since U_1 is dense, there is a point $p_1 \in V(q) \cap U_1$. Since U_1 is open, we can choose $\varepsilon_1 > 0$ such that $\varepsilon_1 < 1$ and

$$\overline{S_{\varepsilon_1}(p_1)} \subset V(q) \cap U_1. \tag{1}$$

Since U_2 is open dense, we can choose a point $p_2 \in S_{\varepsilon_1}(p_1) \cap U_2$ and $\varepsilon_2 > 0$ such that $\varepsilon_2 < \frac{1}{2}$ and

$$\overline{S_{\varepsilon_2}(p_2)} \subset S_{\varepsilon_1}(p_1) \cap U_2.$$

Repeating this process, we can choose points $p_1, p_2, \ldots$ of X and positive numbers $\varepsilon_1, \varepsilon_2, \ldots$ such that

$$\varepsilon_n < 1/n, \quad n = 1, 2, \ldots,$$

$$\overline{S_{\varepsilon_n}(p_n)} \subset S_{\varepsilon_{n-1}}(p_{n-1}) \cap U_n, \quad n = 2, 3, \ldots. \tag{2}$$

Since (2) implies

$$p_n, p_{n+1}, \ldots \in S_{\varepsilon_n}(p_n), \tag{3}$$

$\{p_n \mid n = 1, 2, \ldots\}$ is a Cauchy sequence of points. Hence by F) it converges to a point p of X.

It follows from (3) that

$$p \in \bigcap_{n=1}^{\infty} \overline{S_{\varepsilon_n}(p_n)}. \tag{4}$$

This combined with (2) implies that $p \in \bigcap_{n=1}^{\infty} U_n$. On the other hand, it follows from (4) and (1) that $p \in V(q)$. Therefore

$$V(q) \cap \left[\bigcap_{n=1}^{\infty} U_n \right] \neq \emptyset$$

holds for every nbd $V(q)$ of a given point q of X. This means that $\bigcap_{n=1}^{\infty} U_n$ is dense in X.

Corollary. *Let U_n, $n = 1, 2, \ldots$, be open dense subsets of a locally compact T_2-space X. Then $\bigcap_{n=1}^{\infty} U_n$ is dense in X.*[1]

Proof. The proof is quite similar to that of the theorem, so it is left to the reader.

Let us turn to another important condition for a metric space.

---

[1] We shall later generalize Theorem III.7 and the corollary to Čech complete spaces. J. de Groot [3] established a new class of spaces called subcompact spaces including complete metric spaces and locally compact T_2-spaces as special cases and proved Baire's theorem for subcompact spaces. Generally, a topological space is called a *Baire space* if the intersection of countably many open dense sets is always dense.

Definition III.9. Let X be a metric space. If for every $\varepsilon > 0$ the open covering $\{S_\varepsilon(p) \mid p \in X\}$ contains a finite subcovering, then we call X a *totally bounded metric space*.

G) *Every totally bounded metric space satisfies the 2nd axiom of countability.*

Proof. Using the condition of totally boundedness, for every positive integer n, we can choose $S_{1/n}(p_i^n)$, $i = 1, 2, \ldots, k(n)$, such that

$$\bigcup_{i=1}^{k(n)} S_{1/n}(p_i^n) = X.$$

Now, it is easily seen that $\{S_{1/n}(p_i^n) \mid i = 1, \ldots, k(n);\ n = 1, 2, \ldots\}$ is a countable open basis of X, and hence X satisfies the 2nd axiom of countability.

H) *A metric space X is totally bounded if and only if every maximal filter of X is Cauchy.*

Proof. Let $\mathscr{F}$ be a maximal filter of a totally bounded metric space X. Suppose ε is a given positive number. Then

$$\bigcup_{i=1}^{k} S_\varepsilon(p_i) = X$$

for a finite number of points $p_1, \ldots, p_k$. Now we can assert that $F \subset S_\varepsilon(p_i)$ for some $F \in \mathscr{F}$ and i. For, if we assume the contrary, then for each i

$$(X - S_\varepsilon(p_i)) \cap F \neq \emptyset \quad \text{for all } F \in \mathscr{F}.$$

Therefore

$$X - S_\varepsilon(p_i) \in \mathscr{F}, \quad i = 1, \ldots, k$$

since $\mathscr{F}$ is maximal (II.4.B)). But we also have

$$\bigcap_{i=1}^{k} (X - S_\varepsilon(p_i)) = \emptyset,$$

which contradicts f.i.p. of $\mathscr{F}$. Thus $F \subset S_\varepsilon(p_i)$ for some $F \in \mathscr{F}$ and i, which means that $\mathscr{F}$ is a Cauchy filter.

Conversely, if X is a non-totally bounded metric space, then there is $\varepsilon > 0$ such that any finite number of ε-nbds do not cover X. Therefore

$$\mathscr{S} = \{X - S_\varepsilon(p) \mid p \in X\}$$

has f.i.p., and hence we can construct a maximal filter $\mathscr{F}$ which contains $\mathscr{S}$ as a subcollection (II.4.A)). Now, we can show that $\mathscr{F}$ is not Cauchy. Because, if $F \subset S_\varepsilon(p)$ for some $F \in \mathscr{F}$ and $p \in X$, then

$$F \cap (X - S_\varepsilon(p)) = \emptyset$$

contradicting the fact that

$$F, X - S_\varepsilon(p) \in \mathscr{F}.$$

Hence $F \not\subset S_\varepsilon(p)$ for every $F \in \mathscr{F}$ and $p \in X$, proving that $\mathscr{F}$ is not Cauchy.

Now, combining D) with the definition of a complete metric space and H), we obtain the following important theorem for metric space which shows that compactness of a metric space is a combination of completeness and totally boundedness.

Theorem III.8. *A metric space X is compact if and only if it is complete and totally bounded.*

Example III.12. All bounded subsets of a Euclidean space E^n are totally bounded, and they are the only totally bounded subspaces of E^n. On the other hand the closed sets and only those are complete subspaces of E^n. Generally, we can prove that every generalized Hilbert space $H(A)$ is a complete metric space (but non-compact). To show it we suppose

$$p^{(i)} = \{p_\alpha^{(i)} \mid \alpha \in A\}, \quad i = 1, 2, \dots,$$

form a Cauchy point sequence of $H(A)$. Since for each α, $\{p_\alpha^{(i)} \mid i = 1, 2, \dots\}$ is a Cauchy point sequence of E^1, it converges to a point p_α of E^1, i.e.,

$$\lim_{i\to\infty} p_\alpha^{(i)} = p_\alpha \,.$$

Let $\varepsilon > 0$ be given; then there is i_0 such that for every $i, i' \geqslant i_0$,

$$\left(\sum_{\alpha \in A} (p_\alpha^{(i)} - p_\alpha^{(i')})^2 \right)^{1/2} < \varepsilon \,.$$

Hence for every finite subset A' of A, we obtain

$$\left(\sum_{\alpha \in A'} (p_\alpha^{(i)} - p_\alpha^{(i')})^2 \right)^{1/2} < \varepsilon \,.$$

Letting $i' \to \infty$ in this inequality, we obtain

$$\left(\sum_{\alpha \in A'} (p_\alpha^{(i)} - p_\alpha)^2 \right)^{1/2} \leqslant \varepsilon \,. \tag{1}$$

This implies that

$$\left(\sum_{\alpha \in A'} p_\alpha^2 \right)^{1/2} - \left(\sum_{\alpha \in A'} (p_\alpha^{(i)})^2 \right)^{1/2} \leqslant \left(\sum_{\alpha \in A'} (p_\alpha^{(i)} - p_\alpha)^2 \right)^{1/2} \leqslant \varepsilon \,,$$

i.e.,

$$\left(\sum_{\alpha \in A'} p_\alpha^2 \right)^{1/2} \leqslant \varepsilon + \left(\sum_{\alpha \in A'} (p_\alpha^{(i)})^2 \right)^{1/2} \,.$$

Since $\sum_{\alpha \in A} (p_\alpha^{(i)})^2 < +\infty$, we get $\sum_{\alpha \in A} p_\alpha^2 < +\infty$, i.e.,

$$p = \{p_\alpha \mid \alpha \in A\} \in H(A) \,.$$

On the other hand, (1) implies

$$\rho(p, p^{(i)}) = \left(\sum_{\alpha \in A} (p_\alpha^{(i)} - p_\alpha)^2 \right)^{1/2} \leqslant \varepsilon \,.$$

Therefore $\{p^{(i)} \mid i = 1, 2, \ldots\}$ converges to p, proving that $H(A)$ is complete.

Let $C^*(X)$ be the collection of all bounded continuous functions over a topological space X. We introduce into $C^*(X)$ the metric given in Example III.9. Then we can easily see that $C^*(X)$ is a complete metric space. If X is the unit segment $I = [0, 1]$, then it follows from Weierstrass' theorem[1] that $C^*(I)$ $(= C(I))$ is also separable and therefore it satisfies the 2nd axiom of countability. However, it is not compact.

Exercise III

1. Test separation axioms, compactness and axioms of countability for a cofinite space defined in Exercise II.1.

2. In a T_1-space X, every nbd of a cluster point of a given set A contains infinitely many points of A. The set of cluster points of A is closed.

3. A subset A of a topological space X is called a *retract* if there is a continuous mapping f from X onto A such that $f(x) = x$ for $x \in A$. Every retract of a T_2-space is a closed set.

4. A topological space X is T_2 if and only if every net of X converges to at most one point.

5. Let f be a continuous mapping of a topological space X into a T_2-space Y. Then $\{(x, y) \mid x \in X, y \in Y, f(x) = y\}$ is a closed set of $X \times Y$.

6. A topological space X is T_2 if and only if $\Delta = \{(x, x) \mid x \in X\}$ is closed in $X \times X$.

7. Let f be a continuous mapping of a closed set F of a normal space X into the n-dimensional cube $I^n = \{(x_1, \ldots, x_n) \mid 0 \leq x_i \leq 1, i = 1, \ldots, n\}$. Then f can be continuously extended over X.

8. A T_1-space X is normal if and only if for every finite open covering $\{U_1, \ldots, U_k\}$ there is an open covering $\{V_1, \ldots, V_k\}$ such that $\bar{V}_i \subset U_i$, $i = 1, \ldots, k$ (or a closed covering or a covering by cozero sets satisfying the same condition). (V is called a *cozero set* of X if there is a real-valued continuous function f on X such that $V = \{x \in X \mid f(X) \neq 0\}$.)

9. A T_1-space X is normal if and only if for every finite open covering $\mathcal{U}$, there is an open covering $\mathcal{V}$ with $\mathcal{V}^\Delta < \mathcal{U}$. (This proposition shows that full normality is a natural generalization of normality. The proof is analogous to the last part of the proof of 3.E).)

---

[1] For every $f \in C(I)$ and $\varepsilon > 0$, there is a polynomial g such that $\rho(f, g) < \varepsilon$. Therefore the set of all polynomials with rational coefficients forms a countable dense set of $C(I)$.

10. Let a topological space X be the union of normal closed subspaces $F_1, \ldots, F_k$; then X is normal.

11. Every F_σ-set of a normal space is a normal subspace (Yu. Smirnov's theorem).

12. Give examples of a space which is locally compact but not paracompact and a space which is paracompact but not locally compact.

13. Let S be a locally compact subspace of a T_2-space. Then S is open in $\bar{S}$. Thus a subspace S of a locally compact T_2-space X is locally compact if and only if S is the intersection of a closed set and an open set of X.

14. A real-valued continuous function defined over a compact space X is bounded and takes on its largest and smallest values at some points of X.

15. Let X be a compact space satisfying the 1st axiom of countability. Then every point sequence of X has a convergent subsequence.

14. Let X be a topological space and $\mathscr{D}$ a decomposition of X into closed sets. If X is T_1, then the decomposition space $X(\mathscr{D})$ is T_1. If $\mathscr{D}$ is upper semi-continuous and X is normal, then $X(\mathscr{D})$ is normal. If $\mathscr{D}$ is an upper semi-continuous decomposition into compact sets, then $X(\mathscr{D})$ is T_2, is regular or satisfies the 2nd axiom of countability provided X has the respective property.

17. A topological space X is compact if and only if it satisfies either one of the following conditions:
   (i) every filter basis has a cluster point,
   (ii) every closed filter has a cluster point,
   (iii) every maximal closed filter converges.

18. A topological space X is compact if one of the following conditions is satisfied.
   (i) There is an open base $\mathscr{U}$ of X such that if $\mathscr{U}'$ is a covering of X satisfying $\mathscr{U}' \subset \mathscr{U}$, then $\mathscr{U}'$ has a finite subcovering.
   (ii) There is a closed base $\mathscr{G}$ of X such that if $\mathscr{G}'$ is a subcollection of $\mathscr{G}$ with f.i.p., then $\cap \mathscr{G}' \neq \emptyset$.

19. Let X be a compact T_2-space. Then it is totally disconnected if and only if every point has a nbd base consisting of clopen sets. (To prove 'only if', show that $Q(x) = K(x) = \{x\}$ for every $x \in X$.)

20. We add a new point p_0 to R_4 in Example II.5 and define a nbd basis $\mathscr{U}(p_0)$ of p_0 by

$$\mathcal{U}(p_0) = \left\{ \bigcup_{i=m}^{\infty} U_{1/n_i}(i) \cup \{p_0\} \mid m \text{ and } n_i, \ i = m, \ m+1, \ldots, \text{ are} \right.$$

$$\left. \text{positive integers} \right\}$$

where

$$U_{1/n_i}(i) = S_{1/n_i}((i, 1/n_i)) \cup \{(i, 0)\}.$$

Then the thus obtained space R_4^*, which contains R_4 as a subspace, is separable but satisfies neither the 1st axiom of countability nor the Lindelöf property.

21. Let X be a Lindelöf (separable) space and f a continuous mapping of X onto a topological space Y; then Y is Lindelöf (separable).

22. Each F_σ set of a Lindelöf space is also Lindelöf.

23. Each open subspace of a separable space is separable.

24. Every separable paracompact space is Lindelöf.

25. Let X be a topological space satisfying the 2nd axiom of countability and P a property for closed sets of X such that if $F_1 \supset F_2 \supset \cdots$ and each F_i satisfies P, then $\bigcap_{i=1}^{\infty} F_i$ also satisfies P. Then there is a minimal closed set which satisfies P. (Brouwer's reduction theorem.)

26. $S(A)$ and $N(A)$ in Example III.9 are both complete.

27. A normal space X is called *perfectly normal* if every closed set of X is an F_σ-set. A normal space X is perfectly normal if and only if every closed set F of X is a *zero set*, i.e., there is a real-valued continuous function f on X such that $F = \{x \in X \mid f(x) = 0\}$. Every metric space is perfectly normal. Every perfectly normal space is hereditarily normal.

28. $C(I)$ in Example III.12 is not locally compact.

29. Assume that every totally bounded subset of a metric space X is compact. Then X is complete.

30. Let f be a uniformly continuous mapping from a totally bounded metric space X onto a metric space Y. Then Y is totally bounded. Does the same hold for completeness?

31. Every non-empty open subset of a Baire space is of second category.

32. Let f be a mapping of a complete metric space X into itself, satisfying $\rho(f(p), f(q)) \leq \lambda \rho(p, q)$ for a definite positive number λ with $\lambda < 1$ and any pair p, q of points of X. Then there is one and only one point p_0 of X such that $f(p_0) = p_0$. (S. Banach's theorem. Consider a point sequence $\{p, f(p), f(f(p)), \ldots\}$, noticing that f is continuous.)

33. A T_1-space X is hereditarily normal if and only if for any two sets A and B satisfying $\bar{A} \cap B = \emptyset$, $A \cap \bar{B} = \emptyset$, there are open sets U and V such that $U \supset A$, $V \supset B$, $U \cap V = \emptyset$. A T_1-space X is hereditarily normal if and only if each open set in X is normal.

34. Let f be a closed continuous mapping from a normal space X onto a topological space Y. Then Y is normal.

35. A topological space X is called *extremely disconnected* if the closure of every open set of X is open. Every extremely disconnected T_2-space is totally disconnected, but the converse is not true.

36. Let F and G be disjoint zero sets of a topological space X. Then there is a real-valued continuous function on X such that $f(F) = 0$, $f(G) = 1$, $0 \leq f \leq 1$.

CHAPTER IV

COMPACT SPACES AND RELATED TOPICS

In Section 4 of the last chapter, we learned some elementary properties of compact spaces. The purpose of the present chapter is to give an account of more advanced theories on compact spaces.

1. Product of compact spaces

Theorem IV.1 (Tychonoff's product theorem). *Let* X_γ, $\gamma \in \Gamma$, *be compact spaces. Then their product space* $X = \prod \{X_\gamma \mid \gamma \in \Gamma\}$ *is also a compact space.*

Proof. We denote by π_γ the projection of X onto X_γ. Namely,

$$\pi_\gamma(p) = p_\gamma \quad \text{for } p = \{p_\gamma \mid \gamma \in \Gamma\}.$$

Suppose $\mathscr{F}$ is a given maximal filter of X. Then we can assert that for each $\gamma \in \Gamma$

$$\mathscr{F}_\gamma = \{\pi_\gamma(A) \mid A \in \mathscr{F}\}$$

is a maximal filter of X_γ. First it is clear that $\emptyset \notin \mathscr{F}_\gamma$. Secondly, if $B \supset \pi_\gamma(A)$ for some $A \in \mathscr{F}$ in X_γ, then $\pi_\gamma^{-1}(B) \supset A$ in X, which implies $\pi_\gamma^{-1}(B) \in \mathscr{F}$. Hence

$$B = \pi_\gamma(\pi_\gamma^{-1}(B)) \in \mathscr{F}_\gamma.$$

Thirdly, if $B = \pi_\gamma(A)$, $B' = \pi_\gamma(A')$ for some $A, A' \in \mathscr{F}$, then

$$B \cap B' \supset \pi_\gamma(A \cap A').$$

Since $A \cap A' \in \mathscr{F}$, and hence $\pi_\gamma(A \cap A') \in \mathscr{F}_\gamma$, this implies (as has just been proved)

$$B \cap B' \in \mathcal{F}_\gamma \,.$$

Therefore $\mathcal{F}_\gamma$ is a filter of X_γ.

To show that $\mathcal{F}_\gamma$ is maximal, we suppose B is a subset of X_γ intersecting every member of $\mathcal{F}_\gamma$. Then $\pi_\gamma^{-1}(B)$ is a subset of X intersecting every member of $\mathcal{F}$. Hence by II.4.B), $\pi_\gamma^{-1}(B) \in \mathcal{F}$. Therefore

$$B = \pi_\gamma(\pi_\gamma^{-1}(B)) \in \mathcal{F}_\gamma \,.$$

Applying again II.4.B) to $\mathcal{F}_\gamma$, we conclude that $\mathcal{F}_\gamma$ is a maximal filter.

Now, since X_γ is compact, by Theorem III.5, $\mathcal{F}_\gamma$ converges to a point $q_\gamma \in X_\gamma$. Then we can assert that

$$\mathcal{F} \to q = \{ q_\gamma \mid \gamma \in \Gamma \} \quad \text{in } X.$$

To show it, suppose U is a given nbd of q. Then we can find a nbd U' of q with $U' \subset U$ and of the form

$$U' = \prod_{i=1}^{k} U_{\gamma_i} \times \prod \{ X_\gamma \mid \gamma \neq \gamma_i, i = 1, \ldots, k \} \,,$$

where each U_{γ_i} is a nbd of q_{γ_i} in X_{γ_i}. Note that U' can be expressed as

$$U' = \bigcap_{i=1}^{k} \pi_{\gamma_i}^{-1}(U_{\gamma_i}) \,. \tag{1}$$

On the other hand, it follows from $\mathcal{F}_{\gamma_i} \to q_{\gamma_i}$ that $U_{\gamma_i} \in \mathcal{F}_{\gamma_i}$. Hence $U_{\gamma_i} = \pi_{\gamma_i}(B_i)$ for some $B_i \in \mathcal{F}$. Since $\pi_{\gamma_i}^{-1}(U_{\gamma_i}) \supset B_i$, we obtain $\pi_{\gamma_i}^{-1}(U_{\gamma_i}) \in \mathcal{F}$. Hence, in view of (1), we conclude that $U' \in \mathcal{F}$. Therefore $\mathcal{F}$ converges to q. Thus, by Theorem III.5, we can conclude that X is a compact space.

Corollary. *A topological space X is compact T_2 if and only if it is homeomorphic to a closed set of the product of copies of the unit segment* $[0, 1]$.

Proof. Combine Theorems III.1 and IV.1.

Example IV.1. The Hilbert cube I^ω is a compact metric space, because it is the product of countably many closed segments which are compact.

We can prove Theorem IV.1 using nets instead of filters. Suppose $\pi(\Delta \mid >)$ is a given maximal net of X. Then for each $\gamma \in \Gamma$, $\varphi_\gamma (\Delta \mid >) = (\pi_\gamma \circ \varphi)(\Delta \mid >)$ is easily proved to be a maximal net of X_γ. Therefore $\varphi_\gamma \to q_\gamma$ for some point q_γ of X_γ. Now, it is also easy to prove that $\varphi(\Delta \mid >) \to q = \{q_\gamma \mid \gamma \in \Gamma\}$ in X. This means that X is compact by virtue of the corollary to Theorem III.5.

Theorem IV.2. *For every completely regular space X, there exists a compact T_2-space $\beta(X)$ such that*
    (i) *X is a dense subset of $\beta(X)$,*
    (ii) *every bounded real-valued continuous function over X can be extended to a continuous function over $\beta(X)$.*
    Moreover, such a space $\beta(X)$ is uniquely determined by X in the sense that if $\beta(X)$ and $\beta'(X)$ are compact T_2-spaces satisfying (i), (ii), then $\beta(X)$ and $\beta'(X)$ are homeomorphic by a topological mapping which leaves invariant every point of X.

Proof. By Theorem III.1, X is homeomorphic with a subspace X' of the product space $P = \prod \{I_\alpha \mid \alpha \in A\}$ of the copies of the unit segment $[0, 1]$, where $\{f_\alpha \mid \alpha \in A\}$ is the totality of the continuous functions over X such that $0 \leq f_\alpha \leq 1$. Since X and X' are homeomorphic, we may identify them. Namely we identify each point p of X with $\{f_\alpha(p) \mid \alpha \in A\}$ of X'. Thus we can regard X as a subspace of P. (See the proof of Theorem III.1.) By Theorem IV.1, P is a compact space and hence $\bar{X}$ is also a compact space since it is a closed subset of the compact space P. Putting $\bar{X} = \beta(X)$ we obtain the desired space. (Note that $\beta(X)$ is also T_2 and thus it is a compact, fully normal space by III.3.F).)

First of all, (i) is obviously satisfied by X and $\beta(X)$.

To see (ii) we assume f is a given bounded continuous function defined over X. Then for appropriate positive numbers, ε and k, $0 \leq \varepsilon(f + k) \leq 1$ holds, i.e.

$$\varepsilon(f + k) = f_\alpha$$

for some $\alpha \in A$. At every point $p = \{p_\alpha \mid \alpha \in A\}$ of X, f_α takes the value $f_\alpha(p) = p_\alpha$. Hence putting $g_\alpha(p) = p_\alpha$ for every point $p = \{p_\alpha \mid \alpha \in A\}$ of $\bar{X} = \beta(X)$, we obtain a continuous extension g_α of f_α over $\beta(X)$. Then $(1/\varepsilon)g_\alpha - k$ is a continuous extension of f over $\beta(X)$, which proves (ii).

Finally to show the uniqueness of $\beta(X)$, we assume that $\beta'(X)$ is a given compact T_2-space which satisfies (i), (ii). Let f_α be a given continuous function over X such that $0 \le f_\alpha \le 1$. Then by (ii) there is a continuous extension φ_α of f_α over $\beta'(X)$. Since $\bar{X} = \beta'(X)$ by (i), it is clear that $0 \le \varphi_\alpha \le 1$. Therefore

$$\varphi(p') = \{\varphi_\alpha(p') \mid \alpha \in A\}, \quad p' \in \beta'(X), \tag{1}$$

is a mapping of $\beta'(X)$ into P which leaves each point of X fixed. It follows from the continuity of φ_α that φ is a continuous mapping. Hence

$$\varphi(\beta'(X)) = \varphi(\bar{X}) \subset \overline{\varphi(X)} = \beta(X) \quad \text{in } P$$

On the other hand, since $\beta'(X)$ is compact, $\varphi(\beta'(X))$ is also compact by III.3.B), and hence by III.3.C) it must be a closed subset of $\beta(X)$ containing X. Thus from

$$X \subset \varphi(\beta'(X)) \subset \beta(X)$$

it follows in $\beta(X)$ that

$$\bar{X} \subset \varphi(\beta'(X)) \subset \beta(X).$$

Since $\bar{X} = \beta(X)$, we get

$$\varphi(\beta'(X)) = \beta(X).$$

To prove that φ is one-to-one, we note that by (i) the continuous extension mentioned in (ii) is uniquely determined (Exercise II.31). Suppose that p' and q' are distinct points of $\beta'(X)$. Because $\beta'(X)$ is completely regular, there is a continuous function ψ over $\beta'(X)$ such that

$$\psi(p') = 0, \quad \psi(q') = 1 \quad \text{and} \quad 0 \le \psi \le 1.$$

We denote by f_{α_0} the restriction of ψ to X. Then as noted in the above, the continuous extension φ_{α_0} of f_{α_0} over $\beta'(X)$ is unique and hence $\varphi_{\alpha_0} = \psi$. Therefore

$$\varphi_{\alpha_0}(p') = 0 \quad \text{and} \quad \varphi_{\alpha_0}(p') = 1.$$

This implies that $\varphi(p') \ne \varphi(q')$, and hence φ is one-to-one. Thus it follows from III.3.D) that φ is a homeomorphism of $\beta'(X)$ onto $\beta(X)$, proving the theorem.

Corollary 1. *Let f be a continuous mapping of a completely regular space X into a compact T_2-space Y. Then f can be extended to a continuous mapping of $\beta(X)$ into Y, and the extension is unique.*

Proof. Since Y is completely regular, by Theorem III.1, we can imbed it in the product space P of unit segments I_α, $\alpha \in A$. Therefore we may represent the mapping f by

$$f(p) = \{f_\alpha(p) \mid \alpha \in A\}, \quad p \in X,$$

where each f_α is a real-valued function satisfying $0 \leqslant f_\alpha \leqslant 1$. Since f is continuous, so is f_α. Therefore, by (ii) of Theorem IV.2, we can extend f_α to a continuous function φ_α over $\beta(X)$. It follows from $\beta(X) = \bar{X}$ that $0 \leqslant \varphi_\alpha \leqslant 1$ and hence

$$\varphi(p) = \{\varphi_\alpha(p) \mid \alpha \in A\}, \quad p \in \beta(X),$$

is a continuous mapping of $\beta(X)$ into P. On the other hand, since φ is continuous, we obtain in P

$$\varphi(\beta(X)) = \varphi(\bar{X}) \subset \overline{\varphi(X)} = \overline{f(X)} \subset \bar{Y} = Y.$$

(Note that Y is compact and therefore closed in P.)

Thus φ is the desired continuous extension of f over $\beta(X)$. We can also easily prove the uniqueness of the extension using the fact that $\beta(X) = \bar{X}$.

Corollary 2. *Let $B(X)$ be a compact T_2-space in which X is a dense subset. Then there is a continuous mapping g of $\beta(X)$ onto $B(X)$ which keeps every point of X fixed and maps $\beta(X) - X$ onto $B(X) - X$.*

Proof. Let us denote by f the identity mapping which maps $X \subset \beta(X)$ onto $X \subset B(X)$, namely $f(p) = p$ for every $p \in X$. By use of Corollary 1, we can extend f to a continuous mapping g of $\beta(X)$ into $B(X)$. Since g is continuous, $g(\beta(X))$ is compact and therefore closed in $B(X)$. This combined with $\bar{X} = B(X)$ implies $g(\beta(X)) = B(X)$. To prove $g(\beta(X) - X) \subset B(X) - X$, we suppose that p is a given point of $\beta(X) - X$. Put

$$\mathcal{F} = \{U \cap X \mid U \text{ is a nbd of } p \text{ in } \beta(X)\};$$

then $\mathscr{F}$ is a filter basis in $\beta(X)$ converging to p. Since g is continuous, by II.6.B),

$$g(\mathscr{F}) \to g(p) \quad \text{in } B(X).$$

Observe that $g(\mathscr{F}) = \mathscr{F}$ because each element of $\mathscr{F}$ is a subset of X on which $g = f$. Thus $\mathscr{F} \to g(p)$ in $B(X)$. Since $\mathscr{F}$ converges to no point of X, we conclude that

$$g(p) \in B(X) - X,$$

which proves the assertion.

Definition IV.1. The compact T_2-space $\beta(X)$ obtained in Theorem IV.2 is called the *Čech–(M.H.) Stone's compactification* of X.

A) *Assume that X is a non-compact Tychonoff space. Then every point y of $\beta(X) - X$ has no countable nbd base.*

Proof. Assume to the contrary that y has a countable nbd base U_1, $U_2, \ldots$. Then we may assume

$$\bar{U}_{i+1} \subset U_i, \quad i = 1, 2, \ldots.$$

Observe that $U_i \cap X$ is infinite for every i, because $y \in \bar{X}$. Thus we can select two point sequences

$$P = \{p_1, p_2, \ldots\} \quad \text{and} \quad Q = \{q_1, q_2, \ldots\}$$

such that $P \cap Q = \emptyset$ and

$$p_i, q_i \in (U_{n_i} - \bar{U}_{n_{i+1}}) \cap X, \quad i = 1, 2, \ldots,$$

where

$$n_1 < n_2 < \cdots.$$

Then we can find open nbds V_i of p_i such that

$$V_i \subset (U_{n_i} - \bar{U}_{n_{i+1}} \cup Q) \cap X.$$

Since X is Tychonoff, there are continuous functions f_i from X into $[0, 1]$ such that

$$f_i(p_i) = 1, \qquad f_i(X - V_i) = 0.$$

Then put

$$f = \sum_{i=1}^{\infty} f_i.$$

Observe that $\{V_i \mid i = 1, 2, \ldots\}$ is mutually disjoint and locally finite in X. Thus it is easy to prove that f is a continuous function from X into $[0, 1]$. (The detailed proof is left to the reader.) It is also obvious that

$$f(P) = 1, \qquad f(Q) = 0.$$

Note that $p_i \to y$ and $q_i \to y$ in $\beta(X)$. Thus f cannot be extended over $\beta(X)$, contradicting the property of $\beta(X)$. Thus y has no countable nbd base.

Example IV.2. As indicated by A), $\beta(X)$ is not 1st countable even if X is a metric (non-compact) space. Generally speaking, the structure of $\beta(X)$ is not simple even if X is such a simple space like the real line. A rather exceptional simple example is $\beta(R_5) = R_6$. (See Example III.6.) It is obvious that R_6 is a compact T_2-space containing R_5 as a dense subspace. On the other hand, it is easy to show that for every real-valued, bounded, continuous function f defined on R_5, there is $\alpha \in R_5$ such that

$$f(x) = f(\alpha) \quad \text{for all } x \in R_5 \text{ with } x \geqslant \alpha.$$

Thus we can extend f to the continuous function φ over R_6 by defining that

$$\varphi(\omega_1) = f(\alpha).$$

Therefore R_6 is the Čech–Stone's compactification of R_5.

On the other hand, $(0, 1)$ is a dense subset of $[0, 1]$, but by A) the latter is not the Čech–Stone's compactification of the former, because $[0, 1]$ is 1st countable. Similarly, the 2-dimensional sphere S^2 is not $\beta(E^2)$, though S^2 is a compact T_2-space in which E^2 is dense. (See Example II.12.)

Example IV.3. In Example III.3 we showed that R_4 is completely regular but not normal. On the other hand, $\beta(R_4)$ is compact T_2 and therefore fully normal. Thus a subspace of a fully normal space does not need even to be normal.

We can prove the uniqueness of $\beta(X)$ in a slightly different way from the proof of Theorem IV.2 as follows. Assume that X and X' are homeomorphic and that $\beta(X)$ and $\beta'(X')$ are T_2-compactifications of X and X', respectively, satisfying (i) and (ii) of Theorem IV.2. Let f be a homeomorphism from X onto X' and i and i' the identity mapping of X and X', respectively. By Corollary 1 we extend f and f^{-1} to φ and φ', respectively, where φ is a mapping from $\beta(X)$ into $\beta'(X')$ and φ' from $\beta'(X')$ into $\beta(X)$. Then $\varphi' \circ \varphi$ is a continuous extension of i, and hence $\varphi' \circ \varphi = j$, where j is the identity mapping of $\beta(X)$. In a similar way $\varphi \circ \varphi' = j'$, the identity mapping of $\beta'(X')$. Thus it is easy to see that φ is a homeomorphism from $\beta(X)$ onto $\beta'(X')$ such that $\varphi = f$ on X.

Example IV.4. We gave in Example III.2 a regular space R_3' which is not completely regular. The following classical example due to A. Tychonoff [1] is much more complicated but interesting in its own right. We consider R_6 in Example III.5 and define

$$R_6' = \{0, 1, 2, \ldots, \omega_0\},$$

where ω_0 denotes the first countable ordinal number, and topology of R_6' is the order topology. Let

$$P = R_6 \times R_6';$$

then

$$P = \{(\xi, \eta) \mid 0 \leq \xi \leq \omega_1, 0 \leq \eta \leq \omega_0\},$$

and $\{U_{\alpha\beta}(x_0) \mid 0 \leq \alpha < \alpha_0, 0 \leq \beta < \beta_0\}$ forms a nbd basis of $x_0 = (\alpha_0, \beta_0) \in P$ with $\alpha_0 > 0$, $\beta_0 > 0$, where

$$U_{\alpha\beta}(x_0) = \{(\xi, \eta) \mid \alpha < \xi \leq \alpha_0, \beta < \eta \leq \beta_0\}.$$

It is easily seen that P is a compact T_2-space because both of R_6 and R_6' are compact T_2. Therefore P is regular. Put

$$P' = P - (\omega_1, \omega_0);$$

then P' is also a regular space.

To proceed with the construction of the desired space, we need the following properties of P'.

(a) *For each α with $\alpha < \omega_1$ and n with $n < \omega_0$, we put*

$$X'_\alpha = \{(\xi, \omega_0) \mid \alpha_0 < \xi < \omega_1\},$$

and

$$Y'_n = \{(\omega_1, \eta) \mid n < \eta < \omega_0\}.$$

If U is an open set of P' containing some X'_α, then $\bar{U}$ contains some Y'_n.

Proof. Since U is open, for each $x = (\xi, \omega_0) \in X'_\alpha$, there is a nbd $U_{\xi' n_\xi}(x)$ such that

$$U_{\xi' n_\xi}(x) \subset U.$$

Hence for each ξ with $\alpha < \xi < \omega_1$

$$\{(\xi, \eta) \mid n_\xi < \eta < \omega_0\} \subset U.$$

Since each n_ξ is an integer, there is some n such that $n_\xi = n$ for uncountably many ξ, i.e. $\{\xi \mid n_\xi = n\}$ is cofinal in $\{\xi \mid 0 \leqslant \xi < \omega_1\}$. Then it is easy to see that $Y'_n \subset \bar{U}$.

(b) *If U is an open set of P' containing some Y'_n, then $\bar{U}$ contains some X'_α.*

Proof. Since U is open, for each $x = (\omega_1, \eta) \in Y'_n$, there is a nbd $U_{\alpha_\eta \eta'}(x) \subset U$. Hence

$$\{(\xi, \eta) \mid \alpha_\eta < \xi < \omega_1\} \subset U.$$

Let

$$\sup \{\alpha_\eta \mid (\omega_1, \eta) \in Y'_n\} = \alpha.$$

Then it is easily seen that $X'_\alpha \subset \bar{U}$, because Y'_n and hence $\{\alpha_\eta\}$ is countable which implies $\alpha < \omega_1$.

Now, we denote by R' the union of countably many copies P'^k, $k = 1$, $2, \ldots$, of P'. We consider $\{P'^k \mid k = 1, 2, \ldots\}$ as forming a discrete open covering of R'. Then R' is a regular space. We denote (ξ, η), X'_α, Y'_n in P'^k by $(\xi, \eta)^k$, X'^k_α, Y'^k_n, respectively. We consider a decomposition $\mathscr{D}$ of R' consisting of

$$\{(\omega_1, \eta)^{2n-1}, (\omega_1, \eta)^{2n}\}, \quad 0 \leqslant \eta < \omega_0, n = 1, 2, \ldots,$$

$$\{(\xi, \omega_0)^{2n}, (\xi, \omega_0)^{2n+1}\}, \quad 0 \leqslant \xi < \omega_1, n = 1, 2, \ldots$$

and the other points of R'. Consider the decomposition space

$$R = R'(\mathscr{D}).$$

Since $\mathscr{D}$ is easily verified to be upper semi-continuous, by Exercise III.16, R is a regular space. Let us denote by f the natural mapping of R' onto R and put, for brevity,

$$f(X_\alpha^{k\prime}) = X_\alpha^k,$$

$$f(Y_n^{k\prime}) = Y_n^k$$

$$f(P^{k\prime}) = P^k.$$

Now, we construct a space R_3 by adding a point y_0 to R. For each positive integer l, we put

$$U_l(y_0) = \{y_0\} \cup \left[\bigcup_{k=l+1}^{\infty} P^k \right]. \tag{1}$$

Then we define that $\{U_l(y_0) \mid l = 1, 2, \ldots\}$ is a nbd basis of y_0 in R_3. We can easily verify that R_3 is a regular space. But, now, we can assert that R_3 is not completely regular. For, if R_3 is completely regular, then we can construct Čech–Stone's compactification $\beta(R_3)$ of R_3 and assert that $\beta(R_3) \supsetneqq R_3$, because R_3 is easily seen to be non-compact. (Generally, for a given completely regular space X, $X = \beta(X)$ holds if and only if X is compact.) Hence we can select

$$z_0 \in \beta(R_3) - R_3.$$

Since $\beta(R_3)$ is regular, so is

$$R_0 = R_3 \cup \{z_0\},$$

and hence there are nbds $U_N(y_0)$ of y_0 and $V(z_0)$ of z_0 in R_0 such that

$$\overline{U_N(y_0)} \cap \overline{V(z_0)} = \emptyset. \tag{2}$$

We may suppose without loss of generality that N is odd. Let us continue our discussion in R_0. We choose a sequence $\{V_m(z_0) \mid m = 1, 2, \ldots\}$ of nbds of z_0 such that

$$V_1(z_0) \subset V(z_0) \quad \text{and} \quad \overline{V_{m+1}(z_0)} \subset V_m(z_0). \tag{2'}$$

Put

$$W_m = R_3 - \overline{V_m(z_0)}; \tag{3}$$

then it is clear that

$$\bar{W}_m \subset W_{m+1}, \tag{4}$$

and each W_m is an open set of R_3 containing y_0.

From (1), (2) it is also clear that

$$W_1 \supset U_N(y_0) \supset \bigcup_{k=N+1}^{\infty} P^k, \tag{5}$$

which implies

$$W_1 \supset X_0^{N+1},$$

because

$$P^{N+1} \supset X_0^{N+1}.$$

Putting $\alpha_0 = 0$, we express this relation as

$$W_1 \supset X_{\alpha_0}^{N+1}. \tag{6}$$

Starting here, we proceed with an inductive argument although the detailed proof is left to the reader.

By (a) this implies that

$$\bar{W}_1 \supset Y_{n_1}^{N+1} \quad \text{for some } n_1.$$

Combining this with (6) and (4), we get

$$W_2 \supset X_{\alpha_0}^{N+1} \cup Y_{n_1}^{N+1}. \tag{7}$$

Since N is odd,

$$Y_{n_1}^{N+1} = Y_{n_1}^N$$

follows from the definition of the decomposition $\mathcal{D}$. Therefore

$$W_2 \supset Y_{n_1}^N.$$

By (b) this implies that

$$\bar{W}_2 \supset X_{\alpha_2}^N \quad \text{for some } \alpha_2.$$

Hence

$$W_3 \supset X_{\alpha_2}^N \cup Y_{n_1}^N \tag{8}$$

follows from (4). Since by the definition of $\mathcal{D}$,

$$X_{\alpha_2}^N = X_{\alpha_2}^{N-1},$$

we obtain

$$W_3 \supset X_{\alpha_2}^{N-1}.$$

Now, we can apply the same discussion to W_3 and $X_{\alpha_2}^{N-1}$ as the one applied to W_1 and $X_{\alpha_0}^{N+1}$; then we get n_3 such that

$$W_4 \supset X_{\alpha_2}^{N-1} \cup Y_{n_3}^{N-1}. \tag{9}$$

Repeating this process we get α_N and n_{N+1} for which

$$W_{N+2} \supset X_{\alpha_N}^1 \cup Y_{n_{N+1}}^1. \tag{10}$$

In view of (4), (7), (8), (9), (10), we know that for every k with $1 \leq k \leq N$,

$$W_{N+2} \supset X_{\alpha(k)}^k \cup Y_{n(k)}^k \quad \text{for some } \alpha(k) \text{ and } n(k). \tag{11}$$

On the other hand, from (2), (2'), (3), (5) it follows that

$$V_{N+2}(z_0) \cap R_3 \subset R_3 - W_{N+2} \subset \bigcup \{(R_3 - W_{N+2}) \cap P^k \mid k = 1, \ldots, N\}. \tag{12}$$

(Note that (2), (2'), (3), (5) imply

$$R_3 - W_{N+2} = \overline{V_{N+2}(z_0)} - \{z_0\} \subset \bigcup_{k=1}^{N} P^k .)$$

Let k be a natural number with $1 \leq k \leq N$. Recall that W_{N+2} is an open set of R_3 containing y_0. Then $(R_3 - W_{N+2}) \cap P^k$ is homeomorphic with $P' - W$, where by (11) W is an open set of P' such that

$$W \supset X_{\alpha(k)} \cup Y_{n(k)} . \tag{13}$$

On the other hand,

$$P' - W = P - [W \cup \{(\omega_1, \omega_0)\}] .$$

Modifying the proof of (b), we can derive from (13) that $W \cup \{(\omega_1, \omega_0)\}$ is an open set of P. Since P is compact, $P - [W \cup \{(\omega_1, \omega_0)\}]$ and accordingly $P' - W$ are compact. Thus $(R_3 - W_{N+2}) \cap P^k$ is compact. Therefore from (12) it follows that

$$K = \bigcup \{(R_3 - W_{N+2}) \cap P^k \mid k = 1, \ldots, N\}$$

is a compact subset of R_3 containing $V_{N+2}(z_0) \cap R_3$. Thus

$$z_0 \in \overline{V_{N+2}(z_0) \cap R_3} \subset \bar{K} = K$$

in R_3. On the other hand,

$$z_0 \notin R_3 \supset K ,$$

which is a contradiction. Thus R_3 is not completely regular.[1]

2. Compactification

It is an interesting problem to find a compact space X^* for a given topological space X such that X is homeomorphic with a dense subset X' of X^*. Generally, such a space X^* is called a *compactification* of X. The concept of compactification appears now and then in fields other than

---

[1] As a matter of fact, there is a regular space in which every continuous function is constant! See J. Novak [1], E. Hewitt [1] and A. Mysior [2].

general topology. In fact, we are familiar with a compactification in the theory of functions, where the complex number plane with the point at infinity is a compactification of the plane. We have already studied a compactification $\beta(X)$ in the preceding section. The purpose of the present section is to develop in further detail investigations on compactifications.

Now, let us suppose X is a given topological space; then we consider a space $X^* = X \cup \{p\}$ adding a new point p to X. If we define that p has the only nbd X^*, then X^* turns out to be a compact T-space containing X as a dense subset. Thus we can easily obtain a compactification for every topological space, but such a compactification is not even T_1. What we are concerned with is compactifications with more interesting properties like the Čech–Stone's compactification.

Definition IV.2. Let X be a given topological space. Add a point ∞ of infinity to X and put $\alpha(X) = X \cup \{\infty\}$. We define a topology of $\alpha(X)$ as follows. Each point $x \in X$ has the family of all nbds in X as a nbd base in $\alpha(X)$. On the other hand, $\{U \mid \infty \in U, X - U$ is a compact closed set of $X\}$ is defined as a nbd base of ∞ in $\alpha(X)$. Then it is easy to see that $\alpha(X)$ is a compactification of X. We call $\alpha(X)$ *Alexandroff's one-point compactification* of X.

A) $\alpha(X)$ *is T_2 if and only if X is locally compact and T_2.*

Proof. The 'only if' part is easy and is left to the reader. Let X be locally compact and T_2. Suppose $x \in X$. Then there is a compact nbd G of x in X. Then $\alpha(X) - G$ is a nbd of ∞ which is disjoint from G. Hence $\alpha(X)$ is T_2.

Next, we are going to discuss a general method of H. Wallman and N.A. Shanin to construct compactifications. For that purpose we generalize the concept of filter and its convergence which we originally defined in Definitions II.9 and II.10.

Definition IV.3. Let $\mathscr{B}$ be a collection of subsets of a topological space X such that $B, C \in \mathscr{B}$ implies $B \cap C \in \mathscr{B}$. A subcollection $\mathscr{F}$ of $\mathscr{B}$ is called a $\mathscr{B}$-*filter* if
   (i) $\emptyset \notin \mathscr{F}$,
   (ii) if $B \in \mathscr{F}$ and $B \subset C \in \mathscr{B}$, then $C \in \mathscr{F}$,
   (iii) if $B, C \in \mathscr{F}$, then $B \cap C \in \mathscr{F}$.

A $\mathcal{B}$-filter $\mathcal{F}$ is called *maximal* if for every $\mathcal{B}$-filter $\mathcal{G}$ satisfying $\mathcal{G} \supset \mathcal{F}$, it holds that $\mathcal{G} = \mathcal{F}$.

B) *Every $\mathcal{B}$-filter $\mathcal{F}$ is a base of a filter. Thus $\mathcal{F}$ converges to a point x of X if and only if for every nbd U of x there is $B \in \mathcal{F}$ such that $B \subset U$. A point y of X is a cluster point of $\mathcal{F}$ if and only if $y \in \bar{F}$ for all $F \in \mathcal{F}$.*

Proof. $\mathcal{F}$ is obviously a base of the filter $\mathcal{F}' = \{Y \mid Y$ is a subset of X containing some member of $\mathcal{F}\}$. Thus the rest of the proof follows from the definition of convergence and cluster point of a filter base given after II.4.C).

Example IV.5. If $\mathcal{B}$ is the collection of all closed sets, then a $\mathcal{B}$-filter is called a closed filter as already defined in IV.4. If $\mathcal{B}$ is the collection of all zero sets of X, then we call a $\mathcal{B}$-filter a *zero-filter*.

We can generalize some basic propositions previously proved for filters as follows. (The proofs are left to the reader.)

C) *A $\mathcal{B}$-filter $\mathcal{F}$ is maximal if and only if it satisfies the following condition: If $B \in \mathcal{B}$ and $B \cap F \neq \emptyset$ for all $F \in \mathcal{F}$, then $B \in \mathcal{F}$.*

D) *Let $\mathcal{B}'$ be a subcollection with f.i.p. of $\mathcal{B}$. Then there is a maximal $\mathcal{B}$-filter $\mathcal{G}$ such that $\mathcal{G} \supset \mathcal{B}'$.*

Definition IV.4. Let $\mathcal{B}$ be a closed base of a T_1-space X. Then $\mathcal{B}$ is called a T_1-*base* if it satisfies
  (i) $\emptyset \in \mathcal{B}$,
  (ii) $B_1, B_2 \in \mathcal{B}$ implies $B_1 \cap B_2 \in \mathcal{B}$ and $B_1 \cup B_2 \in \mathcal{B}$,
  (iii) if $x \notin B \in \mathcal{B}$, then there is $B' \in \mathcal{B}$ such that $x \in B'$, $B' \cap B = \emptyset$.
If, moreover, $\mathcal{B}$ satisfies the following condition, then we call it a *normal base*:
  (iv) if $B_1, B_2 \in \mathcal{B}$, and $B_1 \cap B_2 = \emptyset$, then there are $B_1', B_2' \in \mathcal{B}$ such that $B_1 \cap B_1' = \emptyset$, $B_2 \cap B_2' = \emptyset$, $B_1' \cup B_2' = X$.

Theorem IV.3. *Let X be a T_1-space and $\mathcal{B}$ a T_1-base of X. Then let $\sigma(X, \mathcal{B}) =$ the set of all maximal $\mathcal{B}$-filters of X. For each $B \in \mathcal{B}$ we define*

$$\tilde{B} = \{\mathcal{F} \in \sigma(X, \mathcal{B}) \mid B \in \mathcal{F}\}, \qquad \tilde{\mathcal{B}} = \{\tilde{B} \mid B \in \mathcal{B}\}.$$

Introduce a topology into $\sigma(X, \mathcal{B})$ by taking $\tilde{\mathcal{B}}$ as a closed base. Then $\sigma(X, \mathcal{B})$ is a T_1-compactification of X satisfying

(i) $\{\bar{B} \mid B \in \mathcal{B}\}$ *is a closed base of $\sigma(X, \mathcal{B})$,*

(ii) *if B_1, $B_2 \in \mathcal{B}$, then $\bar{B}_1 \cap \bar{B}_2 = \overline{B_1 \cap B_2}$,*

where, both in (i) and (ii), $\bar{B}$ denotes the closure in $\sigma(X, \mathcal{B})$. Furthermore, $\sigma(X, \mathcal{B})$ is T_2 if and only if $\mathcal{B}$ is a normal base.

Proof. We shall prove the theorem step by step as follows.

(1) $\tilde{B}_1 \cup \tilde{B}_2 = (B_1 \cup B_2)^{\tilde{}}$ and $\tilde{B}_1 \cap \tilde{B}_2 = (B_1 \cap B_2)^{\tilde{}}$ hold for every B_1, $B_2 \in \mathcal{B}$.

Let us prove only the first equality. $\tilde{B}_1 \cup \tilde{B}_2 \subset (B_1 \cup B_2)^{\tilde{}}$ is obvious. Assume $\mathcal{F} \not\in \tilde{B}_1 \cup \tilde{B}_2$; then $B_1 \not\in \mathcal{F}$ and $B_2 \not\in \mathcal{F}$. Thus, by C), there are F_1, $F_2 \in \mathcal{F}$ such that

$$F_1 \cap B_1 = \emptyset, \qquad F_2 \cap B_2 = \emptyset.$$

Hence

$$F_1 \cap F_2 \in \mathcal{F} \quad \text{and} \quad (F_1 \cap F_2) \cap (B_1 \cup B_2) = \emptyset.$$

Therefore again by use of C), we obtain $B_1 \cup B_2 \not\in \mathcal{F}$, i.e. $\mathcal{F} \not\in (B_1 \cup B_2)^{\tilde{}}$. This proves $\tilde{B}_1 \cup \tilde{B}_2 \supset (B_1 \cup B_2)^{\tilde{}}$ and eventually (1).

(2) $\tilde{\mathcal{B}}$ satisfies the condition for a closed base given in II.2.C).

This is obvious; in fact it satisfies a stronger condition,

$$\sigma(X, \mathcal{B}) = \tilde{X} \in \tilde{\mathcal{B}}, \qquad \emptyset \in \tilde{\mathcal{B}},$$
$$\tilde{B}_1 \cup \tilde{B}_2 \in \tilde{\mathcal{B}} \quad \text{for every } \tilde{B}_1, \tilde{B}_2 \in \tilde{\mathcal{B}}.$$

Thus $\sigma(X, B)$ turns out to be a topological space with the topology defined by the closed base $\tilde{\mathcal{B}}$.

(3) $\sigma(x, \mathcal{B})$ is a T_1-space.

Let $\mathcal{F}$, $\mathcal{G} \in \sigma(X, \mathcal{B})$, $\mathcal{F} \neq \mathcal{G}$. Then there are $F \in \mathcal{F}$, $G \in \mathcal{G}$ such that $F \cap G = \emptyset$. Now, $\tilde{F} \in \tilde{\mathcal{B}}$ satisfies $\mathcal{F} \in \tilde{F}$ and $\mathcal{G} \not\in \tilde{F}$, which implies that $\sigma(X, \mathcal{B})$ is T_1.

Define a mapping φ from X into $\sigma(X, \mathcal{B})$ by

(4) $\varphi(x) = \mathcal{F}_x = \{B \mid x \in B \in \mathcal{B}\}$.

It follows from (iii) of Definition IV.4 that $\mathcal{F}_x$ is a maximal $\mathcal{B}$-filter. We can show that φ is a topological imbedding. Let x and y be two distinct points of X. Then, since X is T_1, there is $B \in \mathcal{B}$ such that $x \not\in B \ni y$. Now

$$\varphi(x) = \mathcal{F}_x \not\ni B, \qquad \varphi(y) = \mathcal{F}_y \ni B,$$

which proves $\varphi(x) \neq \varphi(y)$, i.e. φ is an injection. Let $\tilde{B} \in \tilde{\mathcal{B}}$. Then

$$\varphi(X) \cap \tilde{B} = \{\mathcal{F}_x \mid x \in B\}.$$

Hence $\varphi^{-1}(\varphi(X) \cap \tilde{B}) = B$. Thus by Exercise II.7 φ is continuous. Similarly we can show that φ is closed. Thus φ is a topological imbedding.

From now on we identify $\varphi(X)$ with X to regard X as a subspace of $\sigma(X, \mathcal{B})$. Note that then $\tilde{B} \cap X = B$ for every $B \in \mathcal{B}$.

(5) $\bar{B} = \tilde{B}$ for every $B \in \mathcal{B}$, where $\bar{B}$ denotes the closure in $\sigma(X, \mathcal{B})$. By Exercise II.7,

$$\bar{B} = \cap\{\tilde{D} \mid D \in \mathcal{B}, \tilde{D} \supset B\} = \cap\{\tilde{D} \mid B \subset D \in \mathcal{B}\} = \tilde{B}.$$

This proves that $\sigma(x, \mathcal{B})$ satisfies the required conditions (i) and (ii).

Thus we obtain

(6) $\bar{X} = \tilde{X} = \sigma(X, \mathcal{B})$.

Hence X is a dense subset of $\sigma(X, \mathcal{B})$.

(7) $\sigma(X, \mathcal{B})$ is compact.

Let $\mathcal{B}' \subset \mathcal{B}$ and suppose that $\tilde{\mathcal{B}}' = \{\tilde{\mathcal{B}} \mid B \in \mathcal{B}'\}$ has f.i.p. Then it follows from (1) that $\mathcal{B}'$ has f.i.p. By D) there is a maximal $\mathcal{B}$-filter $\mathcal{F}$ such that $\mathcal{B}' \subset \mathcal{F}$. Then $\mathcal{F} \in \cap \tilde{\mathcal{B}}'$ holds in $\sigma(X, \mathcal{B})$. Thus by Exercise III.18 $\sigma(X, \mathcal{B})$ is compact. Namely, $\sigma(X, \mathcal{B})$ is a T_1-compactification of X satisfying (i), (ii).

Assume, furthermore, $\mathcal{B}$ is a normal base. Then for any two distinct points $\mathcal{F}$, $\mathcal{G}$ of $\sigma(X, \mathcal{B})$ we find $B_1 \in \mathcal{F}$, $B_2 \in \mathcal{G}$ such that $B_1 \cap B_2 = \emptyset$. By (iv) of Definition IV.4, there are B_1', $B_2' \in \mathcal{B}$ such that

$$B_1 \cap B_1' = \emptyset, \qquad B_2 \cap B_2' = \emptyset, \qquad B_1' \cup B_2' = X.$$

Then $\tilde{B}_1'$, $\tilde{B}_2' \in \tilde{\mathcal{B}}$ satisfy

$$\mathcal{F} \in \sigma - \tilde{B}_1', \qquad \mathcal{G} \in \sigma - \tilde{B}_2', \qquad (\sigma - \tilde{B}_1') \cap (\sigma - \tilde{B}_2') = \emptyset,$$

i.e. $\sigma(X, \mathcal{B})$ (denoted by σ in the above) is T_2.

On the other hand, assume that $\sigma(X, \mathcal{B})$ is normal, and B_1, $B_2 \in \mathcal{B}$, $B_1 \cap B_2 = \emptyset$. Then $\bar{B}_1 \cap \bar{B}_2 = \emptyset$. Choose open sets U_1, U_2 of $\sigma(X, \mathcal{B})$ such that

$$\bar{B}_1 \subset U_1, \qquad \bar{B}_2 \subset U_2, \qquad U_1 \cap U_2 = \emptyset.$$

There are B_1', $B_2' \in \mathcal{B}$ such that

$$\bar{B}_1 \cap \bar{B}_1' = \emptyset, \qquad \bar{B}_1' \supset \sigma(X, B) - U_1, \qquad \bar{B}_2 \cap \bar{B}_2' = \emptyset,$$
$$\bar{B}_2' \supset \sigma(X, \mathscr{B}) - U_2.$$

Then it is obvious that condition (iv) of Definition IV.4 is satisfied by B_1, B_2, B_1', B_2'.

E) $\sigma(X, \mathscr{B})$ *is unique. Namely, if $\sigma'(X, \mathscr{B})$ is another T_1-compactification of X satisfying* (i), (ii) *of Theorem* IV.3, *then there is a homeomorphism between $\sigma(X, \mathscr{B})$ and $\sigma'(X, \mathscr{B})$ which leaves every point of X fixed.*

Proof. Let $x \in \sigma(X, \mathscr{B})$; then x can be regarded as a maximal $\mathscr{B}$-filter and accordingly as a filter base in $\sigma'(X, \mathscr{B})$. Hence it has a cluster point y in $\sigma'(X, \mathscr{B})$, because $\sigma'(X, \mathscr{B})$ is compact. Thus

(1) $y \in \cap\{\bar{F} \mid F \in x\}$, where $\bar{F}$ denotes the closure in $\sigma(X, \mathscr{B})$.

By condition (i) of Theorem IV.3 we have

(2) $\{y\} = \cap\{\bar{B} \mid y \in \bar{B}, B \in \mathscr{B}\}$.

Consider any $B \in \mathscr{B}$ such that $y \in \bar{B}$. Then for any $F \in x$ we obtain $\overline{F \cap B} = \bar{F} \cap \bar{B} \neq \emptyset$, because of (ii) of the theorem. Thus $F \cap B \neq \emptyset$, which implies that $B \in x$. Hence, from (1) and (2) it follows that

$$\{y\} = \cap\{\bar{F} \mid F \in x\}.$$

Now, let U be a given nbd of y in $\sigma'(X, \mathscr{B})$. Then there is $B \in \mathscr{B}$ such that $y \notin \bar{B} \supset \sigma'(X, \mathscr{B}) - U$. Hence there is $F \in x$ such that $\bar{F} \cap \bar{B} = \emptyset$, because otherwise $F \cap B \neq \emptyset$ for all $F \in x$, and accordingly $B \in x$ follows. This implies $y \in \bar{B}$, a contradiction. Thus $\bar{F} \subset U$. Namely x, as regarded as a filter base, converges to y in $\sigma'(X, \mathscr{B})$. Now, we define a mapping φ from $\sigma(X, \mathscr{B})$ into $\sigma'(X, \mathscr{B})$ by

$$\varphi(x) = y.$$

It is obvious that φ leaves each point of X fixed. Assume that x, x' are distinct points of $\sigma(X, \mathscr{B})$. Then there are $F \in x$, $F' \in x'$ such that $F \cap F' = \emptyset$. Hence $\bar{F} \cap \bar{F}' = \emptyset$, where closure is taken in $\sigma'(X, \mathscr{B})$. Since $\varphi(x) \in \bar{F}$ and $\varphi(x') \in \bar{F}'$, we obtain $\varphi(x) \neq \varphi(x')$, i.e. φ is one-to-one.

Next, φ is onto. Because, let $y \in \sigma'(X, \mathscr{B})$; then by (i) of the theorem

$$\{y\} = \cap\{\bar{B}^{\sigma'} \mid B \in \mathscr{B}, y \in \bar{B}^{\sigma'}\},$$

where $\bar{B}^{\sigma'}$ denotes the closure in $\sigma'(X, \mathscr{B})$. Let $\mathscr{F} = \{B \in \mathscr{B} \mid y \in \bar{B}^{\sigma'}\}$.

Then by (ii) of the theorem $\mathcal{F}$ is a $\mathcal{B}$-filter. If $B \in \mathcal{B}$ satisfies $B \cap F \neq \emptyset$ for all $F \in \mathcal{F}$, then $y \in \bar{B}^{\sigma'}$, i.e. $B \in \mathcal{F}$. (Because otherwise from the compactness of $\sigma'(X, \mathcal{B})$ it follows that $\bar{C}^{\sigma'} \cap \bar{B}^{\alpha'} = \emptyset$ for some $C \in \mathcal{B}$ with $y \in \bar{C}^{\sigma'}$, i.e. $C \in \mathcal{F}$. Thus $C \cap B = \emptyset$, a contradiction.) Hence $\mathcal{F}$ is a maximal $\mathcal{B}$-filter by C). Namely, $\mathcal{F}$ can be regarded as a point, say x, of $\sigma(X, \mathcal{B})$. Now, it is obvious that $\varphi(x) = y$, which proves that φ is onto.

It is also easy to see that φ is closed and continuous. Because for each $B \in \mathcal{B}$ it holds that

$$\varphi^{-1}(\bar{B}^{\sigma'}) = \bar{B}^{\sigma}, \tag{3}$$

which proves that φ is closed and continuous (Exercise II.7).

To prove (3), let $x \in \bar{B}^{\sigma} = \tilde{B}$ (the last symbol from the theorem). Then $B \in x$, and hence $\varphi(x) \in \bar{B}^{\sigma'}$. Thus $\varphi^{-1}(\bar{B}^{\sigma'}) \supset \bar{B}^{\sigma}$. Conversely, if $x \notin \bar{B}^{\sigma} = \tilde{B}$, then $F \cap B = \emptyset$ for some $F \in x$. Hence

$$\varphi(x) \in \bar{F}^{\sigma'} \subset \sigma'(X, \mathcal{B}) - \bar{B}^{\sigma'},$$

proving $\varphi^{-1}(\bar{B}^{\sigma'}) \subset \bar{B}^{\sigma}$. Thus φ is the desired homeomorphism.

Definition IV.5. $\sigma(X, \mathcal{B})$ is called *Shanin's* (or *Wallman–Shanin's*) *compactification* of X with respect to $\mathcal{B}$. If $\mathcal{B}$ is the collection of all closed sets of X, then $\sigma(X, \mathcal{B})$ is denoted by $\omega(X)$ and called *Wallman's compactification* of X.[1]

Example IV.6. Let X be a locally compact T_2-space and $\mathcal{B} = \{B \mid \text{(i) } B \text{ is a compact set of } X, \text{ or (ii) } B \text{ is a closed set of } X \text{ such that } B \cup B' = X \text{ for some compact set } B'\}$. Then it is easy to see that $\mathcal{B}$ is a normal base of X. Let B be a closed set of $\alpha(X)$ such that $\infty \notin B$. Then B satisfies (i) in the above, and thus $B \in \mathcal{B}$ and $\bar{B} = B$. Let C be a closed set of $\alpha(X)$ such that $\infty \in C$. Suppose $y \in \alpha(X) - C$. Then there is an open nbd U of y in X such that the closure $\bar{U}$ in X is compact. Then $X - U$ satisfies (ii), and thus $X - U \in \mathcal{B}$. It is easy to see that $y \notin \overline{X - U} \supset C$ in $\alpha(X)$. Thus $\{\bar{B} \mid B \in \mathcal{B}\}$ is a closed base of $\alpha(X)$. Generally $B \in \mathcal{B}$ satisfies $\bar{B} = B$ in $\alpha(X)$ if B satisfies (i), and $\bar{B} = B \cup \{\infty\}$ in $\alpha(X)$ if B satisfies (ii). Hence $\overline{B_1 \cap B_2} = \bar{B}_1 \cap \bar{B}_2$ holds in $\alpha(X)$ whenever $B_1, B_2 \in \mathcal{B}$. Thus it follows from E) that $\alpha(X) = \sigma(X, \mathcal{B})$.

---

[1] H. Wallman [1] defined $\omega(X)$, and N. A. Shanin [1], [2], [3] generalized Wallman's idea to define $\sigma(X, \mathcal{B})$. In fact Shanin discussed his compactification under a more general condition than we do here.

This relation does not hold in general. Suppose X_0 is a cofinite space of infinitely many points. Then $\mathscr{B}_0 =$ the collection of all finite subsets of X_0 is the only T_1-base of X_0. Now, for every $F \in \mathscr{B}_0$, $\infty \notin \bar{F}$ in $\alpha(X_0)$. Thus $\tilde{\mathscr{B}}_0$ is no closed base of $\alpha(X_0)$. Namely, $\alpha(X_0)$ is not Shanin's compactification of X_0.

F) *Let X be a Tychonoff space and $\mathscr{Z}$ the collection of all zero sets of X. Then $\beta(X) = \sigma(X, \mathscr{Z})$.*

Proof. It suffices to prove that $\beta(X)$ satisfies (i) and (ii) of Theorem IV.3. Let $x \in \beta(X) - F$ for a closed set F of $\beta(X)$. Then there is a zero set Z of $\beta(X)$ such that

$$F \subset Z^\circ \subset Z \not\ni x \quad \text{in } \beta(X).$$

Then $Z \cap X \in \mathscr{Z}$, and

$$F \subset \overline{Z \cap X} \subset Z \not\ni x \quad \text{in } \beta(X).$$

Thus condition (i) holds for $\beta(X)$ and $\mathscr{Z}$.

To prove (ii), assume $Z_1, Z_2 \in \mathscr{Z}$, $Z_1 \cap Z_2 = \emptyset$. Then there is a real-valued continuous function f on X such that

$$f(Z_1) = 0, \qquad f(Z_2) = 1, \qquad 0 \leq f \leq 1.$$

We extend f to a continuous function g on $\beta(X)$ and put

$$Z_1' = \{x \in \beta(X) \mid g(x) = 0\}, \qquad Z_2' = \{x \in \beta(X) \mid g(x) = 1\}.$$

Then

$$Z_1' \cap Z_2' = \emptyset, \qquad \bar{Z}_1 \subset Z_1', \qquad \bar{Z}_2 \subset Z_2',$$

where the closure is taken in $\beta(X)$. Thus $\bar{Z}_1 \cap \bar{Z}_2 = \emptyset$. (Namely $\bar{Z}_1 \cap \bar{Z}_2 = \overline{Z_1 \cap Z_2}$ holds.) We shall use this observation to prove (ii) in general. Assume that Z_1 and Z_2 are elements of $\mathscr{Z}$ which are not necessarily disjoint. Suppose $x \in \bar{Z}_1 \cap \bar{Z}_2$, where here and in the following the closure is taken in $\beta(X)$. Let U be a zero-set nbd of x in $\beta(X)$. Then

$$x \in \overline{U \cap Z_1} \cap \overline{U \cap Z_2}.$$

Hence by use of the above observation we obtain

$$U \cap Z_1 \cap U \cap Z_2 = U \cap (Z_1 \cap Z_2) \neq \emptyset \,.$$

Hence $x \in \overline{Z_1 \cap Z_2}$, which proves (ii). Therefore $\beta(X) = \sigma(X, \mathscr{Z})$ follows from E).

G) $\beta(X) = \omega(X)$ *holds if and only if X is a normal space.*

Proof. If $\beta(X) = \omega(X)$, then $\beta(X) = \omega(X) = \sigma(X, \mathscr{C})$, where $\mathscr{C}$ is the collection of all closed sets of X. Since $\beta(X)$ is T_2, by Theorem IV.3 $\mathscr{C}$ is a normal base of X, i.e. it satisfies condition (iv) of Definition IV.4. Hence X is a normal space. The converse can be proved in a similar way as the proof of F).[1]

Example IV.7. Let us study some other aspects of $\omega(X)$. Generally we define (covering) *dimension*, dim X of a topological space X as follows:

$$\dim \emptyset = -1 \,;$$

for a non-negative integer n,

$$\dim X \leq n$$

means that for every finite open covering $\mathscr{U}$ of X there is a finite open covering $\mathscr{V}$ of X such that $\mathscr{V} < \mathscr{U}$, and ord $\mathscr{V} = \max \{i|$ there are i distinct elements of $\mathscr{V}$ whose intersection is non-empty$\} \leq n + 1$.

Then dim $X = n$ means that dim $X \leq n$ holds, but dim $X \leq n - 1$ does not. (dim $X = \infty$ if dim $X \leq n$ does not hold for any integer n.)

Then it is easy to see that dim $\omega(X) = \dim X$ for every T_1-space X. Thus dim $\beta(X) = \dim X$ holds if X is a normal space.

A lattice L is called *distributive* if it satisfies

$$a \vee (b \wedge c) = (a \vee b) \wedge (a \vee c), \qquad a \wedge (b \vee c) = (a \wedge b) \vee (a \wedge c) \,.$$

H. Wallman [1] constructed a compact T_1-space $A(L)$ from a given distributive lattice L with the smallest element 0 and the largest element 1 as follows. We call a subset P of L a *maximal dual idal* if (i) $0 \notin P$, (ii) $b > a \in P$ implies $b \in P$, (iii) $a,\ b \in P$ implies $a \wedge b \in P$. Then, let $A(L)$

[1] It is possible to discuss more general conditions in order that $\sigma(X, \mathscr{B}) = \sigma(X, \mathscr{B}')$. See N. A. Shanin [3].

be the set of all maximal dual idals of L. The topology of $A(L)$ is defined by use of the closed base $\mathcal{D} = \{\bar{a} \mid a \in L\}$, where $\bar{a} = \{P \in A(L) \mid a \in P\}$. As a matter of fact, if L is the lattice of all closed sets of a T_1-space X, then $A(L) = \omega(X)$. Wallman proved that L is isomorphic to the lattice of a closed base of $A(L)$ if and only if L satisfies the *disjunction property*: For every pair of distinct elements a, b of L, there is $c \in L$ such that one of $a \wedge c$ and $b \wedge c$ is 0 and the other is not 0. Thus every distributive lattice with 0, 1 and the disjunction property can be represented as a closed base of a compact T_1-space. The same idea was originated by M. H. Stone [1], who proved that every Boolean algebra can be represented as a closed base of a totally disconnected compact T_2-space, and conversely, the set of all clopen sets of such a space is a Boolean algebra, where a *Boolean algebra* is a distributive lattice with 0 and 1 satisfying the following condition: For every element a there is an element a' for which $a \wedge a' = 0$, $a \vee a' = 1$.

3. More of compactifications

As we saw in the previous section, Čech–Stone's compactification $\beta(X)$ as well as Alexandroff's compactification $\alpha(X)$ (in case that $\alpha(X)$ is T_2) are special cases of Shanin's compactification, which indicates that the method used to construct $\sigma(X, \mathcal{B})$ is quite general. Thus the following question occurs: Is it possible to construct every T_2-compactification $\hat{X}$ of every Tychonoff space X by use of that method? In other words, is it true that $\hat{X} = \sigma(X, \mathcal{B})$ for some normal base $\mathcal{B}$ of X?[1]

The purpose of this section is to discuss this problem, especially the partial positive answers due to J. M. Aarts [1], A. and E. Steiners [1] and C. Bandt [1], and the eventual negative answer due to V. M. Ul'janov [2]. All spaces in this section are at least T_2.

A) *Let $\hat{X}$ be a T_2-compactification of a Tychonoff space X. Then $\hat{X} = \sigma(X, \mathcal{B})$ for some normal base $\mathcal{B}$ of X if and only if X has a closed base $\mathcal{C}$ such that*

   (i) *$C_1, C_2 \in \mathcal{C}$ implies $C_1 \cup C_2 \in \mathcal{C}$ and $C_1 \cap C_2 \in \mathcal{C}$,*
   (ii) *for every $F \in \mathcal{C}$, $F = \overline{F \cap X}$, where the closure is taken in $\hat{X}$.*

---

[1] O. Frink [1] constructed a compactification (probably without knowing Shanin's work) by a method which is essentially the same as Shanin's, and he asked the above question. (He called such a compactification 'of Wallman type'.) So this question is sometimes called *Frink's problem*.

Proof. Assume that $\hat{X} = \sigma(X, \mathscr{B})$ for a normal base $\mathscr{B}$ of X. Then $\mathscr{C} = \{\bar{B} \mid B \in \mathscr{B}\}$ satisfies the conditions. Condition (i) is obvious. To prove (ii), let $F = \bar{B} \in \mathscr{C}$, where $B \in \mathscr{B}$. Then

$$\overline{F \cap X} = \overline{\bar{B} \cap X} = \overline{B \cap X} = \bar{B} = F.$$

Conversely, assume that $\mathscr{C}$ is a closed base of $\hat{X}$ satisfying (i) and (ii). Then put

$$\mathscr{B} = \{F \cap X \mid F \in \mathscr{C}\}.$$

It is obvious that $\mathscr{B}$ is a closed base of X satisfying the conditions (i), (ii) of Definition IV.4. To see (iii), let $x \in X - F \cap X$, where $F \in \mathscr{C}$. Then, since $\hat{X}$ is compact and $\mathscr{C}$ is a closed base of $\hat{X}$, there are $F_1, \ldots, F_k \in \mathscr{C}$ such that

$$x \in F_1 \cap \cdots \cap F_k, \quad F_1 \cap \cdots \cap F_k \cap F = \emptyset.$$

Then $x \in F_1 \cap \cdots \cap F_k \cap X \in \mathscr{B}$, satisfying (iii).

In a similar way we can prove (iv) of Definition IV.4, too, by use of normality and compactness of $\hat{X}$. Thus $\mathscr{B}$ is a normal base of X.

Let $B \in \mathscr{B}$; then $B = F \cap X$ for some $F \in \mathscr{C}$. Hence

$$\bar{B} = \overline{F \cap X} = F,$$

which means that $\{\bar{B} \mid B \in \mathscr{B}\} = \mathscr{C}$. Thus condition (i) of Theorem IV.3 is satisfied. Finally, suppose $B_1, B_2 \in \mathscr{B}$; then

$$B_1 = F_1 \cap X, \qquad B_2 = F_2 \cap X \quad \text{for } F_1, F_2 \in \mathscr{C}.$$

Hence
$$\bar{B}_1 \cap \bar{B}_2 = \overline{F_1 \cap X} \cap \overline{F_2 \cap X} = F_1 \cap F_2 = \overline{F_1 \cap F_2 \cap X} = \overline{B_1 \cap B_2},$$

proving (ii) of Theorem IV.3. Thus from E) it follows that

$$\hat{X} = \sigma(X, \mathscr{B}).$$

Definition IV.6. Let F and G be subsets of a space X. Then

$$\Delta(F, G) = \overline{F - G} \cap \overline{G - F}.$$

A closed base $\mathscr{B}$ of X is called a Δ-*base* if

$$\Delta(F, G) = \emptyset \quad \text{for every } F, G \in \mathscr{B}.$$

A compact T_2-space X is called a Δ-*space* if it has a Δ-base.

Example IV.8. Let I be a closed segment, say $I = [0, \sqrt{2}]$. Let $\mathscr{B}$ be the collection of all finite unions of disjoint closed segments in I with a rational left end and an irrational right end. Then it is easy to see that $\mathscr{B}$ is a Δ-base of I. Thus I is a Δ-space.

B) *If X is a Δ-space and $Y \subset X$, then Y has a Δ-base.*

Proof. The easy proof is left to the reader.

C) *If X is a Δ-space, then it has a Δ-base $\mathscr{B}$ satisfying* (i) *of* A) *and consisting of regular closed sets.*

Proof. Let $\mathscr{C}$ be a Δ-base of X. Put

$$\mathscr{B}_0 = \{B \in \mathscr{C} \mid B = \overline{B^\circ}\}.$$

Then $\mathscr{B}_0$ is a closed base of X. Because, if $x \in X - F$ for a closed set F of X, then there is an open nbd U of X such that $\bar{U} \cap F = \emptyset$. Pick $B \in \mathscr{B}_0$ such that $X - U \subset B \not\ni x$. Then obviously $F \subset \overline{B^\circ} \not\ni x$ and $\overline{B^\circ} \in \mathscr{B}_0$. Thus $\mathscr{B}_0$ is a closed base. Since $\mathscr{B}_0 \subset \mathscr{C}$, $\mathscr{B}_0$ is a Δ-base consisting of regular closed sets.

Let $\mathscr{B}$ be the minimal closed collection satisfying (i) of A) and $\mathscr{B} \supset \mathscr{B}_0$. Then $\mathscr{B}$ is what we want. Obviously it suffices to prove the following: Let $\mathscr{B}_1$ be a Δ-base consisting of regular closed sets, then $\mathscr{B}_2 = \{A \cap B, A \cup B \mid A, B \in \mathscr{B}_1\}$ is also a Δ-base consisting of regular closed sets. Now, assume $A, B \in \mathscr{B}_1$. Then

$$\overline{(A \cup B)^\circ} \supset \overline{A^\circ \cup B^\circ} = \overline{A^\circ} \cup \overline{B^\circ} = A \cup B.$$

Thus $A \cup B$ is a regular closed set.

Let $x \in \overline{A^\circ} \cap \overline{B^\circ}$. Then, since $\Delta(A, B) = \emptyset$, either $x \notin \overline{A - B}$ or $x \notin \overline{B - A}$ holds. Assume, e.g., $x \notin \overline{A - B}$. Consider any open nbd U of x such that $U \cap (A - B) = \emptyset$. Then $U \cap A^\circ \neq \emptyset$. Pick $y \in U \cap A^\circ$. To prove $y \in B^\circ$, assume the contrary; then, because $U \cap A^\circ$ is a nbd of y, we have

$U \cap A^\circ \cap (X - B) \neq \emptyset$. This implies that $U \cap (A - B) \neq \emptyset$, a contradiction. Hence $y \in A^\circ \cap B^\circ$, and thus $U \cap A^\circ \cap B^\circ \neq \emptyset$. This proves $x \in \overline{A^\circ \cap B^\circ}$. Thus

$$A \cap B = \overline{A^\circ} \cap \overline{B^\circ} \subset \overline{A^\circ \cap B^\circ} = \overline{(A \cap B)^\circ},$$

proving that $A \cap B$ is a regular closed set.

Suppose $A, B, C, D \in \mathscr{B}_1$. For brevity we denote $X - A$, $X - B$, etc. by A^c, B^c, etc. Then

$$\Delta(A \cup B, C \cup D) = \overline{A \cup B - C \cup D} \cap \overline{C \cup D - A \cup B}$$

$$= \overline{(A \cup B) \cap C^c \cap D^c} \cap \overline{(C \cup D) \cap A^c \cap B^c}$$

$$= \overline{(A \cap C^c \cap D^c \cup B \cap C^c \cap D^c)}$$

$$\cap \overline{(C \cap A^c \cap B^c \cup D \cap A^c \cap B^c)}$$

$$\subset \overline{(A \cap C^c \cap \overline{C \cap A^c})} \cup \overline{(A \cap D^c \cap \overline{D \cap A^c})}$$

$$\cup \overline{(B \cap C^c \cap \overline{C \cap B^c})} \cup \overline{(B \cap D^c \cap \overline{D \cap B^c})}$$

$$= \Delta(A, C) \cup \Delta(A, D) \cup \Delta(B, C) \cup \Delta(B, D) = \emptyset.$$

In a similar way we can prove

$$\Delta(A \cap B, C \cap D) = \emptyset, \qquad \Delta(A \cup B, C \cap D) = \emptyset.$$

Hence $\mathscr{B}_2$ is a Δ-base consisting of regular closed sets.

Theorem IV.4. *Let X be a Δ-space and Y a dense subset of X. Then X is a Shanin's compactification of Y.*

Proof. By C) X has a Δ-base $\mathscr{C}$ satisfying (i) of A) and consisting of regular closed sets. Let $F \in \mathscr{C}$ and $x \in F$. Then, since F is a regular closed set, $x \in \overline{F^\circ}$. Let U be a given open nbd of x; then $U \cap F^\circ \neq \emptyset$, and hence $U \cap F^\circ \cap Y \neq \emptyset$, because $\overline{Y} = Y$. Thus $x \in \overline{F^\circ \cap Y} \subset \overline{F \cap Y}$. This proves that $F \subset \overline{F \cap Y}$, i.e. $F = \overline{F \cap Y}$. Hence condition (ii) of A) is satisfied by C). Therefore, by A), X is a Shanin's compactification of Y.

Corollary. *If X is a compact T_2-space with a closed base satisfying (i) of A) and consisting of regular closed sets, then X is a Shanin's compactification of its every dense subset.*

D) *Let* $I^\tau = \prod\{I_t \mid t \in T\}$ *denote the product space of* τ *copies of the closed segment* $I = [0, 1]$, *where* τ *is a cardinal number, and* $|T| = \tau$. *If* U *is an open set of* I^τ, *then there is a countable subset* T' *of* T *such that for an open set* W *of* $I^{T'} = \prod\{I_t \mid t \in T'\}$, $V = W \times \prod\{I_t \mid t \in T - T'\}$ *satisfies* $V \subset U$ *and* $\bar{V} = \bar{U}$, *where closures are taken in* I^τ.

Proof. We assume $U \neq \emptyset$, $\bar{U} \neq X$. Let us denote by $\mathscr{B}_t$ a countable base of I_t such that $I_t \in \mathscr{B}_t$. Let $B_i \in \mathscr{B}_{t_i}$, $i = 1, \ldots, k$. Then we call the open set $B = B_1 \times \cdots \times B_k \times \prod\{I_t \mid t \neq t_1, \ldots, t_k\}$ of I^τ a basic rectangle and denote it by $B = (B_1 t_1 B_2 t_2 \ldots B_k t_k)$. Assume $B_i \neq I_{t_i}$ for $i = 1, \ldots, h$, and $B_i = I_{t_i}$ for $i = h + 1, \ldots, k$; then we define that

$$\kappa B = \kappa(B_1 t_1 B_2 t_2 \ldots B_k t_k) = \{t_1, \ldots, t_h\}.$$

Denote by B the collection of all basic rectangles of I^τ and by Γ the collection of all non-empty finite subsets of T. Let $B \in \mathscr{B}$ and $S \subset T$; then we denote by $P(B, S)$ the projection of B in I^S. Further, define that

$$\mathscr{B}' = \{B \in B \mid B \subset U, B \neq \emptyset\}.$$

Then $U = \bigcup\{B \mid B \in \mathscr{B}'\}$. Let

$$\mathscr{P}(S) = \{P(B, S) \mid B \in \mathscr{B}'\} \quad \text{for each } S \in \Gamma.$$

Now, we choose a collection Σ of finite alternating sequences $\{S_1 P_1 S_2 P_2 \ldots S_k P_k\}$ of $S_i \in \Gamma$ and $P_i \in \mathscr{P}(S_i)$ such that $S_i \cap S_j = \emptyset$ for $i \neq j$. First, observe that there is $S_1 \in \Gamma$ such that $S_1 \cap \kappa B \neq \emptyset$ for all $B \in \mathscr{B}'$ unless $\bar{U} = X$. Fix an S_1 satisfying this condition. Then, let

$$\mathscr{P}_1 = \{P(B, S_1) \mid B \in \mathscr{B}'\}.$$

Observe that P_1 is at most countable. For each $P_1 \in \mathscr{P}_1$ we proceed as follows. Put

$$\mathscr{B}(P_1) = \{B \in \mathscr{B}' \mid P(B, S_1) = P_1\}.$$

(i) If there is $S_2 \in \Gamma$ such that $S_1 \cap S_2 = \emptyset$, and $S_2 \cap \kappa B \neq \emptyset$ for all $B \in \mathscr{B}(P_1)$, then we fix such an S_2 and proceed in the same way as in the above (to complete the incomplete sequence $\{S_1 P_1 S_2\}$).

(ii) Otherwise, there is a (at most) countable subset $\mathscr{B}_0$ of $\mathscr{B}(P_1)$ such that for every $B \in \mathscr{B}(P_1)$ there is $B_0 \in \mathscr{B}_0$ satisfying

$$\kappa(B) \cap \kappa(B_0) - S_1 = \emptyset .$$

In this case we define that $\{S_1 P_1\} \in \Sigma$ and put $\mathscr{B}_0 = B(S_1 P_1)$.

This process must end with case (ii) after a finite number of steps, say k. Then we define that

$$\{S_1 P_1 \ldots S_k P_k\} \in \Sigma$$

and assign to it a (at most) countable subset $\mathscr{B}(S_1 P_1 \ldots S_k P_k)$ of $\mathscr{B}(P_1 \ldots P_k) = \{B \in \mathscr{B}' \mid P(B, S_i) = P_i, \ i = 1, \ldots, k\}$ such that for every $B \in \mathscr{B}(P_1 \ldots P_k)$ there is $B_0 \in \mathscr{B}(S_1 P_1 \ldots S_k P_k)$ satisfying

$$(\kappa B \cap \kappa B_0) - (S_1 \cup \cdots \cup S_k) = \emptyset .$$

(A detailed argument is left to the reader.)

Now, it is obvious that Σ is at most countable, and so is

$$\mathscr{B}'' = \bigcup \{\mathscr{B}(S_1 P_1 \ldots S_k P_k) \mid \{S_1 P_1 \ldots S_k P_k\} \in \Sigma\} .$$

Let

$$T' = \bigcup \{\kappa B \mid B \in \mathscr{B}''\} ;$$

then T' is countable. Put

$$W = \bigcup \{P(B, T') \mid B \in \mathscr{B}''\} ;$$

then W is an open set of $I^{T'}$ such that

$$V = W \times \prod \{I_t \mid t \in T - T'\} = \bigcup \{B \mid B \in \mathscr{B}''\} \subset U .$$

Further, it is easy to see that $\bar{V} = \bar{U}$. To see it, let $x \in \bar{U}$. Then for each basic rectangle B containing x, $B \cap B' \neq \emptyset$ for some $B' \in \mathscr{B}'$. Thus $B \cap B' = B'' \in \mathscr{B}'$. We can select $(S_1 P_1 \ldots S_k P_k) \in \Sigma$ such that $P(B'', S_i) = P_i$, $i = 1, \ldots, k$. Thus there is $B_0 \in \mathscr{B}(S_1 P_1 \ldots S_k P_k) \subset \mathscr{B}''$ such that

$$(\kappa B'' \cap \kappa B_0) - (S_1 \cup \cdots \cup S_k) = \emptyset .$$

Thus $B'' \cap B_0 \neq \emptyset$, which implies $B \cap V \neq \emptyset$. Hence $x \in \bar{V}$, proving $\bar{V} = \bar{U}$.

E) *Every regular closed set of I^τ is a zero set.*

Proof. It suffices to show that every regular open set U of I^τ is a cozero set. By D) there is an open set W of $I^{T'}$, where T' is at most countable, such that $V = W \times \prod \{I_t \mid t \in T - T'\}$ satisfies $\bar{V} = \bar{U}$. Since U is regular open, $\overline{U}^\circ = U$. Hence $U = \overline{V}^\circ$. Since $\overline{V}^\circ$ is the product of an open set of the metric space $I^{T'}$ and $\prod \{I_t \mid t \in T - T'\}$, it is a cozero set. Hence U is a cozero set.

Generally, let A be a subset of $I^\tau = \prod \{I_t \mid t \in T\}$. If there are $\{x_t \mid t \in T\} \in A$ and $\{x'_t \mid t \in T\} \notin A$ such that $x_t = x'_t$, then we say that A *depends on the t-coordinate*. The open set V in E) depends only on (at most) countably many coordinates, and so does U. Thus we obtain the following proposition.

F) *Every regular closed set of I^τ depends only on (at most) countably many coordinates.*[1]

G) *Let B and C be non-empty closed sets of $I^2 = I_1 \times I_2$, where $I_i = [0, 1]$, $i = 1, 2$, such that $B = F \times I_2 \neq I^2$, $C \neq I^2$, $\Delta(B, C) = \emptyset$. Then C depends on the first coordinate.*

Proof. Let $x_0 = \inf F$. We assume $x_0 > 0$ and $(x_0, 0) \notin C$, leaving check-up of the other cases to the reader. Put

$$y_0 = \inf \{y \mid (x_0, y) \in C\}.$$

(Observe that if there is no y for which $(x_0, y) \in C$, then the proof is over.) Then $y_0 > 0$, and

$$(x_0, y_0) \in C. \tag{1}$$

Since $(x_0, y) \in B - C$ for every $y < y_0$, $(x_0, y_0) \in \overline{B - C}$. Since $\Delta(B, C) = \emptyset$, this implies $(x_0, y_0) \notin \overline{C - B}$. There is $\varepsilon > 0$ satisfying

$$(x_0 - \varepsilon, x_0 + \varepsilon) \times (y_0 - \varepsilon, y_0 + \varepsilon) \cap (C - B) = \emptyset. \tag{2}$$

---

[1] In fact, as we shall see in VII.1, a closed set of I^τ depends only on (at most) countably many coordinates if and only if it is a zero set.

Since $(x_0 - \varepsilon, x_0) \times \{y_0\} \cap B = \emptyset$, (2) implies

$$(x_0 - \varepsilon, x_0) \times \{y_0\} \cap C = \emptyset.$$

Hence $(x_0 - \varepsilon/2, y_0) \notin C$. Thus from (1) it follows that C depends on the first coordinate.

Now, we can prove C. Bandt's [1] first theorem.

Theorem IV.5. *I^τ is no Δ-space whenever $\tau \geqslant \aleph_2$.*

Proof. Assume that I^τ is a Δ-space. Then by C) X has a Δ-space $\mathscr{B}$ consisting of regular closed sets. Let $I^\tau = \{I_t \mid t \in T\}$. Choose a subset T' of T such that $|T'| = \aleph_1$. Denote by 0 the point of I^τ whose all coordinates are zero and by H_t the closed set of all points of I^τ whose t-coordinate is 1. Then for each $t \in T$ there is $B_t \in \mathscr{B}$ such that $H_t \subset B_t \not\ni 0$. By F) every element of $\mathscr{B}$ depends only on countably many coordinates. Thus $\{B_t \mid t \in T'\}$ depends only on $T'' \subset T$ with $|T''| \leqslant \aleph_1$.[1] Select $t' \in T - T''$. Then for each $t \in T'$, B_t does not depend on the t'-coordinate. Consider a closed subset $I^2 = \prod \{X_{t''} \mid t'' \in T\}$ of I^τ, where $X_t = I_t$, $X_{t'} = I_{t'}$, $X_{t''} = \{0\}$ for $t'' \neq t$, t'. Then $B = B_t \cap I^2$ and $C = B_{t'} \cap I^2$ satisfy the condition of G). Hence C and accordingly $B_{t'}$ depends on the t-coordinate. Thus $B_{t'}$ depends on the uncountably many coordinates T', which is a contradiction. Hence I^τ is no Δ-space.

H) *Let X be a compact T_2-space and $\{B_i \mid i = 1, 2, \ldots\}$ a sequence of closed sets of X such that $\Delta(B_i, B_j) = \emptyset$ whenever $i \neq j$. Then for any disjoint closed sets D and E of X there is a closed set C such that $D \subset C$, $C \cap E = \emptyset$, $\Delta(C, B_i) = \emptyset$ for $i = 1, 2, \ldots$.*

Proof. Let h be a real-valued continuous function on X such that

$$h(D) = 0, \qquad h(E) = 1, \qquad 0 \leqslant h \leqslant 1.$$

Let g_i be a real-valued function on X defined by

$$g_i(x) = \begin{cases} 3^{-i} & \text{if } x \in B_i, \\ 0 & \text{otherwise}. \end{cases}$$

---

[1] Precisely this means that if $x, y \in I^\tau$ have the same t-coordinate for every $t \in T''$ and $x \in B_{t'}$ for some $t' \in T'$, then $y \in B_{t'}$.

Put $g = \sum_{i=1}^{\infty} g_i$, and

$$C = \{x \in X \mid g(x) \geq h(x)\} . \tag{1}$$

Then C is the desired set. First, to see that C is a closed set, let $x \notin C$. Then

$$h(x) - g(x) = \varepsilon > 0 \tag{2}$$

follows. Select a natural number m such that $3^{-m} < \varepsilon$. Put

$$U = h^{-1}(S_{\varepsilon/2}(h(x)) - \bigcup \{B_i \mid i \leq m, \text{ and } x \notin B_i\} .$$

Then obviously U is a nbd of x. We claim that $U \cap C = \emptyset$.
To prove it, we suppose $y \in U$. Then we have

$$h(y) > h(x) - \varepsilon/2 . \tag{3}$$

Also note that $x \notin B_i$ and $i \leq m$ imply $y \notin B_i$, and thus it follows that

$$\sum_{i=1}^{m} g_i(y) \leq \sum_{i=1}^{m} g_i(x). \tag{4}$$

Hence from (4), (2), (3) it follows that

$$g(y) \leq \sum_{i=1}^{m} g_i(y) + 3^{-m}/2 \leq \sum_{i=1}^{m} g_i(x) + 3^{-m}/2 \leq g(x) + 3^{-m}/2$$

$$< g(x) + \varepsilon/2 = h(x) - \varepsilon/2 < h(y) .$$

This means that $y \notin C$. (See (1).) Hence $U \cap C = \emptyset$, proving that C is a closed set. It is obvious that $D \subset C$ and $C \cap E = \emptyset$.
Finally we can prove that $\Delta(C, B_i) = \emptyset$ for each i. Assume the contrary,

$$x \in \Delta(C, B_m) \neq \emptyset \quad \text{for some } m . \tag{5}$$

We claim that there is a nbd U of x satisfying:
(6) If $y \in U \cap (C - B_m)$, $z \in U \cap (B_m - C)$, then $\{i \in N \mid i < m, y \in B_i\} \subset \{i \in N \mid i < m, z \in B_i\}$, where N denotes the set of natural numbers.
Assume the contrary and define $F(U)$ for each nbd U of x by $F(U) =$

$\{i \in N \mid i < m$, there are $y \in U \cap (C - B_m)$ and $z \in U \cap (B_m - C)$ such that $y \in B_i$ and $z \notin B_i\}$. Then $F(U) \neq \emptyset$ for every U follows from the assumption. Since $F(\cap_{i=1}^{k} U_i) \subset \cap_{i=1}^{k} F(U_i)$ for any finite number of nbds U_i, $i = 1, \ldots, k$, of x, $\{F(U) \mid U$ is a nbd of $x\}$ has f.i.p. and accordingly a non-empty intersection. Hence there is a natural number $i < m$ such that for every nbd U of x there are $y \in U \cap (C - B_m)$ and $z \in U \cap (B_m - C)$ such that $y \in B_i$ and $z \notin B_i$. Thus

$$y \in (B_i - B_m) \cap U, \qquad z \in (B_m - B_i) \cap U,$$

and thus

$$U \cap (B_i - B_m) \neq \emptyset, \qquad U \cap (B_m - B_i) \neq \emptyset.$$

Namely $x \in \Delta(B_i, B_m) \neq \emptyset$, which is a contradiction. Hence there is a nbd U of x satisfying (1).

We may assume without loss of generality that

$$U \subset h^{-1}(S_{3^{-m}/4}(h(x))).$$

Now, since x satisfies (5), we can choose $y \in U \cap (C - B_m)$ and $z \in U \cap (B_m - C)$. Then from (1) and (6) it follows that

$$h(y) \le g(y) \le \sum_{i < m} g_i(y) + 3^{-m}/2 \le \sum_{i < m} g_i(z) + 3^{-m}/2$$
$$\le g(z) - g_m(z) + 3^{-m}/2.$$

Since $z \in B_m$, $g_m(z) = 3^{-m}$, and hence

$$g(z) - g_m(z) + 3^{-m}/2 = g(z) - 3^{-m}/2 < h(z) - 3^{-m}/2,$$

because $z \notin C$. (See (1).) Thus we obtain

$$h(y) < h(z) - 3^{-m}/2. \tag{7}$$

On the other hand, since

$$y, z \in U \subset h^{-1}(S_{3^{-m}/4}(h(x))),$$

we have

$$h(y) > h(x) - 3^{-m}/4 > h(z) - 3^{-m}/2,$$

which contradicts (7). Hence we conclude that

$$\Delta(C, B_m) = \emptyset,$$

proving the proposition.

Now, we are in a position to prove the second theorem of Bandt.

Theorem IV.6. I^τ *is a Δ-space for every $\tau \leq \aleph_1$. Thus every compact T_2-space X with $w(X) \leq \aleph_1$ is a Δ-space and accordingly a Shanin's compactification of each dense subset.*

Proof. Let $\mathcal{B}$ be a closed base of I^τ such that $|\mathcal{B}| = \aleph_1$. Well-order the collection

$$\mathcal{B}' = \{(B_1, B_2) \mid B_1, B_2 \in \mathcal{B}, B_1 \cap \overline{B_2^c} = \emptyset\}$$

to put

$$\mathcal{B}' = \{(B_1^\alpha, B_2^\alpha) \mid 0 \leq \alpha < \omega_1\}.$$

Apply H) to get a closed set C_0 for the disjoint closed sets B_1 and $\overline{B_2^c}$, where (B_1, B_2) is the first element of $\mathcal{B}'$. Namely, C_0 is a closed set such that

$$B_1 \subset C_0, \qquad C_0 \cap \overline{B_2^c} = \emptyset.$$

Assume that $0 < \beta < \omega_1$, and for all $\alpha < \beta$ closed sets C_α have been defined to satisfy
  (1) $\Delta(C_\alpha, C_{\alpha'}) = \emptyset$ for every $\alpha, \alpha' < \beta$,
  (2) $B_1^\alpha \subset C_\alpha, C_\alpha \cap \overline{(B_2^\alpha)^c} = \emptyset$.
Then by use of H) we can construct a closed set C_β such that

$$B_1^\beta \subset C_\beta, \qquad C_\beta \cap \overline{(B_2^\beta)^c} = \emptyset, \qquad \Delta(C_\beta, C_\alpha) = \emptyset \quad \text{for all } \alpha < \beta.$$

Now we obtain a collection $\mathcal{C} = \{C_\beta \mid 0 \leq \beta < \tau\}$ of closed sets such that $\Delta(C_\alpha, C_\beta) = \emptyset$ whenever $\alpha \neq \beta$ and such that $B_1^\beta \subset C_\beta, C_\beta \cap \overline{(B_2^\beta)^c} = \emptyset$ for every element (B_1^β, B_2^β) of $\mathcal{B}'$. We claim that $\mathcal{C}$ is a closed base of I^τ. Assume $x \notin F$ for a closed set F and a point x of I^τ. Since $\mathcal{B}$ is a closed base, there is $B_1 \in \mathcal{B}$ such that $F \subset B_1 \not\ni x$. There is an open nbd U of x

such that $\bar{U} \cap B_1 = \emptyset$. There is $B_2 \in \mathscr{B}$ such that $X - U \subset B_2 \not\ni x$. Then $B_1 \cap \overline{B_2^c} = \emptyset$. Thus $(B_1, B_2) = (B_1^\alpha, B_2^\alpha)$ for some $\alpha < \omega_1$. Hence $F \subset B_1^\alpha \subset C_\alpha$, $C_\alpha \cap \overline{(B_2^\alpha)^c} = \emptyset$, $x \in (B_2^\alpha)^c \subset \overline{(B_2^\alpha)^c}$. Therefore $F \subset C_\alpha \not\ni x$, proving that $\mathscr{C}$ is a Δ-base of I^τ. Hence I^τ is a Δ-space.

The following important Corollary 1 was first proved by J. M. Aarts [1] and A. K. Steiner–E. F. Steiner [1] independently; the latter also proved Corollary 2.

Corollary 1. *Every compact metric space is a Shanin's compactification of each of its dense subsets.*

Corollary 2. *Every product of compact metric spaces is a Shanin's compactification of each of its dense subsets.*

Proof. Use the corollary of Theorem IV.4. (It is easy to see that if each of X_α, $\alpha \in A$, has a closed base satisfying (i) of A) and consisting of regular closed sets, then so does $\prod_\alpha X_\alpha$.)

Example IV.9. As indicated by Corollary 2, the converse of Theorem IV.4 is not true. In fact I^τ (for every cardinal number τ) is a Shanin's compactification of each of its dense subsets. However, V. M. Ul'janov [2] proved the following extremely important theorem:

Let $2^\tau \geq \aleph_2$. Then there is a Tychonoff space X such that $|X| = \tau$ and a T_2-compactification $\tilde{X}$ of X such that $w(\tilde{X}) = 2^\tau$, and $\tilde{X}$ is not a Shanin's compactification of X. In his proof Theorem IV.5 plays an essential role. From his result it follows that the continuum hypothesis CH is equivalent with that every T_2-compactification of every separable Tychonoff space is a Shanin's compactification. To see it, assume CH and let $\tilde{X}$ be a T_2-compactification of a separable Tychonoff space X. Then $\tilde{X}$ is also separable, so it has a countable dense subset A. Note that the collection $\mathscr{B}$ of all regular open sets of $\tilde{X}$ is a base of $\tilde{X}$ and that every regular open set U is decided by $U \cap A$. Thus $|\mathscr{B}| \leq 2^\mathfrak{a} = \aleph_1$. Hence by Theorem IV.6, $\tilde{X}$ is a Shanin's compactification.

Conversely, assume that CH does not hold. Then $2^\mathfrak{a} \geq \aleph_2$. Thus by the above-mentioned Ul'janov's theorem there is a Tychonoff space X such that $|X| = \mathfrak{a}$ and a non-Shanin's T_2-compactification $\tilde{X}$ of X such that $w(\tilde{X}) = 2^\mathfrak{a}$.[1]

---

[1] See J. v. Mill–H. Vermeer [1] for further discussions on this aspect.

4. Compact space and the lattice of continuous functions

Let $C(X)$ denote the set of all real-valued continuous functions over a topological space X. Then it can be regarded as a lattice with respect to the order: $f \leq g$ if and only if $f(p) \leq g(p)$ at every point p of X. The supremum and infimum of two elements f, g of $C(X)$ are given by

$$(f \vee g)(p) = \max\{f(p), g(p)\} \quad \text{and} \quad (f \wedge g)(p) = \min\{f(p), g(p)\},$$

respectively. We may also regard $C(X)$ as a ring with respect to the usual sum and product operations. The algebraic structure of $C(X)$ as a lattice or as a ring has a deep connection with the topological structure of X itself. Especially, if X is a compact T_2-space, then X is topologically characterized by the algebraic structure of the lattice $C(X)$. Furthermore this implies that X is topologically characterized by the ring $C(X)$, too. A theory of this kind was first established by I. Gelfand and A. Kolmogoroff [1] regarding $C(X)$ as a ring, and later it was extended to the lattice case by I. Kaplansky [1].[1] In the present section we shall give an account of their theory beginning with the lattice case.

We mean by a *prime ideal* of $C(X)$ the inverse image of 0 under some lattice homomorphism of $C(X)$ onto the two-element lattice $\{0, 1\}$. A prime ideal P of $C(X)$ is said to be *associated* with a point p of X if it satisfies the following condition: If $f \in P$ and $g(p) < f(p)$, then $g \in P$.

A) *A proper non-empty subset J of $C(X)$ is a prime ideal if and only if it satisfies the following conditions*:
  (i) $f, g \in J$ implies $f \vee g \in J$,
  (ii) $f < g \in J$ implies $f \in J$,
  (iii) $f, g \in C(X) - J$ implies $f \wedge g \in C(X) - J$.

Proof. It is left to the reader.

B) *For every point p of X, we can construct a prime ideal which is associated with p.*

Proof. For a fixed real number a, let

---

[1] Some other mathematicians studied similar characterization theories dealing with $C(X)$ as a topological ring or as a similar algebraic system. See G. Silov [1], M. H. Stone [1].

$$P = \{f \mid f \in C(X), f(p) < a\}.$$

Then P is obviously a prime ideal associated with p.

C) *Let X be a compact T_2-space. Then every prime ideal P of $C(X)$ is associated with one and only one point of X.*

Proof. Assume that a prime ideal P is associated with no point of X. Then for every point p of X there exists $f_p \in P$ and $g_p \notin P$ with $g_p(p) < f_p(p)$. Since f_p and g_p are continuous, we can find an open nbd $U(p)$ of p in which $g_p(q) < f_p(q)$ holds. Since X is compact, we can cover X with a finite number of those open nbds, say $U(p_1), \ldots, U(p_k)$. Put

$$f = \bigvee_{i=1}^{k} f_{p_i} \quad \text{and} \quad g = \bigwedge_{i=1}^{k} g_{p_i};$$

then $g < f$, $f \in P$ and $g \notin P$.

This contradicts the condition (ii) of A). Therefore P must be associated with at least one point p_0.

Put

$$s = \sup \{f(p_0) \mid f \in P\}.$$

If $s = +\infty$, then $P = C(X)$ contradicting that P is a prime ideal. Therefore $s < +\infty$. Now, let q_0 be a point of X different from p_0; then choose an element $h \in P$. Since X is completely regular, there is a continuous function h' such that $h'(q_0) < h(q_0)$ and $h'(p_0) > s$ which implies $h' \notin P$. Hence P is not associated with q_0, proving the assertion.

D) *Let P_1 and P_2 be prime ideals of $C(X)$, where X is a compact T_2-space. Then P_1 and P_2 are associated with the same point of X if and only if $P_1 \cap P_2$ contains a prime ideal.*

Proof. Suppose P_1 and P_2 are associated with the same point p_0. Put

$$s_1 = \sup \{f(p_0) \mid f \in P_1\}$$

and

$$s_2 = \sup \{g(p_0) \mid g \in P_2\}.$$

Then

$$P_3 = \{h \mid h \in C(X), h(p_0) < \min \{s_1, s_2\}\}$$

is a prime ideal as seen in B). Since $P_3 \subset P_1 \cap P_2$ is obvious, the condition is necessary.

Conversely, suppose $P_1 \cap P_2$ contains a prime ideal P. Assume that P and P_1 are associated with different points p and p_1 of X respectively, and put

$$s = \sup \{f(p) \,|\, f \in P\} < +\infty .$$

and

$$s_1 = \sup \{g(p_1) \,|\, g \in P_1\} < +\infty.$$

Then we can construct a continuous function f such that

$$f(p) < s \quad \text{and} \quad f(p_1) > s_1 .$$

This implies

$$f \in P \quad \text{and} \quad f \notin P_1 ,$$

contradicting that $P \subset P_1$. Hence P and P_1 are associated with the same point, and so, too, are P and P_2. Consequently P_1 and P_2 are associated with the same point.

We shall call two prime ideals P_1 and P_2 equivalent if $P_1 \cap P_2$ contains a prime ideal. Proposition D) shows that all prime ideals of $C(X)$ can be classified by the equivalence relation. We shall call each of those equivalence classes merely a *prime class* and denote by $\pi(X)$ the totality of the prime classes of $C(X)$. Let B be a subset of $\pi(X)$ and q an element of $\pi(X)$. Then we define that

$$q \in \bar{B}$$

if and only if q and B satisfy the following condition: One can choose a prime ideal $P(q')$ from each $q' \in B$ such that

$$\cap \{P(q') \,|\, q' \in B\} \neq \emptyset$$

and such that

$$\cap \{P(q') \,|\, q' \in B\} \subset P$$

for some $P \in q$.

We can show that this definition makes $\pi(X)$ into a topological space which is homeomorphic to X. Now, we denote by $\pi(p)$ the prime class consisting of the prime ideals associated with $p \in X$. Then by B), C) and D), the correspondence $p \to \pi(p)$ is a one-to-one mapping of X onto $\pi(X)$. We shall denote this mapping also by π.

E) *Let A be a subset and p a point of a compact T_2-space X. Then $p \in \bar{A}$ in X if and only if $\pi(p) \in \overline{\pi(A)}$ in $\pi(X)$.*

Proof. Suppose $p \in \bar{A}$ in X. We consider a set $\{P(a) \mid a \in A\}$ of prime ideals defined by

$$P(a) = \{f \mid f \in C(X), f(a) \le k\},$$

where k is a constant. Then

$$P(a) \in \pi(a), \quad a \in A$$

and

$$k \in \cap \{P(a) \mid a \in A\} \ne \emptyset.$$

Define a prime ideal P by

$$P = \{f \mid f \in C(X), f(p) \le k\}.$$

Then $P \in \pi(p)$. It follows from $p \in \bar{A}$ that

$$\cap \{P(a) \mid a \in A\} \subset P.$$

According to the definition this means that

$$\pi(p) \in \overline{\pi(A)} \quad \text{in } \pi(X).$$

Conversely, suppose that $p \notin \bar{A}$ in X. We consider a given set $\{P(a) \mid a \in A\}$ of prime ideals $P(a) \in \pi(a)$ such that

$$\cap \{P(a) \mid a \in A\} \ne \emptyset$$

and assume that P is a given prime ideal satisying $P \in \pi(p)$. Let

$$f_0 \in \cap \{P(a) \mid a \in A\}$$

and put

$$s(a) = \sup \{f(a) \mid f \in P(a)\}, \quad a \in A,$$
$$s = \sup \{f(p) \mid f \in P\}.$$

Then

$$f_0(a) \leq s(a) \quad \text{at each } a \in A.$$

Since f_0 is bounded,

$$t = \inf \{s(a) \mid a \in A\} \geq \inf \{f_0(a) \mid a \in A\} > -\infty.$$

Since X is completely regular and $p \notin \bar{A}$, we can construct a continuous function f which satisfies

$$f(p) > s \quad \text{and} \quad f(a) < t \quad \text{for all } a \in A.$$

Now, it is obvious that

$$f \in \cap \{P(a) \mid a \in A\},$$

but $f \notin P$, i.e.

$$\cap \{P(a) \mid a \in A\} \not\subset P.$$

This proves that

$$\pi(p) \notin \overline{\pi(A)} \quad \text{in } \pi(X).$$

The following is a direct consequence of E).

F) $\pi(X)$ *is a topological space which is homeomorphic to* X.

Now, we reach the following interesting theorem. The significance of this theorem is in the fact that the topology of a compact T_2-space is characterized by the algebraic structure of the lattice $C(X)$.

Theorem IV.7. *Two compact T_2-spaces X and Y are homeomorphic if and only if the lattices $C(X)$ and $C(Y)$ are isomorphic.*

Proof. The necessity of the condition is obvious. Conversely, suppose $C(X)$ and $C(Y)$ are isomorphic as lattices. Throughout the preceding discussion the topological space $\pi(X)$ was constructed by use of the lattice structure of $C(X)$ only. The topological space $\pi(Y)$ is also constructed by use of the lattice structure of $C(X)$. Therefore the isomorphism between $C(X)$ and $C(Y)$ implies that $\pi(X)$ and $\pi(Y)$ are homeomorphic. This combined with F) proves that X and Y are homeomorphic.

Now, let us discuss the case of non-compact completely regular space X. Let $C^*(X)$ denote the lattice of all bounded continuous functions over X. $C^*(X)$ coincides with $C(X)$ if X is compact. Generally, in the non-compact case $C^*(X)$ does not strictly characterize X as in the compact case, but still there is some connection between X and $C^*(X)$ as seen in the following.

G) *Recall that $\beta(X)$ denotes Čech–Stone's compactification of a completely regular space X. Then two lattices $C^*(X)$ and $C^*(\beta(X))$ are isomorphic.*

Proof. By Theorem IV.2 every continuous function $f \in C^*(X)$ can be uniquely extended to a continuous function $\varphi \in C^*(\beta(X))$. Letting φ correspond to f, we obtain a one-to-one mapping of $C^*(X)$ onto $C^*(\beta(X))$. It is easy to see that this mapping preserves the order between two elements and accordingly is an isomorphism.

The following is a direct consequence of G).

Theorem IV.8. *Let X and Y be completely regular spaces. Then $\beta(X)$ and $\beta(Y)$ are homeomorphic if and only if the lattices $C^*(X)$ and $C^*(Y)$ are isomorphic.*

Although, generally speaking, $C^*(X)$ does not characterize the topology of a non-compact space X, it does so in the special case that X satisfies the first axiom of countability. To prove it we need the following proposition about $\beta(X)$.

H) *Let X and Y be completely regular spaces satisfying the first axiom of countability. If $\beta(X)$ and $\beta(Y)$ are homeomorphic, then X and Y are homeomorphic.*

Proof. By IV.1.A) we know that each point of $\beta(X) - X$ has no countable nbd base in $\beta(X)$. Now, let us prove that each point x of X has a countable nbd base in $\beta(X)$. Let $\{U_1, U_2, \ldots\}$ be a nbd base of x in X. Then for each i

$$\tilde{U}_i = \beta(X) - \overline{(X - U_i)}$$

is an open nbd of x in $\beta(X)$, where the closure is taken in $\beta(X)$. We claim that $\{\tilde{U}_i \mid i = 1, 2, \ldots\}$ is a nbd base of x in $\beta(X)$. To see it, suppose

$$y \in \beta(X), \qquad y \neq x.$$

Then there is a continuous function φ from $\beta(X)$ into $[0, 1]$ such that

$$\varphi(x) = 0, \qquad \varphi(y) = 1.$$
Put
$$V = \varphi^{-1}([0, \tfrac{1}{2})).$$

Then V is an open nbd of x in $\beta(X)$, and hence there is some i for which

$$U_i \subset V \cap X.$$
Now, put
$$W = \varphi^{-1}((\tfrac{1}{2}, 1]).$$

Then W is an open nbd of y in $\beta(X)$ such that

$$W \subset \overline{X - U_i} \quad \text{in } \beta(X).$$

Because, consider any point $z \in W$ and any nbd P of z in $\beta(X)$. Then $P \cap W$ is a nbd of z, and hence

$$(P \cap W) \cap X \neq \emptyset.$$

This implies that

$$P \cap (X - U_i) \neq \emptyset,$$
because
$$W \cap X \subset X - U_i.$$
Hence
$$z \in \overline{X - U_i},$$
proving

$$W \subset \overline{X - U_i} \quad \text{or} \quad W \cap \tilde{U}_i = \emptyset \quad (\text{in } \beta(X)) .$$

Thus we obtain

$$y \notin \bar{\tilde{U}}_i \quad \text{in } \beta(X) .$$

Namely

$$\bigcap_{i=1}^{\infty} \bar{\tilde{U}}_i = \{x\} .$$

Since $\beta(X)$ is compact, as easily seen, this implies that for every nbd U of x in $\beta(X)$, there is i for which

$$\bar{\tilde{U}}_i \subset U .$$

Thus $\{\tilde{U}_i \mid i = 1, 2, \ldots\}$ is a nbd base of x in $\beta(X)$.

Now, suppose φ is a homeomorphism between $\beta(X)$ and $\beta(Y)$. Then

$$\varphi(\beta(X) - X) = \beta(Y) - Y, \qquad \varphi(X) = Y,$$

because, as seen in the above, every point of $\beta(X) - X$ as well as every point of $\beta(Y) - Y$ has no countable nbd basis while every point of X as well as every point of Y does. This proves that X and Y are homeomorphic.

Combining H) with Theorem IV.8, we obtain the following corollary of the same theorem.

Corollary. *Let X and Y be completely regular spaces satisfying the first axiom of countability. Then X and Y are homeomorphic if and only if $C^*(X)$ and $C^*(Y)$ are isomorphic as lattices.*

Now we shall regard the sets $C(X)$ and $C^*(X)$ of all continuous functions and of all bounded continuous functions over X, respectively, as rings with respect to the usual sum and product.

Theorems IV.7, 8 and the Corollary remain true for the rings, too. In fact the theorems for rings are directly derived from the corresponding theorems for lattices. To show this it will be enough to give only the following theorem.

Theorem IV.9. *Two compact T_2-spaces X and Y are homeomorphic if and only if the rings $C(X)$ and $C(Y)$ are algebraically isomorphic.*

Proof. It suffices to prove only the 'if' part. Let f, g be given two elements of $C(X)$. Then $f \geqslant g$ if and only if $f - g = h^2$ for some $h \in C(X)$. This means that the lattice structure of $C(X)$ is completely determined by its ring structure. Therefore if $C(X)$ and $C(Y)$ are ring-isomorphic, then they are also lattice-isomorphic. Hence this theorem is derived directly from Theorem IV.7.

Definition IV.7. Let B be a linear space (vector space) with scalar multiplication by the real numbers. Suppose that to each $x \in B$ a real number $\|x\|$ is assigned satisfying

$$\|x\| \geqslant 0 \,,$$

$$\|x\| = 0 \text{ if and only if } x = 0 \,,$$

$$\|rx\| = |r| \cdot \|x\| \,,$$

$$\|x + y\| \leqslant \|x\| + \|y\| \,,$$

where r denotes a real number and x, y elements of B. Then we call $\|x\|$ the *norm* of x. A linear space with norm is called a *normed linear space*. Every normed linear space is a metric space with the metric

$$\rho(x, y) = \|x - y\| \,.$$

A normed linear space is called a *Banach space* if it is a complete metric space. A subset V of a linear space is called *convex* if x, $y \in V$ implies $\alpha x + \beta y \in V$ whenever α, β are positive numbers satisfying $\alpha + \beta = 1$. A subset W of B is called *symmetric* if

$$-W = \{x \mid -x \in W\} = W \,.$$

Let B and B' be normed linear spaces. Then they are called *isomorphic* if there is an isomorphism between the linear spaces B and B' which preserves the norm.

We continue to denote by B a normed linear space. Let a, b be fixed points of B such that $a \neq 0$. Then we call $\{ra + b \mid -1 \leqslant r \leqslant 1\}$ ($\subset X$) a *segment* and each point $ra + b$ of the segment an *interior point* if $-1 < r < 1$. An element e of B is called an *extreme point* if $\|e\| = 1$ and if e is no interior point of any segment in

$$\overline{S_1(0)} = \{x \in B \mid \|x\| \leqslant 1\} \,.$$

Example IV.10. An important example of a Banach space is a generalized Hilbert space $H(A)$, where the operations are defined by

$$\{x_\alpha \mid \alpha \in A\} + \{x'_\alpha \mid \alpha \in A\} = \{x_\alpha + x'_\alpha \mid \alpha \in A\},$$
$$r\{x_\alpha \mid \alpha \in A\} = \{rx_\alpha \mid \alpha \in A\},$$

and the norm by

$$\|\{x_\alpha \mid \alpha \in A\}\| = \left(\sum_{\alpha \in A} x_\alpha^2\right)^{1/2}.$$

Let X be a compact T_2-space. Then it is also easy to see that the linear space $C(X)$ of all real-valued continuous functions on X is a Banach space with respect to the norm

$$\|f\| = \max\{|f(x)| \mid x \in X\}, \quad f \in C(X).$$

Let e be an extreme point of $C(X)$. Then it is easy to see that f cannot take on any value but $+1$ and -1. To prove it, assume the contrary: $e(x) = \alpha$ for some $x \in X$ and $-1 < \alpha < +1$. Select $\varepsilon > 0$ such that $-1 + \varepsilon < \alpha < 1 - \varepsilon$. Let U be an open nbd of x such that $-1 + \varepsilon < e(y) < 1 - \varepsilon$ whenever $y \in U$. Let $f \in C(X)$ satisfy

$$f(x) = \varepsilon, \quad f(X - U) = 0, \quad 0 \le f \le \varepsilon.$$

Then the segment $\{rf + e \mid -1 \le r \le 1\}$ contains e as an interior point and lies in $\overline{S_1(0)} = \{g \in C(X) \mid \|g\| \le 1\}$. This contradicts that e is an extreme point.

It is also easy to prove that 1 is an extreme point of $C(X)$. To see it, assume that 1 is an interior point of a segment $\{ra + b \mid -1 \le r \le 1\} \subset \overline{S_1(0)}$ ($a \ne 0$). Assume $r_0 a + b = 1$ for some r_0 with $-1 < r_0 < 1$. Then

$$ra + b = ra + (1 - r_0 a) = 1 + (r - r_0)a.$$

Since $a \ne 0$, assume, e.g., $a(x) > 0$ for some $x \in X$. Then

$$1 + (r - r_0)a(x) > 1 \quad \text{for } r > r_0,$$

which contradicts that the segment lies in $\overline{S_1(0)}$. Now, we can prove the following proposition:

I) *Let X and Y be compact T_2-spaces. Then X and Y are homeomorphic if and only if the Banach spaces $C(X)$ and $C(Y)$ are isomorphic.*

Let e be a fixed extreme point of $C(X)$. Then X is decomposed into the sum of two disjoint clopen sets G and H defined by

$$G = \{x \in X \mid e(x) = 1\}, \qquad H = \{x \in X \mid e(x) = -1\}.$$

Define that $f \geqslant 0$ in $C(X)$ if and only if

$$\left\| \frac{f}{\|f\|} - e \right\| \leqslant 1.$$

Then $f \geqslant 0$ holds if and only if

$$f(x) \geqslant 0 \quad \text{for all } x \in G,$$
$$f(x) \leqslant 0 \quad \text{for all } x \in H.$$

Thus the proposition follows by use of an argument similar to the proof of Theorem IV.7.[1]

Example IV.11. In the preceding discussion we have derived the ring case from the lattice case to avoid repeating similar arguments. However, if we consider the former case alone, we can enjoy a simpler discussion. Let X be a compact T_2-space and $C(X)$ the ring of the continuous functions over X. We mean by an *ideal* of $C(X)$ a non-empty subset J of $C(X)$ such that f, $g \in J$ and $h \in C(X)$ imply $f + g \in J$ and $hf \in J$. We call an ideal J of $C(X)$ a *maximal ideal* if J is a proper subset of $C(X)$ and $J \subset J' \neq C(X)$ for an ideal J', implies $J = J'$.

Denoting by $I(X)$ the totality of maximal ideals of $C(X)$ we define the closure $\bar{B}$ of a subset B of $I(X)$ as follows: $J \in I(X)$ belongs to $\bar{B}$, i.e. $J \in \bar{B}$, if and only if

$$\cap \{J' \mid J' \in B\} \subset J.$$

---

[1] S. Banach [1] showed that $C(X)$ as a linear topological space fails to characterize the compact T_2-space X. C. Bessaga–A. Pelczyński [1] proved that for any two infinite compact metric spaces X_1 and X_2 the metric spaces $C(X_1)$ and $C(X_2)$ are homeomorphic. (See Example III.9.) It is interesting that the algebraic structure of $C(X)$ seems to characterize the topology of X better than the topological structure of $C(X)$ does.

Now, we can easily show that

$$I(p) = \{f \mid f(p) = 0, f \in C(X)\}$$

is a maximal ideal for each point p of X and conversely that every maximal ideal is equal to $I(p)$ for some $p \in X$. Thus we obtain a one-to-one mapping $I(p)$ of X onto $I(X)$. Moreover, we can prove that $p \in X$ and $A \subset X$ satisfy

$$p \in \bar{A} \quad \text{in } X$$

if and only if

$$I(p) \in \overline{I(A)} \quad \text{in } I(X).$$

This means that $I(X)$ with the above definition of closure is a topological space which is homeomorphic to X itself. Pursuing this discussion we can verify all of Theorems IV.7, 8 and the Corollary for the rings of continuous functions.

Suppose X is a (not necessarily compact) completely regular space. Then we denote by $I^*(X)$ the topological space of the maximal ideals of $C^*(X)$. We can easily prove that $I^*(X)$ is homeomorphic to $\beta(X)$. Thus this gives another definition of Čech–Stone's compactification.

5. Extensions of the concept of compactness

Nowadays, there is a tremendous number of extensions and modifications of the very important concept 'compact space'. We have already learned that paracompact and locally compact spaces are significant examples of these extensions and modifications. 'Lindelöf space' may also be regarded as an extension of 'compact space'. Generally speaking, most of the other miscellaneous extensions and modifications are not so important as paracompactness, but still some of them are certainly interesting enough for our attention. In the present and following sections we shall give a brief account of a few of them.

Definition IV.8. Let X be a topological space. If every sequence of points of X contains a convergent subsequence, then it is called a *sequentially compact space*. If every countable open covering of X has a finite subcovering, then it is called a *countably compact space*. If every real-valued continuous function over X is bounded, then it is called a *pseudo-compact space*.

Example IV.12. Let N be the discrete space of all natural numbers. Then $\beta(N)$ is compact but not sequentially compact, because the sequence N contains no convergent subsequence.

The other two conditions in Definition IV.8 are properly weaker than compactness. For example, R_5 in Example II.1 is countably compact, sequentially compact and pseudo-compact but not compact.

Now let us study relations among the new concepts given in Definition IV.8. Each of the following assertions is obvious or is easy to verify. Some of the proofs are left to the reader.

A) *Every compact space is countably compact.*

B) *Every sequentially compact space is countably compact.*

Proof. If X is not countably compact, then there is a countable open covering $\{U_i \mid i = 1, 2, \ldots\}$ which contains no finite subcovering. Choose points

$$p_i \in X - \bigcup_{j=1}^{i} U_j, \quad i = 1, 2, \ldots;$$

then we obtain a point sequence $\{p_i \mid i = 1, 2, \ldots\}$ which contains no convergent subsequence.

C) *Every countably compact space is pseudo-compact.*

Proof. Let f be a given continuous function over a countably compact space X. Putting

$$U_i = \{p \mid |f(p)| < i\},$$

we get a countable open covering $\{U_i \mid i = 1, 2, \ldots\}$ of X. We can cover X with finitely many of U_i, say $U_{i_1}, \ldots, U_{i_k}$. Then $|f(p)| < \max\{i_1, \ldots, i_k\}$ proving that X is pseudo-compact.

D) *A topological space X is countably compact if and only if every point sequence of X has a cluster point.*

Proof. Assume that $\{p_i \mid i = 1, 2, \ldots\}$ is a point sequence which has no cluster point. Then put

$$U_i = \bigcup\{U \mid U \text{ is an open set of } X \text{ such that } x_j \notin U \text{ for all } j > i\}.$$

It is obvious that $\{U_i \mid i = 1, 2, \ldots\}$ is an open cover of X. Observe that this cover has no finite subcover. Hence X is not countably compact. Conversely assume that X is not countably compact. Then it has a countable open cover $\{U_i \mid i = 1, 2, \ldots\}$ which has no finite subcover. Select a point sequence $\{p_i \mid i = 1, 2, \ldots\}$ such that $p_i \in X - U_1 \cup U_2 \cup \cdots \cup U_i$. Now it is obvious that $\{p_i\}$ has no cluster point.

E) *A Lindelöf space is compact if and only if it is countably compact. A paracompact space is compact if and only if it is countably compact.*

Proof. Let us prove only the 'if' part of the second statement. Suppose X is a countably compact paracompact space. To prove the compactness of X, it suffices to show that a given locally finite open cover $\mathcal{U}$ of X has a finite subcover. Assume that $\mathcal{U}$ has no finite subcover. Then we can select a point sequence $\{p_i \mid i = 1, 2, \ldots\}$ such that $p_i \in X - S(p_1, \mathcal{U}) \cup \cdots \cup S(p_{i-1}, \mathcal{U})$. Now it is obvious that $\{p_i\}$ has no cluster point, which contradicts D).

F) *A first countable space X is countably compact if and only if it is sequentially compact.*

Proof. Let $\{p_i \mid i = 1, 2, \ldots\}$ be a point sequence of X. Then it has a cluster point p. It is obvious that a subsequence of $\{p_i\}$ converges to p.

G) *A normal space is countably compact if and only if it is pseudo-compact.*

Proof. Suppose X is not countably compact. Then using the argument in the proof of B), there is a point sequence $\{p_i \mid i = 1, 2, \ldots\}$ which has no cluster point. Therefore $P = \{p_i \mid i = 1, 2, \ldots\}$ is a closed set of X. The function f defined by

$$f(p_i) = i, \quad i = 1, 2, \ldots,$$

is a continuous function on P. Hence by Tietze's extension theorem, f can be extended to a continuous function φ over X. φ is clearly unbounded, i.e. X is not pseudo-compact.

Combining E), F), G) and III.3.E), we obtain:

H) *For paracompact T_2-spaces satisfying the 1st axiom of countability, the four conditions: compactness, sequential compactness, countable compactness and pseudo-compactness coincide.*[1]

Now, let us turn to the properties of those spaces. Unfortunately, for sequentially compact, countably compact and pseudo-compact spaces, there holds no product theorem like Tychonoff's product theorem for compact spaces. For example, the product space of two countably compact spaces need not be countably compact.[2] As for hereditability, we can easily verify that each closed set of a sequentially compact space or a countably compact space is also sequentially compact or countably compact, respectively; a closed set of a pseudo-compact normal space is also pseudo-compact.

In the following are some other properties.

I) *A T_1-space X is countably compact if and only if every infinite subset of X has an accumulation point.*

Proof. Left to the reader.

J) *A Tychonoff space X is pseudo-compact if and only if every locally finite open covering of X has a finite subcovering.*[3]

Proof. Assume that X is not pseudo-compact. Then there is a real-valued, unbounded, continuous function f defined over X. Put

$$U_n = \{x \in X \mid n - 1 < |f(x)| < n + 1\}, \quad n = 0, 1, 2, \ldots.$$

Then $\{U_n \mid n = 0, 1, \ldots\}$ is a locally finite open covering of X which has no finite subcovering.

---

[1] It is no wonder that in calculus these conditions are sometimes not distinguished, because spaces used there are often metric.

[2] See H. Terasaka [1] or J. Novak [2]. See also S. Mrowka [2], Z. Frolik [1], [2], C. T. S. Scarborough and A. H. Stone [1].

[3] Due to Yu. M. Smirnov [5]. For further characterizations of pseudo-compact spaces, see I. Glicksberg [1], S. Mardésic–P. Papic [1], J. Colmez [1], K. Iseki [1], R. W. Bagley–E. H. Connell–J. D. McKnight Jr. [1]. The following theorem of I. Glicksberg [2] is especially interesting: *Let X_α, $\alpha \in A$, be Tychonoff spaces and suppose that the set $\prod \{X_\alpha \mid \alpha \neq \alpha_0\}$ is infinite for every $\alpha_0 \in A$. Then $\prod \{X_\alpha \mid \alpha \in A\}$ is pseudo-compact if and only if*

$$\beta(\prod\{X_\alpha \mid \alpha \in A\}) = \prod\{\beta(X_\alpha) \mid \alpha \in A\}.$$

Conversely, assume that $\mathcal{U}$ is a locally finite open covering of X which has no finite subcovering. Let $U_1, U_2, \ldots$ be distinct non-empty members of $\mathcal{U}$. Select $x_i \in U_i$, $i = 1, 2, \ldots$, and define real-valued continuous functions f_i such that

$$f_i(x_i) = i, \qquad f_i(X - U_i) = 0, \qquad 0 \leqslant f_i \leqslant i.$$

Put

$$f = \max \{f_i \mid i = 1, 2, \ldots\}.$$

Then, since $\mathcal{U}$ is locally finite, f is continuous and besides it is unbounded. Thus X is not pseudo-compact.

Finally, let us discuss spaces whose topologies are determined by the compact subsets.[1]

Definition IV.9. A topological space is called a *k-space* if a subset is closed if and only if its intersection with every compact closed set is closed.

So in a Hausdorff k-space a subset is closed if and only if its intersection with every compact set is compact. It is interesting to see that the collection of Hausdorff k-spaces contains a considerably wide class of spaces, as implied by the following propositions. It is also of interest that the topologies of such general spaces are completely determined by such a special class of sets as the compact sets.

K) *Every locally compact T_2-space is a k-space. Every T_2-space satisfying the 1st axiom of countability is a k-space.*

Proof. The proof of the first assertion is left to the reader. Let X be a T_2-space satisfying the 1st axiom of countability and A a non-closed subset of X. Then there is a point $p \in \bar{A} - A$. Since X satisfies the 1st axiom, we can choose a point sequence $\{p_i \mid i = 1, 2, \ldots\}$ from A such that $p_i \to p$. Then

$$P = \{p_i \mid i = 1, 2, \ldots\} \cup \{p\}$$

is a compact subset of X, but

$$P \cap A = \{p_i \mid i = 1, 2, \ldots\}$$

[1] See R. Arens [1]. In this connection A. Arhangelski [3], [4] got interesting results.

is non-compact. On the other hand it is obvious that if F is a closed set of X, then its intersection with each compact set of X is compact. Thus X is a k-space.[1]

It is known that a subspace of a k-space does not need to be a k-space and that the product of two k-spaces may fail to be a k-space. However, we can prove that the product of a T_2 k-space and a locally compact T_2-space is a k-space. The following proposition shows a relationship between k-spaces and general topological spaces:

For an arbitrary T_2 topological space X, there is a k-space X' and a continuous, one-to-one mapping f of X' onto X such that f is a homeomorphism on every compact set of either X' or X. Moreover, such a space X' is uniquely determined.

The following proposition[2] is also interesting.

L) *A T_2-space X is a k-space if and only if it is homeomorphic to a decomposition space of a locally compact space.*

Proof. Suppose a T_2-space X is a decomposition space of a locally compact space Z with the natural (quotient) mapping φ. Let F be a subset of X whose intersection with each compact (and accordingly closed) set is closed. To prove that $\varphi^{-1}(F)$ is closed we assume the contrary; then there is a point

$$q \in \overline{\varphi^{-1}(F)} - \varphi^{-1}(F).$$

Take a compact nbd V of q. Then from the continuity of φ we can easily prove that

$$\varphi(q) \in \overline{\varphi(V) \cap F} - \varphi(V) \cap F,$$

i.e. $\varphi(V) \cap F$ is not closed, but this contradicts the property of F since $\varphi(V)$ is compact (and accordingly closed). Therefore $\varphi^{-1}(F)$ is closed in Z, and hence F is closed in X. This proves that X is a k-space.

---

[1] A remarkable property of 1st countable spaces is that their topology is determined in terms of the convergence of point sequences. Namely, a set of a 1st countable space is open if and only if every point sequence converging to a point in the set is residual in it. But conversely this condition does not imply 1st axiom; spaces satisfying the condition and similar spaces were investigated by S. Franklin and A. V. Arhangelskii. S. Franklin [1], for example, defined a *sequential space* as a space in which a set U is open if and only if every point sequence converging to a point of U is, itself, residual in U. He proved that a space is sequential if and only if it is the quotient space of a metric space.

[2] Due to D. E. Cohen [1].

Conversely, let X be a k-space. We denote by $\{K_\alpha \mid \alpha \in A\}$ the collection of all compact closed sets of X. Let H_α be a compact space homeomorphic to K_α and denote by φ_α the topological mapping of H_α onto K_α. We assume $H_\alpha \cap H_{\alpha'} = \emptyset$ for $\alpha \neq \alpha'$. Let Z be the discrete sum of H_α, $\alpha \in A$. Then Z turns out to be a locally compact space. Define a mapping φ of Z onto X by $\varphi(q) = \varphi_\alpha(q)$ if $q \in H_\alpha$. Then we can easily see that a subset F of X is closed if and only if $\varphi^{-1}(F)$ is closed in Z, because X is a k-space. Thus X is a decomposition space of the locally compact space Z.

Example IV.13. Let X be a completely regular space. If X is compact, then it is closed in every completely regular space which contains X as a subspace. Conversely, if X is not compact, then X is a non-closed subset of the completely regular space $\beta(X)$. In other words, a completely regular compact space is characterized as an absolutely closed space in completely regular spaces. This fact suggests to us another way to modify compactness. Generally, suppose (P) is a property for topological spaces. If a topological space X with (P) is always closed in any topological space which satisfies (P) and contains X as a subset, then X is called (P)-*closed or absolutely* (P)-*closed*. Thus completely regular-closed means completely regular and compact. T_2-*closed* and *regular-closed* are conditions somewhat different from compactness. In Example IV.3, we have nearly shown that R_3 is regular-closed, though it is non-compact. (Do not mix up 'regular-closed space' with 'regular closed set' in Exercise II.5.)

We can further extend these concepts to define absolutely F_σ-space and absolutely Borel-space, which were investigated for metric spaces by C. Kuratowski and by A. H. Stone [5], [6], [7]. For example, Stone proved that a metric space is absolutely F_σ (for metric spaces) if and only if it is a countable union of locally compact spaces.[1]

6. Realcompact space

In the present section we give a brief account of realcompact space, a considerably important generalization of compact space.[2]

---

[1] Another direction of generalization of compact space is to consider a condition like: Every open covering $\mathcal{U}$ of X with $|\mathcal{U}| \leq b$ has a sub-covering $\mathcal{U}'$ with $|\mathcal{U}'| < a$, where a and b are any cardinal numbers such that $a \leq b$. J. E. Vaughan [1] gives a good survey to results in this aspect.

[2] We owe the definition and early developments of realcompact space to E. Hewitt [3].

Definition IV.10. A topological space X is called a *realcompact space* (or *Q-space*) if it is homeomorphic to a closed set of the product space of copies of the real line E^1.

In the following are properties of realcompact spaces which follow directly from the definition.

A) *Every compact T_2-space is realcompact.*

Proof. See the corollary to Theorem IV.1.

B) *Every realcompact space is a Tychonoff space.*

C) *Every closed set of a realcompact space is realcompact.*

D) *The product space of arbitrarily many realcompact spaces is realcompact.*

Definition IV.11. Let Y be a subset of a topological space X. If every real-valued (bounded) continuous function on Y can be extended to a continuous function on X, then we say that Y is *C-embedded* (*C*-embedded*) in X.

Example IV.14. Every Tychonoff space X is C^*-embedded in $\beta(X)$. Every closed set F of a normal space X is C-embedded in X.

Let X be a given Tychonoff space. (In the rest of this section all spaces are at least Tychonoff.) We denote by $C(X)$ and $C^*(X)$ the ring of all real-valued continuous functions on X and the ring of all real-valued bounded continuous functions on X, respectively. We also let $\tilde{C}(X) = C(X) - C^*(X)$ in this section. Namely $\tilde{C}(X)$ is the set of all unbounded continuous functions on X.

Now, we are going to construct a realcompactification $\gamma(X)$ of X, by use of a method similar to the construction of $\beta(X)$. This time we use the product of real lines in place of the product of unit segments used for $\beta(X)$.

Let $C(X) = \{f_\alpha \mid \alpha \in A\}$. Then we define a mapping f from X into $Q = \prod \{R_\alpha \mid \alpha \in A\}$, where each R_α is a copy of the real line E^1, by

$$f(x) = \{f_\alpha(x) \mid \alpha \in A\}, \quad x \in X.$$

(Note that f_α is regarded as a mapping from X into R_α.) Then it is easy to see that f is a homeomorphism between X and $f(X) \subset Q$. Now identify X and $f(X)$ to regard X as a subset of Q. Then each point x of X can be expressed as

$$x = \{f_\alpha(x) \mid \alpha \in A\}.$$

Then $\gamma(X) = \bar{X}$ (the closure of X in Q) is obviously a realcompact space and contains X as a dense subset. It is also easy to see that X is C-embedded in $\gamma(X)$ (actually in Q).

Let

$$C^*(X) = \{f_\alpha \mid \alpha \in A^*\}, \qquad \tilde{C}(X) = \{f_\alpha \mid \alpha \in \tilde{A}\}.$$

To each $\alpha \in \tilde{A}$ we assign Alexandroff's one-point compactification $S_\alpha = R_\alpha \cup \{\infty_\alpha\}$ of R_α. (Namely, S_α is a circle.) To each $\alpha \in A^*$ we assign a closed segment I_α such that $I_\alpha \supset f_\alpha(X)$. Then denote by $\hat{X}$ the closure of X in $Q' = \prod_{\alpha \in A^*} I_\alpha \times \prod_{\alpha \in \tilde{A}} S_\alpha$. Now, it is easy to see that

$$\gamma(X) \subset \hat{X} = \beta(X).$$

The last equality follows from the uniqueness of Čech–Stone's compactification, because $\hat{X}$ is a compact T_2-space containing X as a dense subset, and X is C^*-embedded in $\hat{X}$. (The detailed proof is left to the reader.)

Let $f \in C(X)$; then f can be regarded as a continuous mapping from X into the circle $S^1 = E^1 \cup \{\infty\}$. Then by Corollary 1 to Theorem IV.2 we can extend f to a continuous mapping from $\beta(X)$ into S^1. In the rest of this section we denote the (unique) extension of f by f^*.

E) $\gamma(X) = \{x \in \beta(X) \mid f^*(x) \neq \infty \text{ for every } f \in C(X)\}$.

Proof. Let $x \in \gamma(X)$ and $f \in C(X)$. Then, since X is C-embedded in $\gamma(X)$, $f^*(x) \neq \infty$.

Conversely, let $x \in \beta(X) - \gamma(X)$. Then $x \in Q' - Q$, because $\gamma(X)$ and $\beta(X)$ are closures of X in Q and in Q', respectively. Since

$$x \in Q' = \prod_{\alpha \in A^*} I_\alpha \times \prod_{\alpha \in \tilde{A}} S_\alpha,$$

we can express it as $x = \{x_\alpha \mid \alpha \in A\}$, where $A = A^* \cup \tilde{A}$. Since $x \notin Q$, $x_\alpha \in S_\alpha - R_\alpha$ for some $\alpha \in \tilde{A}$. This means that $f_\alpha^*(x) = \infty$.

F) *Let X be a Tychonoff space and $\gamma'(X)$ a realcompact space such that X is dense and C-embedded in $\gamma'(X)$. Let f be a continuous mapping from X into a realcompact space Y. Then f can be extended to a continuous mapping from $\gamma'(X)$ into Y.*

Proof. By Definition IV.10 Y may be regarded as a closed set of $\Pi\{R_\beta \mid \beta \in B\}$, where each R_β is a copy of the real line E^1. Denote by π_β the projection from Y to R_β. We can extend $\pi_\beta \circ f$ to a continuous function g_β from $\gamma'(X)$ into R_β. Now, define a mapping g from $\gamma'(X)$ into ΠR_β by

$$g(x) = \{g_\beta(x) \mid \beta \in B\}, \quad x \in \gamma'(X).$$

Then

$$g(\gamma'(X)) = g(\bar{X}) \subset \overline{g(X)} = \bar{Y} = Y.$$

Thus g is the desired extension of f.

Theorem IV.10.[1] *Let X be a Tychonoff space. Then there is a realcompact space $\gamma(X)$ such that*

(i) *X is a dense subset of $\gamma(X)$,*

(ii) *every real-valued continuous function over X can be extended to a continuous function over $\gamma(X)$.*

Such a space $\gamma(X)$ is uniquely determined by X and called the real-compactification of X.

Proof. To prove the uniqueness of $\gamma(X)$, let X and X' be homeomorphic Tychonoff spaces and $\gamma(X)$ and $\gamma'(X')$ realcompact spaces satisfying conditions (i) and (ii) for X and X', respectively. Let ι be a homeomorphism from X onto X' and denote by i, i', $\hat{\imath}$, $\hat{\imath}'$ the identity mappings of X, X', $\gamma(X)$ and $\gamma'(X')$, respectively. Now by F) we extend ι and ι^{-1} to $\hat{\iota}$ and $(\iota^{-1})\hat{\ }$, respectively, where $\hat{\iota}$ is a continuous mapping from $\gamma(X)$ into $\gamma'(X')$ and $(\iota^{-1})\hat{\ }$ from $\gamma'(X')$ into $\gamma(X)$. Then $(\iota^{-1})\hat{\ } \circ \hat{\iota}$ and $\hat{\iota} \circ (\iota^{-1})\hat{\ }$ are continuous extensions of i and i', respectively. Thus

$$(\iota^{-1})\hat{\ } \circ \hat{\iota} = \hat{\imath}, \qquad \hat{\iota} \circ (\iota^{-1})\hat{\ } = \hat{\imath}',$$

[1] Due to E. Hewitt [3].

which proves that $\hat{\iota}$ is a topological mapping from $\gamma(X)$ onto $\gamma'(X')$ such that $\hat{\iota} = \iota$ on X.

Corollary 1. $\gamma(X) = X$ *holds if and only if* X *is realcompact.*

Corollary 2. *A* T_2-*space* X *is compact if and only if it is pseudo-compact and realcompact.*

Proof. Assume that X is pseudo-compact and realcompact. Then, since $C(X) = C^*(X)$, from E) it follows that $\gamma(X) = \beta(X)$. From Corollary 1 we obtain $X = \beta(X)$, i.e. X is compact.

Example IV.15. The space R_5 in Example II.1 is pseudo-compact but not compact. Thus it is not realcompact either. On the other hand, a product space of arbitrarily many copies of E^1 and especially E^1 itself are realcompact but neither compact nor pseudo-compact.

Definition IV.12. A maximal zero-filter $\mathscr{Z}$ of a space X is called a *real filter* if it has *c.i.p.*, i.e. $\bigcap_{i=1}^{\infty} Z_i \neq \emptyset$ for every $Z_i \in \mathscr{Z}$, $i = 1, 2, \ldots$. Let $f \in C(X)$; then we define

$$Z(f) = \{x \in X \mid f(x) = 0\}, \qquad P(f) = \{x \in X \mid f(x) \geq 0\}.$$

Let J be a maximal ideal of $C(X)$; then we define

$$\mathscr{Z}(J) = \{Z(f) \mid f \in J\}.$$

Let $\mathscr{Z}$ be a maximal zero-filter of X; then we define

$$J(\mathscr{Z}) = \{f \in C(X) \mid Z(f) \in \mathscr{Z}\}.$$

G) *Let* J *be a maximal ideal of* $C(X)$. *If* $f \in J$, $g \in C(X)$ *and* $Z(f) = Z(g)$, *then* $g \in J$.

Proof. The easy proof is left to the reader.

H) *Let* J *be a maximal ideal of* $C(X)$. *Then* $\mathscr{Z}(J)$ *is a maximal zero-filter of* X.

Proof. It is easy to see that $\mathscr{Z}(J)$ is a zero-filter, so the proof is left to the reader. Suppose $\mathscr{Z}'$ is a zero-filter such that $\mathscr{Z}' \supsetneq \mathscr{Z}(J)$. Then select $Z \in \mathscr{Z}' - \mathscr{Z}(J)$. Let $Z = Z(f)$ and $f \in C(X)$. Then $f \notin J$. Put

$$J' = \{h + \varphi f \mid h \in J, \varphi \in C(X)\}.$$

Then J' satisfies $f \in J' \supset J$. Thus $J' \neq J$. Since J is a maximal ideal, this implies $J' = C(X)$. Thus

$$1 = h + \varphi f \quad \text{for some } h \in J \text{ and } \varphi \in C(X).$$

Hence

$$Z(h) \cap Z(f) = \emptyset,$$

which contradicts that $Z(h), Z(f) \in \mathscr{Z}'$. Therefore $\mathscr{Z}' = \mathscr{Z}(J)$, proving that $\mathscr{Z}(J)$ is maximal.

I) *Let $\mathscr{Z}$ be a maximal zero-filter of X. Then $J(\mathscr{Z})$ is a maximal ideal of $C(X)$.*

Proof. It is easy to prove that $J(\mathscr{Z})$ is an ideal, so it is left to the reader. To see that it is maximal, let J' be an ideal such that $J' \supset J(\mathscr{Z})$. Suppose $f \in J'$ and $Z \in \mathscr{Z}$. Then $Z = Z(g)$ for some $g \in J(\mathscr{Z})$. Thus

$$Z(f) \cap Z = Z(f) \cap Z(g) \neq \emptyset,$$

because otherwise $J' = C(X)$ would follow. Since $\mathscr{Z}$ is maximal, $Z(f) \in \mathscr{Z}$ follows, i.e. $f \in J(\mathscr{Z})$. Hence $J' = J(\mathscr{Z})$, proving that $J(\mathscr{Z})$ is maximal.

J) $\mathscr{Z}(J(\mathscr{Z})) = \mathscr{Z}$ *for every maximal zero-filter $\mathscr{Z}$ of X. $J(\mathscr{Z}(J)) = J$ for every maximal ideal J of $C(X)$.*

Proof. Left to the reader.

Now, we consider a maximal ideal J of $C(X)$ and its quotient ring $C(X)/J$. Then we define an order $\geqslant$ in $C(X)/J$ by

$$\bar{f} \geqslant \bar{g} \quad \text{if and only if} \quad \bar{f} - \bar{g} = \bar{h}^2 \text{ for some } \bar{h} \in C(X)/J.$$

Generally we denote by $\bar{f}$ the element of $C(X)/J$ represented by $f \in C(X)$.

K) $\bar{f} \geqslant \bar{0}$ *holds in $C(X)/J$ if and only if $f - |f| \in J$.*

Proof. First observe that $|g| \leq |f|$ and $g \in J$ imply $f \in J$ because of G) and H). Let $\bar{f} \geq \bar{0}$; then

$$\bar{f} = \bar{h}^2 = \overline{h^2} \quad \text{for some } h \in C(X),$$

i.e. $f - h^2 \in J$. Hence

$$|f - h^2| \geq ||f| - h^2|.$$

Thus from the above observation it follows that $|f| - h^2 \in J$. Thus

$$f - |f| \in J.$$

The converse is obvious.

L) $\bar{f} \geq \bar{0}$ *holds in* $C(X)/J$ *if and only if* $P(f) \in \mathcal{L}(J)$.

Proof. $\bar{f} \geq \bar{0}$ if and only if $f - |f| \in J$ if and only if

$$P(f) = Z(f - |f|) \in \mathcal{L}(J).$$

M) $C(X)/J$ *is totally ordered.*

Proof. It is easy to prove that $C(X)/J$ is partially ordered by the above defined order, so it is left to the reader. Now, let $f \in C(X)$; then

$$(f - |f|)(f + |f|) = 0 \in J.$$

Thus

$$Z(f - |f|) \cup Z(f + |f|) = Z((f - |f|)(f + |f|)) \in \mathcal{L}(J).$$

Thus

$$Z(f - |f|) \in \mathcal{L}(J) \quad \text{or} \quad Z(f + |f|) \in \mathcal{L}(J),$$

because $\mathcal{L}(J)$ is maximal. Namely,

$$f - |f| \in J \quad \text{or} \quad f + |f| \in J.$$

Hence by K)

$$\bar{f} \geq \bar{0} \quad \text{or} \quad -\bar{f} \geq \bar{0},$$

proving that $C(X)/J$ is totally ordered.

N) *The mapping* $\theta(r) = \bar{r}$, $r \in R$, *is an order-preserving, one-to-one homomorphism from the field R of all real numbers into the ring* $C(X)/J$.

Proof. Assume $\theta(r) = \bar{r} = \bar{0}$ in $C(X)/J$. Then $r \in J$. Since $J \subsetneqq C(X)$, $r = 0$ follows. Namely, θ is one-to-one. It is obvious that θ is order-preserving.

Definition IV.13. Generally, let A be a totally ordered field such that the set N of all natural numbers is contained in A. If there is $r \in A$ such that $r \geq n$ for all $n \in N$, then A is called *non-Archimedean*. Otherwise A is called *Archimedean*.

We need the following facts in algebra, whose easy proofs are left to the reader.

Remarks. (1) Let A be a commutative ring with 1 and J its maximal ideal; then the quotient ring A/J is a field. (Thus $C(X)/J$ is a field.)

(2) Every Archimedean field is isomorphic to a subfield of the real field R.

(3) Let φ be a one-to-one homomorphism from the real field R into itself. Then φ is the identity mapping of R.

Definition IV.14. A maximal ideal J of $C(X)$ is called a *real ideal* if $C(X)/J$ is isomorphic to the real field R. An ideal J of $C(X)$ is called *fixed* to a point x of X if

$$x \in \cap \{Z(f) \mid f \in J\}.$$

Example IV.16. Let x be a point of a Tychonoff space X. Then

$$J_x = \{f \in C(X) \mid f(x) = 0\}$$

is obviously a real ideal of $C(X)$ and fixed to x.

On the other hand, consider the spaces R_5 and R_6 in Example IV.2. Put

$$J = \{f \in C(R_5) \mid \hat{f}(\omega_1) = 0\},$$

where $\hat{f}$ denotes the extension of f over R_6. Then it is easy to see that J is a real ideal, because for each $f \in C(R_5)$ there is a real number $r(f)$ such that

$$f([\alpha, \omega_1)) = r(f) \quad \text{for some } \alpha < \omega_1,$$

and the mapping which maps $\bar{f}$ to $r(f)$ is an isomorphism between $C(R_5)/J$ and R. But J is fixed to no point of R_5.

O) *A maximal ideal J of $C(X)$ is real if and only if $C(X)/J$ is Archimedean.*

Proof. The necessity of the condition is obvious.

Assume that $C(X)/J$ is Archimedean. By Remark (2) there is an isomorphism φ from $C(X)/J$ to a subfield of R. Let ψ be the natural homomorphism from R into $C(X)/J$. Then $\varphi \circ \psi$ is a one-to-one homomorphism from R into R. Thus by Remark (3) $\varphi \circ \psi$ is onto, and so is φ, proving that J is a real ideal.

P) *A maximal ideal J of $C(X)$ is real if and only if $\mathscr{Z}(J)$ is real.*

Proof. Assume that $\mathscr{Z}(J)$ is not real. Then there are $f_n \in J$, $n = 1, 2, \ldots$, satisfying

$$\bigcap_{n=1}^{\infty} Z(f_n) = \emptyset .$$

Define

$$g_n = \min\{|f_n|, 1/2^n\}, \qquad g = \sum_{n=1}^{\infty} g_n .$$

Then $g \geq 0$, $g \in C(X)$, and hence $\bar{g} \geq \bar{0}$ in $C(X)/J$.

On the other hand,

$$Z(g) = \bigcap_{n=1}^{\infty} Z(f_n) = \emptyset ,$$

and hence $\bar{g} > \bar{0}$. For each natural number n, we obtain

$$g(x) \leq 2^{-n} \quad \text{on } Z(f_1) \cap \cdots \cap Z(f_n) \in \mathscr{Z}(J) .$$

Thus by L) $\bar{g} \leq (\bar{2})^{-n}$, i.e. $(\bar{g})^{-1} \leq \bar{2}^n$, $n = 1, 2, \ldots$, in $C(X)/J$. Hence $C(X)/J$ is non-Archimedean.

Conversely, assume that J is not real, i.e. $C(X)/J$ is non-Archimedean because of O). Thus there is $f \in C(X)$ such that

$$\bar{f} \geqslant \bar{n}, \quad n = 1, 2, \ldots .$$

Then
$$\bigcap_{n=1}^{\infty} P(f - n) = \emptyset,$$

while $P(f - n) \in \mathscr{L}(J)$, $n = 1, 2, \ldots$, because of L). Thus $\mathscr{L}(J)$ is not real.

Q) *Let $y \in \beta(X)$. Then, since $\beta(X) = \sigma(X, \mathscr{L})$, y may be regarded as a maximal zero-filter of X, which we denote by $\mathscr{L}_y$. Then $y \in \gamma(x)$ if and only if $\mathscr{L}_y$ is real.*

Proof. Suppose $\mathscr{L}_y$ is not real. Then there are $Z_i \in \mathscr{L}_y$, $i = 1, 2, \ldots$, such that $\bigcap_{i=1}^{\infty} Z_i = \emptyset$. Let $Z_i = Z(f_i)$, where $f_i \in C(X)$. Then construct $g \in C(X)$ as in the proof of P). Then $1/g \in C(X)$, and for every natural number m

$$1/g \geqslant 2^m \quad \text{on } Z_1 \cap \cdots \cap Z_m .$$

Since $y \in \overline{Z_1 \cap \cdots \cap Z_m}$ holds in $\beta(X)$, we obtain

$$(1/g)^*(y) = \infty,$$

proving that $y \notin \gamma(X)$.

Conversely, suppose that

$$y \in \beta(X) - \gamma(X).$$

Then there is $f \in C(X)$ such that $f^*(y) = \infty$. Put

$$P_n = \{x \in X \mid |f(x)| \geqslant n\}, \quad n = 1, 2, \ldots .$$

Then $P_n \in \mathscr{L}_y$ is obvious, and so is $\bigcap_{n=1}^{\infty} P_n = \emptyset$. Thus $\mathscr{L}_y$ is not real.

The following theorem is due to E. Hewitt [3] and T. Shirota [1].[1]

Theorem IV.11. *For a Tychonoff space X the following conditions are equivalent:*

　(i) *X is realcompact,*

　(ii) *every real filter $\mathscr{L}$ of X converges,*

　(iii) *every real ideal J of $C(X)$ is fixed to a point of X.*

---

[1] To be precise, Hewitt proved (ii) ⇔ (iii), and Shirota (i) ⇔ (ii).

Proof. X is realcompact if and only if $\gamma(X) = X$ because of Corollary 1 to Theorem IV.10. Now, assume the condition (ii); then from Q) it follows that $\gamma(X) = X$, i.e. (i) follows. Conversely, if $\gamma(X) = X$, and if $\mathscr{X}$ is a real filter of X, then $\mathscr{X} = \mathscr{X}_y$ for some $y \in \beta(X)$. By Q),

$$y \in \gamma(X) = X.$$

Thus $\mathscr{X} \to y$ in X. Therefore (i) and (ii) are equivalent.

From J) and P) it follows that (ii) and (iii) are also equivalent.

Corollary 1. *Two realcompact spaces X and Y are homeomorphic if and only if $C(X)$ and $C(Y)$ are ring isomorphic.*

Proof. It follows from (i)⇔(iii) in the above theorem, by use of the method given in Example IV.11.

Corollary 2. *Every regular Lindelöf space is realcompact.*

Proof. It follows from (i)⇔(ii) in the above theorem.

Example IV.17. Let S be the Sorgenfrey line. Then, as proved before, S is Lindelöf while $S \times S$ is not Lindelöf. Thus by D) $S \times S$ is realcompact.

M. Katětov [2] proved that a paracompact T_2-space X is realcompact if and only if every closed discrete subspace of X is realcompact.[1] A discrete space Y is realcompact if and only if $|Y|$ is *non-measurable*. A cardinal number Y is called non-measurable if every countably additive (at most) two-valued measure m on 2^Y satisfying $m(\{y\}) = 0$ for every $y \in Y$ is the trivial measure 0. It is easy to see that $\mathfrak{a}$ and $\mathfrak{c}$, e.g. are non-measurable. It is also known that the statement 'Every cardinal number is non-measurable' is consistent with the ordinary axioms of set theory. Thus the discrete space R_d of all real numbers is realcompact, while there is a continuous one-to-one mapping from R_d onto R_S, which is not realcompact.[2]

---

[1] A proof will be given later in VIII.4.

[2] It was proved by V. Ponomarev [7] and Z. Frolik [4] that the image of a realcompact space by a perfect mapping is not necessarily realcompact. Thus realcompactness is quite different in this respect from compactness, which is preserved by a continuous mapping.

For further results on realcompact spaces, see e.g. I. Glicksberg [1], M. Henriksen–L. Gillman [1], G. Aquaro [1], T. Ishiwata [1], W. Comfort [3], M. Hušek [1], W. McArthur

Exercise IV

1. A topological space X is compact if and only if every maximal closed filter of X converges.

2. Let $\mathscr{H}$ be a closed subbasis of a topological space X. Then X is compact if and only if every subcollection of $\mathscr{H}$ with f.i.p. has a non-empty intersection (J. Alexander's theorem). Use this theorem to prove Theorem IV.1.

3. The product space $P = \Pi\{D_\alpha \mid \alpha \in A\}$ of the two point discrete spaces D_α is compact, but it is not sequentially compact unless the cardinal number of A is less than $\mathfrak{c}$.

4. (i) If X is dense and C^*-embedded in a Tychonoff space Y, then $Y \subset \beta(X)$.
   (ii) Let F be a closed subset of a normal space X. Then $\beta(F) = \bar{F}$, where $\bar{F}$ denotes the closure of F in $\beta(X)$.

5. $\beta(R_5 \times R_5) = R_6 \times R_6$ (cf. Example IV.2).

6. $\alpha(X)$ is metrizable (i.e., it is homeomorphic to a metric space) if and only if X is a locally compact T_2-space satisfying the second axiom of countability.

7. Let X be a Tychonoff space. Then there is a normal base $\mathscr{B}$ of X such that $w(\sigma(X, \mathscr{B})) = w(X)$.

8. $\beta(X)$ is connected if and only if X is connected.

9. $\beta(X)$ is extremely disconnected if and only if X is extremely disconnected.

10. A completely regular space X is locally compact if and only if it is open in every T_2-compactification if and only if it is open in some T_2-compactification.

(footnote continued from p. 184)

[1], H. Ohta [1], [2], A. Mysior [1]. R. Engelking–S. Mrowka [1] generalized the concept of realcompact space as follows. Let E be a space; then a space is called an *E-compact space* if it is homeomorphic to a closed subset of the product of copies of E. See S. Mrowka [4], [5] on this aspect. If $E = I$, E^1, $\{0, 1\}$ with the discrete topology, then E-compact means compact T_2, realcompact, totally disconnected and compact T_2, respectively. One of the other interesting cases is N-compact space, where N is the discrete space of all natural numbers. See e.g. P. Nyikos [1]. To the reader who is interested in more extensive study of realcompact spaces, we also recommend L. Gillman–M. Jerison's [1] book and J. v. d. Slot's [1] survey article.

11. Let f be a bounded continuous function defined over a T_1-space X. Then f can be continuously extended over the Wallman's compactification $\omega(X)$ of X.

12. Let X be a T_1-space, A a subset of $\omega(X)$ and p a point of $\omega(X)$. Then $p \in \bar{A}$ in $\omega(X)$ if and only if $\cap\{p' \mid p' \in A\} \subset p$. In the last inequality we regard p and p' as closed collections in X (i.e. maximal closed filters).

13. Let X be a T_1-space and p a point of $\omega(X)$. If F is a closed set of X satisfying $F \notin p$, then

$$U(F) = \{q \mid q \in \omega(X); F \notin q\}$$

is an open nbd of p and $\{U(F) \mid F$ is a closed set of X satisfying $F \notin p\}$ forms a nbd basis of p in $\omega(X)$.

14. Reprove 2.F) by use of Theorem IV.2.

15. Let X be a cofinite space with infinitely many points. Then X is a compactification of each of its infinite subsets Y but no Shanin's compactification unless $X = Y$.

16. Let X be a compact T_2-space and $C(X)$ the ring of all real-valued continuous functions over X. Then every maximal ideal I of $C(X)$ has the form

$$I(p) = \{f \mid f(p) = 0, f \in C(X)\}$$

for some point p of X. (See Example IV.11. Hint: If the assertion is false, then for every point $p \in X$, there is $f_p \in I$ such that $f_p(p) \neq 0$. Thus we can find an open nbd $U(p)$ of p in which $f_p^2 > 0$.)

17. Discuss a method to define $\beta(X)$ by use of the ring $C^*(X)$ of all real-valued, bounded, continuous functions on the Tychonoff space X.

18. Let $C'(X)$ denote the ring of all complex-valued continuous functions on a Tychonoff space X. Then two compact T_2-spaces X and Y are homeomorphic if and only if $C'(X)$ and $C'(Y)$ are ring isomorphic.

19. The product space of countably many sequentially compact spaces is sequentially compact.

20. X is pseudo-compact if and only if every countable cover of X by cozero sets has a finite subcover.

21. A Tychonoff space X is pseudo-compact if and only if every maximal ideal of $C(X)$ is real.

22. Let X and Y be T_1-spaces. If X is compact or sequentially compact, and Y is countably compact, then $X \times Y$ is countably compact.

23. Let X be compact or sequentially compact and Y pseudo-compact. Then $X \times Y$ is pseudo-compact.

24. A sequential T_1-space is sequentially compact if and only if it is countably compact.

25. A mapping of a k-space X into a topological space Y is continuous if and only if it is continuous over every compact set of X.

26. The product space of a locally compact T_2-space and a k-T_2-space is a k-space.

27. Let X be the quotient space obtained from the real line by identifying all integers. Then X is sequential but not first countable (due to S. Franklin).

28. Every sequential T_2-space is a k-space.

29. Let f be a perfect map from a Tychonoff space X onto Y. If Y is realcompact, then so is X. (It is known that the perfect image of a realcompact space is not necessarily realcompact.)

30. Let X and Y be Tychonoff spaces in which every singleton is a G_δ-space. Then X and Y are homeomorphic if and only if $C(X)$ and $C(Y)$ are ring isomorphic.

CHAPTER V

PARACOMPACT SPACES AND RELATED TOPICS

The purpose of this chapter is to give a rather detailed account of the theory of paracompact spaces. Since the concept of paracompactness was invented by J. Dieudonné [1] and the fundamental theorem by A. H. Stone [1] was established soon after, there has been a great development in the theory of this class of spaces, and now it is a major condition, frequently used in topology. A significance of this newer category of spaces is in the fact that it contains two important classical categories, compact spaces and metric spaces, as special cases and still it is concrete enough to allow fruitful theories in itself.

1. Fundamental theorem

In the present chapter we need some new terminologies for coverings and collections which recently have proved to be powerful tools for study not only of paracompact spaces but also of metric spaces.

Definition V.1. Let $\mathcal{U}$ be a collection in a topological space X. If every point p of X has a nbd which intersects at most one member of $\mathcal{U}$, then $\mathcal{U}$ is called *discrete*. If every member U of $\mathcal{U}$ intersects at most finitely many (countably many) members of $\mathcal{U}$, then $\mathcal{U}$ is called *star-finite* (*star-countable*). If for every subcollection $\mathcal{U}'$ of $\mathcal{U}$,

$$\bigcup\{\bar{U} \mid U \in \mathcal{U}'\} = \overline{\bigcup\{U \mid U \in \mathcal{U}'\}},$$

then $\mathcal{U}$ is called *closure-preserving*.

Let $\mathcal{U}$ and $\mathcal{V}$ be collections; then we say that $\mathcal{V}$ is *cushioned in* $\mathcal{U}$ if we can assign to each $V \in \mathcal{V}$ a $U(V) \in \mathcal{U}$ such that for every subcollection $\mathcal{V}'$ of $\mathcal{V}$,

$$\overline{\bigcup\{V \mid V \in \mathcal{V}'\}} \subset \bigcup\{U(V) \mid V \in \mathcal{V}'\}.$$

We often use the above terminologies for coverings, i.e., we use terminologies like star-finite covering, closure-preserving covering, etc. In particular, we call a covering $\mathcal{U}$ a *cushioned refinement* of a covering $\mathcal{V}$ if $\mathcal{U}$ is cushioned in $\mathcal{V}$.

Let $\mathcal{U}$ be a collection (covering) such that

$$\mathcal{U} = \bigcup_{i=1}^{\infty} \mathcal{U}_i, \quad i = 1, 2, \ldots,$$

where each $\mathcal{U}_i$ is a locally finite collection. Then $\mathcal{U}$ is called a *σ-locally finite collection* (*covering*). In a similar way we can define *σ-discrete, σ-closure-preserving, σ-star-finite*, etc.

Example V.1. It is clear that discrete is the strongest among the conditions in Definition V.1, and that relationships between them are given by the following scheme (see II.5.C)):

discrete $\Rightarrow$ star-finite $\Rightarrow$ locally finite $\Rightarrow$ closure preserving $\Rightarrow$ cushioned.

In this scheme the implication 'star-finite $\Rightarrow$ locally finite' is valid only for open coverings, and 'closure-preserving $\Rightarrow$ cushioned' should be understood to mean the following: If $\mathcal{U}$ is a closure-preserving collection, then $\mathcal{U}$ is cushioned in $\bar{\mathcal{U}}$. We can easily show that those conditions are essentially different from each other. For example, in E^1 $\{(-n, n) \mid n = 1, 2, \ldots\}$ is a closure-preserving, but non-locally finite open covering.

To prove A. H. Stone's fundamental theorem in an efficient manner, we need the following Theorem V.1, due to E. Michael [1], which is interesting in its own right. Let us begin with Proposition A) which is a lemma for Theorem V.1.

A) *A regular space X is paracompact if and only if every open covering of X has a locally finite (not necessarily open) refinement.*

Proof. It suffices to prove only the sufficiency of the condition. Let X be a regular space satisfying the condition, and $\mathcal{U}$ a given open covering of X. Then by use of the hypothesis, we can find a locally finite (not necessarily open) covering $\mathcal{A}$ such that $\mathcal{A} < \mathcal{U}$. Since $\mathcal{A}$ is locally finite, there is an open covering $\mathcal{P}$ each of whose members intersects only finitely many members of $\mathcal{A}$. Since X is regular, we can construct an open covering $\mathcal{Q}$ with $\bar{\mathcal{Q}} < \mathcal{P}$. Again by use of the hypothesis, we can construct a locally finite covering $\mathscr{B}$ with $\mathscr{B} < \mathscr{D}$. Then $\bar{\mathscr{B}}$ is easily seen to be a locally finite covering satisfying $\bar{\mathscr{B}} < \mathscr{P}$. Thus $\mathscr{C} = \bar{\mathscr{B}}$ is a locally finite closed covering of X each of whose members intersects only finitely many members of $\mathscr{A}$.

For each $A \in \mathscr{A}$, we obtain an open set A', defined by

$$A' = X - \bigcup\{C \mid C \cap A = \emptyset, C \in \mathscr{C}\}.$$

In fact, we note that A' is an open set containing A (see II.5.D)). Furthermore, note that $C \in \mathscr{C}$ intersects A' if and only if C intersects A; this means that each member of $\mathscr{C}$ intersects only finitely many A'. Hence, since $\mathscr{C}$ is a locally finite covering, $\{A' \mid A \in \mathscr{A}\}$ is locally finite. For each $A \in \mathscr{A}$, choose a member $U(A)$ of $\mathscr{U}$ such that $A \subset U(A)$. Put

$$\mathscr{V} = \{A' \cap U(A) \mid A \in \mathscr{A}\};$$

then $\mathscr{V}$ is a locally finite open covering of X such that $\mathscr{V} < \mathscr{U}$. This proves that X is paracompact.

B) *For every countable, open covering $\{V_i \mid i = 1, 2, \ldots\}$ of a topological space X, there is a locally finite covering $\{A_i \mid i = 1, 2, \ldots\}$ such that $A_i \subset V_i$, $i = 1, 2, \ldots$.*

Proof. Putting

$$A_1 = W_1$$

and

$$A_i = W_i - \bigcup_{j=1}^{i-1} W_j, \quad i = 2, 3, \ldots,$$

we obtain a desired covering $\{A_i \mid i = 1, 2, \ldots\}$.

Theorem V.1. *A regular space X is paracompact if and only if for every open covering $\mathscr{U}$ of X, there is a σ-locally finite open covering $\mathscr{V}$ with $\mathscr{V} < \mathscr{U}$.*

Proof. Since the necessity of the condition is clear, we shall only prove the sufficiency. Let $\mathscr{U}$ be a given open covering of X. Then there is an open refinement $\mathscr{V} = \bigcup_{i=1}^{\infty} \mathscr{V}_i$ of $\mathscr{U}$, where each $\mathscr{V}_i$ is a locally finite open collection. Put

$$V_i = \bigcup \{V \mid V \in \mathcal{V}_i\}, \quad i = 1, 2, \ldots ;$$

then by B) there is a locally finite covering $\{A_i \mid i = 1, 2, \ldots\}$ such that $A_i \subset V_i$. Further put

$$\mathcal{B} = \{A_i \cap V \mid V \in \mathcal{V}_i ; i = 1, 2, \ldots\}.$$

Then it follows from the local finiteness of $\mathcal{V}_i$ and $\{A_i \mid i = 1, 2, \ldots\}$ that $\mathcal{B}$ is a locally finite covering of X. It follows from $\mathcal{V} < \mathcal{U}$ that $\mathcal{B} < \mathcal{U}$. Though $\mathcal{B}$ may not be open, we conclude by A) that X is paracompact.

Corollary. *Every Lindelöf regular space is paracompact, and accordingly fully normal.*

Theorem V.2 (A. H. Stone's coincidence theorem). *A T_2-space X is paracompact if and only if it is fully normal.*

Proof. Since the 'only if' part has already been proved in III.3.E), it suffices to prove only the 'if' part. Let X be a fully normal space and $\mathcal{U}$ a given open covering of X. By virtue of Zermelo's theorem, we may regard $\mathcal{U}$ as a well-ordered collection

$$\mathcal{U} = \{U_\alpha \mid 0 \leq \alpha < \tau\},$$

where τ denotes a fixed ordinal number. Since X is fully normal, we can find open coverings $\mathcal{U}_i$, $i = 1, 2, \ldots$, such that

$$\mathcal{U} > \mathcal{U}_1^\Delta > \mathcal{U}_1 > \mathcal{U}_2^\Delta > \cdots .$$

Put

$$V_\alpha^1 = X - S(X - U_\alpha, \mathcal{U}_1)$$

and

$$V_\alpha^n = S(V_\alpha^{n-1}, \mathcal{U}_n), \quad n = 2, 3, \ldots .$$

Then V_α^n, $n = 2, 3, \ldots$, are open sets satisfying

$$\overline{V_\alpha^{n-1}} \subset V_\alpha^n \subset U_\alpha$$

because $\mathcal{U}_n^\Delta < \mathcal{U}_{n-1}$.

Put

$$V_\alpha = \bigcup_{n=1}^\infty V_\alpha^n \,;$$

then $V_\alpha \subset U_\alpha$ and $\bigcup \{V_\alpha \mid \alpha < \tau\} = X,$[1] because $\bigcup \{V_\alpha^1 \mid \alpha < \tau\} = X$ which follows from $\mathcal{U}_1^4 < \mathcal{U}$. Further put

$$W_\alpha^n = V_\alpha^n - \overline{\bigcup \{V_\beta^{n+1} \mid \beta < \alpha\}}$$

and

$$\mathcal{W}_n = \{W_\alpha^n \mid \alpha < \tau\}, \quad n = 2, 3, \ldots.$$

Then $\bigcup_{n=2}^\infty \mathcal{W}_n$ is an open covering of X. For, let p be a given point of X and

$$\alpha = \min\{\beta \mid p \in V_\beta\}\,.$$

Then, since $p \in V_\alpha$, $p \in V_\alpha^n$ for some n. Since $p \notin V_\beta$ for every $\beta < \alpha$, $p \notin V_\beta^{n+2}$. Hence in view of the definition of V_β^{n+2}, we obtain

$$S(p, \mathcal{U}_{n+2}) \cap V_\beta^{n+1} = \emptyset\,,$$

i.e.,

$$S(p, \mathcal{U}_{n+2}) \cap (\bigcup \{V_\beta^{n+1} \mid \beta < \alpha\}) = \emptyset\,.$$

Thus

$$p \notin \overline{\bigcup \{V_\beta^{n+1} \mid \beta < \alpha\}}\,,$$

which combined with $p \in V_\alpha^n$ implies $p \in W_\alpha^n$.

Next, each $\mathcal{W}_n$ is locally finite. (As a matter of fact it is discrete.) To prove this, it suffices to show that each member of $\mathcal{U}_{n+1}$ meets at most one member of $\mathcal{W}_n$. Suppose that W_α^n and W_β^n, $\beta < \alpha$, are two given members of $\mathcal{W}_n$ and U is a member of $\mathcal{U}_{n+1}$. If $U \cap W_\beta^n \neq \emptyset$, then $U \cap V_\beta^n \neq \emptyset$, and hence

$$U \subset V_\beta^{n+1} = S(V_\beta^n, \mathcal{U}_{n+1})\,.$$

Therefore $U \cap W_\alpha^n = \emptyset$, because $V_\beta^{n+1} \cap W_\alpha^n = \emptyset$ by the definition of W_α^n. Since $\mathcal{U}_{n+1}$ is an open covering of X, this proves that $\mathcal{W}_n$ is locally finite. It follows from $W_\alpha^n \subset V_\alpha^n \subset U_\alpha$ that

---

[1] We consider ordinal numbers beginning at 0 and therefore we often write $\alpha < \tau$ meaning $0 \leq \alpha < \tau$.

$$\bigcup_{n=2}^{\infty} \mathcal{W}_n < \mathcal{U} .$$

Therefore $\bigcup_{n=2}^{\infty} \mathcal{W}_n$ is a σ-locally finite open refinement of $\mathcal{U}$. On the other hand, the regularity of X is implied by its full normality and hence by Theorem V.1 X is paracompact.[1]

Corollary 1.[2] *Every metric space is paracompact.*

Proof. Combine III.5.A) and Theorem V.2.

Corollary 2. *Let $\mathcal{U}$ be a normal covering of a topological space X. Then $\mathcal{U}$ has a σ-discrete open refinement.*

Proof. This proposition is practically proved in the proof of Theorem V.2.

Corollary 3. *A regular space X is paracompact if and only if every open covering of X has a σ-discrete open refinement.*

Proof. This proposition is a direct consequence of Theorem V.1 and the above corollary.

Corollary 4. *Let $\mathcal{U}$ be a locally finite open covering of a normal space X. Then $\mathcal{U}$ has a σ-discrete open refinement and a σ-discrete closed refinement.*

Proof. Let $\mathcal{U} = \{U_\alpha \mid \alpha \in A\}$. Then by III.2.C) there is an open covering $\mathcal{V} = \{V_\alpha \mid \alpha \in A\}$ of X such that $\bar{V}_\alpha \subset U_\alpha$ for all $\alpha \in A$. Then by use of the argument of the proof of III.3.E), there is a locally finite open covering $\mathcal{W}_1$ of X such that $\mathcal{W}_1^\Delta < \mathcal{V}$. By repeating the same argument we obtain a normal sequence

$$\mathcal{W}_1 > \mathcal{W}_2^\Delta > \mathcal{W}_2 > \mathcal{W}_3^\Delta > \cdots .$$

Thus $\mathcal{V}$ is a normal covering. Hence by Corollary 2 there is a σ-discrete open refinement $\mathcal{P}$ of $\mathcal{V}$. $\bar{\mathcal{P}}$ is a σ-discrete closed refinement of $\mathcal{U}$.

---

[1] This proof is a revision of the original proof of A. H. Stone [1]. The same paper is important not only in its results but also in its method of proof which originated highly refined techniques of analysing coverings, although we can find a primitive idea of this kind of technique in D. Montgometry [1].

[2] M. E. Rudin [1] gave a direct proof of this proposition.

We have obtained the concept of paracompactness as a generalization of the concept of compactness. On the other hand, the concept of full normality can be regarded as a strong normality and thus it belongs to the separation axioms. The meaning of A. H. Stone's theorem is that those two concepts, belonging to different categories, coincide in every T_2-space. We can summarize implications between conditions for T_2-spaces in the following diagram:

$$\text{metrizable} \Rightarrow \text{fully normal} \Rightarrow \text{normal} \Rightarrow \text{completely regular} \Rightarrow \text{regular} \Rightarrow T_2$$
$$\Updownarrow$$
$$\text{compact} \Rightarrow \text{paracompact}$$

2. Further properties of paracompact spaces [1]

A) *Every F_σ subset of a paracompact space is paracompact.*

Proof. Let X' be an F_σ-set of a paracompact space X, and $\mathcal{U}'$ an open covering of X'. Then for every member $U' \in \mathcal{U}'$, we can find an open set U of X such that $U \cap X' = U'$. We put

$$\mathcal{U} = \{U \mid U' \in \mathcal{U}'\}.$$

Suppose

$$X' = \bigcup_{i=1}^{\infty} F_i \quad \text{for closed sets } F_i, \ i = 1, 2, \ldots,$$

and put

$$\mathcal{U}_i = \{X - F_1\} \cup \mathcal{U}, \quad i = 1, 2, \ldots.$$

Then each $\mathcal{U}_i$ is an open covering of X. Since X is paracompact, for each i there is a locally finite open refinement $\mathcal{W}_i$ of $\mathcal{U}_i$. Put

$$\mathcal{W}'_i = \{W \mid W \in \mathcal{W}_i, \ W \cap F_i \neq \emptyset\};$$

then $\mathcal{W}'_i$ is a locally finite open collection which satisfies $\mathcal{W}'_i < \mathcal{U}$ and which covers F_i. Thus $\mathcal{V}' = \bigcup_{i=1}^{\infty} \mathcal{V}'_i$, where $\mathcal{V}'_i = \{W \cap X' \mid W \in \mathcal{W}'_i\}$ is a σ-locally finite open covering of the subspace X', satisfying $\mathcal{V}' < \mathcal{U}'$. On

---

[1] The contents of this section are due chiefly to E. Michael [1], [4], [5]. Example V.2 is due to E. Michael [6].

the other hand, X' is regular as a subspace of the regular space X. Therefore by Theorem V.1 we can conclude that X' is paracompact.

B) *Let X be a T_1-space and $\{F_\alpha \mid \alpha \in A\}$ a locally finite closed covering of X whose members are paracompact T_2. Then X is paracompact T_2.*[1]

Proof. Let p be a point of X and U a given nbd of p. Suppose

$$p \in F_{\alpha_i}, \quad i = 1, \ldots, k,$$

and

$$p \notin F_\alpha \quad \text{for } \alpha \neq \alpha_i, \ i = 1, \ldots, k.$$

Put

$$M = X - \bigcup \{F_\alpha \mid \alpha \neq \alpha_i, \ i = 1, \ldots, k\};$$

then M is an open set such that

$$M \subset \bigcup_{i=1}^{k} F_{\alpha_i}.$$

Since each F_{α_i} is regular, there is an open nbd V_i of p in F_{α_i} such that $\bar{V}_i \subset U$. Since V_i is open in F_{α_i}, $V_i = W_i \cap F_{\alpha_i}$ for some open set W_i in X. Put

$$W = M \cap \left(\bigcap_{i=1}^{k} W_i \right);$$

then we can easily see that W is an open nbd of p satisfying $\bar{W} \subset U$. Therefore X is regular.

Now, suppose $\mathcal{U}$ is a given open covering of X. Since each F_α is paracompact, there is a locally finite open covering $\mathcal{V}_\alpha$ of F_α such that $\mathcal{V}_\alpha < \mathcal{U}$. Then $\mathcal{V} = \bigcup \{\mathcal{V}_\alpha \mid \alpha \in A\}$ is a locally finite covering of X satisfying $\mathcal{V} < \mathcal{U}$ because $\{F_\alpha \mid \alpha \in A\}$ is locally finite, and each F_α is closed. Thus from 1.A) it follows that X is paracompact.

Example V.2. As we saw in Example III.8, the product space of paracompact spaces need not be paracompact. As a matter of fact even the product of a paracompact T_2-space and a metric space may fail to be normal as shown in the following example.

---

[1] M. Katětov [4] proved a similar theorem for normal spaces.

Let R be the subspace consisting of the irrational points in the real line E^1 and let M be the real line but with a modified topology, i.e., every irrational point q of M has $\{\{q\}\}$ as its nbd basis while every rational point p of M has the usual nbd basis $\{S_{1/n}(p) \mid n = 1, 2, \ldots\}$, where $S_{1/n}(p) = \{x \in M \mid |x - p| < 1/n\}$. Now it is obvious that R is metric and M is T_2. To see the full normality of M, we suppose $\mathcal{U}$ is a given open covering of M. Then there is an open refinement $\mathcal{V}$ of $\mathcal{U}$ such that $\mathcal{V} = \{\{q\}, S_{\varepsilon(p)}(p) \mid q$ is irrational; p is rational$\}$, where $\varepsilon(p)$ is a positive number assigned to each rational p such that $\varepsilon(p) \leqslant 1$. Putting

$$\mathcal{W} = \{\{q\}, S_{\varepsilon(p)/4}(p) \mid q \text{ is irrational}; p \text{ is rational}\},$$

we get an open covering $\mathcal{W}$ satisfying $\mathcal{W}^\Delta < \mathcal{U}$, proving that M is fully normal and, accordingly, paracompact.

We denote by M' and M'' the set of the rational points and the irrational points in M respectively. Then $F = R \times M'$ and $G = \{(q, q) \mid q \in M''\}$ are obviously disjoint closed sets of $R \times M$. Now we claim that F and G cannot be separated by open sets. To see this, assume the contrary that there are open sets U and V of $R \times M$ for which

$$U \supset F, \qquad V \supset G \quad \text{and} \quad U \cap V = \emptyset.$$

Put

$$V_n = \{q \mid q \in M'' \text{ and } S_{1/n}(q) \times \{q\} \subset V\};$$

then

$$\bigcup_{n=1}^{\infty} V_n = M''.$$

Note that M'' is not F_σ in M. For, if $M'' = \bigcup_{n=1}^{\infty} K_n$ for closed K_n, then $M'' = \bigcup_{n=1}^{\infty} \bar{K}_n$ for the closures $\bar{K}_n$ with respect to the usual topology of E^1. Since $R - \bar{K}_n \supset M'$, each $\bar{K}_n$ is a nowhere dense set of E^1, which contradicts the fact that M'' is of the second category in E^1. (See Example II.8.) Now, let us turn to the topology of M. Since M'' is not F_σ,

$$\bigcup_{n=1}^{\infty} \bar{V}_n \underset{\neq}{\supsetneq} M'',$$

i.e.,

$$\bigcup_{n=1}^{\infty} \bar{V}_n \cap M' \neq \emptyset.$$

Hence $\bar{V}_n \cap M' \neq \emptyset$ for some n. Let

$$p \in \bar{V}_n \cap M',$$

and then choose an irrational number q such that

$$|p - q| < 1/2n.$$

Then since $(q, p) \in F$, there is some $m \geqslant 2n$ such that

$$S_{1/m}(q) \times S_{1/m}(p) \subset U.$$

Since $p \in \bar{V}_n$, we can choose

$$q' \in V_n \cap S_{1/m}(p).$$

Then
$$(q, q') \in S_{1/m}(q) \times S_{1/m}(p) \subset U.$$

On the other hand,

$$|p - q'| < 1/m \leqslant 1/2n,$$

which combined with $|p - q| < 1/2n$ implies

$$|q - q'| < 1/n.$$

Since $q' \in V_n$, it follows from the definition of V_n that $(q, q') \in V$, contradicting $U \cap V = \emptyset$. Therefore $R \times M$ is not normal. This space M is called *Michael line*.

Now, let us proceed to improve Theorem V.1 using the concept of cushioned refinement. To this end we make the convention that an indexed covering $\{B_\alpha \mid \alpha \in A\}$ is a *cushioned refinement* of an indexed covering $\{U_\alpha \mid \alpha \in A\}$ if for every $A' \subset A$

$$\overline{\bigcup \{B_\alpha \mid \alpha \in A'\}} \subset \bigcup \{U_\alpha \mid \alpha \in A'\}.$$

C) *Let a covering $\mathcal{U} = \{U_\alpha \mid \alpha \in A\}$ have a cushioned refinement. Then it has an indexed cushioned refinement.*

Proof. Let $\mathscr{C}$ be a cushioned refinement of $\mathcal{U}$; then for each $C \in \mathscr{C}$, there is $U(C) \in \mathcal{U}$ satisfying the condition of Definition V.1. Now, put

$$B_\alpha = \cup \{C \mid U(C) = U_\alpha, C \in \mathscr{C}\}.$$

Then $\{B_\alpha \mid \alpha \in A\}$ is an indexed cushioned refinement of $\mathscr{U}$.

Theorem V.3. *A T_1-space X is paracompact if and only if every open covering of X has a cushioned (not necessarily open) refinement.*

Proof. Since the 'only if' part is clear, we shall prove only the 'if' part. First we can assert that

(1) X is normal.

For, if F_1 and F_2 are disjoint closed sets of X, then $\{X - F_1, X - F_2\}$ is an open covering of X. Therefore by C) it has an indexed refinement $\{B_1, B_2\}$ such that

$$\bar{B}_1 \subset X - F_1 \quad \text{and} \quad \bar{B}_2 \subset X - F_2.$$
Putting
$$U_1 = X - \bar{B}_1, \qquad U_2 = X - \bar{B}_2,$$

we obtain disjoint open sets U_1 and U_2 such that

$$U_1 \supset F_1 \quad \text{and} \quad U_2 \supset F_2$$

proving that X is normal.

Now, let $\mathscr{U}$ be a given open covering of X. By use of Zermelo's theorem we may regard $\mathscr{U}$ as a well-ordered collection

$$\mathscr{U} = \{U_\alpha \mid 0 \leqslant \alpha < \tau\},$$

where τ is a fixed ordinal number. Then for each natural number i we can construct an indexed cushioned refinement $\mathscr{B}_i = \{B_{\alpha i} \mid \alpha < \tau\}$ of $\mathscr{U}$ such that

$$\overline{\cup \{B_{\beta i} \mid \beta < \alpha\}} \cap B_{\alpha i+1} = \emptyset \tag{2$_i$}$$
and
$$B_{\alpha i} \cap \overline{\cup \{B_{\beta i+1} \mid \beta > \alpha\}} = \emptyset. \tag{3$_i$}$$

We shall define $\mathscr{B}_i$ by induction on the number i. For $i = 1$, by C) we choose an arbitrary indexed cushioned refinement $\mathscr{B}_1 = \{B_{\alpha 1} \mid \alpha < \tau\}$ of $\mathscr{U}$. Suppose that desired refinements $\mathscr{B}_i = \{B_{\alpha i} \mid \alpha < \tau\}$ have been defined for $i = 1, \ldots, n$ and let us construct $\mathscr{B}_{n+1} = \{B_{\alpha n+1} \mid \alpha < \tau\}$. To do so we put

$$U_{\alpha n+1} = U_\alpha - \overline{\bigcup\{B_{\beta n} \mid \beta < \alpha\}}, \quad \alpha < \tau. \tag{4}$$

Then $\mathcal{U}_{n+1} = \{U_{\alpha n+1} \mid \alpha < \tau\}$ is an open covering of X. To see that $\mathcal{U}_{n+1}$ covers X, we suppose p is a given point of X and α is the first ordinal number such that $p \in U_\alpha$. Then

$$p \in U_{\alpha n+1} = U_\alpha - \overline{\bigcup\{B_{\beta n} \mid \beta < \alpha\}}$$

because $\mathcal{B}_n$ is an indexed cushioned refinement of $\mathcal{U}$, i.e.,

$$\overline{\bigcup\{B_{\beta n} \mid \beta < \alpha\}} \subset \bigcup\{U_\beta \mid \beta < \alpha\}.$$

This proves that $\mathcal{U}_{n+1}$ covers X. Using C) we construct an indexed cushioned refinement $\mathcal{B}_{n+1} = \{B_{\alpha n+1} \mid \alpha < \tau\}$ of $\mathcal{U}_{n+1}$. It is obvious that $\mathcal{B}_{n+1}$ is an indexed cushioned refinement of $\mathcal{U}$.

To see that $\mathcal{B}_{n+1}$ satisfies $(2)_n$, we suppose $p \in B_{\alpha n+1}$. Then $p \in U_{\alpha n+1}$ which, because of the definition (4) of $U_{\alpha n+1}$, implies that

$$p \notin \overline{\bigcup\{B_{\beta n} \mid \beta < \alpha\}},$$

proving $(2)_n$. To see $(3)_n$, we suppose $p \in B_{\alpha n}$. Then $p \notin U_{\beta n+1}$ for every $\beta > \alpha$, by the definition (4) of $U_{\beta n+1}$. This implies

$$p \notin \overline{\bigcup\{B_{\beta n+1} \mid \beta > \alpha\}}$$

because $\mathcal{B}_{n+1}$ is an indexed cushioned refinement of $\mathcal{U}_{n+1}$. Thus $(3)_n$ is proved.

Now, using the thus defined $\mathcal{B}_i$, we can construct an open covering

$$\mathcal{V} = \{V_{\alpha i} \mid \alpha < \tau; i = 1, 2, \ldots\}$$

of X such that for each i,

$$V_{\alpha i} \subset U_\alpha, \quad \alpha < \tau, \tag{5}$$

$$V_{\alpha i} \cap V_{\beta i} = \emptyset \quad \text{whenever } \alpha \neq \beta. \tag{6}$$

To this end we put

$$V_{\alpha i} = X - \overline{\bigcup\{B_{\beta i} \mid \beta \neq \alpha\}}. \tag{7}$$

Since $\mathcal{B}_i = \{B_{\alpha i} \mid \alpha < \tau\}$ is a covering of X and a cushioned refinement of the indexed covering $\mathcal{U}$, we obtain

$$V_{\alpha i} \subset B_{\alpha i} \subset U_{\alpha},$$

which proves (5). This relation combined with the definition (7) of $V_{\alpha i}$ also implies (6).

To prove that $\mathcal{V}$ is a covering, we suppose p is a given point of X. Then we put

$$\alpha_i = \min\{\alpha \mid p \in B_{\alpha i}\}, \quad i = 1, 2, \ldots,$$

recalling that each $\mathcal{B}_i$ is a covering. Then it holds that

$$p \in B_{\alpha_i i}, \quad i = 1, 2, \ldots. \tag{8}$$

Let

$$\alpha_k = \min\{\alpha_i \mid i = 1, 2, \ldots\}.$$

Then $p \in B_{\alpha_k k}$ follows from (8). This combined with $(3)_k$ implies

$$p \notin \overline{\bigcup \{B_{\beta k+1} \mid \beta > \alpha_k\}}. \tag{9}$$

On the other hand, from (8) we obtain $p \in B_{\alpha_{k+2} k+2}$, where it should be kept in mind that $\alpha_{k+2} \geqslant \alpha_k$. Hence it follows from $(2)_k$ that

$$p \notin \overline{\bigcup \{B_{\beta k+1} \mid \beta < \alpha_{k+2}\}}$$

which implies

$$p \notin \overline{\bigcup \{B_{\beta k+1} \mid \beta < \alpha_k\}}. \tag{10}$$

Thus combining (9), (10) with the definition (7) of $V_{\alpha i}$, we obtain

$$p \in V_{\alpha_k k+1},$$

proving that $\mathcal{V}$ is an open covering of X.

To complete the proof, by use of C) we construct a cushioned refinement $\{C_{\alpha i} \mid \alpha < \tau; i = 1, 2, \ldots\}$ of the indexed covering $\mathcal{V}$. Then for each i

$$\overline{\bigcup \{C_{\alpha i} \mid \alpha < \tau\}} \subset \bigcup \{V_{\alpha i} \mid \alpha < \tau\}.$$

Since X is normal as proved in (1), there is an open set W_i such that

$$\overline{\bigcup\{C_{\alpha i} \mid \alpha < \tau\}} \subset W_i \subset \bar{W}_i \subset \bigcup\{V_{\alpha i} \mid \alpha < \tau\}. \tag{11}$$

Put

$$\mathscr{W}_i = \{V_{\alpha i} \cap W_i \mid \alpha < \tau\}, \quad i = 1, 2, \ldots.$$

Then each $\mathscr{W}_i$ is a locally finite (as a matter of fact, discrete) open collection. For, let p be a given point of X. If $p \notin \bigcup\{V_{\alpha i} \mid \alpha < \tau\}$, then by (11) $p \notin \bar{W}_i$; hence $X - \bar{W}_i$ is a nbd of p which intersects no member of $\mathscr{W}_i$. If $p \in \bigcup\{V_{\alpha i} \mid \alpha < \tau\}$, then $p \in V_{\alpha i}$ for some α. Therefore $V_{\alpha i}$ is a nbd of p which intersects no $V_{\beta i} \cap W_i$ for $\beta \neq \alpha$, because by virtue of (6) $V_{\alpha i} \cap V_{\beta i} = \emptyset$. Thus $\mathscr{W}_i$ is locally finite and further, by virtue of (5), it satisfies $\mathscr{W}_i < \mathscr{U}$.

Finally to prove that $\bigcup_{i=1}^{\infty} \mathscr{W}_i$ covers X, we take a given point p of X. Since $\{C_{\alpha i} \mid \alpha < \tau; i = 1, 2, \ldots\}$ is a covering and, moreover, a cushioned refinement of the indexed covering $\mathscr{V}$, $p \in C_{\alpha i}$ for some α, i, and consequently $p \in V_{\alpha i}$. It follows from (11) that $p \in W_i$. Thus

$$p \in V_{\alpha i} \cap W_i \in \mathscr{W}_i$$

proving that $\bigcup_{i=1}^{\infty} \mathscr{W}_i$ covers X. Therefore $\bigcup_{i=1}^{\infty} \mathscr{W}_i$ is a σ-locally finite open refinement of $\mathscr{U}$, and hence we can conclude from Theorem V.1 that X is paracompact.

This theorem practically implies 1.A) and the following Corollary 1 as direct consequences, because the 'only if' parts of those propositions are obvious.

Corollary 1. *A regular space X is paracompact if and only if every open covering of X has a closure-preserving (not necessarily open) refinement.*

Corollary 2. *Let X be a paracompact T_2-space and f a closed continuous mapping of X onto a topological space Y. Then Y is also paracompact T_2.*

Proof. First of all Y is regular. For, let q and G be a point and a closed set of Y such that $q \notin G$. Then $f^{-1}(q)$ and $f^{-1}(G)$ are disjoint closed sets of X. Since X is normal, there are disjoint open sets U_1, U_2 of X for which

$$U_1 \supset f^{-1}(q) \quad \text{and} \quad U_2 \supset f^{-1}(G).$$

Put

$$V_1 = Y - f(X - U_1) \quad \text{and} \quad V_2 = Y - f(X - U_2).$$

Then V_1 and V_2 are open sets of Y and are easily seen to satisfy

$$V_1 \ni q, \qquad V_2 \supset G \quad \text{and} \quad V_1 \cap V_2 = \emptyset .$$

Therefore Y is regular.

We observe here that generally a closed collection $\mathscr{F}$ is closure-preserving if and only if for every subcollection $\mathscr{F}'$ of $\mathscr{F}$, $\cup \{F \mid F \in \mathscr{F}'\}$ is a closed set.

Suppose $\mathscr{V}$ is a given open covering of Y. Then $f^{-1}(\mathscr{V}) = \{f^{-1}(V) \mid V \in \mathscr{V}\}$ is an open covering of X and therefore it has a closure-preserving refinement $\mathscr{F} = \{F_\alpha \mid \alpha \in A\}$. Since X is regular, we may assume that $\mathscr{F}$ is a closed cover. Now, put

$$f(\mathscr{F}) = \{f(F_\alpha) \mid \alpha \in A\} .$$

Then obviously $f(\mathscr{F})$ is a closed refinement of $\mathscr{V}$ in Y. Furthermore $f(\mathscr{F})$ is closure-preserving. Because for any subset A' of A

$$\cup \{f(F_\alpha) \mid \alpha \in A'\} = f(\cup \{F_\alpha \mid \alpha \in A'\}) ,$$

and the closedness of the last set follows from the fact that f is closed, and $\mathscr{F}$ is a closure-preserving closed cover. Thus, by Corollary 1, Y is paracompact.

It is obvious that we cannot interchange the domain and range in the above Corollary 2. However, we have the following.

D) *Let f be a closed continuous mapping from X onto Y such that $f^{-1}(y)$ is compact for every $y \in Y$. (Such a mapping is called a perfect mapping.) If Y is paracompact, then so is X.*[1]

Proof. Let $\mathscr{U}$ be a given open cover of X. Then for each $y \in Y$, $f^{-1}(y)$ is covered by a finite subcollection of $\mathscr{U}$, which we denote by $\mathscr{U}_y$. Put

$$V_y = Y - f(X - \cup \{U \mid U \in \mathscr{U}_y\}) .$$

Then V_y is a nbd of y satisfying $f^{-1}(V_y) \subset \cup \mathscr{U}_y$. Since Y is paracompact, there is a locally finite open refinement $\mathscr{W}$ of $\{V_y \mid y \in Y\}$. For each $W \in \mathscr{W}$ we assign $y = y(W)$ such that $W \subset V_y$ and thus

---

[1] Due to S. Hanai [2].

$$f^{-1}(W) \subset \cup \mathcal{U}_y.$$

Then put

$$\mathcal{U}(W) = \{U \cap f^{-1}(W) \mid U \in \mathcal{U}_y\}.$$

Now, it is easy to see that $\cup \{\mathcal{U}(W) \mid W \in \mathcal{W}\}$ is a locally finite open refinement of $\mathcal{U}$. Therefore X is paracompact.

Theorem V.4. *A T_1-space X is paracompact if and only if every open covering of X has a σ-cushioned open refinement.*

Proof. Since the necessity of the condition is clear, we shall prove only the sufficiency. Let $\mathcal{U}$ be a given open covering of X. Then there is a σ-cushioned open refinement $\mathcal{V} = \cup_{n=1}^{\infty} \mathcal{V}_n$ of $\mathcal{U}$, where each $\mathcal{V}_n$ is cushioned in $\mathcal{U}$. We denote by $U(V)$ the member of $\mathcal{U}$ assigned to $V \in \mathcal{V}_n$ satisfying the condition of Definition V.1. Now, for each $p \in X$, let

$$n(p) = \inf\{n \mid p \in \cup \{V \mid V \in \mathcal{V}_n\}\}. \tag{1}$$

Then we denote by $V(p)$ a member of $\mathcal{V}_{n(p)}$ which contains p. Put

$$U(p) = U(V(p)). \tag{2}$$

Furthermore, for each $q \in X$, we put

$$W(q) = \cup \{V \mid V \in \mathcal{V}_{n(q)}\} - \cup \left\{ V \mid V \in \bigcup_{k=1}^{n(q)} \mathcal{V}_k, q \notin U(V) \right\}. \tag{3}$$

Then $W(q)$ is a nbd of q, because

$$q \notin \cup \left\{ U(V) \mid V \in \bigcup_{k=1}^{n(q)} \mathcal{V}_k, q \notin U(V) \right\} \supset \overline{\cup \left\{ V \mid V \in \bigcup_{k=1}^{n(q)} \mathcal{V}_k, q \notin U(V) \right\}}$$

which implies that

$$X - \cup \left\{ V \mid V \in \bigcup_{k=1}^{n(q)} \mathcal{V}_k, q \notin U(V) \right\}$$

is a nbd of q, and thus $W(q)$ is a nbd of q.

Now we claim that

$$q \notin U(p) \text{ implies } p \notin W(q). \tag{4}$$

For, if $n(p) \leqslant n(q)$, then it follows from $q \notin U(p)$ and (2) that

$$p \in V(p) \subset \cup \left\{ V \mid V \in \bigcup_{k=1}^{n(q)} \mathscr{V}_k, q \notin U(V) \right\},$$

because $V(p)$ satisfies the condition for V in the above. This implies that $p \notin W(q)$ because of the definition (3) of $W(q)$. If $n(p) > n(q)$, then

$$p \notin \cup \{V \mid V \in \mathscr{V}_{n(q)}\}$$

by the definition (1) of $n(p)$, and hence $p \notin W(q)$ follows from (3).
  Finally, we put

$$M(U) = \{p \mid U(p) = U\}$$

for each $U \in \mathscr{U}$. Then

$$\mathscr{M} = \{M(U) \mid U \in \mathscr{U}\}$$

is a cushioned refinement of $\mathscr{U}$. Because, if

$$q \notin \cup \{U \mid U \in \mathscr{U}'\}$$

for a subcollection $\mathscr{U}'$ of $\mathscr{U}$, then for every $U \in \mathscr{U}'$ and for each point $p \in M(U)$, we obtain

$$U(p) = U \in \mathscr{U}'$$

which implies

$$q \notin U(p) = U.$$

Therefore it follows from (4) that $p \notin W(q)$. Hence

$$W(q) \cap M(U) = \emptyset$$

for every $U \in \mathscr{U}'$. This means that

i.e.,
$$q \notin \overline{\bigcup \{M(U) \mid U \in \mathcal{U}'\}},$$

$$\overline{\bigcup \{M(U) \mid U \in \mathcal{U}'\}} \subset \bigcup \{U \mid U \in \mathcal{U}'\}$$

proving that $\mathcal{M}$ is cushioned in $\mathcal{U}$. Since $\mathcal{M}$ is obviously a covering of X, it is a cushioned refinement of $\mathcal{U}$. Thus it follows from Theorem V.3 that X is paracompact.

Theorem V.4 practically implies Theorem V.1 and the following corollary.

Corollary. *A regular space X is paracompact if and only if every open covering of X has a σ-closure preserving open refinement.*[1]

3. Countably paracompact space and collectionwise normal space

In the present section we shall define countably paracompact space and collectionwise normal space and study their basic properties. These conditions are somewhat weaker than paracompactness and full normality, respectively, and considerably important in general topology. We shall discuss more of their properties later in this book.

Definition V.2. A topological space X is called *countably paracompact* if every countable open covering of X has a locally finite open refinement.

Example V.3. It is clear that every paracompact space is countably paracompact, but the converse is not true. The space R_5 in Example II.1 is as shown in Example IV.3, normal but not fully normal, and therefore non-paracompact. On the other hand, R_5 is countably compact and therefore countably paracompact.

First we shall give some conditions (due to C. H. Dowker [3], M. Katětov [3]) for a normal space to be countably paracompact.

Theorem V.5. *The following properties of a normal space X are equivalent*:
   (i) *X is countably paracompact,*

---

[1] See H. Tamano [3], [4], Y. Katuta [1], [3], B. H. McCandless [1] and J. Mack [2] for other interesting characterizations of paracompact space. See also H. Tamano–J. Vaughan [1].

(ii) *every countable open covering of X has a point-finite open refinement,*

(iii) *every countable open covering* $\{U_i \mid i = 1, 2, \ldots\}$ *has an open refinement* $\{V_i \mid i = 1, 2, \ldots\}$ *with* $\bar{V}_i \subset U_i$,

(iv) *for every countable open covering* $\{U_i \mid i = 1, 2, \ldots\}$ *with* $U_i \subset U_{i+1}$, *there is a closed covering* $\{F_i \mid i = 1, 2, \ldots\}$ *such that* $F_i \subset U_i$.

(v) *for every sequence* $\{F_i \mid i = 1, 2, \ldots\}$ *of closed sets with* $F_i \supset F_{i+1}$, $\bigcap_{i=1}^{\infty} F_i = \emptyset$, *there is a sequence* $\{U_i \mid i = 1, 2, \ldots\}$ *of open sets such that* $U_i \supset F_i$, $\bigcap_{i=1}^{\infty} U_i = \emptyset$.

Proof. (i) $\Rightarrow$ (ii) is clear because every locally finite covering is point-finite.

(ii) $\Rightarrow$ (iii). Let $\mathcal{W}$ be a point-finite open refinement of $\{U_i \mid i = 1, 2, \ldots\}$. Then putting

$$W_i = \bigcup \{W \mid W \in \mathcal{W}, \, W \subset U_i, \, W \not\subset U_j \text{ for } j = 1, \ldots, i-1\},$$

we get a point-finite open covering $\{W_i \mid i = 1, 2, \ldots\}$ with $W_i \subset U_i$. Hence by III.2.C) we can construct an open covering $\{V_i \mid i = 1, 2, \ldots\}$ with $\bar{V}_i \subset W_i$. This implies $\bar{V}_i \subset U_i$.

(iii) $\Rightarrow$ (iv) is clear.

(iv) $\Rightarrow$ (i). Let $\{U_i \mid i = 1, 2, \ldots\}$ be a given countable open covering of X with $U_i \subset U_{i+1}$. Then by (iv) there is a closed covering $\{F_i \mid i = 1, 2, \ldots\}$ such that $F_i \subset U_i$. Since X is normal, for each i there is an open set V_i such that

$$F_i \subset V_i \subset \bar{V}_i \subset U_i.$$

Let $W_i = U_i - \bigcup_{j=1}^{i-1} \bar{V}_j$. Then $\{W_i \mid i = 1, 2, \ldots\}$ is a locally finite open refinement of $\{U_i\}$. Now, suppose $\{U_i' \mid i = 1, 2, \ldots\}$ is a given countable open covering of X which does not necessarily satisfy $U_i' \subset U_{i+1}'$. Then we put

$$U_i = \bigcup_{j=1}^{i} U_j', \quad i = 1, 2, \ldots,$$

to obtain an increasing open covering $\{U_i \mid i = 1, 2, \ldots\}$.

Construct a locally finite open refinement $\{W_i \mid i = 1, 2, \ldots\}$ of $\{U_i \mid i = 1, 2, \ldots\}$ using the process mentioned in the above and put

$$W_i' = \left(\bigcup_{j=1}^{\infty} W_j \right) \cap U_i'.$$

Then $\{W_i' \mid i = 1, 2, \ldots\}$ is a locally finite open refinement of $\{U_i' \mid i = 1, 2, \ldots\}$, because its local finiteness follows from that of $\{W_i\}$. On the other hand, each point p of X belongs to W_i' for the first number i satisfying $p \in U_i'$. Thus X is countably paracompact.

It is obvious that (iv) and (v) are equivalent.

Corollary. *Every perfectly normal space X is countably paracompact.*

Proof. Let $\{U_i \mid i = 1, 2, \ldots\}$ be an increasing open covering of X. Since X is perfectly normal, for each i there are closed sets F_{is}, $s = 1, 2, \ldots$, such that

$$U_i = \bigcup_{s=1}^{\infty} F_{is} \quad \text{and} \quad F_{is} \subset F_{is+1}.$$

Put

$$G_i = \bigcup_{j=1}^{i} F_{ji}.$$

Then $G_i \subset U_i$ and $\{G_i \mid i = 1, 2, \ldots\}$ is a closed covering of X. Hence by (iv) of the theorem, X is countably paracompact.

Without assuming normality, we can characterize countably paracompact spaces as follows.[1]

Theorem V.6. *A topological space X is countably paracompact if and only if for any decreasing sequence $\{F_i \mid i = 1, 2, \ldots\}$ of closed sets with $\bigcap_{i=1}^{\infty} F_i = \emptyset$, there is a decreasing sequence $\{U_i \mid i = 1, 2, \ldots\}$ of open sets satisfying $U_i \supset F_i$ and $\bigcap_{i=1}^{\infty} \bar{U}_i = \emptyset$.*

Proof. Assume the condition is satisfied by X. Let $\mathcal{V} = \{V_i \mid i = 1, 2, \ldots\}$ be a given countable open covering of X. Then put

$$F_i = X - V_1 \cup \cdots \cup V_i, \quad i = 1, 2, \ldots,$$

to obtain a decreasing sequence $\{F_i \mid i = 1, 2, \ldots\}$ of closed sets such that $\bigcap_{i=1}^{\infty} F_i = \emptyset$. Let $\{U_i \mid i = 1, 2, \ldots\}$ be a decreasing sequence of open sets satisfying the condition. Then put

---

[1]Due to F. Ishikawa [1].

$$W_i = X - \bar{U}_i \; ;$$

then $\{W_i \mid i = 1, 2, \ldots\}$ is an open covering of X such that

$$\bar{W}_i \subset X - F_i = V_1 \cup \cdots \cup V_i .$$

Put

$$P_i = V_i - \bar{W}_1 \cup \cdots \cup \bar{W}_{i-1} \; ;$$

then $\{P_i \mid i = 1, 2, \ldots\}$ is a locally finite open refinement of $\mathscr{V}$, and hence X is countably paracompact.

Conversely, assume that X is countably paracompact. Let $\{F_i\}$ be a decreasing sequence of closed sets with $\cap_{i=1}^{\infty} F_i = \emptyset$. Then there is a locally finite open covering $\mathscr{W}$ of X such that $\mathscr{W} < \{X - F_i \mid i = 1, 2, \ldots\}$. Now put

$$U_i = \cup \{W \in \mathscr{W} \mid W \not\subset X - F_j \text{ for } j = 1, \ldots, i\}, \quad i = 1, 2, \ldots . \quad (1)$$

Then $F_i \subset U_i$ is obvious. It is also obvious that $U_i \supset U_{i+1}$, because $F_i \supset F_{i+1}$. Finally, to prove $\cap_{i=1}^{\infty} \bar{U}_i = \emptyset$, let $x \in X$. Then there is a nbd V of x which intersects only finitely many members of $\mathscr{W}$, say $W_1, \ldots, W_k$. Suppose

$$W_j \subset X - F_{i(j)}, \quad j = i, \ldots, k ,$$

and

$$i_0 = \max\{i(1), \ldots, i(k)\} .$$

Then $\mathscr{W} \ni W \not\subset X - F_{i_0}$ implies $W \neq W_1, \ldots, W_k$ and accordingly $V \cap W = \emptyset$. Thus from (1) it follows that

$$V \cap U_{i_0} = \emptyset .$$

Namely $x \notin \bar{U}_{i_0}$, proving that $\cap_{i=1}^{\infty} \bar{U}_i = \emptyset$.

Definition V.3. A T_1-space X is called *collectionwise normal* if for every discrete closed collection $\{F_\alpha \mid \alpha \in A\}$ in X there is a disjoint open collection $\{U_\alpha \mid \alpha \in A\}$ such that $F_\alpha \subset U_\alpha$ for all $\alpha \in A$.

A) *Every fully normal space is collectionwise normal.*

Proof. Let $\{F_\alpha \mid \alpha \in A\}$ be a discrete closed collection in a fully normal space X. Then there is an open covering $\mathcal{V}$ of X each of whose members intersects at most one member of $\{F_\alpha\}$. Then let $\mathcal{W}$ be an open covering of X such that $\mathcal{W}^\Delta < \mathcal{V}$. Put

$$U_\alpha = S(F_\alpha, \mathcal{W}), \quad \alpha \in A.$$

Then $\{U_\alpha \mid \alpha \in A\}$ satisfies the required condition.

Example V.4. R_5 in Example II.1 is collectionwise normal but not fully normal.[1]

B) *Let X be a collectionwise normal space and $\{F_\alpha \mid \alpha \in A\}$ a discrete closed collection in X. Then there is a discrete open collection $\{V_\alpha \mid \alpha \in A\}$ such that $F_\alpha \subset U_\alpha$ for all $\alpha \in A$.*

Proof. Select a disjoint open collection $\{U_\alpha \mid \alpha \in A\}$ such that $F_\alpha \subset U_\alpha$. Since X is normal, there is an open set W such that

$$\bigcup\{F_\alpha \mid \alpha \in A\} \subset W \subset \bar{W} \subset \bigcup\{U_\alpha \mid \alpha \in A\}.$$

Now put

$$V_\alpha = W \cap U_\alpha, \quad \alpha \in A.$$

Then $\{V_\alpha \mid \alpha \in A\}$ satisfies the desired condition.

C) *A T_1-space X is collectionwise normal if and only if for every closed set F of X and every locally finite open covering $\mathcal{U} = \{U_\alpha \mid \alpha \in A\}$ of F, there is a locally finite open covering $\mathcal{V} = \{V_\alpha \mid \alpha \in A\}$ of X such that $F \cap \bar{V}_\alpha \subset U_\alpha$ for all $\alpha \in A$.*

Proof. Assume that X satisfies the said condition and that $\{F_\alpha \mid \alpha \in A\}$ is a discrete closed collection in X. Then observe that $F = \bigcup\{F_\alpha \mid \alpha \in A\}$ is a closed set of X and that $\{F_\alpha \mid \alpha \in A\}$ forms a discrete open covering of F. Thus there is a locally finite open covering $\mathcal{V} = \{V_\alpha \mid \alpha \in A\}$ of X such that

$$F \cap \bar{V}_\alpha \subset F_\alpha \quad \text{for all } \alpha \in A.$$

Thus it follows that

[1] See R. H. Bing [1] for an example of a normal but non-collectionwise normal space.

$$F \cap V_\alpha = F \cap \bar{V}_\alpha = F_\alpha \,.$$

Put

$$U_\alpha = V_\alpha - \bigcup \{\bar{V}_\beta \mid \beta \neq \alpha\} \,.$$

Then $\{U_\alpha \mid \alpha \in A\}$ is a disjoint open collection such that $F_\alpha \subset U_\alpha$. Hence X is collectionwise normal.

Conversely, suppose that $\mathscr{U} = \{U_\alpha \mid \alpha \in A\}$ is a locally finite open covering of a closed set F of a collectionwise normal space X. Then by Corollary 4 to Theorem V.2 there is a σ-discrete closed covering $\mathscr{F} = \bigcup_{i=1}^\infty \mathscr{F}_i$ of F such that $\mathscr{F} < \mathscr{U}$, where each $\mathscr{F}_i$ is discrete in F and accordingly in X, too. We may assume, without loss of generality, that $\mathscr{F}_i$ can be expressed as

$$\mathscr{F}_i = \{F_{i\alpha} \mid \alpha \in A\}, \quad \text{with } F_{i\alpha} \subset U_\alpha \text{ for all } \alpha \in A \,.$$

Since X is collectionwise normal, by B) there is a discrete open collection $\{U_{i\alpha} \mid \alpha \in A\}$ in X such that

$$F_{i\alpha} \subset U_{i\alpha} \subset U_\alpha \cup (X - F) \quad \text{for all } \alpha \in A \,. \tag{1}$$

For each (i, α) select an open set $W_{i\alpha}$ of X satisfying

$$F_{i\alpha} \subset W_{i\alpha} \subset \bar{W}_{i\alpha} \subset U_{i\alpha} \,.$$

Put

$$W = \bigcup \{W_{i\alpha} \mid \alpha \in A, i = 1, 2, \ldots\} \,.$$

Then, since $F \subset W$, by use of Urysohn's lemma we can choose a cozero set P of X such that

$$X - W \subset P \subset X - F \,. \tag{2}$$

Let

$$P = \bigcup_{i=1}^\infty Q_i = \bigcup_{i=1}^\infty \bar{Q}_i \quad \text{for open sets } Q_i \text{ of } X, i = 1, 2, \ldots \,.$$

For a technical reason we define that

$$A' = A \cup \{\alpha_0\}, \quad \alpha_0 \notin A \,,$$
$$W_{i\alpha_0} = Q_i, \quad i = 1, 2, \ldots, \tag{3}$$
$$U_i\alpha_0 = P, \quad i = 1, 2, \ldots \,.$$

Then $\{U_{i\alpha} \mid \alpha \in A'\}$ is a locally finite open collection for each fixed i,

$$\bar{W}_{i\alpha} \subset U_{i\alpha} \quad \text{for each } i \text{ and } \alpha \in A',$$

and

$$\bigcup \{W_{i\alpha} \mid i = 1, 2, \ldots, \alpha \in A'\} = X.$$

Put

$$U'_i = \bigcup \{U_{i\alpha} \mid \alpha \in A'\}, \qquad W_i = \bigcup \{W_{i\alpha} \mid \alpha \in A'\};$$

then

$$\bar{W}_i \subset U'_i \quad \text{and} \quad \bigcup_{i=1}^{\infty} W_i = X.$$

Further, put

$$W'_i = U'_i - \bigcup_{j=1}^{i-1} \bar{W}_j, \quad i = 1, 2, \ldots.$$

Then $\{W'_i \mid i = 1, 2, \ldots\}$ is a locally finite open covering of X. Put

$$V_{i\alpha} = I_{i\alpha} \cap W'_i \quad \text{for each } i \text{ and } \alpha \in A'.$$

Then $\{V_{i\alpha} \mid i = 1, 2, \ldots, \alpha \in A'\}$ is a locally finite open covering of X satisfying $V_{i\alpha} \subset U_{i\alpha}$. Thus

$$V_{i\alpha} \subset U_\alpha \cup (X - F) \quad \text{for every } \alpha \in A,$$

and

$$V_{i\alpha_0} \subset P \subset X - F$$

(see (1), (2), (3)). Select a fixed element $\alpha_1 \in A$. Now we define open sets V'_α, $\alpha \in A$, as follows:

$$V'_{\alpha_1} = \bigcup \{V_{i\alpha_0} \cup V_{i\alpha_1} \mid i = 1, 2, \ldots\},$$
$$V'_\alpha = \bigcup \{V_{i\alpha} \mid i = 1, 2, \ldots\} \quad \text{for } \alpha \in A \text{ with } \alpha \neq \alpha_1.$$

Then $\{V'_\alpha \mid \alpha \in A\}$ is a locally finite open covering of X such that

$$F \cap V'_\alpha \subset U_\alpha \quad \text{for all } \alpha \in A.$$

Since X is normal, there is an open covering $\{V_\alpha \mid \alpha \in A\}$ satisfying $\bar{V}_\alpha \subset V'_\alpha$ for all $\alpha \in A$. Then $\{V_\alpha \mid \alpha \in A\}$ satisfies the required condition.

The following proposition[1] is interesting in comparison with B).

D) *A normal space X is collectionwise normal and countably paracompact if and only if for every locally finite closed collection $\{F_\alpha \mid \alpha \in A\}$ in X, there is a locally finite open collection $\{U_\alpha \mid \alpha \in A\}$ such that $F_\alpha \subset U_\alpha$ for all $\alpha \in A$.*

Proof. Assume the condition is satisfied. Let $\{F_\alpha \mid \alpha \in A\}$ be a given discrete closed collection in X. We well-order the elements of the collection to put

$$\{F_\alpha \mid \alpha \in A\} = \{F_\alpha \mid 0 \leq \alpha < \tau\},$$

where τ is a fixed ordinal number. Then select a locally finite open collection $\{U_\alpha \mid 0 \leq \alpha < \tau\}$ satisfying $F_\alpha \subset U_\alpha$. Then for each α there is an open set V_α such that

$$F_\alpha \subset V_\alpha \subset \bar{V}_\alpha \subset U_\alpha.$$

Put

$$W_\alpha = U_\alpha - \bigcup \{\bar{V}_\beta \mid 0 \leq \beta < \tau\}, \quad 0 \leq \alpha < \tau.$$

Then $\{W_\alpha \mid 0 \leq \alpha < \tau\}$ is a discrete open collection satisfying $F_\alpha \subset W_\alpha$, because $\{\bar{V}_\alpha\}$ is locally finite. Thus X is collectionwise normal. Now, let $\{F_i \mid i = 1, 2, \ldots\}$ be a decreasing sequence of closed sets of X with $\bigcap_{i=1}^\infty F_i = \emptyset$. Then $\{F_i\}$ is a locally finite closed collection. Hence there is a locally finite open collection $\{U_i \mid i = 1, 2, \ldots\}$ such that $U_i \supset F_i$. Since $\{U_i\}$ is locally finite, $\bigcap_{i=1}^\infty \bar{U}_i = \emptyset$. Thus by Theorem V.6 X is countably paracompact.

Conversely, assume that $\mathscr{F} = \{F_\alpha \mid \alpha \in A\}$ is a locally finite closed collection in a countably paracompact, collectionwise normal space X. We define the *order* of $\mathscr{F}$ at a point $x \in X$ (denoted by $\text{ord}_x \mathscr{F}$) as follows:

$\text{ord}_x \mathscr{F}$ = the number of the elements of F which contain x,

$\text{ord}\, \mathscr{F} = \sup\{\text{ord}_x \mathscr{F} \mid x \in X\}$.

(ord $\mathscr{F}$ for a covering was defined before.)

Now, if ord $\mathscr{F} < \infty$, then we can construct an open collection satisfying the desired condition as follows.

---

[1] Due to C. H. Dowker [5] and M. Katětov [5].

Since in case of ord $\mathscr{F} = 1$ the assertion follows directly from the collectionwise normality, we assume it in case of ord $\mathscr{F} \leq n - 1$ to use an induction argument. Then suppose that ord $\mathscr{F} = n$. Put

$$\mathscr{F}' = \{F_{\alpha_1} \cap \cdots \cap F_{\alpha_n} \mid \alpha_1, \ldots, \alpha_n \text{ are distinct members of } A\}.$$

Then $\mathscr{F}'$ is a discrete closed collection in X. Thus there is a discrete open collection $\{V(\alpha_1, \ldots, \alpha_n) \mid \alpha_1, \ldots, \alpha_n \in A\}$ such that

$$F_{\alpha_1} \cap \cdots \cap F_{\alpha_n} \subset V(\alpha_1, \ldots, \alpha_n) \subset X - \bigcup\{F_\alpha \mid \alpha \neq \alpha_1, \ldots, \alpha_n\}.$$

Put

$$V = \bigcup\{V(\alpha_1, \ldots, \alpha_n) \mid \alpha_1, \ldots, \alpha_n \in A\}.$$

Then $\{F_\alpha \cap V^c \mid \alpha \in A\}$ has order $\leq n - 1$, where $V^c = X - V$. Thus by use of the induction hypothesis we can find a locally finite open collection $\{V_\alpha \mid \alpha \in A\}$ such that

$$F_\alpha \cap V^c \subset V_\alpha.$$

Then put

$$U_\alpha = V_\alpha \cup (\bigcup\{V(\alpha_1, \ldots, \alpha_n) \mid \alpha \in \{\alpha_1, \ldots, \alpha_n\}\}).$$

Now it is obvious that the collection $\{U_\alpha \mid \alpha \in A\}$ satisfies the desired condition. Hence the assertion is proved in the special case that ord $\mathscr{F} < \infty$.

Now, assume ord $\mathscr{F} = \infty$. Then put

$$G_n = \{x \in X \mid \mathrm{ord}_x\mathscr{F} \geq n\}, \quad n = 1, 2, \ldots.$$

Then, since $\{G_n\}$ is a decreasing sequence of closed sets with $\bigcap_{n=1}^\infty G_n = \emptyset$, there is a decreasing sequence $\{V_n\}$ of open sets such that

$$G_n \subset V_n, \quad \bigcap_{n=1}^\infty \bar{V}_n = \emptyset,$$

because X is countably paracompact. For each n $\{F_\alpha \cap V_n^c \mid \alpha \in A\}$ has order $\leq n - 1$. Hence, as proved in the above, there is a locally finite open collection $\{U_{\alpha n} \mid \alpha \in A\}$ such that

$$F_\alpha \cap V_{n+1}^c \subset U_{\alpha n} \subset V_1 \cup \cdots \cup V_n \quad \text{for every } \alpha \in A. \tag{1}$$

Now, put

$$U_\alpha = \bigcup \{U_{\alpha n} \mid n = 1, 2, \ldots\}, \quad \alpha \in A.$$

Then $F_\alpha \subset U_\alpha$. Because, if $x \in F_\alpha$ and $\mathrm{ord}_x \mathscr{F} = n$, then $x \in U_{\alpha n} \subset U_\alpha$ follows from (1). To prove that $\{U_\alpha \mid \alpha \in A\}$ is locally finite, let $x \in X$. Then $x \notin \bar{V}_n$ for some n. Thus $X - \bar{V}_n$ is a nbd of x which is disjoint from all $U_{\alpha m}$ for $m \geq n$ (because of (1)). On the other hand $\{U_{\alpha m} \mid m < n, \alpha \in A\}$ is locally finite. Thus x has a nbd W which intersects at most finitely many members of this collection. Hence $(X - \bar{V}_n) \cap W$ is a nbd of x which meets at most finitely many of U_α. Therefore $\{U_\alpha \mid \alpha \in A\}$ is locally finite, proving our assertion.

Example V.5. It was a long unsolved question if every normal space is countably paracompact. This problem was completely solved by M. E. Rudin [3], who gave an example of a collectionwise normal but non-countably paracompact space. Her example requires a rather lengthy proof, but we can easily construct a non-countably paracompact topological space in which every two disjoint closed sets can be separated by open sets. For example, the real line $R = (-\infty, \infty)$ with the nbd basis

$$\mathscr{U}(p) = \{(-\infty, q) \mid q > p\}, \quad p \in R,$$

is such a space. (The countable open covering $\{U_i \mid i = 1, 2, \ldots\}$ where $U_i = (-\infty, i)$ has no locally finite open refinement.) But this space R is not T_1 and hence not normal in our sense. It is also easy to see that Niemytzki space R_4 is a non-countably paracompact Tychonoff space.

4. Modifications of the concept of paracompactness

There are various modifications of the concept of paracompactness or of full-normality though probably not so many as for the concept of compactness. We have already learned two of them, countable paracompactness and collection-wise normality. In this section we shall give a brief account of some other considerably important modifications. One of them is a little stronger than paracompactness, and the others weaker.

Definition V.4. A topological space X is said to be *strongly paracompact* or to have the *star-finite property* if every open covering of X has a star-finite open refinement.

From the above definition we obtain the following obvious proposition.

A) *Every compact space is strongly paracompact. Every strongly paracompact space is paracompact.*

Example V.6. The definition of strong paracompactness seems similar to that of paracompactness. But an essential difference between the two is that a metric space is not necessarily strongly paracompact while it is paracompact. For example, a metric star-space $S(A)$ with an uncountable index set A (Example III.9) is a metric space which is not strongly paracompact.

On the other hand, a discrete metric space X with metric

$$\rho(p, q) = \begin{cases} 1 & \text{if } p \neq q, \\ 0 & \text{if } p = q, \end{cases}$$

is a strongly paracompact metric space, but not compact if X contains infinitely many points.

Although not every metric space is strongly paracompact, every separable metric space is strongly paracompact, as implied by the following proposition.

B) *Every regular Lindelöf space is strongly paracompact.*

Proof. Let $\mathcal{U}$ be a given open covering. Then we can choose a countable open refinement $\mathcal{V} = \{V_i \mid i = 1, 2, \ldots\}$ of $\mathcal{U}$. Since by III.4.C), X is paracompact, we may assume that $\mathcal{V}$ is locally finite. Furthermore, there is an open covering $\mathcal{W}$ with $\mathcal{W}^* < \mathcal{V}$. Put

$$W_i = \bigcup \{W \mid W \in \mathcal{W}, S(W, \mathcal{W}) \subset V_i\};$$

then $\mathcal{W}' = \{W_i \mid i = 1, 2, \ldots\}$ is an open covering of X satisfying $\bar{W}_i \subset V_i$. By Urysohn's lemma, for each i we construct a continuous function f_i such that

$$f_i(\bar{W}_i) = i, \qquad f_i(X - V_i) = 0 \quad \text{and} \quad 0 \leq f_i \leq i.$$

Put

$$f = \sup\{f_i \mid i = 1, 2, \ldots\};$$

then f is a continuous function over X because $\mathcal{V}$ is locally finite. Putting

$$\mathcal{M} = \{M_i \mid i = 0, 1, 2, \ldots\},$$

where

$$M_i = \{p \mid i - 1 < f(p) < i + 1\},$$

we obtain an open covering $\mathcal{M}$ of X. Then $\mathcal{N} = \mathcal{M} \wedge \mathcal{M}'$ is easily verified to be a star-finite open refinement of $\mathcal{V}$, because each M_i meets at most finitely many members of $\mathcal{M}$ and $\mathcal{W}'$. Therefore $\mathcal{N}$ is a star-finite open refinement of $\mathcal{U}$ proving that X is strongly paracompact.

C) *A connected regular space X is strongly paracompact if and only if it is Lindelöf.*[1]

Proof. Assume that X is connected, regular and strongly paracompact. Let $\mathcal{U}$ be a given open cover of X. Select a star-finite open refinement $\mathcal{V} = \{V_\alpha \mid \alpha \in A\}$ of $\mathcal{U}$. Fix $\alpha \in A$ assuming $V_\alpha \neq \emptyset$. Define

$$W = \bigcup_{n=1}^{\infty} S^n(V_\alpha, \mathcal{V}).$$

Then W is open. We can see that W is closed, too. Because, if $x \notin W$, then $S(x, \mathcal{V}) \cap W = \emptyset$; hence $x \notin \bar{W}$, proving that W is closed. Since X is connected $W = X$. Since $\mathcal{V}$ is star-finite, W is a sum of countably many members of $\mathcal{V}$. Thus $\mathcal{U}$ has a countable subcover, and hence X is Lindelöf.

It is easy to see that every closed set of a strongly paracompact space is strongly paracompact. Neither the product of two strongly paracompact spaces nor the sum of two strongly paracompact closed sets need be strongly paracompact. Example V.2 can be modified to give a metric space X and a Lindelöf space Y such that $X \times Y$ is not normal.[2]

Definition V.5. A topological space X is called *metacompact* or *subparacompact* if every open cover $\mathcal{U}$ of X has a point-finite open refinement or a σ-discrete closed refinement, respectively. X is called *θ-refinable* if for every open cover $\mathcal{U}$ of X there is a sequence $\{V_n \mid n = 1, 2, \ldots\}$ of open covers such that $\mathcal{V}_n < \mathcal{U}$, $n = 1, 2, \ldots$, and such that for each $x \in X$ and for some n, $\mathrm{ord}_x \mathcal{V}_n$ is finite.

---

[1] Due to K. Morita [1].
[2] See E. Michael [6]. See V. Trnkova [1], Y. Yasui [1] for the sum of strongly paracompact spaces.

D) *Every paracompact space is metacompact. Every paracompact T_2-space is subparacompact. Every metacompact space as well as every subparacompact space is θ-refinable.*

Proof. Let us prove only that every subparacompact space X is θ-refinable. Suppose $\mathcal{U}$ is a given open cover of X. Then there is a closed refinement $\mathcal{G} = \bigcup_{n=1}^{\infty} \mathcal{G}_n$ of $\mathcal{U}$, where each $\mathcal{G}_n$ is discrete. Let $\mathcal{G}_n = \{G_\alpha \mid \alpha \in A_n\}$. To each G_α assign a member $U(\alpha)$ of $\mathcal{U}$ satisfying $U(\alpha) \supset G_\alpha$. Then put

$$V_\alpha = (X - \bigcup\{G_{\alpha'} \mid \alpha' \neq \alpha\}) \cap U(\alpha), \quad \alpha \in A_n,$$

$$\mathcal{V}_n = \{(X - \bigcup\{G_\alpha \mid \alpha \in A_n\}) \cap U \mid U \in \mathcal{U}\} \cup \{V_\alpha \mid \alpha \in A_n\}.$$

Then it is obvious that $\text{ord}_x \mathcal{V}_n \leq 1$ for each $x \in \bigcup\{G_\alpha \mid \alpha \in A_n\}$, and $\mathcal{V}_n < \mathcal{U}$. Hence X is θ-refinable.

Theorem V.7.[1] *The following conditions are equivalent for a topological space X:*

(i) *X is a subparacompact space,*

(ii) *every open cover $\mathcal{U}$ of X has a σ-locally finite closed refinement,*

(iii) *every open cover $\mathcal{U}$ of X has a σ-closure-preserving closed refinement,*

(iv) *for every open cover $\mathcal{U}$ of X there is a sequence $\{V_n \mid n = 1, 2, \ldots\}$ of open covers such that for each $x \in X$ there is n satisfying $S(x, \mathcal{V}_n) \subset U$ for some $U \in \mathcal{U}$.*

Proof. It is obvious that (i) implies (ii), and (ii) implies (iii).

Now, suppose that X is subparacompact. To prove (iv), let $\mathcal{F} = \bigcup_{n=1}^{\infty} \mathcal{F}_n$ be a closed refinement of a given open cover $\mathcal{U}$ such that $\mathcal{F}_n = \{F_\alpha \mid \alpha \in A_n\}$ is discrete for every n. To each F_α we assign $U(\alpha) \in \mathcal{U}$ satisfying $U(\alpha) \supset F_\alpha$. Put

$$V_\alpha = (X - \bigcup\{U_{\alpha'} \mid \alpha' \neq \alpha\}) \cap U(\alpha),$$

$$V = X - \bigcup\{U_\alpha \mid \alpha \in A_n\},$$

$$\mathcal{V}_n = \{V_\alpha \mid \alpha \in A_n\} \cup \{V\}.$$

---

[1] Due to M. M. Čoban [1], D. K. Burke–R. A. Stoltenberg [1] and D. K. Burke [2]. The class of subparacompact spaces was first considered by L. F. McAuley [2] and A. V. Arhangelskii [6]. The concept of θ-refinable space was first introduced by J. M. Worrell–H. H. Wicke [1]. See also R. Hodel [4] and J. R. Boone [1] for subparacompactness and θ-refinability.

Then each $\mathscr{V}_n$ is an open cover of X. Let $x \in X$. Then there is n such that $x \in F_\alpha$ for some $\alpha \in A_n$. Thus $S(x, \mathscr{V}_n) = V_\alpha \subset U(\alpha)$. Therefore (iv) holds for X.

Now, assume that (iv) holds. Then well-order the elements of a given open cover $\mathscr{U}$ as $\mathscr{U} = \{U_\alpha \mid 0 \leq \alpha < \tau\}$. We choose a sequence $\{\mathscr{V}_i \mid i = 1, 2, \ldots\}$ of open covers satisfying the condition (iv). For each $\mathscr{V}_i$ we further choose a sequence $\{\mathscr{V}_{ij} \mid i, j = 1, 2, \ldots\}$ of open covers satisfying the condition (iv) for $\mathscr{V}_i$. Namely, for each $x \in X$ there is j such that $S(x, \mathscr{V}_{ij}) \subset V$ for some $V \in \mathscr{V}_i$. We continue the same process. To be precise, assume that $\{\mathscr{V}_{i_1 i_2 \ldots i_k} \mid i_1, i_2, \ldots, i_k = 1, 2, \ldots\}$ has been defined. Then we choose a sequence $\{\mathscr{V}_{i_1 i_2 \ldots i_k j} \mid j = 1, 2, \ldots\}$ satisfying the condition (iv) for $\mathscr{V}_{i_1 \ldots i_k}$. Then reorder $\{\mathscr{V}_{i_1 \ldots i_k} \mid i_1, \ldots, i_k = 1, 2, \ldots; \ k = 1, 2, \ldots\}$ as $\{\mathscr{V}_j' \mid j = 1, 2, \ldots\}$. Now, we put

$$\mathscr{W}_j = \mathscr{V}_1' \wedge \cdots \wedge \mathscr{V}_j'.$$

Then $\{\mathscr{W}_j \mid j = 1, 2, \ldots\}$ is a sequence of open covers of X such that $\mathscr{W}_{j+1} < \mathscr{W}_j < \mathscr{U}$, for each $x \in X$ there is j satisfying $S(x, \mathscr{W}_j) \subset U$ for some $U \in \mathscr{U}$, and such that for each j and each $x \in X$ there is $j' > j$ satisfying $S(x, \mathscr{W}_{j'}) \subset W$ for some $W \in \mathscr{W}_j$. Then we define

$$F_{\alpha n} = \{x \in X \mid S(x, \mathscr{W}_n) \subset U_\alpha\}, \quad 0 \leq \alpha < \tau, \ n = 1, 2, \ldots. \quad (1)$$

$F_{\alpha n}$ is obviously a closed set contained in U_α. Further we put

$$G_{\alpha n m} = F_{\alpha n} - \cup \{S(F_\beta, \mathscr{W}_m) \mid \beta < \alpha\}, \quad 0 \leq \alpha < \tau; \ n, m = 1, 2, \ldots, \quad (2)$$

$$\mathscr{G}_{nm} = \{G_{\alpha n m} \mid 0 \leq \alpha < \tau\}.$$

Then each $G_{\alpha n m}$ is a closed set contained in U_α. It is obvious that each $\mathscr{G}_{nm}$ is discrete. Let us prove that $\cup_{n,m=1}^{\infty} \mathscr{G}_{nm} = \mathscr{G}$ covers X. Let $x \in X$ be given. Then there is n such that $S(x, \mathscr{W}_n) \subset U_\alpha$ for some α. Put

$$\alpha = \min\{\alpha' \mid S(x, \mathscr{W}_n) \subset U_{\alpha'} \text{ for some } n\}. \quad (3)$$

Assume $S(x, \mathscr{W}_n) \subset U_\alpha$. Then by (1) we obtain

$$x \in F_{\alpha n}. \quad (4)$$

There is $m > n$ such that $S(x, \mathscr{W}_m) \subset W$ for some $W \in \mathscr{W}_n$. We claim that

$$S(x, \mathcal{W}_m) \cap F_{\beta n} = \emptyset \quad \text{for all } \beta < \alpha.$$

Because, if $S(x, \mathcal{W}_m) \cap F_{\beta n} \neq \emptyset$ for some $\beta < \alpha$, then select $W \in \mathcal{W}_n$ satisfying $W \supset S(x, \mathcal{W}_m)$. Then $W \cap F_{\beta n} \neq \emptyset$. Pick $y \in W \cap F_{\beta n}$. Then, because of the definition (3) of α, $S(x, \mathcal{W}_m) \not\subset U_\beta$. Hence $W \not\subset U_\beta$. Thus $S(y, \mathcal{W}_n) \not\subset U_\beta$ while $y \in F_{\beta n}$, which contradicts (1). Hence $S(x, \mathcal{W}_m) \cap F_{\beta n} = \emptyset$ for all $\beta < \alpha$. This implies that $x \notin S(F_{\beta n}, \mathcal{W}_m)$ for all $\beta < \alpha$. Thus from (2) and (4) it follows that $x \in G_{\alpha n m}$, proving that $\mathcal{G}$ is a σ-discrete closed cover of X such that $\mathcal{G} < \mathcal{U}$. Hence X is subparacompact.

Finally, let us prove that (iii) implies (i). Assume that X satisfies (iii) and $\mathcal{U} = \{U_\alpha \mid 0 \leq \alpha < \tau\}$ is a given open cover of X. Generally, let $\mathcal{F}$ be a closure-preserving closed collection in X. Then we define that

$$F_\alpha = \bigcup \{F \in \mathcal{F} \mid F \subset U_\alpha\},$$

$$\mathcal{V}(\mathcal{U}, \mathcal{F}) = \left\{ U_\alpha - \bigcup_{\beta < \alpha} F_\beta \mid 0 \leq \alpha < \tau \right\}.$$

Then $\mathcal{V}(\mathcal{U}, \mathcal{F})$ is an open cover of X. Now, we use this construction in the following. Let $\bigcup_{n=1}^{\infty} \mathcal{F}_n$ be a σ-closure-preserving closed refinement of $\mathcal{U}$. Let $\bigcup_{m=1}^{\infty} \mathcal{F}_{nm}$ be a σ-closure-preserving closed refinement of $\mathcal{V}(\mathcal{U}, \mathcal{F}_n)$. Define $\bigcup_{p=1}^{\infty} \mathcal{F}_{nmp}$ in a similar way. Continue the same process to define closure-preserving closed collections

$$\{\mathcal{F}_{n_1 \ldots n_k} \mid n_1, n_2, \ldots, n_k = 1, 2, \ldots; k = 1, 2, \ldots\}.$$

Renumber them to put

$$\{\mathcal{F}_{n_1 \ldots n_k}\} = \{\mathcal{G}_i \mid i = 1, 2, \ldots\}.$$

This sequence of closed collections has the following property: If $x \in X$, then $x \in G \subset U_\alpha$ for some $\mathcal{G}_i$ and $G \in \mathcal{G}_i$ and α. For every i and $x \in X$, there is j such that $x \in G' \subset V$ for some $G' \in \mathcal{G}_j$ and some $V \in \mathcal{V}(\mathcal{U}, \mathcal{G}_i)$.

Now, we define G_α^{ij} for each pair (i, j) of natural numbers and α by

$$G_\alpha^{ij} = \left[\bigcup \left\{ G \in \mathcal{G}_j \mid G \cap \left(\bigcup_{\beta < \alpha} G_\beta^i \right) = \emptyset, G \subset U_\alpha \right\} \right] \cap G_\alpha^i, \tag{5}$$

where

$$G_\alpha^i = \bigcup \{G \in \mathcal{G}_i \mid G \subset U_\alpha\}.$$

Further we put

$$\mathcal{G}_{ij} = \{G_\alpha^{ij} \mid 0 \leqslant \alpha < \tau\}, \quad i, j = 1, 2, \dots.$$

Then each $\mathcal{G}_{ij}$ is a discrete closed collection such that $\mathcal{G}_{ij} < \mathcal{U}$, because $\mathcal{G}_i$ and $\mathcal{G}_j$ are closure-preserving closed collections. (The detail is left to the reader.) Now, we claim that $\mathcal{G} = \bigcup_{i,j=1}^\infty \mathcal{G}_{ij}$ covers X. Let $x \in X$. Then there is n such that $x \in G \subset U_{\alpha'}$ for some $G \in \mathcal{G}_n$ and for some α'. Put

$$\alpha = \min\{\alpha' \mid \text{there is } n \text{ and } G \in \mathcal{G}_n \text{ such that } x \in G \subset U_{\alpha'}\}. \quad (6)$$

Suppose

$$x \in G \subset U_\alpha, \quad G \in \mathcal{G}_i.$$

Select j for which there is $G' \in \mathcal{G}_j$ satisfying $x \in G' \subset V$ for some $V \in \mathcal{V}(\mathcal{U}, \mathcal{G}_i)$. Let

$$V = U_{\alpha'} - \bigcup_{\beta < \alpha'} G_\beta^i.$$

For $\beta < \alpha$, $G' \not\subset U_\beta$ because of (6). Hence $\alpha' < \alpha$ cannot happen, i.e. $\alpha' \geqslant \alpha$. On the other hand, for $\gamma > \alpha$ we have

$$G' \not\subset U_\gamma - \bigcup_{\beta < \gamma} G_\beta^i,$$

because

$$G' \cap \left(\bigcup_{\beta < \gamma} G_\beta^i\right) \supset G' \cap G_\alpha^i \supset G' \cap G \ni x,$$

and hence

$$G' \cap \left(\bigcup_{\beta < \gamma} G_\beta^i\right) \neq \emptyset.$$

Thus $\alpha' > \alpha$ cannot happen, i.e. $\alpha' = \alpha$. Hence

$$G' \subset U_\alpha - \bigcup_{\beta < \alpha} G_\beta^i \subset U_\alpha.$$

Hence $x \in G' \cap G \subset G_\alpha^{ij}$ (see (5)). Thus $\mathcal{G}$ covers X, and hence (i) holds.

Example V.7. It is easy to see that Niemytzki space R_4 is subparacompact but not metacompact. On the other hand, let $X = \{(x, y) \mid x$ and y are real numbers, and $y \geq 0\}$ be a topological space defined by the base $\{\{(x, y)\}, V_n(x) \mid y > 0, -\infty < x < +\infty\}$, where

$$V_n(x) = \{(u, v) \in X \mid u - v = x \text{ and } 0 \leq v < 1/n, \text{ or } u + v = x \text{ and } 0 \leq v < 1/n\}.$$

Then it is easy to see that X is metacompact but not subparacompact.

It is not so difficult to prove that the product $S \times S$ of the Sorgenfrey line with itself is not metacompact. (Actually it is subparacompact.) Thus metacompactness is not preserved by a finite product. On the other hand, J. M. Worrell [2] proved that a T_2-space is metacompact if it is the image of a metacompact space by a closed continuous map. K. Alster–R. Engelking [1] gave an example of a paracompact T_2-space X such that $X \times X$ is not subparacompact. T. Przymusiński [1] gave a Lindelöf regular space X and a separable metric space Y such that $X \times Y$ is not θ-refinable.

E) *Every collectionwise normal θ-refinable space X is paracompact.*

Proof. Let $\mathcal{U}$ be a given open cover of X. Then there are open covers $\{\mathcal{V}_n \mid n = 1, 2, \ldots\}$ satisfying the condition of Definition V.5. Suppose $\mathcal{V}_n = \{V_\alpha \mid \alpha \in A_n\}$. For a fixed n and for an arbitrary choice $\alpha_1, \ldots, \alpha_k$ of finitely many distinct elements of A_n, we put

$$F_n(\alpha_1, \ldots, \alpha_k) = \{x \in X \mid x \in V_{\alpha_1} \cap \cdots \cap V_{\alpha_k} - \cup \{V_\alpha \mid \alpha \neq \alpha_1, \ldots, \alpha_k\}\}.$$

Then $\{F_n(\alpha) \mid \alpha \in A_n\}$ is a discrete closed collection. Hence there is a discrete open collection $\{U_n(\alpha) \mid \alpha \in A_n\}$ such that $U_n(\alpha) \supset F_n(\alpha)$. We can select $U_n(\alpha)$ contained in some element of $\mathcal{U}$ because $\mathcal{V}_n < \mathcal{U}$. Put

$$U_n^1 = \cup \{U_n(\alpha) \mid \alpha \in A_n\}.$$

Now $\{F_n(\alpha_1, \alpha_2) - U_n^1 \mid \alpha_1, \alpha_2 \in A_n, \alpha_1 \neq \alpha_2\}$ is a discrete closed collection. Because, if $x \in X - U_n^1$, then $\text{ord}_x \mathcal{V}_n \geq 2$. Hence there are distinct α_1, $\alpha_2 \in A_n$ such that $x \in V_{\alpha_1} \cap V_{\alpha_2}$. Now $V_{\alpha_1} \cap V_{\alpha_2}$ intersects $F_n(\alpha_1', \alpha_2')$ only if $\{\alpha_1, \alpha_2\} = \{\alpha_1', \alpha_2'\}$. Hence the concerned collection is discrete. It is also obvious that each $F_n(\alpha_1, \alpha_2) - U_n^1$ is closed. Thus there is a discrete open collection $\{U_n(\alpha_1, \alpha_2) \mid \alpha_1, \alpha_2 \in A_n, \alpha_1 \neq \alpha_2\}$ such that

$U_n(\alpha_1, \alpha_2) \supset F_n(\alpha_1, \alpha_2) - U_n^1$ and such that $U_n(\alpha_1, \alpha_2)$ is contained in some element of $\mathcal{V}_n$. Then put

$$U_n^2 = \bigcup \{ U_n(\alpha_1, \alpha_2) \mid \alpha_1, \alpha_2 \in A_n, \alpha_1 \neq \alpha_2 \}.$$

Continue the same process to define a discrete open collection

$$U_{nk} = \{ U_n(\alpha_1, \ldots, \alpha_k) \mid \alpha_1, \ldots, \alpha_k \in A_n \}$$

such that

$$U_n(\alpha_1, \ldots, \alpha_k) \supset F_n(\alpha_1, \ldots, \alpha_k) \quad \text{and} \quad \mathcal{U}_{nk} < \mathcal{V}_n.$$

Now $\bigcup_{m, k=1}^{\infty} \mathcal{U}_{nk}$ is obviously a σ-discrete open refinement of $\mathcal{U}$. Hence by Michael's theorem, X is paracompact.

In a θ-refinable space compactness and countable compactness coincide as it was the case for a paracompact space.

F) *Every countably compact θ-refinable space X is compact.*

Proof. First observe that every discrete collection of non-empty closed sets of X is finite because X is countably compact. Let $\mathcal{U}$ be a given open cover of X and $\{ \mathcal{V}_n \mid n = 1, 2, \ldots \}$ open refinements of $\mathcal{U}$ satisfying the condition of Definition V.5. We use the same symbol $F_n(\alpha_1, \ldots, \alpha_k)$ as in the proof of D). Then $\{ F_n(\alpha) \mid \alpha \in A \}$ is finite because of the above observation. Hence it is covered by a finite subcollection $\mathcal{V}_{n1}$ of $\mathcal{V}_n$. Let $V_{n1} = \bigcup \mathcal{V}_{n1}$. Then $\{ F_n(\alpha_1, \alpha_2) - V_{n1} \mid \alpha_1, \alpha_2 \in A_n, \alpha_1 \neq \alpha_2 \}$ is a discrete closed collection and hence finite. Thus it is covered by a finite subcover $\mathcal{V}_{n2}$ of $\mathcal{V}_n$. Continue the same process; then we obtain a sequence $\{ \mathcal{V}_{ni} \mid i = 1, 2, \ldots \}$ of finite subcollections of $\mathcal{V}_n$ such that $\bigcup \{ \mathcal{V}_{ni} \mid i = 1, 2, \ldots \}$ covers $\bigcup \{ F_n(\alpha_1, \ldots, \alpha_k) \mid \alpha_1, \ldots, \alpha_k \in A_n ; k = 1, 2, \ldots \}$. Hence $\bigcup \{ \mathcal{V}_{ni} \mid i = 1, 2, \ldots ; n = 1, 2, \ldots \}$ is a countable subcover of $\bigcup_{n=1}^{\infty} \mathcal{V}_n$. Hence $\mathcal{U}$ has a countable subcover and accordingly a finite subcover. Thus X is compact.[1]

---

[1] See P. Zenor [1], Y. Yasui [3] for another modification of paracompactness called $\mathcal{B}$-property.

5. Characterization by product spaces

As seen in Example V.2, the product space of two normal spaces may fail to be normal even if one of them is metric and the other is paracompact. This fact leads us to the following problem: Let P be a class of normal spaces; then what is the necessary and sufficient condition for a normal space X in order that the product space of X with every space belonging to P be normal? We shall discuss this problem in cases that P is compact metric (due to C. H. Dowker [3]), compact (due to H. Tamano [1]) and metric (due to K. Morita [3]). The former two cases will be the topic of the present section while the last will be left to Section 7 of the next chapter. It is interesting that in the former cases we obtain a new characterization of countably paracompact normal space and of paracompact T_2-spaces, respectively. On the other hand, the last case (that P is metric) leads us to a new class of spaces.

In the present section we use K. Morita's [4] method.

A) *Let X be a paracompact T_2-space and Y a compact T_2-space. Then $X \times Y$ is paracompact T_2.*

Proof. It suffices to prove that $X \times Y$ is paracompact. Suppose $\mathcal{W}$ is a given open covering of $X \times Y$. Then for each $x \in X$ we can select open nbds $U_i(x)$, $i = 1, \ldots, k(x)$, of x in X and an open covering $\{V_{ix} \mid i = 1, \ldots, k(x)\}$ of Y such that each $U_i(x) \times V_{ix}$ is contained in some member of $\mathcal{W}$, because Y is compact. Put

$$U(x) = \bigcap_{i=1}^{k(x)} U_i(x).$$

Then, since X is paracompact, there is a locally finite open covering $\mathcal{P}$ of X such that

$$\mathcal{P} < \{U(x) \mid x \in X\}.$$

To each $P \in \mathcal{P}$ we assign $x(P) \in X$ such that

$$P \subset U(x(P)).$$

Then put

$$\mathscr{W}' = \{U(x(P)) \times U_i(x(P)) \mid i = 1, \ldots, k(x(P)); P \in \mathscr{P}\}.$$

Now, it is easy to see that $\mathscr{W}'$ is a locally finite open covering of $X \times Y$ such that $\mathscr{W}' < \mathscr{W}$. Thus $X \times Y$ is paracompact.

Definition V.6. Let ρ be a real-valued continuous function on $X \times X$, where X is a topological space. Then ρ is called a *pseudo-metric* of X if it satisfies $\rho(x, y) \geqslant 0$, $\rho(x, x) = 0$, $\rho(x, y) = \rho(y, x)$ and $\rho(x, y) + \rho(y, z) \geqslant \rho(x, z)$ for every $x, y, z \in X$. (Namely, a pseudo-metric is not required to satisfy $\rho(x, y) > 0$ for $x \neq y$ like a metric.) Let μ be an infinite cardinal number and X a topological space. If every open covering $\mathscr{U}$ of X with $|\mathscr{U}| \leqslant \mu$ has a locally finite open refinement, then X is called μ-*paracompact*. (Thus X is paracompact if and only if it is μ-paracompact for every μ.)

We are going to modify some corollaries of Theorem V.2 to use them in the following arguments.

B) *Let ρ be a pseudo-metric of a topological space X. Define that $x \sim y$ for $x, y \in X$ if and only if $\rho(x, y) = 0$. Then denote by Y the set of the equivalence classes and by φ the natural mapping from X onto Y, i.e. for each $x \in X$ $\varphi(x)$ is the class which contains x. Define a real-valued function ρ' on $Y \times Y$ by*

$$\rho'(y, y') = \rho(x, x'), \quad x \in \varphi^{-1}(y), x' \in \varphi^{-1}(y').$$

Then ρ' is a metric of Y, i.e. Y is a metric space, and φ is a continuous mapping.

Proof. It is easy to verify that ρ' is well-defined and satisfies the conditions for a metric. So the detailed check-up is left to the reader. Now, let $y \in Y$ and $\varepsilon > 0$. Then $S_\varepsilon(y) = \{y' \in Y \mid \rho'(y, y') < \varepsilon\}$ is a basic nbd of y in the metric space Y. Suppose $x \in \varphi^{-1}(y)$. It is obvious that

$$\varphi(B_\varepsilon(x)) \subset S_\varepsilon(y),$$

where $B_\varepsilon(x) = \{x' \in X \mid \rho(x, x') < \varepsilon\}$. Since $B_\varepsilon(x)$ is a nbd of x because of the continuity of ρ, φ is continuous.

C) *Let $\mathscr{U}$ be a normal covering of a topological space X. Then $\mathscr{U}$ has a locally finite open refinement which consists of cozero sets.*

Proof. There is a normal sequence $\{\mathcal{U}_i \mid i = 1, 2, \ldots\}$ such that

$$\mathcal{U} > \mathcal{U}_1 > \mathcal{U}_2^* > \mathcal{U}_2 > \mathcal{U}_3^* > \cdots .$$

We define a real-valued function ρ on $X \times X$ as follows. For every rational number of the form $k/2^n$, $k = 1, 2, \ldots, 2^n - 1$; $n = 1, 2, \ldots$, we define an open covering $\mathcal{V}(k/2^n)$ as follows:

$$\mathcal{V}\left(\frac{1}{2}\right) = \mathcal{U}_1,$$

$$\mathcal{V}\left(\frac{1}{2^2}\right) = \mathcal{U}_2, \qquad \mathcal{V}\left(\frac{3}{2^2}\right) = \left\{S(V_1, \mathcal{U}_2) \mid V \in \mathcal{V}\left(\frac{1}{2}\right)\right\},$$

$$\mathcal{V}\left(\frac{1}{2^3}\right) = \mathcal{U}_3, \qquad \mathcal{V}\left(\frac{3}{2^3}\right) = \left\{S(V, \mathcal{U}_3) \mid V \in \mathcal{V}\left(\frac{1}{2^2}\right)\right\},$$

$$\mathcal{V}\left(\frac{5}{2^3}\right) = \left\{S(V, \mathcal{U}_3) \mid V \in \mathcal{V}\left(\frac{1}{2}\right)\right\},$$

$$\mathcal{V}\left(\frac{7}{2^3}\right) = \left\{S(V, \mathcal{U}_3) \mid V \in \mathcal{V}\left(\frac{3}{2^2}\right)\right\}, \ldots .$$

Generally, assume we have defined $\mathcal{V}(k/2^n)$, $k = 1, \ldots, 2^n - 1$. Then we define $\mathcal{V}(k/2^{n+1})$ by

$$\mathcal{V}\left(\frac{1}{2^{n+1}}\right) = \mathcal{U}_{n+1},$$

$$\mathcal{V}\left(\frac{k}{2^{n+1}}\right) = \mathcal{V}\left(\frac{2k'}{2^{n+1}}\right) = \mathcal{V}\left(\frac{k'}{2^n}\right) \quad \text{if } k = 2k',$$

and

$$\mathcal{V}\left(\frac{k}{2^{n+1}}\right) = \mathcal{V}\left(\frac{2k'+1}{2^{n+1}}\right) = \left\{S(V, \mathcal{U}_{n+1}) \mid V \in \mathcal{V}\left(\frac{k'}{2^n}\right)\right\}$$

if $k = 2k' + 1$, where $1 \leq k' \leq 2^n - 1$.

Moreover, we put

$$\mathcal{V}(1) = \{X\}.$$

We can derive from the definition that

$$\mathcal{V}\left(\frac{1}{2^{n+1}}\right) = \mathcal{U}_{n+1}, \tag{1}$$

$$\mathcal{V}\left(\frac{k}{2^n}+\frac{1}{2^{n+1}}\right) = \left\{S(V, \mathcal{U}_{n+1}) \mid V \in \mathcal{V}\left(\frac{k}{2^n}\right)\right\} \quad \text{if } 1 \le k \le 2^n - 1, \quad (2)$$

$$\mathcal{V}(\nu) < \mathcal{V}(\nu') \quad \text{if } \nu < \nu'. \tag{3}$$

The validity of (1) and (2) is clear. To see (3), we shall show, by induction on the number n, that

$$S(V, \mathcal{U}_n) \subset V' \quad \text{for every } V \in \mathcal{V}(k/2^n) \text{ and for some}$$

$$V' \in \mathcal{V}\left(\frac{k+1}{2^n}\right), \quad 1 \le k \le 2^n - 1. \tag{4}_n$$

It is easily seen that $(4)_n$, $n = 1, 2, \ldots$, imply (3), because ν and ν' can be expressed as fractions with a common denominator.

First, $\mathcal{V}(\frac{1}{2}) = \mathcal{U}_1$ and $\mathcal{V}(\frac{2}{2}) = \{X\}$. Hence $(4)_1$ is obviously true. Assuming the validity of $(4)_n$, we can show that $(4)_{n+1}$ is also true. To do so, we note that

$$\mathcal{V}\left(\frac{k'}{2^n}\right) = \mathcal{V}\left(\frac{2k'}{2^{n+1}}\right), \qquad \mathcal{V}\left(\frac{k'}{2^n}+\frac{1}{2^{n+1}}\right) = \mathcal{V}\left(\frac{2k'+1}{2^{n+1}}\right)$$

and

$$\mathcal{V}\left(\frac{k'+1}{2^n}\right) = \mathcal{V}\left(\frac{2k'+2}{2^{n+1}}\right).$$

We divide the proof into the two cases, $k = 2k'$ and $k = 2k'+1$. If $V \in \mathcal{V}(2k'/2^{n+1})$, then by (2)

$$V' = S(V, \mathcal{U}_{n+1}) \in \mathcal{V}\left(\frac{2k'+1}{2^{n+1}}\right),$$

and hence the assertion $(4)_{n+1}$ is true in the case $k = 2k'$. If $V \in \mathcal{V}((2k'+1)/2^{n+1})$ and $k' \ge 1$, then by (2)

$$V = S(V_0, \mathcal{U}_{n+1}) \quad \text{for some } V_0 \in \mathcal{V}\left(\frac{k'}{2^n}\right).$$

Therefore by the induction hypothesis, we obtain

$$S(V_0, \mathcal{U}_n) \subset V' \quad \text{for some } V' \in \mathcal{V}\left(\frac{k'+1}{2^n}\right).$$

This combined with $\mathcal{U}_{n+1}^* < \mathcal{U}_n$ implies

$$S(V, \mathcal{U}_{n+1}) = S(S(V_0, \mathcal{U}_{n+1}), \mathcal{U}_{n+1}) \subset S(V_0, \mathcal{U}_n) \subset V'$$

$$\text{for } V' \in \mathcal{V}\left(\frac{2k'+2}{2^{n+1}}\right),$$

which proves $(4)_{n+1}$ in the case $k = 2k' + 1$. Finally, in case $k = 1$, $(4)_{n+1}$ is a direct consequence of (1) and $\mathcal{U}_{n+1}^* < \mathcal{U}_n$. Thus in any case $(4)_{n+1}$ is proved, i.e. $(4)_n$ is true for every n. Consequently, we can define open coverings $\mathcal{V}(k/2^n)$, $k = 1, \ldots, 2^n$; $n = 1, 2, \ldots$, which satisfy (1), (2) and (3).

Now, we define real-valued functions $\varphi(p, q)$ and $\rho(p, q)$ with two variables $p, q \in X$ by

$$\varphi(p, q) = \inf\left\{\frac{k}{2^n} \,\middle|\, q \in S\left(p, \mathcal{V}\left(\frac{k}{2^n}\right)\right)\right\}$$

and

$$\rho(p, q) = \sup\{|\varphi(p, x) - \varphi(q, x)| \mid x \in X\}.$$

Then we can prove that ρ is a pseudo-metric of X. Among the conditions which should be satisfied by ρ, $\rho(p, q) = \rho(q, p) \geq 0$ and $\rho(p, p) = 0$ are obvious. To prove the triangle inequality for $\rho(p, q)$, let p, a, q be given points of X. Then

$$\rho(p, q) \leq \sup\{|\varphi(p, x) - \varphi(a, x)| + |\varphi(a, x) - \varphi(q, x)| \mid x \in X\}$$

$$\leq \sup\{|\varphi(p, x) - \varphi(a, x)| \mid x \in X\}$$

$$+ \sup\{|\varphi(a, x) - \varphi(q, x)| \mid x \in X\}$$

$$= \rho(p, a) + \rho(a, q)$$

To prove that ρ is continuous on $X \times X$ it suffices to verify that for a given $\varepsilon > 0$ and a given point p of X, $B_\varepsilon(p) = \{q \mid \rho(p, q) < \varepsilon\}$ is a nbd of p. Take a positive integer n such that

$$3/2^{n+1} < \varepsilon.$$

Then

$$U(p) = S(p, \mathcal{U}_{n+1}) = S\left(p, \mathcal{V}\left(\frac{1}{2^{n+1}}\right)\right)$$

is a nbd of p. Consider a given point q of $U(p)$; then

$$p, q \in V \quad \text{for some } V \in \mathscr{V}\left(\frac{1}{2^{n+1}}\right). \tag{5}$$

Take an arbitrary point x of X; then

$$(k-1)/2^n \leqslant \varphi(q, x) < k/2^n \tag{6}$$

for some integer k with $2 \leqslant k \leqslant 2^n - 1$, or

$$\varphi(q, x) < 1/2^n$$

or

$$\varphi(q, x) \geqslant (2^n - 1)/2^n .$$

In case of (6) we obtain

$$q \in S\left(x, \mathscr{V}\left(\frac{k}{2^n}\right)\right), \tag{7}$$

because, otherwise, $\varphi(q, x) \geqslant k/2^n$ follows from the definition of $\varphi(q, x)$, contradicting (7). From (5) and (7) we obtain

$$p, x \in S\left(V', \mathscr{V}\left(\frac{1}{2^{n+1}}\right)\right) \quad \text{for some } V' \in \mathscr{V}\left(\frac{k}{2^n}\right).$$

Hence it follows from (2) that

$$p \in S\left(x, \mathscr{V}\left(\frac{k}{2^n} + \frac{1}{2^{n+1}}\right)\right).$$

Therefore

$$\varphi(p, x) \leqslant \frac{k}{2^n} + \frac{1}{2^{n+1}},$$

which combined with (7) implies

$$\varphi(p, x) \leqslant \varphi(q, x) + \frac{3}{2^{n+1}}.$$

In a similar way we can prove that

$$\varphi(q, x) \leqslant \varphi(p, x) + \frac{3}{2^{n+1}}.$$

Therefore we conclude that

$$|\varphi(p, x) - \varphi(q, x)| \leq \frac{3}{2^{n+1}}.$$

In the other two cases we can also easily show that $|\varphi(p, x) - \varphi(q, x)| \leq 3/2^{n+1}$. Thus

$$\rho(p, q) = \sup\{|\varphi(p, x) - \varphi(q, x)| \mid x \in X\} \leq \frac{3}{2^{n+1}} < \varepsilon,$$

i.e. $q \in B_\varepsilon(p)$. This means that

$$U(p) \subset B_\varepsilon(p),$$

and hence $B_\varepsilon(p)$ is a ndb of p.

Now, by B) there is a metric space Y with a metric ρ and a continuous mapping φ from X onto Y such that

$$\rho(x, x') = \rho'(\varphi(x), \varphi(x')).$$

Thus

$$\varphi^{-1}(S_\varepsilon(\varphi(x))) = B_\varepsilon(x) \quad \text{for each } \varepsilon > 0 \text{ and } x \in X,$$

where we use the same symbols as in the previous proof. Note that

$$B_{1/4}(x) \subset S(x, \mathcal{U}_2).$$

Thus

$$\{\varphi^{-1}(S_{1/4}(y)) \mid y \in Y\} < \mathcal{U}_1 < \mathcal{U}.$$

Since Y is metric and accordingly paracompact, there is a locally finite open covering $\mathcal{W}$ of Y such that

$$\mathcal{W} < \{S_{1/4}(y) \mid y \in Y\}.$$

Hence $\varphi^{-1}(\mathcal{W}) = \{\varphi^{-1}(W) \mid W \in \mathcal{W}\}$ is a locally finite open refinement of $\mathcal{U}$ in X. Since we can choose $\mathcal{W}$ consisting of cozero sets of Y, $\varphi^{-1}(\mathcal{W})$ consists of cozero sets of X.

D) *If a covering $\mathcal{U}$ of a topological space X has a σ-locally finite open refinement consisting of cozero sets, then $\mathcal{U}$ is a normal covering.*

Proof. Let $\mathcal{V}$ be a σ-locally finite open covering such that

$$\mathcal{V} = \bigcup_{n=1}^{\infty} \mathcal{V}_n < \mathcal{U},$$

where each $\mathcal{V}_n$ is a locally finite open collection whose elements are cozero sets. Let

$$\mathcal{V}_n = \{V_{\alpha n} \mid \alpha \in A_n\}, \quad V_{\alpha n} = \{p \mid f_{\alpha n}(p) > 0\}$$

for a real-valued continuous function $f_{\alpha n}$ over X satisfying $0 \leqslant f_{\alpha n} \leqslant 1$. Since $\mathcal{V}_n$ is locally finite,

$$f_n(p) = \sum_{\alpha \in A_n} f_{\alpha n}(p), \quad p \in X,$$

is a continuous function. Putting

$$f(p) = \sum_{n=1}^{\infty} \frac{f_n(p)}{2^n(1 + f_n(p))}, \quad p \in X,$$

we obtain a continuous function such that $f(p) > 0$ at every $p \in X$. Further, we put

$$g_{\alpha n}(p) = \frac{f_{\alpha n}(p)}{2^n f(p)(1 + f_n(p))}, \quad p \in X.$$

Then $g_{\alpha n}$, $\alpha \in A_n$, $n = 1, 2, \ldots$, are continuous functions satisfying

$$\sum_{n=1}^{\infty} \sum_{\alpha \in A_n} g_{\alpha n}(p) = 1, \quad p \in X, \tag{1}$$

and

$$V_{\alpha n} = \{p \mid g_{\alpha n}(p) > 0\}.$$

Let us consider a metric space Y whose points are the points $\{x_{\alpha n} \mid \alpha \in A_n, \; n = 1, 2, \ldots\}$ of the cartesian product of the copies $I_{\alpha n}$, $\alpha \in A_n$, $n = 1, 2, \ldots$, of the unit segment $[0, 1]$, satisfying

$$\sum x_{\alpha n} = 1$$

and whose metric is defined by

$$\rho(x, y) = \sum \{|x_{\alpha n} - y_{\alpha n}| \mid \alpha \in A_n, n = 1, 2, \ldots\}$$

for $x = \{x_{\alpha n}\}$ and $y = \{y_{\alpha n}\}$.

To every point p of X, we assign a point

$$g(p) = \{g_{\alpha n}(p) \mid \alpha \in A_n, n = 1, 2, \ldots\}$$

of Y (see (1)). Then g is seen to be a continuous mapping of X into Y. For, suppose p_0 is a given point of X, and ε is a positive number. Then we can choose $(\alpha_1, n_1), \ldots, (\alpha_k, n_k)$ and a nbd $U(p_0)$ of p_0 such that

$$\sum \{g_{\alpha n}(p_0) \mid (\alpha, n) \neq (\alpha_i, n_i), i = 1, \ldots, k\} < \varepsilon$$

and

$$\sum_{i=1}^{k} |g_{\alpha_i n_i}(p) - g_{\alpha_i n_i}(p_0)| < \varepsilon \quad \text{for every } p \in U(p_0).$$

Then, in view of the fact that

$$\sum g_{\alpha n}(p) = \sum g_{\alpha n}(p_0) = 1,$$

we obtain

$$\sum_{(\alpha, n) \neq (\alpha_i, n_i)} g_{\alpha n}(p) = \sum_{i=1}^{k} (g_{\alpha_i n_i}(p_0) - g_{\alpha_i n_i}(p)) + \sum_{(\alpha, n) \neq (\alpha_i, n_i)} g_{\alpha n}(p_0)$$

$$\leq \sum_{i=1}^{k} |g_{\alpha_i n_i}(p_0) - g_{\alpha_i n_i}(p)| + \sum_{(\alpha, n) \neq (\alpha_i, n_i)} g_{\alpha n}(p_0)$$

$$< 2\varepsilon \quad \text{for every } p \in U(p_0).$$

This implies that

$$\rho(g(p), g(p_0)) = \sum_{i=1}^{k} |g_{\alpha_i n_i}(p) - g_{\alpha_i n_i}(p_0)| + \sum_{(\alpha, n) \neq (\alpha_i, n_i)} |g_{\alpha n}(p) - g_{\alpha n}(p_0)|$$

$$< \varepsilon + \sum_{(\alpha, n) \neq (\alpha_i, n_i)} (g_{\alpha n}(p) + g_{\alpha n}(p_0))$$

$$< 4\varepsilon$$

for every $p \in U(p_0)$, which proves the continuity of g. Put

$$W_{\alpha n} = \{\{x_{\alpha n}\} \mid x_{\alpha n} > 0, \{x_{\alpha n}\} \in g(X)\} \, ;$$

then

$$\mathcal{W} = \{W_{\alpha n} \mid \alpha \in A_n, n = 1, 2, \ldots\}$$

is an open covering of $g(X)$ satisfying

$$g^{-1}(\mathcal{W}) = \{g^{-1}(W) \mid W \in \mathcal{W}\} < \mathcal{V},$$

because

$$g^{-1}(W_{\alpha n}) = V_{\alpha n} \, .$$

Since $g(X)$ is a metric space, we can construct open coverings $\mathcal{W}_1, \mathcal{W}_2, \ldots$ of $g(X)$ such that

$$\mathcal{W} > \mathcal{W}_1^{\Delta} > \mathcal{W}_1 > \mathcal{W}_2^{\Delta} > \cdots .$$

Now

$$\mathcal{V}_i = f^{-1}(\mathcal{W}_i), \quad i = 1, 2, \ldots,$$

are open coverings of X satisfying

$$\mathcal{U} > \mathcal{V} > \mathcal{V}_1^{\Delta} > \mathcal{V}_1 > \mathcal{V}_2^{\Delta} \cdots .$$

Hence $\mathcal{U}$ is a normal covering.

E) *Let $\mathcal{U}$ be an open covering of a topological space X. If there is a normal covering $\mathcal{V}$ of X such that for each $V \in \mathcal{V}$ the restriction $\{V \cap U \mid U \in \mathcal{U}\}$ of $\mathcal{U}$ to V is a normal covering of V, then $\mathcal{U}$ is a normal covering of X.*

Proof. Since $\mathcal{V}$ is normal, by D) there is a locally finite open refinement $\mathcal{V}'$ of $\mathcal{V}$ consisting of cozero sets. We may assume $\bar{\mathcal{V}}' < \mathcal{V}$. Let

$$\mathcal{V}' = \{V_\alpha \mid \alpha \in A\}, \quad V_\alpha = \{x \in X \mid f_\alpha(x) > 0\},$$

where f_α is a continuous function over X with $0 \leqslant f_\alpha \leqslant 1$. To each V_α we assign a member $V(V_\alpha)$ of $\mathcal{V}$ containing $\bar{V}_\alpha$. Since $\{V(V_\alpha) \cap U \mid U \in \mathcal{U}\}$ is a normal covering of $V(V_\alpha)$, we can construct a locally finite cozero covering $\mathcal{V}_\alpha$ of the subspace $V(V_\alpha)$ such that

$$\mathcal{V}_\alpha < \{V(V_\alpha) \cap U \mid U \in \mathcal{U}\}.$$

Now put

$$\mathscr{W}_\alpha = \{V_\alpha \cap V \mid V \in \mathscr{V}_\alpha\}.$$

Then it easily follows from the local finiteness of $\mathscr{V}'$ that $\mathscr{W} = \cup \{\mathscr{W}_\alpha \mid \alpha \in A\}$ is a locally finite open covering of X satisfying $\mathscr{W} < \mathscr{U}$.

On the other hand, for each $V \in \mathscr{V}_\alpha$ there is a continuous function g over $V(V_\alpha)$ such that

$$0 \leqslant g \leqslant 1, \qquad V = \{x \in V(V_\alpha) \mid g(x) > 0\}.$$

Put

$$h(x) = \begin{cases} g(x)f_\alpha(x) & \text{if } x \in V_\alpha, \\ 0 & \text{if } x \in X - V_\alpha. \end{cases}$$

Then h is a continuous function on X satisfying

$$V_\alpha \cap V = \{x \in X \mid h(x) > 0\}.$$

Namely, $\mathscr{W}$ consists of cozero sets of X. Hence by D) $\mathscr{U}$ is a normal covering.

F) *Let $\mathscr{U}$ be an open covering of a normal space X. If there is a normal open covering $\mathscr{V}$ of X such that each $V \in \mathscr{V}$ is covered by finitely many elements of $\mathscr{U}$, then $\mathscr{U}$ is a normal covering.*

Proof. Select a normal covering $\mathscr{W}$ of X such that $\bar{\mathscr{W}} < \mathscr{V}$. Then for each $W \in \mathscr{W}$ $\bar{W}$ is a normal space covered by finitely many elements of $\mathscr{U}$. Thus the restriction of $\mathscr{U}$ to $\bar{W}$ is a normal covering of $\bar{W}$. Hence by C) $\mathscr{U}$ is normal.

G) *Let ω_α be the least ordinal number with cardinality $\aleph_\alpha$. Denote by W_α the topological space of all ordinal numbers $\leqslant \omega_\alpha$ with the order topology. (See the remark of Example I.10 and Example II.1.)*

Assume that $X \times W_\alpha$ is a normal space for a topological space X. Then each open covering $\mathscr{U}$ of X with $|\mathscr{U}| \leqslant \aleph_\alpha$ has a closed refinement $\mathscr{G}$ with $|\mathscr{G}| \leqslant \aleph_\alpha$.

Proof. The proof will be done by use of transfinite induction on α. Let $\alpha > 0$ and assume that the proposition is true for all ordinal numbers β with $0 \leqslant \beta < \alpha$. Let $\{U_\lambda \mid \lambda < \omega_\alpha\}$ be an open covering of X. Put

$$V_\lambda = \cup\{U_{\lambda'} \mid \lambda' < \lambda\}, \quad \lambda < \omega_\alpha, \tag{1}$$

$$F = X \times W_\alpha - \cup\{V_\lambda \times (\lambda, \omega_\alpha] \mid \lambda < \omega_\alpha\}, \tag{2}$$

$$G = X \times \{\omega_\alpha\}. \tag{3}$$

Then F and G are disjoint closed sets of $X \times W_\alpha$. Since $X \times W_\alpha$ is normal, there are open sets P and Q such that

$$P \supset F, \quad Q \supset G, \quad P \cap Q = \emptyset.$$

Let

$$F_\lambda = \{x \in X \mid (x, \lambda) \notin P\},^1 \quad \lambda < \omega_\alpha.$$

Then F_λ is obviously a closed set of X. We claim that

$$X = \cup\{F_\lambda \mid \lambda < \omega_\alpha\}, \tag{4}$$

$$F_\lambda \subset V_\lambda \quad \text{for all } \lambda < \omega_\alpha. \tag{5}$$

To prove (4), let $x \in X$. Then $(x, \omega_\alpha) \in Q$. Since Q is open,

$$\{x\} \times (\lambda', \omega_\alpha] \subset Q \quad \text{for some } \lambda' < \omega_\alpha.$$

Hence $(x, \lambda) \notin P$ for λ satisfying $\lambda' < \lambda < \omega_\alpha$. Namely $x \in F_\lambda$, proving (4).
To prove (5), let $x \in F_\lambda$; then

$$(x, \lambda) \notin P \supset F.$$

Hence by (2)

$$(x, \lambda) \in V_{x'} \times (\lambda', \omega_\alpha] \quad \text{for some } \lambda'.$$

Hence $\lambda' < \lambda$, i.e. by (1) $x \in V_{\lambda'} \subset V_\lambda$, proving (5).
Now observe that W_β is a closed set of W_α for every $\beta < \alpha$. Hence $F_\lambda \times W_\beta$ is normal. Let $\lambda \in W_\alpha$ and $\|[0, \lambda]\| = \aleph_\beta$, where $[0, \lambda] = \{\lambda' \in W_\alpha \mid 0 \le \lambda' \le \lambda\}$. Put

$$\mathcal{U}_{\lambda\lambda'} = \{F_\lambda \cap U_{\lambda'} \mid \lambda' \le \lambda\}.$$

---

[1] (x, λ) denotes a point of the product space $X \times W_\alpha$. The same symbol is not used in this proof to mean an interval, while $(\lambda, \omega_\alpha]$ denotes an interval in W_α, i.e. $(\lambda, \omega_\alpha] = \{\lambda \in W_\alpha \mid \lambda < \lambda' \le \omega_\alpha\}$.

Then $\mathscr{U}_{\lambda\lambda'}$ is an open covering of F_λ with cardinality $\aleph_\beta$ (because of (1) and (5)). Hence by the induction hypothesis and the above observation there is a closed covering $\mathscr{F}_\lambda$ of F_λ such that

$$\mathscr{F}_\lambda < \mathscr{U}_{\lambda\lambda'}, \qquad |\mathscr{F}_\lambda| \le \aleph_\beta \,.$$

Put

$$\mathscr{G} = \bigcup \{\mathscr{F}_\lambda \mid \lambda < \omega_\alpha\} \,.$$

Then by (4), $\mathscr{G}$ is a closed covering of X such that

$$\mathscr{G} < \mathscr{U}, \qquad |\mathscr{G}| \le \aleph_\alpha^2 = \aleph_\alpha \,.$$

The above argument is true without the induction hypothesis if $\alpha = 0$, because every finite open covering of a normal space can be shrunk to a closed covering. (The details are left to the reader.) Thus the proposition is proved.

H) *Let X be a topological space and I^μ the product space of μ copies of the unit segment $[0, 1]$, where μ is an infinite cardinal number. If $X \times I^\mu$ is normal, then X is μ-paracompact and normal.*

Proof. It is obvious that X is normal.

Let $\mu = \aleph_\alpha$; then W_α denotes the space in G). Observe that W_α is a closed subset of I^μ because of the corollary to Theorem III.6 and III.3.C). Thus $X \times W_\alpha$ is normal. Let $\mathscr{W}$ be a given open covering of X with $|\mathscr{W}| \le \mu$. Then, by use of G), we can find an open covering $\mathscr{G} = \{G_\lambda \mid \lambda < \omega_\alpha\}$ and a closed covering $\mathscr{F} = \{F_\lambda \mid \lambda < \omega_\alpha\}$ such that

$$\mathscr{G} \subset \mathscr{W} \quad \text{and} \quad F_\lambda \subset G_\lambda \quad \text{for all } \lambda \,,$$

where ω_α denotes the least ordinal number with cardinality $\aleph_\alpha$. Since X is normal, for each λ there is a continuous function φ_λ from X into $I_\lambda = [0, 1]$ such that

$$\varphi_\lambda(F_\lambda) = 0, \qquad \varphi_\lambda(X - G_\lambda) = 1 \,.$$

Suppose

$$I^\mu = \{I_\lambda \mid 0 \le \lambda < \omega_\alpha\} \,;$$

then define a continuous mapping φ from X into I^μ by

$$\varphi(x) = \{\varphi_\lambda(x) \mid 0 \leq \lambda < \omega_\alpha\}, \quad x \in X.$$

Put

$$V_\lambda = \{\{y_{\lambda'} \mid \lambda' < \omega_\alpha\} \in I^\mu \mid y_\lambda < 1\}, \quad \lambda < \omega_\alpha,$$

$$H = \cup\{V_\lambda \mid \lambda < \omega_\alpha\}. \tag{1}$$

Then H is an open set of I^μ. Observe that $\varphi(X) \subset H$ because $\{F_\lambda\}$ covers X. Also note that

$$\varphi^{-1}(V_\lambda) \subset G_\lambda. \tag{2}$$

Because if $x \in X - G_\lambda$, then $\varphi_\lambda(x) = 1$, and hence $\varphi(x) \notin V_\lambda$, i.e. $x \notin \varphi^{-1}(V_\lambda)$. Further put

$$D = \{(x, \varphi(x)) \mid x \in X\}.$$

Then D is a closed set of $X \times I^\mu$ such that

$$D \subset X \times H \quad \text{(Exercise III.5)}.$$

Now, define a continuous mapping Φ from $X \times I^\mu$ into $I = [0, 1]$ such that

$$\Phi(D) = 0, \qquad \Phi(X \times I^\mu - X \times H) = 1.$$

Define a real-valued function ρ on $X \times X$ by

$$\rho(x, x') = \max\{|\Phi(x, y) - \Phi(x', y)| \mid y \in I^\mu\}, \quad x, x' \in X.$$

Then ρ is easily seen to be a pseudo-metric of X. Let

$$B(x) = \{x' \in X \mid \rho(x, x') < \tfrac{1}{2}\},$$

$$\mathcal{U} = \{B(x) \mid x \in X\}.$$

Then $\mathcal{U}$ is a normal covering of X. Let $x' \in B(x)$; then

$$\Phi(x, \varphi(x')) = \Phi(x, \varphi(x')) - \Phi(x', \varphi(x')) \leq \rho(x, x') < \tfrac{1}{2}.$$

Hence

$$\varphi(B(x)) \subset \{y \in I^\mu \mid \Phi(x, y) < \tfrac{1}{2}\},$$

and thus

$$\overline{\varphi(B(x))} \subset \{y \in I^\mu \mid \varPhi(x, y) \leqslant \tfrac{1}{2}\} \subset H \quad \text{in } I^\mu. \tag{3}$$

Because if $y \in I^\mu - H$, then $(x, y) \in X \times I^\mu - X \times H$, and hence $\varPhi(x, y) = 1 > \tfrac{1}{2}$. Note that $\overline{\varphi(B(x))}$ is a closed set of I^μ and accordingly compact. Thus from (3) and (1) it follows that $\varphi(B(x))$ is covered by finitely many of V_λ's. Hence by (2), $B(x)$ is covered by finitely many of G_λ's. Since $\{G_\lambda\}$ is a subcovering of $\mathcal{W}$, each member $B(x)$ of $\mathcal{U}$ is covered by finitely many members of $\mathcal{W}$. Hence by F) $\mathcal{W}$ is a normal covering. Thus by C) $\mathcal{W}$ has a locally finite open refinement, proving that X is μ-paracompact.

Theorem V.8 (Tamano's theorem). *Let X be a topological space. Then X is paracompact T_2 if and only if the product space of X with every compact T_2-space Y is normal.*

Proof. The 'only if' part is the already proved proposition A).

Conversely, if the condition is satisfied, then $X \times I^\mu$ is normal for every cardinal number μ. Thus by H) X is μ-paracompact for every μ. Hence X is paracompact. X is obviously T_2, too.[1]

I) *Let φ be a perfect mapping from a topological space Y onto a topological space Z. If $X \times Y$ is normal, then so is $X \times Z$.*[2]

Proof. Define a continuous mapping ψ from $X \times Y$ onto $X \times Z$ by

$$\psi(x, y) = (x, \varphi(y)), \quad (x, y) \in X \times Y.$$

[1] H. Tamano's [1] original method of proof is also quite interesting. He proved that if $X \times \beta(X)$ is normal for a Tychonoff space X, then X is paracompact. Thus, if $X \times Y$ is normal for every compact T_2-space Y, then X is paracompact. The above result of Tamano also implies that if $X \times Y$ is normal for every paracompact T_2-space Y, then $X \times Y$ is paracompact for every paracompact T_2-space Y, because then $X \times Y \times \beta(X \times Y)$ is normal. K. Morita [4] and H. Tamano [2] proved that if $X \times B(X)$ is normal for a Tychonoff space X and its T_2-compactification $B(X)$, then X is paracompact. Y. Katuta [2] obtained a necessary and sufficient condition in order that $X \times Y$ be normal for every paracompact T_2-space Y. M. E. Rudin and M. Atsuji proved that $X \times Y$ is normal for every normal space Y if and only if X is a discrete space (see M. Atsuji [4]). See also T. Ishii [1], J. Suzuki [1] and R. Telgársky [1] for interesting results and R. Telgársky [2] for a survey of this aspect.

Tamano characterized more properties of a Tychonoff space X in terms of the product of X with a T_2-compactification. For example, X is second countable if and only if $X \times B(X)$ is perfectly normal for some T_2-compactification $B(X)$; X is collectionwise normal if and only if for every closed set F of X, $F \times \beta(X)$ is C^*-embedded in $X \times \beta(X)$.

O. T. Alas [1] proved that a regular space X is countably paracompact and normal if and only if $X \times \alpha(X_0)$ is normal where X_0 is the discrete space with $|X_0| = |X|$.

[2] M. E. Rudin [5] proved that this proposition is true if φ is merely a closed continuous mapping, and X is compact T_2.

Then it is easy to see that ψ is a closed mapping (actually a perfect mapping). To prove it, assume that F is a closed set of $X \times Y$ and

$$(x, z) \in X \times Z - \psi(F).$$

Then $\psi^{-1}(x, z) = \{x\} \times \varphi^{-1}(z)$ is disjoint from F. Since $\varphi^{-1}(z)$ is compact, there are an open nbd U of x in X and an open set V of Y satisfying

$$V \supset \varphi^{-1}(z), \qquad U \times V \cap F = \emptyset.$$

Since φ is closed, $U \times (Z - \varphi(Y - V))$ is an open nbd of (x, z), which is disjoint from $\psi(F)$. Thus

$$(x, z) \notin \overline{\psi(F)} \quad \text{in } X \times Z,$$

proving that $\psi(F)$ is closed. Hence ψ is a closed mapping. Thus $X \times Z$ is normal (Exercise III.33).

J) *Let X be a μ-paracompact, normal space, and Y a compact T_2-space with $w(Y) \le \mu$. Then $X \times Y$ is normal.*

Proof. Suppose that F and G are disjoint closed sets of $X \times Y$ and $\{V_\lambda \mid \lambda \in \Lambda\}$ a base of Y such that $|\Lambda| \le \mu$. Let Γ denote the collection of all finite subsets of Λ. For each $\gamma \in \Gamma$ we define

$$H_\gamma = \bigcup \{V_\lambda \mid \lambda \in \gamma\}.$$

Further we define, for each $x \in X$,

$$F(x) = \{y \in Y \mid (x, y) \in F\}, \qquad G(x) = \{y \in Y \mid (x, y) \in G\},$$

to put

$$U_\gamma = \{x \in X \mid F(x) \subset H_\gamma, \bar{H}_\gamma \subset Y - G(x)\}. \tag{1}$$

Then U_γ is an open set of X. To prove it, let $x_0 \in U_\gamma$. Let $\{W_\delta \mid \delta \in \Delta\}$ be the set of all nbds of x_0 in X, where we regard Δ as a directed set with the order defined by

$$\delta > \delta' \quad \text{if and only if} \quad W_\delta \subset W_{\delta'}.$$

Assume that x_0 is no interior point of U_γ, i.e. $W_\delta \not\subset U_\gamma$ for all $\delta \in \Delta$.

Then select $\varphi(\delta) \in W_\delta - U_\gamma$ for each $\delta \in \Delta$. Then, because of the definition (1) of U_γ, there are two possible cases.

(i) If $F(\varphi(\delta)) \not\subset H_\gamma$ for a cofinal subset Δ' of Δ and for all $\delta \in \Delta'$, then pick a point

$$\psi(\delta) \in F(\varphi(\delta)) - H_\gamma \quad \text{for each } \delta \in \Delta'.$$

Since Y is compact, the net $\psi(\Delta' \,|\, >)$ has a cluster point $y_0 \in Y - H_\gamma$. Then we can prove that

$$(x_0, y_0) \notin F, \tag{2}$$

because $(x_0, y_0) \in F$ implies

$$y_0 \in F(x_0) \subset H_\gamma.$$

(Recall that $x_0 \in U_\gamma$ and accordingly $F(x_0) \subset H_\gamma$ follow from (1).) But this is impossible.

On the other hand, from $\psi(\delta) \in F(\varphi(\delta))$ it follows that $(\varphi(\delta), \psi(\delta)) \in F$ for every $\delta \in \Delta'$. Since y_0 is a cluster point of $\psi(\Delta' \,|\, >)$ and $\varphi(\Delta' \,|\, >) \to x_0$, (x_0, y_0) is a cluster point of the net $\theta(\Delta' \,|\, >)$ defined by

$$\theta(\delta) = (\varphi(\delta), \psi(\delta)), \quad \delta \in \Delta'. \tag{3}$$

Since F is closed, $(x_0, y_0) \in F$, which contradicts (2).

(ii) If $\bar{H}_\gamma \cap G(x_\delta) \neq \emptyset$ for a cofinal subset Δ' of Δ and all $\delta \in \Delta'$, then select

$$\psi(\delta) \in \bar{H}_\gamma \cap G(x_\delta) \quad \text{for every } \delta \in \Delta'.$$

Let y_0 be a cluster point of the net ψ. Then

$$y_0 \in \bar{H}_\gamma \subset X - G(x_0).$$

(Recall (1) and $x_0 \in U_\gamma$.) Thus we obtain

$$(x_0, y_0) \notin G. \tag{4}$$

On the other hand, from $\psi(\delta) \in G(\varphi(\delta))$ it follows that

$$(\varphi(\delta), \psi(\delta)) \in G \quad \text{for all } \delta \in \Delta'.$$

Since θ defined by (3) has (x_0, y_0) as a cluster point, and G is closed,

$$(x_0, y_0) \in G,$$

which contradicts (4). Thus U_γ is proved to be open.

We note that

$$\cup \{U_\gamma \mid \gamma \in \Gamma\} = X,$$

because $\{V_\lambda \mid \lambda \in \Lambda\}$ is a base of the compact T_2-space Y, and hence

$$F(x) \subset H_\gamma \subset \bar{H}_\gamma \subset Y - G(x)$$

holds for each $x \in X$ and some $\gamma \in \Gamma$. Since $|\Gamma| \leq \mu$, and X is μ-paracompact, there is a locally finite open covering $\{P_\gamma \mid \gamma \in \Gamma\}$ such that

$$\bar{P}_\gamma \subset U_\gamma \quad \text{for all } \gamma \in \Gamma.$$

Put

$$P = \cup \{P_\gamma \times H_\gamma \mid \gamma \in \Gamma\};$$

then, since $\{P_\gamma \times H_\gamma \mid \gamma \in \Gamma\}$ is locally finite, we obtain

$$\bar{P} = \cup \{\bar{P}_\gamma \times \bar{H}_\gamma \mid \gamma \in \Gamma\}. \tag{5}$$

Hence it follows that

$$F \subset P \subset \bar{P} \subset X \times Y - G. \tag{6}$$

Since P is open, this proves that $X \times Y$ is normal. To show (6), let $(x, y) \in F$ and $x \in P_\gamma$; then $x \in U_\gamma$, and hence $y \in F(x) \subset H_\gamma$ because of (1). Thus

$$(x, y) \in P_\gamma \times H_\gamma \subset P.$$

Let $(x, y) \in G$ and $x \in \bar{P}_\gamma$; then $x \in U_\gamma$, and hence $y \notin \bar{H}_\gamma$, because $\bar{H}_\gamma \cap G(x) = \emptyset$ follows from (1). Thus

$$(x, y) \notin \bar{P}_\gamma \times \bar{H}_\gamma \quad \text{for all } \gamma \in \Gamma,$$

which implies $(x, y) \notin \bar{P}$ because of (5).

Now, the following important theorem due to K. Morita [4] and its corollary due to C. H. Dowker [3] easily follow.

Theorem V.9. *The following conditions are equivalent for a topological space X:*

(i) *X is μ-paracompact and normal,*

(ii) *$X \times I^\mu$ is normal,*

(iii) *$X \times D^\mu$ is normal, where D^μ is the product space of μ copies of the discrete space D consisting of two points.*

Proof. By J), (i) implies (iii). By I) and IV.6.A), (iii) implies (ii). By H), (ii) implies (i).

Corollary. *The following conditions are equivalent for a topological space X:*

(i) *X is countably paracompact and normal,*

(ii) *$X \times Y$ is normal for every compact metric space Y,*

(iii) *$X \times I$ is normal, where I denotes the segment $[0, 1]$.*

Proof. Assume that (i) holds. Then by Theorem V.9, $X \times I^\omega$ is normal, where I^ω is the Hilbert cube. Suppose Y is a given compact metric space. Then it may be regarded as a closed set of I^ω. Hence $X \times Y$ is a closed set of $X \times I^\omega$, and thus it is normal, proving (ii).

It is obvious that (ii) implies (iii). Assume that (iii) holds. Since $D^{\aleph_0}$ is a closed subset of I, $X \times D^{\aleph_0}$ is a closed set of $X \times I$. Hence $X \times D^{\aleph_0}$ is normal. Therefore by Theorem V.9 X is countably paracompact and normal.

Exercise V

1. Let $\mathscr{U}$, $\mathscr{V}$ be collections in a topological space X. If $\mathscr{U}$ is closure-preserving and satisfies $\bar{\mathscr{U}} < \mathscr{V}$, then $\mathscr{U}$ is cushioned in $\mathscr{V}$.

2. A T_1-space X is paracompact if and only if for every open covering $\mathscr{U}$ of X, there is a sequence $\{\mathscr{U}_n \mid n = 1, 2, \ldots\}$ of open coverings such that for each point p of X there is a nbd V of p for which $S(V, \mathscr{U}_n) \subset U$ for some n and some $U \in \mathscr{U}$ (A. Arhangelskii's theorem. Apply Theorem V.4.)

3. Let X be a countably paracompact normal space and $\{U_i \mid i = 1, 2, \ldots\}$ an open covering of X such that each U_i is paracompact. Then X is paracompact.

4. A T_1-space X is countably paracompact and normal if and only if for every countable open covering $\mathcal{U}$ of X, there is an open covering $\mathcal{V}$ such that $\mathcal{V}^* < \mathcal{U}$.

5. Let X be a countably paracompact normal space and f a closed continuous mapping of X onto a topological space Y. Then Y is countably paracompact and normal.

6. Every star space with an uncountable index set A is not strongly paracompact.

7. The closed sets of a strongly paracompact space are strongly paracompact. The same holds for collectionwise normality and the other modifications of paracompactness defined in Section 4.

8. Every locally compact, fully normal space X (and especially every locally compact, metric space) is strongly paracompact. (Hint: Take an open covering $\mathcal{U}$, the closures of whose elements are compact, and an open covering $\mathcal{V}$ with $\mathcal{V}^* < \mathcal{U}$. Then for each $V \in \mathcal{V}$, $S^\infty(V, \mathcal{V}) = \bigcup_{n=1}^\infty S^n(V, \mathcal{V})$ is σ-compact and therefore Lindelöf. Now, apply 4.B) to each $S^\infty(V, \mathcal{V})$ noting that $\{S^\infty(V, \mathcal{V}) \mid V \in \mathcal{V}\}$ is a discrete open covering of X.)

9. $N(A)$ is strongly paracompact while $N(A) \times (0, 1)$ is not, where A is an uncountable set.

10. A metric space X is strongly paracompact if and only if it is homeomorphic to a subset of $I^\omega \times N(A)$, where $w(X) = |A|$.

11. A regular space X is strongly paracompact if and only if every open cover of X has a star-countable open refinement. (Yu. Smirnov's theorem.)

12. The product of a strongly paracompact space and a compact space is strongly paracompact.

13. Let f be a perfect map from a topological space X onto a strongly paracompact space Y. Then X is strongly paracompact.

14. The image of a collectionwise normal (subparacompact) space X by a closed continuous map is collectionwise normal (subparacompact). The same does not hold for strong paracompactness.

15. Let f be a perfect map from a regular space X onto a subparacompact space Y. Then X is subparacompact.

16. Every F_σ-set of a collectionwise normal (subparacompact) space is collectionwise normal (subparacompact).

17. A T_1-space X is hereditarily collectionwise normal if and only if every open set of X is collectionwise normal.

18. Every perfectly normal paracompact space is hereditarily paracompact.

19. Niemytzki space is not metacompact.

20. The space X in Example V.7 is not subparacompact.

21. The product $N^{\aleph_1}$ of $\aleph_1$ copies of the countable discrete space N is not normal. (Hint: Prove that every regular open set of $N^{\aleph_1}$ depends only on countably many coordinates, in view of IV.3.D). Define disjoint closed sets F and G by $F = \{\{x_\alpha \mid \alpha \in A\} \in N^{\aleph_1} \mid x_\alpha \neq x_{\alpha'}$ whenever $\alpha \neq \alpha'$ and $x_\alpha \neq 1 \neq x_{\alpha'}\}$, $G = \{\{x_\alpha \mid \alpha \in A\} \in N^{\aleph_1} \mid x_\alpha \neq x_{\alpha'}$ whenever $\alpha \neq \alpha'$ and $x_\alpha \neq 2 \neq x_{\alpha'}\}$. Show that F and G cannot be separated by disjoint regular open sets.)

22. The product $\prod_{\alpha \in A} X_\alpha$ of metric spaces X_α, $\alpha \in A$, is normal if and only if at most countably many of X_α, $\alpha \in A$, are non-compact. (A. H. Stone's [1] theorem.)

23. The product of a μ-paracompact space and a compact space is μ-paracompact.

24. Show by an example that the product of a normal space and a compact T_2-space need not be normal.

25. Every countably compact normal space is collectionwise normal.

26. The product $S \times S$ of Sorgenfrey line with itself is not countably paracompact.

27. Every countably paracompact, 1st countable space is regular. (Due to C. Aull [1].)

CHAPTER VI

METRIZABLE SPACES AND RELATED TOPICS

The purpose of this chapter is to give a considerably detailed account of theories on metrizable spaces and of related topics. We shall first investigate conditions for metrizability, and then generalize the concept of metric space. Especially, we shall deal with a new category of spaces which includes metric spaces, compact spaces and other important spaces as special cases.

1. Metrizability

In the present section we shall discuss the problem of finding necessary and sufficient conditions for a topological space to be homeomorphic with a metric space. There are various theorems which have been established in the field of metrization theory, but still they do not seem to be exhausted. We shall give a quick view of some remarkable theorems, chiefly using the method due to J. Nagata [3]. In later sections we shall discuss newer results in metrization theory.

Definition VI.1. A topological space X is called *metrizable* if it is homeomorphic with a metric space.

A) *A topological space X is metrizable if and only if we can define a metric $\rho(p, q)$ of X satisfying* (i)–(iv) *of Definition* III.6 *such that* $\{S_{1/n}(p) \mid n = 1, 2, \ldots\}$ *is a nbd basis of each point p of X, where* $S_{1/n}(p) = \{q \mid \rho(p, q) < 1/n\}$. *We call such a metric $\rho(x, y)$ compatible* with the topology of X.

Proof. It is clearly a direct consequence of Definition VI.1.

Example VI.1. A simple example of a metrizable space is a discrete space X, where we can define a metric ρ by

$$\rho(x, y) = 1 \quad \text{if } x \neq y, \qquad \rho(x, x) = 0 \quad \text{for every } x \in X.$$

Then ρ is obviously compatible with the discrete topology of X. On the other hand, a metrizable space must have all topological properties possessed by a metric space. Thus the 1st countable normal space R_5 in Example II.1 is not metrizable, because it is not fully normal. (See Example III.3.)

Let $\{I_\alpha \mid \alpha \in A\}$ be an infinite collection of segments $I_\alpha = [0, 1]$. Denote by X the discrete sum of I_α, $\alpha \in A$. Then we define a decomposition $\mathcal{D}$ of X by identifying all zeros in X while leaving the other points as they are (as singletons). Then the quotient space $S'(A) = X/\mathcal{D}$ is called a (non-metrizable) *star-space* or (non-metrizable) *hedgehog*. Obviously $S'(A)$ has a stronger topology than the metric topology of $S(A)$ in Example III.9. The quotient mapping φ from X onto $S'(A)$ is obviously a closed mapping. Hence by Corollary 2 to Theorem V.3, $S'(A)$ is paracompact T_2. But it is not metrizable, because the center point 0 has no countable nbd base, and accordingly $S'(A)$ is not 1st countable.

In metrization theory the following classical theorem is still fundamental.

Theorem VI.1 (Alexandroff–Urysohn's metrization theorem). *A T_1-space X is metrizable if and only if there is a sequence $\mathcal{U}_1, \mathcal{U}_2, \ldots$ of open coverings such that*
 (i) $\mathcal{U}_1 > \mathcal{U}_2^* > \mathcal{U}_2 > \mathcal{U}_3^* > \cdots,$[1]
 (ii) $\{S(p, \mathcal{U}_n) \mid n = 1, 2, \ldots\}$ *is a nbd base at each point p of X.*

Proof. Necessity. Let X be a metric space. Put

$$\mathcal{U}_n = \{S_{1/3^n}(p) \mid p \in X\};$$

then all conditions are obviously satisfied.

Sufficiency. It was nearly proved in the proof of V.5.C). We use here the same technique and symbols to define a pseudo-metric ρ of X by use of the normal sequence $\{U_n\}$. Then in the present case we can prove that ρ is a metric compatible with the topology of X. Assume that p and q are distinct points of X. Then

---

[1] Namely $\{\mathcal{U}_i\}$ is a normal sequence. Obviously (i) may be replaced with the condition
(i') $\mathcal{U}_1 > \mathcal{U}_2^\Delta > \mathcal{U}_2 > \mathcal{U}_3^\Delta > \cdots.$

$$\rho(p, q) = \sup\{|\varphi(p, x) - \varphi(q, x)| \mid x \in X\}$$
$$\geq |\varphi(p, q) - \varphi(q, q)| = \varphi(p, q).$$

Since X is T_1, by the condition (ii) we can choose an n for which

$$q \not\in S(p, \mathcal{U}_n) = S(p, \mathcal{V}(1/2^n)).$$

Hence $\varphi(p, q) \geq 1/2^n$, and hence $\rho(p, q) \geq 1/2^n$, i.e. $\rho(p, q) \neq 0$. Thus ρ is a metric of X.

Now, let us turn to the proof of that ρ is compatible with the topology of X. Since ρ is continuous on $X \times X$ and $\rho(p, p) = 0$, for each natural number i, there is a nbd $U(p)$ of p such that

$$U(p) \subset S_{1/i}(p).$$

Conversely, let $U(p)$ be a given nbd of p. Then by (ii) there is an n such that

$$S(p, \mathcal{U}_n) \subset U(p).$$

Take a natural number i with $1/i < 1/2^n$ and suppose $\rho(p, q) < 1/i$ for a point q of X. Then

$$\varphi(p, q) \leq \rho(p, q) < 1/2^n.$$

Therefore by the definition of φ

$$q \in S(p, \mathcal{V}(1/2^n)) = S(p, \mathcal{U}_n) \subset U(p),$$

i.e.,

$$S_{1/i}(p) \subset U(p).$$

This proves that $\{S_{1/i}(p) \mid i = 1, 2, \ldots\}$ forms a nbd base at p. Thus ρ is a metric compatible with the topology of X, and hence by A) X is a metrizable space.

Corollary. *A T_2-space X is metrizable if and only if it is paracompact and developable, where a topological space is called* developable *if it has a sequence $\{\mathcal{U}_1, \mathcal{U}_2, \ldots\}$ of open covers such that $\{S(p, \mathcal{U}_n) \mid n = 1, 2, \ldots\}$ is a nbd base of each point p of X, and $\{\mathcal{U}_n \mid n = 1, 2, \ldots\}$ is called a* development *of X.*

Proof. The easy proof is left to the reader. (Use full-normality of a paracompact T_2-space.)

Theorem VI.2. *A T_1-space X is metrizable if and only if for each point p of X, there exist two sequences $\{U_n(p) \mid n = 1, 2, \ldots\}$ and $\{V_n(p) \mid n = 1, 2, \ldots\}$ of nbds of p such that*

(i) *$\{U_n(p) \mid n = 1, 2, \ldots\}$ is a nbd basis of p,*
(ii) *$q \not\in U_n(p)$ implies $V_n(q) \cap V_n(p) = \emptyset$,[1]*
(iii) *$q \in V_n(p)$ implies $V_n(q) \subset U_n(p)$.*

Proof.[2] The necessity is almost clear if we put

$$U_n(p) = S_{1/2^n}(p), \qquad V_n(p) = S_{1/2^{n+1}}(p).$$

Therefore we shall only prove the sufficiency. First of all we can prove that X is paracompact if it satisfies the conditions of the theorem. To show this we consider a given open covering $\mathcal{U} = \{U_\alpha \mid \alpha \in A\}$. Put

$$V_{n\alpha} = \bigcup \{(V_n(p))^\circ \mid U_n(p) \subset U_\alpha\},$$
$$\mathcal{V}_n = \{V_{n\alpha} \mid \alpha \in A\}. \tag{1}$$

Then by (i) $\mathcal{V} = \bigcup_{n=1}^{\infty} \mathcal{V}_n$ is an open covering of X such that $\mathcal{V} < \mathcal{U}$. We can prove, moreover, that each $\mathcal{V}_n$ is cushioned in $\mathcal{U}$. To see it, we suppose that A' is a given subset of A. Take a point $q \not\in \bigcup\{U_\alpha \mid \alpha \in A'\}$ and a point p satisfying $U_n(p) \subset U_\alpha$ for some $a \in A'$. Then $q \not\in U_n(p)$, which combined with (ii) implies that

$$V_n(q) \cap V_n(p) = \emptyset,$$

i.e.,

$$V_n(q) \cap (V_n(p))^\circ = \emptyset.$$

Therefore

$$V_n(q) \cap [\bigcup\{V_{n\alpha} \mid \alpha \in A'\}] = \emptyset$$

follows from (1), and hence

$$q \not\in \overline{\bigcup\{V_{n\alpha} \mid \alpha \in A'\}}.$$

---

[1] This condition implies that $\overline{V_n(p)} \subset U_n(p)$ for each p.
[2] We owe the idea of deriving Theorem VI.2 from Theorem V.4 to E. Michael [5].

Thus we have verified that

$$\overline{\cup \{V_{n\alpha} \mid \alpha \in A'\}} \subset \cup \{U_\alpha \mid \alpha \in A'\},$$

proving that $\mathscr{V}_n$ is cushioned in $\mathscr{U}$. Hence $\mathscr{V}$ is a σ-cushioned open refinement of $\mathscr{U}$. Thus by Theorem V.4, we can conclude that X is paracompact. Since X is T_1, the conditions (i) and (ii) obviously imply that X is T_2.

To complete the proof, let us show that if we put

$$\mathscr{W}_m = \{(V_m(q))^\circ \mid q \in X\}, \quad m = 1, 2, \dots, \tag{2}$$

then $\{S(p, \mathscr{W}_m) \mid m = 1, 2, \dots\}$ is a nbd basis of each point p of X. To do so we first note that we can assume $m \geq n$ implies $U_m(p) \subset U_n(p)$ and $V_m(p) \subset V_n(p)$. Because if not, we replace $U_n(p)$ and $V_n(p)$ with $\cap_{i=1}^n U_i(p)$ and $\cap_{i=1}^n V_i(p)$, respectively. Then they satisfy the desired conditions besides the conditions (i)–(iii). Let $U(p)$ be an arbitrary nbd of $p \in X$, and choose n for which $U_n(p) \subset U(p)$. Then by use of (i) we can find $m \geq n$ such that $U_m(p) \subset V_n(p)$. If $p \in (V_m(q))^\circ$ for some point q, then

$$V_m(q) \cap V_m(p) \neq \emptyset,$$

which combined with (ii) implies

$$q \in U_m(p) \subset V_n(p).$$

Hence it follows from (iii) that

$$V_m(q) \subset V_n(q) \subset U_n(p).$$

This of course implies

$$(V_m(q))^\circ \subset U_n(p).$$

Therefore we have

$$S(p, \mathscr{W}_m) \subset U_n(p) \subset U(p) \quad \text{(see (2))}.$$

This means that $\{S(p, \mathscr{W}_m) \mid m = 1, 2, \dots\}$ is a nbd basis of p. Note that each $\mathscr{W}_m$ is an open covering of X.

Thus by the Corollary of Theorem VI.1, X is metrizable.

Corollary. *A T_1-space X is metrizable if and only if it satisfies one of the following conditions*:

(i) *There exists a nbd basis $\{W_n(p) \mid n = 1, 2, \ldots\}$ for each point p of X such that for every n and $p \in X$, there exists an $m = m(n, p)$ for which $W_m(p) \cap W_m(q) \neq \emptyset$ implies $W_m(q) \subset W_n(p)$.*

(ii) *There exists a sequence $\{\mathscr{F}_n \mid n = 1, 2, \ldots\}$ of closure-preserving closed coverings of X such that for every nbd $U(p)$ of each point p of X, there is an n for which $S(p, \mathscr{F}_n) \subset U(p)$.*

(iii) *There exists a sequence $\{\mathscr{U}_n \mid n = 1, 2, \ldots\}$ of open coverings of X such that $\{S^2(p, \mathscr{U}_n) \mid n = 1, 2, \ldots\}$ is a nbd basis of each point p of X.*[1]

Proof. The necessity of the conditions is almost obvious. We shall verify only the sufficiency.

The sufficiency of (i). We first note that we may assume without loss of generality that $m \geqslant n$ implies $W_m(p) \subset W_n(p)$ and that $m(n, p) \geqslant n$. Putting

$$\mathscr{W}_n = \{W_n(p) \mid p \in X\},$$

we define nbds $U_n(p)$, $V_n(p)$ of each point p of X by

$$U_n(p) = S(p, \mathscr{W}_n), \qquad V_n(p) = W_{m(n,p)}(p).$$

Then we can assert that

$$V_n(p) \cap V_n(q) \neq \emptyset \text{ implies } V_n(q) \subset U_n(p). \tag{1}$$

To see it, suppose that $V_n(p) \cap V_n(q) \neq \emptyset$; then

---

[1] We originally owe (i) to A. H. Frink [1] and (ii) and (iii) to K. Morita [3]. We can derive several other conditions for metrizability from Theorem VI.2. For example, a T_1-space X is metrizable if and only if it satisfies either of the following conditions:

(i)′ There is a non-negative valued function $\varphi(p, q)$ over $X \times X$ such that (a) $\varphi(p, q) = \varphi(q, p)$, (b) for each fixed closed set F, $d(p, F) = \inf\{\varphi(p, q) \mid q \in F\}$ is a continuous function of p, (c) $\{S'_n(p) \mid n = 1, 2, \ldots\}$ where $S'_n(p) = \{q \mid \varphi(p, q) < 1/n\}$ is a nbd basis of each point p of X.

(ii)′ There is a collection $\{f_\alpha \mid \alpha \in A\}$ of real-valued functions over X such that (a) for every subset B of A, $\sup\{f_\beta \mid \beta \in B\}$ and $\inf\{f_\beta \mid \beta \in B\}$ are continuous, (b) for every nbd U of each point p of X, there is $\alpha \in A$ and a real number ε for which $f_\alpha(p) < \varepsilon$, $f_\alpha(X - U) \geqslant \varepsilon$. See J. Nagata [3].

In (i)′ $d(p, F)$ may be regarded as a map from the closed sets of X into $C(X)$. P. Zenor [2] gave a metrizability condition in terms of such maps.

$$W_{m(n,p)}(p) \cap W_{m(n,q)}(q) \neq \emptyset .$$ (2)

In the case that $m(n, p) \leq m(n, q)$ we have

$$W_{m(n,q)}(q) \subset W_{m(n,p)}(q) \subset W_n(p),$$

because, as implied by (2),

$$W_{m(n,p)}(p) \cap W_{m(n,p)}(q) \neq \emptyset .$$

Therefore

$$V_n(q) = W_{m(n,q)}(q) \subset W_n(p) \subset U_n(p).$$

In the case that $m(n, p) \geq m(n, q)$ we have

$$W_{m(n,p)}(p) \subset W_n(q)$$

by an argument quite similar to that of the previous case. Thus $p \in W_n(q)$ which, combined with the definition of $U_n(p)$, implies $W_n(q) \subset U_n(p)$. Therefore

$$V_n(q) = W_{m(n,q)}(q) \subset W_n(q) \subset U_n(p)$$

(note that $m(n, q) \geq n$).

   Thus the validity of (1) is proved. It is clear that (1) implies (ii) and (iii) of Theorem VI.2. To prove (i) of Theorem VI.2, we suppose $U(p)$ is a given nbd of a point p. Then $W_n(p) \subset U(p)$ for some p. Now, from the hypothesis of the theorem we obtain

$$U_{m(n,p)}(p) = S(p, \mathcal{W}_{m(n,p)}) \subset W_n(p) \subset U(p),$$

proving (i). Thus by Theorem VI.2 we can conclude the metrizability of X.

   The sufficiency of (ii).

   Since $\mathcal{F}_n$ is closure-preserving,

$$N_n(p) = \cap \{X - F \mid p \in X - F, F \in \mathcal{F}_n\}$$ (3)

$$= X - \cup \{F \mid p \notin F \in \mathcal{F}_n\}$$

is an open nbd of each point p of X. For each pair (n, m) of natural numbers, we define nbds $U_{nm}(p)$ and $V_{nm}(p)$ of p as follows: If $S(p, \mathscr{F}_m) \subset N_n(p)$, then

$$U_{nm}(p) = N_n(p) \quad \text{and} \quad V_{nm}(p) = N_m(p).$$

If $S(p, \mathscr{F}_m) \not\subset N_n(p)$, then

$$U_{nm}(p) = X \quad \text{and} \quad V_{nm}(p) = N_m(p).$$

Since $\mathscr{F}_m$ covers X, we obtain

$$N_m(p) \subset S(p, \mathscr{F}_m) \quad \text{for each } p \in X \text{ (see (3))}. \tag{4}$$

Suppose $q \notin U_{nm}(p)$; then $U_{nm}(p) \neq X$, i.e., it is the case that

$$S(p, \mathscr{F}_m) \subset N_n(p) = U_{nm}(p), \tag{5}$$

which implies

$$q \notin S(p, \mathscr{F}_m).$$

Therefore it follows from (3) and (4) that

$$V_{nm}(q) = N_m(q) \subset X - S(p, \mathscr{F}_m) \subset X - N_m(p) = X - V_{nm}(p).$$

On the other hand, it follows from (3)–(5) that $q \in V_{nm}(p)$ implies

$$V_{nm}(q) = N_m(q) \subset N_m(p) \subset S(p, \mathscr{F}_m) \subset U_{nm}(p).$$

By the hypothesis on $\{\mathscr{F}_n \mid n = 1, 2, \ldots\}$, $\{U_{nm}(p) \mid n, m = 1, 2, \ldots\}$ is easily seen to be a nbd basis of p. Thus $\{U_{nm}(p)\}$ and $\{V_{nm}(p)\}$ satisfy all the conditions required in Theorem VI.2. Hence by the same theorem, X is metrizable.

The sufficiency of (iii).

Put
$$U_n(p) = S^2(p, \mathscr{U}_n) \quad \text{and} \quad V_n(p) = S(p, \mathscr{U}_n).$$

Then all the conditions required in Theorem VI.2 are clearly satisfied, and hence X is metrizable.

Theorem VI.3 (Nagata–Smirnov's metrization theorem). *A regular space* X *is metrizable if and only if it has a σ-locally finite open basis.*

Proof. Necessity. Suppose X is a metric space and put

$$\mathscr{S}_n = \{S_{1/n}(p) \mid p \in X\}.$$

Then, since X is paracompact by Corollary 1 to A. H. Stone's coincidence theorem, there is a locally finite open refinement $\mathscr{V}_n$ of $\mathscr{S}_n$. It is clear that $\bigcup_{n=1}^{\infty} \mathscr{V}_n$ is a σ-locally finite open basis of X.

Sufficiency. Let $\mathscr{V} = \bigcup_{n=1}^{\infty} \mathscr{V}_n$ be a σ-locally finite open basis of a regular space X, where each $\mathscr{V}_n$ is a locally finite open collection. For every pair (n, m) of natural numbers and each point p of X, we define nbds $U_{nm}(p)$ and $V_{nm}(p)$ as follows. Put

$$V_n(p) = \cap\{V \mid p \in V \in \mathscr{V}_n\}.$$

Then $V_n(p)$ is an open nbd of p because of the local finiteness of $\mathscr{V}_n$. Now, if

$$p \in U \subset \bar{U} \subset V_n(p) \quad \text{for some } U \in \mathscr{V}_m,$$

then we put

$$U_{nm}(p) = V_n(p)$$

and

$$V_{nm}(p) = U \cap [\cap\{X - \bar{V} \mid p \notin \bar{V}, V \in \mathscr{V}_m\}];$$

otherwise we put

$$U_{nm}(p) = X$$

and

$$V_{nm}(p) = V_n(p) \cap [\cap\{X - \bar{V} \mid p \notin \bar{V}, V \in \mathscr{V}_m\}].$$

Since $\mathscr{V}_m$ is locally finite, $V_{nm}(p)$ is an open nbd of p. Suppose $W(p)$ is a given nbd of a point p of X; then since $\mathscr{V}$ is an open basis and X is regular, we can choose n, m and $U \in \mathscr{V}_m$ such that

$$p \in U \subset \bar{U} \subset V_n(p) \subset W(p).$$

Hence by the definition of $U_{nm}(p)$,

$$U_{nm}(p) = V_n(p) \subset U(p),$$

proving that $\{U_{nm}(p) \mid n, m = 1, 2, \ldots\}$ is a nbd basis of p.

On the other hand, let $q \not\in U_{nm}(p)$; then this implies that $U_{nm}(p) \neq X$, and therefore that there is some $U \in \mathscr{V}_m$ such that

$$p \in U \subset \bar{U} \subset V_n(p) = U_{nm}(p) \quad \text{and} \quad V_{nm}(p) \subset U.$$

Hence $q \not\in \bar{U}$, which combined with the definition of $V_{nm}(q)$ implies

$$V_{nm}(q) \subset X - \bar{U}.$$

Since $V_{nm}(p) \subset U$, we get

$$V_{nm}(q) \cap V_{nm}(p) = \emptyset.$$

Finally, suppose $q \in V_{nm}(p)$. Since $V_{nm}(p) \subset V_n(p)$ follows from the definition of $V_{nm}(p)$,

$$q \in V_n(p) = \cap\{V \mid p \in V \in \mathscr{V}_n\}.$$

Hence
$$V_n(q) = \cap\{V \mid q \in V \in \mathscr{V}_n\} \subset V_n(p).$$

Since $V_{nm}(q) \subset V_n(q)$, we obtain

$$V_{nm}(q) \subset V_n(p) = U_{nm}(p)$$

in the case that $U_{nm}(p) = V_n(p)$. If $U_{nm}(p) = X$, then $V_{nm}(q) \subset U_{nm}(p)$ is clear. Thus $\{U_{nm}(p) \mid n, m = 1, 2, \ldots\}$ and $\{V_{nm}(p) \mid n, m = 1, 2, \ldots\}$ satisfy the conditions (i), (ii) and (iii) of Theorem VI.2, and hence X is metrizable.[1]

As a corollary of this theorem we obtain the following which gives a complete answer to the metrization problem in the separable case and is a variation of Theorem III.6.

Corollary 1 (Urysohn's metrization theorem). *A topological space satisfying the 2nd axiom of countability is metrizable if and only if it is regular.*

---

[1] As for other proofs of this theorem, see J. Nagata [2] or Yu. M. Smirnov [2]. A sketch of the former proof will be found in Example VI.1.

Corollary 2. *A compact T_2-space X is metrizable if and only if in the product space $X \times X$ the diagonal set $\Delta = \{(p, p) \mid p \in X\}$ is G_δ.*[1]

Proof. The necessity of the condition is clear, because $X \times X$ is metrizable. To see the sufficiency, we use the following proposition B) to get a sequence $\mathcal{U}_1, \mathcal{U}_2, \ldots$ of open covers such that $\bigcap_{n=1}^{\infty} S(p, \mathcal{U}_n) = \{p\}$ for each $p \in X$. Since X is compact and regular, there are finite open covers $\mathcal{V}_n$, $n = 1, 2, \ldots$, such that $\bar{\mathcal{V}}_n < \mathcal{U}_n$. Then $\bigcap_{n=1}^{\infty} \overline{S(p, \mathcal{V}_n)} = \{p\}$ for every $p \in X$. Let $\mathcal{B} = \{X - \bar{V}_1 \cup \cdots \cup \bar{V}_k \mid V_1, \ldots, V_k \in \bigcup_{n=1}^{\infty} \mathcal{V}_n; k = 1, 2, \ldots\}$. Then $\mathcal{B}$ is a countable open collection. Now we claim that $\mathcal{B}$ is a base for X. To prove it, let p be a point of X and U an open nbd of p. Then for each $x \in X - U$, there is $V(x) \in \bigcup_{n=1}^{\infty} \mathcal{V}_n$ such that $p \notin \overline{V(x)}$. Cover $X - U$ by finitely many of $V(x)$'s, say $V(x_1), \ldots, V(x_k)$. Then $X - \overline{V(x_1)} \cup \cdots \cup \overline{V(x_k)} \in \mathcal{B}$, and

$$p \in X - \overline{V(x_1)} \cup \cdots \cup \overline{V(x_k)} \subset U.$$

Thus $\mathcal{B}$ is a countable base for X, and hence by Corollary 1 X is metrizable.

B) *Let X be a topological space. Then $\Delta = \{(p, p) \mid p \in X\}$ is G_δ in $X \times X$ if and only if X has a sequence $\mathcal{U}_1, \mathcal{U}_2, \ldots$ of open covers such that $\bigcap_{n=1}^{\infty} S(p, \mathcal{U}_n) = \{p\}$ for every $p \in X$.*

Proof. Assume that Δ is G_δ. Then, let $\Delta = \bigcap_{n=1}^{\infty} W_n$, where each W_n is an open set of $X \times X$. To each $x \in X$ we can assign an open nbd $U_n(x)$ such that $U_n(x) \times U_n(x) \subset W_n$. Put

$$\mathcal{U}_n = \{U_n(x) \mid x \in X\}.$$

Then it is easy to see that $\mathcal{U}_1, \mathcal{U}_2, \ldots$ satisfy the above condition. Conversely, if $\mathcal{U}_1, \mathcal{U}_2, \ldots$ is a sequence satisfying the condition, then put $W_n = \bigcup \{U \times U \mid U \in \mathcal{U}_n\}$. It is easy to see that W_n, $n = 1, 2, \ldots$, are open sets of $X \times X$ such that

$$\bigcap_{n=1}^{\infty} W_n = \Delta.$$

---

[1] For brevity such a space X is said to have a G_δ-*diagonal*.

Example VI.2. We can derive Theorem VI.3 directly from Alexandroff–Urysohn's metrization theorem as follows. Suppose $\bigcup_{n=1}^{\infty} \mathscr{V}_n$ is a σ-locally finite open basis of X, where we assume each $\mathscr{V}_n$ is a locally finite open covering of X. First, X is normal. For, let F and G be disjoint closed sets of X. Put

$$U_n = \bigcup \{V \mid V \in \mathscr{V}_n, \; \bar{V} \cap G = \emptyset\}$$

and

$$V_n = \bigcup \{V \mid V \in \mathscr{V}_n, \; \bar{V} \cap F = \emptyset\}.$$

Then U_n and V_n are open sets satisfying

$$\bar{U}_n \cap G = \emptyset, \qquad \bar{V}_n \cap F = \emptyset,$$

$$\bigcup_{n=1}^{\infty} U_n \supset F \quad \text{and} \quad \bigcup_{n=1}^{\infty} V_n \supset G.$$

Therefore

$$U = \bigcup_{n=1}^{\infty} \left(U_n - \bigcup_{i=1}^{n} \bar{V}_i \right) \quad \text{and} \quad V = \bigcup_{n=1}^{\infty} \left(V_n - \bigcup_{i=1}^{n} \bar{U}_i \right)$$

are disjoint open sets containing F and G respectively. Secondly, X is perfectly normal. For using the notation in the above, we obtain

$$F = \bigcap_{n=1}^{\infty} (X - \bar{V}_n),$$

which proves that F is G_δ.

Thirdly, it follows from the normality of X that for every locally finite open covering $\mathscr{U}$ of X, there is a locally finite open covering $\mathscr{N}$ such that $\mathscr{N}^\Delta < \mathscr{U}$. The method of proof is quite analogous to the last part of the proof of III.3.E) and therefore it is left to the reader. Now, let $\mathscr{V}_n = \{V_\alpha \mid \alpha \in A_n\}$; then using the perfect normality of X, we express $X - V_\alpha$ as

$$X - V_\alpha = \bigcap_{k=1}^{\infty} V_\alpha^k,$$

where V_α^k are open sets. For each finite subset A' of A_n we put

Then
$$W(k, A') = [\cap\{V_\alpha \mid \alpha \in A'\}] \cap [\cap\{V_\alpha^k \mid \alpha \in A_n - A'\}] .$$

$$\mathcal{W}_{nk} = \{W(k, A') \mid A' \text{ is a finite subset of } A_n\}$$

is a locally finite open covering of X. On the other hand, for every nbd $W(p)$ of each point p of X, there are n and k for which

$$S(p, \mathcal{W}_{nk}) \subset W(p) .$$

Thus using the third assertion in the above and an argument analogous with that in the last part of the proof of Theorem VI.2 we can construct a sequence of open coverings satisfying the condition of Theorem VI.1. Therefore X is metrizable.

Theorem VI.4 (Bing's metrization theorem). *A regular space X is metrizable if and only if it has a σ-discrete open basis.*

Proof. The sufficiency of the condition is a direct consequence of Theorem VI.3 because every discrete collection is locally finite. As for the necessity, by Corollary 2 of Theorem V.2, for each n the open covering $\mathcal{S}_n = \{S_{1/n}(p) \mid p \in X\}$ has a σ-discrete open refinement $\mathcal{U}_n$. Thus $\bigcup_{n=1}^\infty \mathcal{U}_n$ is easily seen to be a σ-discrete open basis of X.[1]

Corollary. *A topological space X is metrizable if and only if it satisfies one of the following conditions*:
   (i) *X is regular, and there exists a sequence $\{\mathcal{U}_n \mid n = 1, 2, \ldots\}$ of closure-preserving open collections of X such that $\{S(p, \mathcal{U}_n) \mid n = 1, 2, \ldots, S(p, \mathcal{U}_n) \neq \emptyset\}$ is a nbd basis of each point p of X.*
   (ii) *X is collectionwise normal and developable.*

Proof. Since the necessity of either condition is clear, we shall only prove the sufficiency.
   The sufficiency of (i). Using Zermelo's theorem, we regard each $\mathcal{U}_n$ as a well-ordered collection,

$$\mathcal{U}_n = \{U_{\alpha n} \mid 0 \leqslant \alpha < \tau_n\} ,$$

where τ_n denotes a fixed ordinal number. For every triplet (l, m, n) of

[1] See R. H. Bing [1] for a direct proof.

natural numbers, we define an open collection $\mathcal{W}_{lmn}$ as follows:

$$V(U_{\alpha n}) = \cup \{U \mid U \in \mathcal{U}_m, \bar{U} \subset U_{\alpha n}\},$$

$$W(U_{\alpha n}) = \cup \{U \mid U \in \mathcal{U}_l, \bar{U} \subset V(U_{\alpha n})\},$$

$$W_{\alpha n} = W(U_{\alpha n}) - \cup \{\overline{V(U_{\beta n})} \mid \beta < \alpha\}, \tag{1}$$

and

$$\mathcal{W}_{lmn} = \{W_{\alpha n} \mid 0 \leq \alpha < \tau_n\}.$$

Note that $\{W(U_{\alpha n}) \mid 0 \leq \alpha < \tau_n\}$ is closure-preserving, because $\mathcal{U}_l$ is so. Then $\mathcal{W}_{lmn}$ is discrete. For if $p \notin \overline{W(U_{\alpha n})}$ for every α, then $X - \cup \{\overline{W(U_{\alpha n})} \mid 0 \leq \alpha < \tau_n\}$ is an open nbd of p intersecting no member of $\mathcal{W}_{lmn}$. Otherwise, we denote by α the first ordinal number for which

$$p \in \overline{W(U_{\alpha n})}.$$

Then $[X - \cup \{\overline{W(U_{\beta n})} \mid \beta < \alpha\}] \cap V(U_{\alpha n})$ is an open nbd of p which intersects no member of $\mathcal{W}_{lmn}$ except $W_{\alpha n}$. To see that $\cup \{\mathcal{W}_{lmn} \mid l, m, n = 1, 2, \ldots\}$ is an open basis of X, we suppose $N(p)$ is a given nbd of a point p of X. Then there is an n for which

$$\emptyset \neq S(p, \mathcal{U}_n) \subset N(p). \tag{2}$$

Let α be the first ordinal number for which $p \in U_{\alpha n}$. Now, since X is regular, there are m, l, $U_1 \in \mathcal{U}_m$ and $U_2 \in \mathcal{U}_l$ such that

$$p \in U_2 \subset \bar{U}_2 \subset U_1 \subset \bar{U}_1 \subset U_{\alpha n}.$$

Therefore $p \in W(U_{\alpha n})$ for the triplet (l, m, n). Since by the definition of α,

$$p \notin U_{\beta n} \supset \overline{V(U_{\beta n})} \quad \text{for every } \beta \text{ with } \beta < \alpha,$$

it follows from the definition (1) of $W_{\alpha n}$ that

$$p \in W_{\alpha n} \subset U_{\alpha n} \subset N(p) \quad (\text{see (2)}).$$

This proves that $\{\mathcal{W}_{lmn} \mid l, m, n = 1, 2, \ldots\}$ is an open basis. Thus by Theorem VI.4, X is metrizable.

The sufficiency of (ii). In view of the corollary of Theorem VI.1, all we have to show is that X is paracompact. Let $\mathcal{W} = \{W_\alpha \mid 0 \le \alpha < \tau\}$ be a given open covering of X. Then for each natural number n and ordinal number α with $0 \le \alpha < \tau$, we put

$$F_{\alpha n} = [X - S(X - W_\alpha, \mathcal{U}_n)] - \bigcup \{W_\beta \mid \beta < \alpha\}.$$

Then $\{F_{\alpha n} \mid 0 \le \alpha < \tau\}$ is easily seen to be a discrete closed collection. Hence by virtue of V.3.A), there is a discrete open collection $\{U_{\alpha n} \mid 0 \le \alpha < \tau\}$ for which $U_{\alpha n} \supset F_{\alpha n}$. We can choose $U_{\alpha n}$ such that $U_{\alpha n} \subset W_\alpha$, because $F_{\alpha n} \subset W_\alpha$. Now

$$\mathcal{U} = \{U_{\alpha n} \mid 0 \le \alpha < \tau, n = 1, 2, \ldots\}$$

is a σ-discrete and accordingly σ-locally finite open refinement of $\mathcal{W}$. Because for a given point p of X, we can determine the first ordinal number α such that $p \in W_\alpha$ and a natural number n such that $S(p, \mathcal{U}_n) \subset W_\alpha$; then $p \in F_{\alpha n} \subset U_{\alpha n}$ is obvious, i.e. $\mathcal{W}$ covers X. Therefore by Theorem V.1, X is paracompact. This proves the metrizability of X.

A regular developable space is called a *Moore space*. A famous question on Moore space is: Is every normal Moore space metrizable? This question is not completely settled yet. In the following is an important theorem of F. B. Jones [1] (Theorem VI.5) which partially answers this question.

C) *Assume the continuum hypothesis (CH). Then every separable normal Moore space X is $\aleph_1$-compact, i.e. every uncountable subset Y of X has a cluster point.*

Proof. Suppose D is a countable dense subset of X. Assume that Y is an uncountable set with no cluster point. Then, because of CH, we may assume $|Y| = \mathfrak{c}$. Suppose Z is a given subset of Y. Then Z and $Y - Z$ are disjoint closed sets. Hence we can select an open set $U(Z)$ such that

$$Z \subset U(Z), \qquad \overline{U(Z)} \cap (Y - Z) = \emptyset.$$

If Z_1 and Z_2 are distinct subsets of Y, then either $U(Z_1) - \overline{U(Z_2)} \ne \emptyset$ or $U(Z_2) - \overline{U(Z_1)} \ne \emptyset$ holds. Hence

$$U(Z_1) \cap D \ne U(Z_2) \cap D.$$

Consider the map which associates Z $(\subset Y)$ with $U(Z) \cap D$ $(\subset D)$. Then this map is one-to-one, which implies that

$$2^c = |2^Y| \leqslant |2^D| = c,$$

a contradiction. Hence X is $\aleph_1$-compact.

Theorem VI.5. *Assume CH. Then every separable normal Moore space X is metrizable.*

Proof. Let $\mathcal{U}_1, \mathcal{U}_2, \ldots$ be a sequence of open covers of X such that $\{S(x, \mathcal{U}_n) \mid n = 1, 2, \ldots\}$ is a nbd base of each $x \in X$. Now, it suffices to show that X is Lindelöf. Because then each $\mathcal{U}_n$ has a countable subcover $\mathcal{U}'_n$ and thus $\bigcup_{n=1}^{\infty} \mathcal{U}'_n$ is a countable base for X. Hence metrizability of X follows from Urysohn's metrization theorem.

To prove that X is Lindelöf, assume that $\mathcal{V}$ is a given open cover of X. For a fixed n we put

$$X_n = \{x \in X \mid S(x, \mathcal{U}_n) \subset V \text{ for some } V \in \mathcal{V}\};$$

well-order the points of X_n to express $X_n = \{x_\gamma \mid 0 \leqslant \gamma < \tau\}$. We define a sequence $\{\gamma_i\}$ of ordinal numbers with $0 \leqslant \gamma_i < \tau$ and a sequence $\{V_i\}$ of elements of $\mathcal{V}$, by use of induction as follows. Let $\gamma_0 = 0$. Select $V_0 \in \mathcal{V}$ such that $S(x_0, \mathcal{U}_n) \subset V_0$. Suppose γ_i have been defined for all $\alpha < \beta$, and also $V_\alpha \in \mathcal{V}$ such that $S(x_{\gamma_\alpha}, \mathcal{U}_n) \subset V_\alpha$. Then

$$\gamma_\beta = \min \{\gamma \mid x_\gamma \in X_n - \bigcup_{\alpha < \beta} V_\alpha\} \quad \text{if } X_n \not\subset \bigcup_{\alpha < \beta} V_\alpha.$$

Obviously the sequence $\{x_{\gamma_i}\}$ has no cluster point, and hence by C) $X_n \subset \bigcup_{\alpha < \beta} V_\alpha$ for some countable ordinal number β. Since $X = \bigcup_{n=1}^{\infty} X_n$, X is covered by a countable subcover of $\mathcal{V}$. Thus X is Lindelöf, and accordingly metrizable.

Recently, some problems are studied in relation to axiomatic set theory. In this aspect, different answers may be obtained to the same question, according to what axioms one may assume, and this is the case in the above *normal Moore space question*. It is known that the usual set theory axioms (Zermelo–Fraenkel's set theory axioms including the axiom of choice) + the negation of CH + *Martin's axiom* (= If X is a compact T_2-space which satisfies *CCC* (= *countable chain condition* = Any disjoint open collection is at most countable), then the intersection of fewer than $2^{\aleph_0}$ dense, open sets is dense) are consistent. Assuming these axioms[1] it is proved that there is a separable, normal, non-metrizable Moore space. (See F. D. Tall [2].) T. Przymusiński–F. Tall [1] proved that the statement: A normal Moore space satisfying CCC is metrizable, is independent of the set theory axioms. W. Fleissner [1], [2] constructed a normal, non-metrizable Moore space assuming CH and proved that the consistency of the statement: All normal Moore spaces are metrizable, implies the consistency of the existence of a measurable cardinal number.[2]

Example VI.3. In view of Theorem VI.3 and (i) of the corollary to Theorem VI.4, the reader may wonder whether or not the existence of a σ-closure-preserving open basis is sufficient for a regular space to be metrizable. However, it is not sufficient, as seen in the following example. Let $X = \{(x, y) \mid (x, y) \in E^2, 0 < x < 1, y \geq 0\}$. For $(p, 0) \in X$ we define

$$U_n(p, 0) = (p, 0) \cup \left\{(x, y) \mid (x, y) \in X, \ y < n - (n^2 - (x - p)^2)^{1/2}, \right.$$

$$\left. |x - p| < \frac{1}{n}\right\}$$

and define $\{U_n(p, 0) \mid n = 1, 2, \ldots\}$ as a nbd basis of $(p, 0)$. We define nbd bases for the other points of X as usual. Then, as easily seen, X is a regular space satisfying the 1st axiom of countability, but non-metrizable because it does not satisfy the 2nd axiom of countability, although it is separable. We can also prove that X has a σ-closure-preserving open basis.[3] Spaces with a σ-closure-preserving base will be studied later.

---

[1] Also proved under the same assumption is that every perfectly normal, compact space is separable (I. Juhász). Another interesting example in this aspect is *Souslin's conjecture*: Let L satisfy the following conditions; then L is homeomorphic to the real line: (i) L is a totally ordered set without first or last element, (ii) L is a topological space with the interval topology, (iii) L is connected and satisfies CCC. It is known that both affirmative and negative answers to Souslin's conjecture are consistent with the usual set theory axioms. Furthermore, if the negation of CH and Martin's axiom (besides the set theory axioms) are assumed, then Souslin's conjecture is true (R.M. Solovay–S. Tennenbaum).

[2] See also the following survey articles and research papers in this aspect: D. Traylor [1], M. E. Rudin [2], [7], F. D. Tall [1], [3], and P. Nyikos [2].

[3] As for the proof, see J. Ceder [1].

The existence of a σ-point-finite open basis is not sufficient for metrizability of a regular space either. For example, Michael line M is a regular space with a σ-point-finite open basis, but non-metrizable. On the other hand, not every metric space has a σ-*star-finite open basis* (an open basis which is the union of countably many star-finite open coverings). Every star space $S(A)$ with an uncountable index set A is a metric space which has no σ-star-finite open basis. The existence of a σ-star-finite open basis is not sufficient for a metric space to be strongly paracompact. Let $N(A)$ be a generalized Baire's zero-dimensional space with an uncountable index set A. Then the product space $N(A) \times \{x \mid 0 < x < 1\}$ has a σ-star-finite open basis, but is not strongly paracompact.[1]

Now, let us study other types of metrization theorems which were obtained by P. S. Alexandroff [2] and A. Arhangelskii [1].

An open basis $\mathcal{U}$ of a topological space X is called a *regular open basis* if for every nbd V of each point p of X, there is a nbd W of p such that at most finitely many members U of $\mathcal{U}$ satisfy $U \cap W \neq \emptyset$ and $U \cap (X - V) \neq \emptyset$ at the same time.

An open basis $\mathcal{U}$ is called a *uniform open basis* if for every nbd V of each point p of X, there are at most finitely many members U of $\mathcal{U}$ for which $p \in U \not\subseteq V$.

It is clear that the former is a stronger condition than the latter.

Theorem VI.6. *A collectionwise normal space X is metrizable if and only if it has a uniform open basis $\mathcal{U}$.*[2]

Proof. To see the necessity, for every n we choose a locally finite open refinement $\mathcal{U}_n$ of $\{S_{1/n}(p) \mid p \in X\}$. Then $\mathcal{U} = \bigcup_{n=1}^{\infty} \mathcal{U}_n$ is a uniform open basis.

Conversely, to prove the sufficiency of the theorem, we first note that every non-empty member of the uniform open basis $\mathcal{U}$ is contained in at most finitely many members of $\mathcal{U}$. For, if $U \in \mathcal{U}$ is contained in infinitely many distinct members $U_1, U_2, \ldots$ of $\mathcal{U}$, then suppose every U_i is different from U and take a point p of U. Then each U_i contains p and

---

[1] Exercise V.9. See J. Nagata [4].

[2] A. Arhangelskii [9] called a base $\mathcal{U}$ '*of countable order*' if for any sequence $U_1 \supsetneq U_2 \supsetneq \cdots$ of elements of $\mathcal{U}$ with $x \in \bigcup_{i=1}^{\infty} U_i$, $\{U_i\}$ is a nbd base of x. He proved that a paracompact T_2-space is metrizable if and only if it has a base of countable order. J. Worrell–H. Wicke [1] studied spaces with a base of countable order and especially proved that a T_1-space is developable if and only if it is θ-refinable and has a base of countable order.

intersects the complement $X - U$ of the nbd U of p. This contradicts the fact that U is a uniform open basis.

For every $U \in \mathcal{U}$, we denote by $r(U)$ the number of the members of $\mathcal{U}$ which contain U. Put

$$\mathcal{U}_n = \{U \mid U \in \mathcal{U}, r(U) \geq n\} \cup \{\{p\} \mid p \text{ is an isolated point of } X\}.$$

Then each $\mathcal{U}_n$ is easily seen to be an open covering of X. Let us prove that $\{S(p, \mathcal{U}_n) \mid n = 1, 2, \ldots\}$ is a nbd basis of each point p of X. To do so, we suppose V is a given nbd of p. Since $\mathcal{U}$ is a uniform open basis, at most finitely many members of $\mathcal{U}$, say $U_1, \ldots, U_n$, meet both p and $X - V$. Then

$$r(U_i) \leq n, \quad i = 1, \ldots, n,$$

and hence

$$U_i \notin \mathcal{U}_{n+1}, \quad i = 1, \ldots, n.$$

This means that every member of $\mathcal{U}_{n+1}$ containing p is contained in V. Hence $S(p, \mathcal{U}_{n+1}) \subset V$ proving our assertion. Thus by (ii) of the corollary to Theorem VI.4, X is metrizable.

Corollary. *A topological space X is metacompact and developable if and only if X has a uniform open basis $\mathcal{U}$.*

Proof. Necessity of the condition is easy to prove. So assume the condition. It follows from the proof of the theorem that X is developable. Let $\mathcal{V}$ be a given open cover of X. Then to each $x \in X$ we assign $U(x) \in \mathcal{U}$ such that (i) $x \in U(x) \subset V$ for some $V \in \mathcal{V}$, (ii) for any $U' \in \mathcal{U}$ with $r(U') < r(U(x))$ and $x \in U'$, U' is contained in no element of $\mathcal{V}$. Now it is obvious that $\mathcal{U}' = \{U(x) \mid x \in X\}$ is a point-finite open refinement of $\mathcal{V}$. Hence X is metacompact.

Theorem VI.7. *A T_1-space X is metrizable if and only if it has a regular open basis $\mathcal{U}$.*

Proof. See the proof of the necessity of Theorem VI.6, where $\mathcal{U}$ is obviously a regular open basis. Thus we shall prove only the sufficiency. First we can prove that X is regular. For, let V be a given open nbd of a point p of X.

Then there is a nbd W of p such that at most finitely many members of $\mathcal{U}$ intersect both W and $X - V$. We may assume that W is an open nbd. Now, suppose $q \notin V$. Then, since X is T_1, we know that $\mathcal{U}$ has infinitely many members containing q unless q is an isolated point. Therefore, there is a member U' of $\mathcal{U}$ such that $q \in U'$ and $U' \cap W = \emptyset$. If q is an isolated point, then $\{q\}$ is a nbd of q which does not intersect W. This means that $q \notin \overline{W}$, i.e. $\overline{W} \subset V$. Thus X is regular.

To show the metrizability of X, we shall prove in general that every regular open basis is a σ-locally finite open basis. In fact, since $\mathcal{U}$ is regular, every member U of $\mathcal{U}$ is contained in at most finitely many members of $\mathcal{U}$ as shown in the proof of Theorem VI.6. We call $U \in \mathcal{U}$ a maximal element of $\mathcal{U}$ if U is contained in no other member of $\mathcal{U}$. Denote by $\mathcal{U}_1$ the collection of all maximal elements of $\mathcal{U}$. Then $\mathcal{U}_1$ is locally finite. For, let p be a given point of X. Take a member U_0 of $\mathcal{U}_1$ which contains p. If U_0 meets infinitely many members of $\mathcal{U}_1$, then all of them but U_0 itself is not contained in U_0, because they are maximal elements. Hence there must be a nbd of p which intersects only finitely many of them, because $\mathcal{U}$ is regular. This means that $\mathcal{U}_1$ is locally finite.

Now, denote by $\mathcal{U}_2$ the collection of the maximal elements of $\mathcal{U} - \mathcal{U}_1$. Then we can verify in a similar way that $\mathcal{U}_2$ is locally finite. To prove the local finiteness of $\mathcal{U}_2$, we must consider such a special case as p is an isolated point and $\{p\} \in \mathcal{U}_1$. In this case, precisely speaking, the argument for $\mathcal{U}_1$ is not valid for $\mathcal{U}_2$, but $\{p\}$ is a nbd of p which meets no member of $\mathcal{U}_2$, and hence it is still true that $\mathcal{U}_2$ is locally finite. Next we denote by $\mathcal{U}_3$ the collection of the maximal elements of $\mathcal{U} - (\mathcal{U}_1 \cup \mathcal{U}_2)$; then $\mathcal{U}_3$ is also locally finite. Repeating this process, we obtain locally finite open collections $\mathcal{U}_1, \mathcal{U}_2, \mathcal{U}_3, \ldots$. Since each member of $\mathcal{U}$ is contained in at most finitely many members of $\mathcal{U}$, we have

$$\mathcal{U} = \bigcup_{i=1}^{\infty} \mathcal{U}_i.$$

Thus $\mathcal{U}$ is a σ-locally finite open basis, and hence by Nagata–Smirnov's metrization theorem, X is metrizable.[1]

---

[1] Another example of an interesting metrization theorem of A. Arhangelskii [3]: A T_2-space X is metrizable if and only if it has a sequence $\mathcal{U}_1, \mathcal{U}_2, \ldots$ of open coverings such that for every compact set C and its open nbd V, there is i for which $S(C, \mathcal{U}_i) \subset V$. Nowadays, *metrization theory* is growing to a huge compound of various results. R. E. Hodel's nice survey article [3] is recommended in this aspect. See also A. Arhangelskii [12], H. Martin [1], H. Hung [1], [2] for newer results on metrization. R. Hodel [6] extended some metrization theorems to higher cardinality. See also P. Nyikos–H.-C. Reichel [1] in this aspect.

2. Complete metrizability

This section concerns the problem to find conditions in order that a metric satisfying certain conditions can be introduced into a given topological space. For example, the following proposition will be almost clear in view of the fact that every regular space with the 2nd axiom of countability is a subset of a Hilbert cube, which is totally bounded.

A) *A regular space X can be introduced with a compatible totally bounded metric if and only if X satisfies the 2nd axiom of countability.*

Definition VI.2. We call a topological space X *completely metrizable* if we can introduce in X a complete metric. To study a condition for complete metrizability, we give the following definition. A completely regular space X is called *Čech complete* if it is G_δ in its Čech–Stone's compactification $\beta(X)$.

B) *A completely regular space X is Čech complete if and only if it is G_δ in every T_2-compactification if and only if it is G_δ in some T_2-compactification.*

Proof. It is a direct consequence of Corollary 2 to Theorem IV.2.

C) *Every locally compact T_2-space X is Čech complete.*

Proof. Let $p \in X$. Then there is a compact nbd U of p in X. We can assert that

$$\overline{X - U} \supset \beta(X) - X \quad \text{in } \beta(X).$$

If it is not true, then there is

$$q \in \beta(X) - \overline{X - U}.$$

Since

$$\overline{X - U} \cup \bar{U} = \bar{X} = \beta(X),$$

$$q \in \bar{U} \cap (\beta(X) - X).$$

But this is impossible, because U is closed in $\beta(X)$ because of its compactness. Hence,

$$\overline{X - U} \supset \beta(X) - X,$$

i.e.,

$$\beta(X) - \overline{X - U} \subset X.$$

Since $p \notin \overline{X - U}$ is clear, $V = \beta(X) - \overline{X - U}$ is a nbd of p (in $\beta(X)$) which is contained in X. Thus X is an open set of $\beta(X)$, proving C).

D) *Let X be a non-compact Tychonoff space; then X is Čech complete if and only if there is a sequence $\mathscr{F}_i$, $i = 1, 2, \ldots$, of collections of zero sets with f.i.p. such that*

(i) $\cap \{F \mid F \in \mathscr{F}_i\} = \emptyset$, *and*

(ii) *for every non-convergent maximal zero filter $\mathscr{F}$ of X, there is an i for which $\mathscr{F}_i \subset \mathscr{F}$.*[1]

Proof. The necessity. First we note that $\beta(X) = \sigma(X, \mathscr{Z})$ for the collection $\mathscr{Z}$ of all zero sets of X (see IV.2). Therefore $X = \cap_{i=1}^{\infty} U_i$ for some open sets U_i, $i = 1, 2, \ldots$, of $\sigma(X, \mathscr{Z})$ which we denote by $\sigma(X)$. Then each $\sigma(X) - U_i$ is a closed set of $\sigma(X)$. Hence

$$\sigma(X) - U_i = \cap \{\bar{F}_\alpha \mid \alpha \in A_i\}$$

for some zero sets F_α of X, where we denote by $\bar{F}_\alpha$ the closure of F_α in $\sigma(X)$. Since X is non-compact, $X \neq \sigma(X)$ and therefore we may assume that

$$\sigma(X) - U_i \neq \emptyset, \quad i = 1, 2, \ldots.$$

Putting

$$\mathscr{F}_i = \{F_\alpha \mid \alpha \in A_i\},$$

we get a sequence $\{\mathscr{F}_i \mid i = 1, 2, \ldots\}$ of collections of zero sets with f.i.p. For if $\cap_{j=1}^{k} F_{\alpha_j} = \emptyset$ for some $F_{\alpha_j} \in \mathscr{F}_i$, then

$$\bigcap_{j=1}^{k} \bar{F}_{\alpha_j} = \overline{\bigcap_{j=1}^{k} F_{\alpha_j}} = \emptyset \quad \text{(see (ii') of Theorem IV.3)},$$

contradicting $\sigma(X) - U_i \neq \emptyset$. Furthermore, each $\mathscr{F}_i$ satisfies

$$\cap \{F \mid F \in \mathscr{F}_i\} = \emptyset$$

since

$$\cap \{F \mid F \in \mathscr{F}_i\} \subset (\sigma(X) - U_i) \cap X = \emptyset.$$

[1] Due to N. A. Shanin [3].

To prove condition (ii) for $\{\mathcal{F}_i \mid i = 1, 2, \ldots\}$, we suppose $\mathcal{F}$ is a given maximal zero filter which does not converge. We may regard $\mathcal{F}$ as a point p of $\sigma(X) - X$. Hence there is an i for which $p \notin U_i$. Then $p \in \bar{F}$ for every $F \in \mathcal{F}_i$. Hence $F \in \mathcal{F}$ for every $F \in \mathcal{F}_i$, which implies $\mathcal{F}_i \subset \mathcal{F}$, proving (ii).

The sufficiency. Put

$$F_i = \cap \{\bar{F} \mid F \in \mathcal{F}_i\}, \tag{1}$$

where we denote by $\bar{F}$ the closure of F in $\sigma(X)$. Then $U_i = X - F_i$ is an open set of $\sigma(X)$ containing X because of the condition (i). (Note that $\bar{F} \cap X = F$ since $F \in \mathcal{Z}$.) Now, we can show that $X = \cap_{i=1}^{\infty} U_i$. For, if $p \in \sigma(X) - X$, then we may regard p as a non-convergent maximal zero filter $\mathcal{F}$ in X. Hence, by (ii), $\mathcal{F}_i \subset \mathcal{F}$ for some i. This means that $p \in \bar{F}$ for every $F \in \mathcal{F}_i$. Therefore it follows from the definition (1) of F_i that $p \in F_i$, i.e. $p \notin U_i$. This proves that $X = \cap_{i=1}^{\infty} U_i$.

Now, we can prove the following theorem to give a necessary and sufficient condition for complete metrizability.

Theorem VI.8 (Čech's theorem). *A metrizable space X is completely metrizable if and only if it is Čech complete.*

Proof. The necessity. Suppose X is a complete metric space. If X is compact, then it is clearly Čech complete. If X is not compact, then neither is it totally bounded, because a totally bounded complete metric space is compact. Hence there exists some natural number i_0 such that for every i with $i \geqslant i_0$, $\{S_{1/i}(p) \mid p \in X\}$ has no finite subcovering. Therefore

$$\mathcal{F}_i = \{X - S_{1/i}(p) \mid p \in X\}, \quad i = i_0, i_0 + 1, \ldots,$$

are collections of zero sets with f.i.p. $\mathcal{F}_i$ clearly satisfies (i) of D). To prove (ii), we suppose that $\mathcal{F}$ is a given maximal zero filter which does not converge. Since X is complete, there is a positive integer $i \geqslant i_0$ for which no member of $\mathcal{F}$ is contained in any $S_{1/i}(p)$. Therefore

$$(X - S_{1/i}(p)) \cap F \neq \emptyset \quad \text{for all } F \in \mathcal{F}.$$

This implies

$$X - S_{1/i}(p) \in \mathcal{F}, \quad p \in X.$$

Thus $\mathcal{F}_i \subset \mathcal{F}$, proving (ii).

The sufficiency. Since X is metrizable, by Alexandroff–Urysohn's metrization theorem, there is a sequence $\mathcal{V}_1, \mathcal{V}_2, \ldots$ of open coverings of X such that

$$\mathcal{V}_1 > \mathcal{V}_2^* > \mathcal{V}_2 > \mathcal{V}_3^* > \cdots$$

and $\{S(p, \mathcal{V}_i) \mid i = 1, 2, \ldots\}$ is a nbd basis of each $p \in X$. On the other hand, by D) there is a sequence $\mathcal{F}_i$, $i = 1, 2, \ldots$, of collections of zero sets with f.i.p. satisfying (i), (ii) of D). Put

$$\mathcal{U}_i = (X - F \mid F \in \mathcal{F}_i); \tag{1}$$

then we get another sequence $\mathcal{U}_1, \mathcal{U}_2, \ldots$ of open coverings. Since X is fully normal as a metric space, we can construct open coverings $\mathcal{W}_1, \mathcal{W}_2, \ldots$ such that

$$\mathcal{W}_1 < \mathcal{V}_1 \wedge \mathcal{U}_1,$$
$$\mathcal{W}_2^* < \mathcal{W}_1 \wedge \mathcal{V}_2 \wedge \mathcal{U}_2,$$
$$\mathcal{W}_3^* < \mathcal{W}_2 \wedge \mathcal{V}_3 \wedge \mathcal{U}_3,$$
$$\vdots$$
$$\mathcal{W}_i^* < \mathcal{W}_{i-1} \wedge \mathcal{V}_i \wedge \mathcal{U}_i,$$
$$\vdots$$

Then

$$\mathcal{W}_1 > \mathcal{W}_2^* > \mathcal{W}_2 > \mathcal{W}_3^* > \cdots,$$
$$\mathcal{W}_i < \mathcal{U}_i, \qquad \mathcal{W}_i < \mathcal{V}_i, \quad i = 1, 2, \ldots.$$

Therefore $\{S(p, \mathcal{W}_i) \mid i = 1, 2, \ldots\}$ is a nbd basis of each $p \in X$. Using $\mathcal{W}_1, \mathcal{W}_2, \ldots$ in place of $\mathcal{U}_1, \mathcal{U}_2, \ldots$ in the proof of Alexandroff–Urysohn's metrization theorem, we can introduce a metric ρ into X. To show the completeness of this metric, we consider a filter $\mathcal{G}$ which has no cluster point. Then $\bar{\mathcal{G}}$ is a zero filter which has no cluster point and satisfies $\bar{\mathcal{G}} \subset \mathcal{G}$. Construct a maximal zero filter $\mathcal{F}$ with $\mathcal{F} \supset \bar{\mathcal{G}}$. Then $\mathcal{F}$ has no cluster point either. Therefore it does not converge. Hence, by (ii) of D), $\mathcal{F}_i \subset \mathcal{F}$ for some i. Suppose F is a given member of $\mathcal{F}$. Then

$$F \cap F' \neq \emptyset \quad \text{for every } F' \in \mathcal{F}_i, \tag{2}$$

because F, $F' \in \mathcal{F}$. On the other hand, since $\mathcal{W}_i < \mathcal{U}_i$, for every $W \in \mathcal{W}_i$ there is $U \in \mathcal{U}_i$ such that $U \supset W$. Since $U = X - F'$ for some $F' \in \mathcal{F}_i$ (see (1)), this implies that $W \cap F' = \emptyset$. Hence it follows from (2) that $F \not\subset W$ for every $W \in \mathcal{W}_i$. We can easily see from the proof of Alexandroff–Urysohn's metrization theorem that

$$\{S_{1/j}(p) \mid p \in X\} < \mathcal{W}_i, \quad \text{where } j = 2^{i+2}.$$

Therefore $F \not\subset S_{1/j}(p)$ for every $F \in \mathcal{F}$ and every $p \in X$. Thus $\mathcal{F}$ is not Cauchy. Consequently $\bar{\mathcal{G}}$ and $\mathcal{G}$ are not Cauchy either. Hence we can conclude that every Cauchy filter of X has a cluster point and therefore converges. This proves that X is complete with respect to the introduced metric.[1]

We can extend Baire's theorem to a Čech complete space as follows.

Theorem VI.9. *Let U_n, $n = 1, 2, \ldots$, be dense open sets of a Čech complete space X. Then $U = \bigcap_{n=1}^{\infty} U_n$ is dense in X.*

Proof. Let F_i, $i = 1, 2, \ldots$, be a sequence of collections of zero sets of X with f.i.p. satisfying the conditions (i), (ii) of D). Suppose V is a given open nbd of a given point x of X. Then, since $V \cap U_1 \neq \emptyset$, we can select $x_1 \in V \cap U_1$. Since $\bigcap\{F \mid F \in \mathcal{F}_1\} = \emptyset$, there is $F_1 \in \mathcal{F}_1$ and a zero set nbd V_1 of x_1 such that

$$V_1 \subset V \cap U_1 \cap (X - F_1). \tag{1}$$

Since $V_1^{\circ} \cap U_2 \neq \emptyset$, we can select $x_2 \in V_1^{\circ} \cap U_2$. Since $\bigcap\{F \mid F \in \mathcal{F}_2\} = \emptyset$, there is $F_2 \in \mathcal{F}_2$ and a zero set nbd V_2 of x_2 such that

$$V_2 \subset V_1 \cap U_2 \cap (X - F_2).$$

Repeating the same process, we obtain a sequence $F_i \in \mathcal{F}_i$, $i = 1, 2, \ldots$, and a sequence V_i, $i = 1, 2, \ldots$, of non-empty zero sets such that

$$V_i \subset V_{i-1} \cap U_i \cap (X - F_i), \quad i = 2, 3, \ldots. \tag{2}$$

Now, expand $\{V_i \mid i = 1, 2, \ldots\}$ to a maximal zero-filter $\mathcal{F}$. Then from (2) it follows that

---

[1] As for more conditions for complete metrizability, see, for example, J. Nagata [2], [3].

$$F_i \notin \mathscr{F}, \quad i = 1, 2, \ldots,$$

and hence

$$\mathscr{F}_i \not\subset \mathscr{F}, \quad i = 1, 2, \ldots.$$

Thus by (ii) of D) $\mathscr{F} \to p$ for some $p \in X$. Then $p \in \bigcap_{i=1}^{\infty} V_i$ and hence by (2) we obtain

$$p \in \bigcap_{i=1}^{\infty} U_i = U.$$

On the other hand, from (1) it follows that

$$p \in V_1 \subset V.$$

Thus

$$V \cap U \neq \emptyset,$$

i.e. $x \in \bar{U}$. This proves that U is dense in X.

Example VI.4. Let Q and Q' denote the subspaces of E^1 consisting of the rationals and the irrationals, respectively. Then Q' is a G_δ-set of E^1. Since E^1 is homeomorphic to the open interval $(0, 1)$, the closed interval $[0, 1]$ is a T_2-compactification of E^1 and also of Q'. Since Q' is G_δ in $[0, 1]$, it is Čech complete, while it is not locally compact. On the other hand, Q is not Čech complete. Because, if we put $Q = \{x_1, x_2, \ldots\}$, then $U_n = Q - \{x_1, \ldots x_n\}$ is a dense open set of Q satisfying

$$\bigcap_{n=1}^{\infty} U_n = \emptyset.$$

Therefore by Theorem VI.9 Q is not Čech complete.

3. Imbedding

In this section we shall concern ourselves with the problem of topologically imbedding a given metrizable space in a concrete space. Generally, a topological space X is said to be *topologically imbedded* in a topological space Y if X is homeomorphic with a subspace of Y. As a matter of fact, we have already dealt with such a type of problem in Theorem III.6, i.e., imbedding separable metric spaces in the Hilbert cube. Our concern in the present section will be chiefly aimed at general metric spaces. Following C. H. Dowker [1] and H. J. Kowalsky [1], we can topologically imbed a given metrizable space in a generalized Hilbert space and in the product of star-spaces (see Example III.9).

Theorem VI.10. *Let A be a set of power μ where μ is an infinite cardinal number. Then a topological space X with weight $\leq \mu$ is metrizable if and only if it is homeomorphic with a subset of the generalized Hilbert space $H(A)$.*

Proof. Since the sufficiency of the condition is obvious, we shall prove only the necessity. Suppose X is a metric space with weight $\leq \mu$. By Theorem VI.4, there is a σ-discrete open basis $\mathcal{U} = \bigcup_{n=1}^{\infty} \mathcal{U}_n$ where each $\mathcal{U}_n$ is a discrete open collection. Suppose that $\mathcal{U}_n = \{U_\alpha \mid \alpha \in A_n\}$. Then we can express each U_α as $U_\alpha = \bigcup_{i=1}^{\infty} F_{\alpha i}$, for closed sets $F_{\alpha i}$, $i = 1, 2, \ldots$. This means that we may assume, without loss of generality, that there are discrete open collections $\mathcal{U}_n = \{U_\alpha \mid \alpha \in A_n\}$, $n = 1, 2, \ldots$, and closed collections $\{F_\alpha \mid \alpha \in A_n\}$, $n = 1, 2, \ldots$, such that $F_\alpha \subset U_\alpha$ and such that for every nbd V of each point p of X, there is an α for which

$$P \in F_\alpha \subset U_\alpha \subset V.$$

We note that $|A_n| \leq \mu$ is derived from the discreteness of $\mathcal{U}_n$ and consequently $|\bigcup A_n| \leq \mu$ (see Corollary 3 to Theorem I.3). Now, we put $\bigcup_{n=1}^{\infty} A_n = A$. For each $\alpha \in A_n$, we define a continuous function f_α such that

$$f_\alpha(F_\alpha) = \frac{1}{n}, \qquad f_\alpha(X - U_\alpha) = 0 \quad \text{and} \quad 0 \leq f_\alpha \leq \frac{1}{n}.$$

Then, since each $\mathcal{U}_n$ is discrete, $f(p) = \{f_\alpha(p) \mid \alpha \in A\}$ is a continuous mapping of X into the generalized Hilbert space $H(A)$. On the other hand, for every nbd V of each point p of X, there is an α for which

$$p \in F_\alpha \subset U_\alpha \subset V.$$

Hence if $q \in X - V$, then

$$f_\alpha(p) = \frac{1}{n} \quad \text{and} \quad f_\alpha(q) = 0,$$

where n is the number for which $\alpha \in A_n$. Thus

$$\rho(f(p), f(q)) \geq \frac{1}{n} \quad \text{in } H(A),$$

which proves that f is one-to-one, and f^{-1} is continuous. Thus f is a homeomorphic mapping of X onto a subset of $H(A)$.

Theorem VI.10 is a generalization of Theorem III.6, because if $|A| = \mathfrak{a}$, then $H(A)$ turns out to be the separable Hilbert space.

Theorem VI.11. *Let A be a set with power μ where μ is an infinite cardinal number. Then a topological space X with weight $\leq \mu$ is metrizable if and only if it is homeomorphic with a subspace of the product space of countably many star-spaces with index set A.*

Proof. It suffices to prove only the necessity. First pursuing the same discussion and using the same notation as in the previous proof, we define F_α and U_α, $\alpha \in A = \bigcup_{n=1}^{\infty} A_n$. Then for each α we define a continuous function f_α such that

$$f_\alpha(F_\alpha) = 1, \qquad f_\alpha(X - U_\alpha) = 0 \quad \text{and} \quad 0 \leq f_\alpha \leq 1.$$

Now, for each A_n, we construct a star-space $S(A_n)$ and note that $S(A_n)$ is the sum of the unit segments I_α, $\alpha \in A_n$, whose zeros are identified. We put

$$P = \prod_{n=1}^{\infty} S(A_n).$$

Now, for each natural number n, we define a mapping f_n of X into $S(A_n)$ by

$$f_n(p) = \begin{cases} f_\alpha(p) \in I_\alpha & \text{if } p \in U_\alpha, \\ 0 & \text{if } p \notin \bigcup \{U_\alpha \mid \alpha \in A_n\}. \end{cases}$$

Since $\mathcal{U}_n = \{U_\alpha \mid \alpha \in A_n\}$ is discrete, this uniquely defines a mapping f_n

over X. Moreover, it follows from the discreteness of $\mathcal{U}_n$ that f_n is continuous. Hence the mapping

$$f(p) = \{f_n(p) \mid n = 1, 2, \ldots\}$$

is a continuous mapping of X into P. We can also prove in the same way as for Theorem VI.10 that f is a homeomorphism. Thus X is homeomorphic with a subspace of P. Since each $S(A_n)$ is a subspace of $S(A)$, X is homeomorphic with a subspace of the product space of countably many copies of the star-space $S(A)$.[1]

Example VI.5. We have seen in Theorem III.6 that a separable metric space is characterized as a subset of the Hilbert cube. On the other hand, by use of Theorem VI.10 (Theorem VI.11), a completely metrizable space can be characterized as a G_δ subset of a generalized Hilbert space (the countable product of star-spaces). Because, if X is a completely metrizable space, then it is topologically imbedded in a generalized Hilbert space H. Since the closure $\bar{X}$ in $\beta(H)$ is a T_2-compactification of X, X is G_δ in $\bar{X}$. Observe that $\bar{X} \cap H$ is G_δ in H. Thus X is G_δ in H. On the other hand, by virtue of 2.B), it is easy to see that a G_δ-set in a Čech complete space is Čech complete. Now, let us give an imbedding theorem due to C. Kuratowski for a separable completely metrizable space X. By Theorem III.6, X is homeomorphic with a subset X_0 of the Hilbert cube I^ω. Since X_0 is Čech complete, by 2.B) it is G_δ in $\bar{X}_0$ and accordingly in I^ω, too. We consider the product space E^∞ of countably many copies of E^1. Then I^ω is a closed subset of E^∞, and hence X_0 is a G_δ set of E^∞.

We express X_0 as $X_0 = \bigcap_{i=1}^\infty U_i$, where each U_i is an open set of E^∞. Putting

$$f_i(p) = 1/\rho(p, E^\infty - U_i),$$

we get a continuous function f_i defined over U_i. Define a mapping f of X_0 into $X_0 \times E^\infty$ by

---

[1] As for imbedding theory for metric spaces of weight $\leq \mu$ and dimension $\leq n$, see e.g. J. Nagata [5]. In the field of imbedding theory for non-metrizable spaces A. Zarelua [1] and B. Pasynkov [2] obtained for given μ and n a compact T_2-space X such that every completely regular space Y with weight $\leq \mu$ and dimension $\leq n$ is homeomorphic with a subset of X. Another aspect to be noted is representation of mappings between abstract spaces as concrete mappings. M. Edelstein [1] obtained some remarkable results in this aspect.

$$f(p) = \{p, f_1(p), f_2(p), \ldots\}, \quad p \in X_0.$$

It is easily seen that f is a homeomorphism. We may regard f as a mapping of X_0 into $E^\infty \times E^\infty$, because X_0 is a subset of E^∞. Then we can prove that $f(X_0)$ is closed in $E^\infty \times E^\infty$.

To prove that, suppose

$$p_n \in X_0, \quad n = 1, 2, \ldots,$$

i.e. $f(p_n) \in f(X_0)$, and $f(p_n) \to q$ in $E^\infty \times E^\infty$. Let

$$q = (q_0, q_1, q_2, \ldots), \quad \text{where } q_0 \in E^\infty, q_i \in E^1, i = 1, 2, \ldots;$$

then, since $f(p_n) = (p_n, f_1(p_n), f_2(p_n), \ldots)$, we know that

$$p_n \to q_0 \quad \text{in } E^\infty$$

and

$$\lim_{n \to \infty} f_i(p_n) = q_i, \quad i = 1, 2, \ldots.$$

Thus, if $q_0 \in X_0$, then by virtue of the continuity of f_i, we obtain

$$f_i(q_0) = q_i, \quad i = 1, 2, \ldots,$$

i.e.,

$$q = (q_0, f_1(q_0), f_2(q_0), \ldots) = f(q_0) \in f(X_0).$$

On the contrary, if we assume $q_0 \in E^\infty - X_0$, then $q_0 \notin U_i$ for some i. This implies that $\rho(q_0, E^\infty - U_i) = 0$, and hence

$$\lim_{n \to \infty} f_i(p_n) = \lim_{n \to \infty} 1/\rho(p_n, E^\infty - U_i) = \infty.$$

This, however, contradicts the fact that

$$\lim_{n \to \infty} f_i(p_n) = q_i.$$

Therefore we can conclude that $q \in f(X_0)$. This implies that $f(X_0)$ is closed in $E^\infty \times E^\infty$, which is homeomorphic to E^∞. Therefore X is homeomorphic with a closed set of E^∞. Conversely, it is clear that every closed set of E^∞ is a complete, separable metric space because E^∞ is a complete, separable metric space. It was proved by R. D. Anderson [1] that E^{∞} and Hilbert space H are homeomorphic.[1] Hence a topological space is separable and completely metrizable if and only if it is homeomorphic with a closed subset of Hilbert space.

Imbedding theorems help us to visually understand abstract spaces. From another point of view, they topologically characterize subsets of concrete spaces. E.g., subsets of a Hilbert space are topologically characterized as separable metrizable spaces. Then is it possible to characterize the Hilbert space itself in terms of simple topological properties? J. de Groot [4], [5] has obtained amazing results in this aspect. According to De Groot a subbase $\mathcal{U}$ for a space X is *binary* if every cover of X by members of $\mathcal{U}$ has a subcover consisting of two members. $\mathcal{U}$ is *comparable* if for any U, U_1, $U_2 \in \mathcal{U}$ such that $U \cup U_1 = X$, $U \cup U_2 = X$, either $U_1 \subset U_2$ or $U_1 \supset U_2$ is true. J. de Groot's theorem: A topological space X is homeomorphic to the n-cube I^n (or Hilbert cube I^{ω}) if and only if (i) X is metrizable, (ii) X is compact, (iii) X is connected, (iv) dim $X = n$ (or dim $X = \infty$), (v) X has a comparable and binary subbase. He also gave characterizations of E^n, S^n and manifolds. (A topological space X is called a *manifold* if each point of X has a nbd which is homeomorphic to an open set of E^n.)

Example VI.6. Sometimes we consider an imbedding which satisfies a stronger condition. Suppose X and Y are metric spaces. If f is a one-to-one mapping of X onto Y which preserves the metric, then we call f an *isometry*. If there is an isometry which maps X onto a subspace of X, then X is said to be *isometrically imbedded* in Y. Let X be a metric space. Then we denote by $C^*(X)$ the metric space of all bounded continuous functions over X. $C^*(X)$ is not only a complete metric space but also a Banach space. Considering a fixed point p_0 of X, we let each point p of X correspond to a member f_p of $C^*(X)$, where

$$f_p(q) = \rho(p, q) - \rho(p_0, q) .$$

Then it is easy to see that this correspondence is an isometry mapping of X into $C^*(X)$. Thus every metric space can be isometrically imbedded in a Banach space.[2]

---

[1] Since his work, study of topological properties of Hilbert space and related spaces is becoming active under the name of *infinite dimensional topology*. See e.g. R. D. Anderson [2].

[2] This is a result essentially due to C. Kuratowski [2].

4. Union and image of metrizable spaces

It is obvious that a subspace of a metrizable space is also metrizable. We also know that the product space of countably many metrizable spaces is metrizable[1] (Example III.10). But as for union and image of metric spaces, there are interesting problems which are not so easy to answer. Is the union of metrizable spaces also metrizable under some additional condition? Is the image of a metrizable space by a continuous mapping also metrizable with some condition on the mapping? The purpose of the present section is to answer these questions.

Theorem VI.12. *Let $\mathscr{F} = \{F_\alpha \mid \alpha \in A\}$ be a locally finite closed covering of a topological space X. If each subspace F_α is metrizable, then X is also metrizable.*[2]

Proof. First we note that X is T_1 because each F_α is closed and T_1. By Corollary (ii) of Theorem VI.2, there is a sequence $\{\mathscr{F}_{\alpha n} \mid n = 1, 2, \ldots\}$ of locally finite closed coverings of F_α such that for every nbd U of $p \in F_\alpha$, there is an n for which

$$S(p, \mathscr{F}_{\alpha n}) \subset U.$$

We may assume without loss of generality that $\mathscr{F}_{\alpha n+1} < \mathscr{F}_{\alpha n}$ for each $\alpha \in A$. Now, putting

$$\mathscr{F}_n = \bigcup \{\mathscr{F}_{\alpha n} \mid \alpha \in A\}$$

for each n, we obtain a closure-preserving closed covering of X because $\{F_\alpha \mid \alpha \in A\}$ is locally finite. Let U be a given nbd of a point p of X. Then p is contained in only finitely many members of $\mathscr{F}$, say $F_{\alpha_1}, \ldots, F_{\alpha_k}$. For each α_i we can find n_i for which

$$S(p, \mathscr{F}_{n_i \alpha_i}) \subset U.$$

---

[1] We, of course, mean by product space the one with weak topology. It is obvious that the product space of countably many metric spaces with box topology (box product) is not necessarily metrizable. But it is an interesting question if such a box product is paracompact. H. Tamano announced an affirmative answer to this question without giving a proof. M. E. Rudin [4] has proved that continuum hypothesis implies that the box product of countably many locally compact, σ-compact, metric spaces is paracompact.

[2] This theorem and its corollary are due to J. Nagata [2] and Yu. M. Smirnov [2] respectively.

Let $n = \max\{n_1, \ldots, n_k\}$; then $S(p, \mathscr{F}_n) \subset U$. Hence by the same corollary we have used in the above, we can conclude that X is metrizable.

Corollary. *A paracompact T_2-space X is metrizable if it is locally metrizable.*

Proof. Since X is locally metrizable, there is an open covering $\mathscr{U}$ of X each of whose elements is metrizable. Since X is paracompact T_2, there is a closed, locally finite refinement $\mathscr{F}$ of $\mathscr{U}$. Since each member of $\mathscr{F}$ is metrizable, by the theorem X is metrizable.

Now, let us consider a similar problem, replacing a locally finite covering with a countable covering. It may be rather unexpected for the reader that in the countable case we need somewhat more complicated additional conditions. First we need some lemmas to prove Theorem VI.13.

A) *A topological space X which is the union of countably many separable metric spaces is hereditarily Lindelöf.*

Proof. Let $X = \bigcup_{i=1}^{\infty} X_i$, where each X_i is a separable metric space. Suppose A is a given subset of X; then A is the union of the countably many separable metric spaces $A_i = A \cap X_i$, $i = 1, 2, \ldots$. Let $\mathscr{U}$ be a given open covering of A; then for each i, there are countably many members U_{ij}, $j = 1, 2, \ldots$, of $\mathscr{U}$ covering A_i. Hence $\{U_{ij} \mid i, j = 1, 2, \ldots\}$ is a countable subcovering of $\mathscr{U}$, which means that A is Lindelöf.

B) *Every closed set F of a hereditarily Lindelöf, regular space X is G_δ, i.e., X is perfectly normal.*

Proof. Since X is regular, there is an open covering $\mathscr{U}$ of $X - F$ such that $\bar{U} \subset X - F$ for every $U \in \mathscr{U}$. Since $X - F$ is Lindelöf, we can cover $X - F$ with countably many members $U_1, U_2, \ldots$ of $\mathscr{U}$. Thus

$$F = \bigcap_{i=1}^{\infty} (X - \bar{U}_i),$$

proving that F is G_δ.

C) *If a locally, countably compact T_2-space X is the union of countably many separable metric spaces X_i, $i = 1, 2, \ldots$, then X is metrizable.*

Proof. We shall first prove this proposition in the case that X is compact. Then $X \times X$ is the union of countably many separable metric subsets $X_i \times X_j$, i, $j = 1, 2, \ldots$. Therefore by A) and B) the diagonal $\Delta = \{(p, p) \mid p \in X\}$ of $X \times X$ is G_δ, and hence by Corollary 2 to Theorem VI.3, X is metrizable.

Now, let us prove C) in the general case. Since by A) X is hereditarily Lindelöf and locally countably compact, X is locally compact. Hence by use of the compact case in the above, we can assert that X is locally metrizable. On the other hand, X is regular by III.3.G). Thus by III.4.C), X is paracompact. Hence it follows from the corollary to Theorem VI.12 that X is metrizable.

Theorem VI.13. *Let X be a collectionwise normal, locally countably compact space. If X is the union of countably many closed subspaces F_n, $n = 1, 2, \ldots$, each of which is metrizable, then X is metrizable.*[1]

Proof. To begin with, let us show that X is paracompact. Suppose $\mathcal{U}$ is a given open covering of X. Then, for each n the open covering $\{U \cap F_n \mid U \in \mathcal{U}\}$ of F_n has a σ-discrete closed refinement $\mathcal{G}_n$, by Corollary 2 to Theorem V.2 and the fact that F_n is paracompact. (The corollary asserts only that there is a σ-discrete open refinement, but by using the regularity of F_n we can easily find a σ-discrete closed refinement, too.) Each $\mathcal{G}_n$ is a σ-discrete closed collection in X satisfying $\mathcal{G}_n < \mathcal{U}$. Since X is collectionwise normal, using V.3.A) we can construct a σ-discrete open collection $\mathcal{V}_n$ such that

$$\mathcal{G}_n < \mathcal{V}_n < \mathcal{U} .$$

Now $\mathcal{V} = \bigcup_{n=1}^{\infty} \mathcal{V}_n$ is a σ-discrete open covering of X such that $\mathcal{V} < \mathcal{U}$. Therefore by Corollary 3 to Theorem V.2, X is paracompact. Each point p of X has a countably compact nbd $N(p)$. $N(p)$ is a countable sum of the separable metric spaces $N(p) \cap F_n$, $n = 1, 2, \ldots$. Hence by C), $N(p)$ is metrizable, i.e., X is locally metrizable and paracompact. Therefore the metrizability of X follows from the corollary to Theorem VI.12.

---

[1] This theorem was first obtained by A. H. Stone [3] after Yu. M. Smirnov [7] proved C). The present proof is due to H. Corson. It is not known if collectionwise normality can be replaced by normality, but complete regularity is known to be insufficient. As for further conditions on metrizability of unions, see, for example, H. H. Corson and E. Michael [1], S. Hanai [3], Y. Tanaka [1], Y. Tanaka–Zhou Hao-xuan [1].

Now, let us turn to the problem concerning metrizability of the image of a metrizable space under a closed continuous mapping. In this aspect, the following theorem due to A. H. Stone [2] and K. Morita and S. Hanai [1] is the most interesting.[1]

Theorem VI.14. *Let f be a closed continuous mapping of a metrizable space X onto a topological space Y. Then Y is metrizable if and only if the boundary $Bf^{-1}(q)$ of the inverse image $f^{-1}(q)$ of each point q of Y is compact.*

Proof. Necessity. Assume $Bf^{-1}(q)$ is non-compact while Y is metrizable. Then, since X is metrizable, $Bf^{-1}(q)$ is not countably compact and hence there is a sequence $p_1, p_2, \ldots$ of points of $Bf^{-1}(q)$ which has no cluster point. On the other hand, since Y is metrizable, q has a countable nbd basis $U_1, U_2, U_3, \ldots$, where we may suppose U_i, $i = 1, 2, \ldots$, are open sets. Since f is continuous, $f^{-1}(U_1), f^{-1}(U_2), \ldots$ are open sets of X containing $f^{-1}(q)$. For each i, since $p_i \in Bf^{-1}(q)$, we can choose a point p_i' for which

$$p_i' \in (f^{-1}(U_i) - f^{-1}(q)) \cap S_{1/i}(p_i).$$

Then, since $\{p_i \mid i = 1, 2, \ldots\}$ has no cluster point, $\{p_i' \mid i = 1, 2, \ldots\}$, too, has no cluster point, and consequently it is a closed set of X. Hence $\{f(p_i') \mid i = 1, 2, \ldots\}$ must be a closed set of Y because f is a closed mapping. But $f(p_i') \in U_i$, which implies that $f(p_i') \to q$, while $f(p_i') \neq q$, $i = 1, 2, \ldots$, follows from $p_i' \notin f^{-1}(q)$. This means that $\{f(p_i') \mid i = 1, 2, \ldots\}$ is not closed in Y, which is a contradiction. Thus we can conclude that $Bf^{-1}(q)$ must be compact.

Sufficiency. We shall apply Theorem VI.2 to prove the metrizability of Y. It is clear that Y is T_1, because each point of Y is the image of a point of X, and accordingly closed.

Now, let

$$X' = \cup \{B(f^{-1}(y)) \mid y \in Y\}.$$

Then X' is a closed set of X, and the restriction of f to X' is a perfect mapping from X' onto Y. Namely, it is closed and continuous, and the

---

[1] For other similar results, see A. H. Stone [2]. There is the inverse problem of finding the conditions insuring metrizability of the inverse image of a metrizable space. V. V. Proizvolov [1] and F. Slaughter–J. Atkins [1] obtained some results in this area.

inverse image of each point of Y is compact. Thus we may assume without loss of generality that f itself is a perfect mapping.

We are going to use Corollary (ii) of Theorem VI.2. Let $\mathcal{F}_i$, $i = 1, 2, \ldots$, be a sequence of coverings of X which satisfies the conditions of the said corollary. We may assume without loss of generality that

$$\mathcal{F}_{i+1} < \mathcal{F}_i, \quad i = 1, 2, \ldots.$$

Then

$$f(\mathcal{F}_i) = \{f(F) \mid F \in \mathcal{F}_i\}$$

is obviously a closure-preserving closed covering of Y. Let $y \in Y$ and V an open nbd of y. Then

$$f^{-1}(y) \subset f^{-1}(V) \quad \text{in } X.$$

We claim that

$$S(f^{-1}(y), \mathcal{F}_i) \subset f^{-1}(V) \quad \text{for some } i.$$

Because, otherwise, there is a point sequence $\{x_i \mid i = 1, 2, \ldots\}$ in $f^{-1}(y)$ such that

$$S(x, \mathcal{F}_i) \not\subset f^{-1}(V), \quad i = 1, 2, \ldots. \tag{1}$$

Since $f^{-1}(y)$ is compact, $\{x_i\}$ has a cluster point $x \in f^{-1}(y)$. Then

$$S(x, \mathcal{F}_n) \subset f^{-1}(V) \quad \text{for some } n, \tag{2}$$

because $f^{-1}(V)$ is a nbd of x. On the other hand, since $\mathcal{F}_n$ is closure-preserving,

$$W = X - \bigcup\{F \mid x \notin F \in \mathcal{F}_n\} \tag{3}$$

is an open nbd of x. Hence we can select $i \geq n$ for which $x_i \in W$. Then the definition (3) of W implies that $x \in F$ for every element $F \in \mathcal{F}_n$ satisfying $x_i \in F$. Thus

$$S(x_i, \mathcal{F}_n) \subset S(x, \mathcal{F}_n),$$

while

$$S(x_i, \mathcal{F}_n) \supset S(x_i, \mathcal{F}_i) \not\subset f^{-1}(V)$$

by (1). Hence

$$S(x, \mathscr{F}_n) \not\subset f^{-1}(V),$$

which contradicts (2). Therefore our claim is proved. Now it is obvious that

$$S(y, f(\mathscr{F}_i)) \subset V \quad \text{in } Y.$$

Hence by Corollary (ii) of Theorem VI.2, Y is metrizable.

Example VI.7. As we saw in Example VI.1, $S'(A)$ is the image of a metrizable space X by a closed continuous mapping, but it is not metrizable. It is also easy to verify that the image of a metrizable space by an open continuous mapping is not necessarily metrizable. In fact we shall prove in the next chapter that every 1st countable T_1-space is the image of a metric space by an open continuous mapping.[1]

Corollary 1. *Let f be a closed continuous mapping of a metrizable space X onto a topological space Y. Then Y is metrizable if and only if it satisfies the 1st axiom of countability.*

Proof. The necessity is obvious. Conversely, if Y satisfies the 1st axiom, then for each $q \in Y$, $Bf^{-1}(q)$ is compact, because in the proof of the necessity of Theorem VI.14 we have practically used only the 1st countability of Y. Thus by the same theorem, Y is metrizable.

Corollary 2. *Let A be a set of power μ, where μ is an infinite cardinal number. Then a topological space X is a metrizable space of weight $\leqslant \mu$ if and only if there exists a subspace Y of Baire's zero-dimensional space $N(A)$ and a perfect mapping f of Y onto X.*

Proof. Necessity. Let X be a metrizable space of weight $\leqslant \mu$. Since X is paracompact and regular, for each natural number n we can choose a locally finite closed covering $\mathscr{F}_n$ such that

$$\mathscr{F}_n < \{S_{1/n}(p) \mid p \in X\}. \tag{1}$$

---

[1] See D. Hyman [2] and P. Zenor [3] for metrizability of closed continuous images of metric spaces.

We can easily see that the local finiteness of $\mathscr{F}_n$ implies $|\mathscr{F}_n| < \mu$, and hence we may put

$$\mathscr{F}_n = \{F(\alpha, n) \mid \alpha \in A\},$$

where some $F(\alpha, n)$ may be empty.[1] We define a subset Y of $N(A)$ by

$$Y = \left\{ (\alpha_1, \alpha_2, \ldots) \mid \alpha_i \in A, i = 1, 2, \ldots ; \bigcap_{n=1}^{\infty} F(\alpha_n, n) \neq \emptyset \right\}.$$

We define a mapping f of Y into X by

$$f(\alpha) = \bigcap_{n=1}^{\infty} F(\alpha_n, n) \quad \text{for } \alpha = (\alpha_1, \alpha_2, \ldots) \in Y. \tag{2}$$

It follows from (1) that f defines a unique mapping of Y into X. We can assert, moreover, that f is the desired mapping.

From the definition (2) it is clear that f is a continuous mapping of Y onto X. To see that $f^{-1}(p)$ is compact for each $p \in X$, we consider a point sequence $\alpha^i \in f^{-1}(p)$, $i = 1, 2, \ldots$, putting

$$\alpha^i = (\alpha_1^i, \alpha_2^i, \ldots).$$

Then $f(\alpha^i) = p$, which implies that $p \in F(\alpha_n^i, n)$, $i = 1, 2, \ldots$, for each n. We let n be fixed for a while. Since $\mathscr{F}_n$ is locally finite, for each n we can choose a subsequence $i_1, i_2, \ldots$ of $1, 2, \ldots$ such that

$$F(\alpha_n^{i_1}, n) = F(\alpha_n^{i_2}, n) = \cdots,$$

i.e.,

$$\alpha_n^{i_1} = \alpha_n^{i_2} = \cdots.$$

Using this observation for $n = 1$, we denote by $\{i_{11}, i_{12}, \ldots\}$ a subsequence of $\{1, 2, \ldots\}$ such that

$$\alpha_1^{i_{11}} = \alpha_1^{i_{12}} = \alpha_1^{i_{13}} = \cdots.$$

We denote by $\{i_{21}, i_{22}, \ldots\}$ a subsequence of $\{i_{11}, i_{12}, \ldots\}$ such that

$$\alpha_2^{i_{21}} = \alpha_2^{i_{22}} = \alpha_2^{i_{23}} = \cdots.$$

---

[1] We assume $\alpha \neq \alpha'$ implies $F(\alpha, n) \neq F(\alpha', n)$ unless $F(\alpha, n)$ is empty.

Repeating this process we obtain sequences $\{i_{k1}, i_{k2}, \ldots\}$, $k = 1, 2, \ldots$. Then putting $i_{kk} = j_k$, $k = 1, 2, \ldots$, we obtain a subsequence $\{j_1, j_2, \ldots\}$ of $\{1, 2, \ldots\}$ such that

$$\alpha_k^{j_k} = \alpha_k^{j_{k+1}} = \alpha_k^{j_{k+2}} = \cdots, \quad k = 1, 2, \ldots. \tag{3}$$

We put, for brevity,

$$\beta_k = \alpha_k^{j_k}, \quad k = 1, 2, \ldots.$$

Then it is obvious that

$$\beta = (\beta_1, \beta_2, \ldots) \in Y$$

and

$$f(\beta) = \bigcap_{n=1}^{\infty} F(\beta_n, n) = p.$$

On the other hand, the sequence $\{\alpha^{j_1}, \alpha^{j_2}, \ldots\}$ clearly converges to β in Y. For, by (3) α^{j_k} and β are equal up to the kth coordinate. Hence $f^{-1}(p)$ is sequentially compact. Since $f^{-1}(p)$ is a metric space, by IV.4.H) $f^{-1}(p)$ is compact.

To see that f is closed, we consider a closed set G of Y. Let p be a point of X which is not in $f(G)$. Then

$$f^{-1}(p) \cap G = \emptyset \quad \text{in } Y.$$

For each finite sequence $(\alpha_1, \ldots, \alpha_k)$ of elements of A, we define a subset $U(\alpha_1, \ldots, \alpha_k)$ of $N(A)$ by

$$U(\alpha_1, \ldots, \alpha_k) = \{(\alpha_1', \alpha_2', \ldots) \mid \alpha_i' = \alpha_i, i = 1, \ldots, k\}.$$

It is clear that $\{U(\alpha_1, \ldots, \alpha_k) \cap Y \mid \alpha_i \in A, i = 1, \ldots, k; k = 1, 2, \ldots\}$ forms a σ-discrete open basis of Y. We put

$$\mathcal{V}_k = \{U(\alpha_1, \ldots, \alpha_k) \cap Y \mid \alpha_i \in A, i = 1, \ldots, k\}.$$

Then by use of the compactness of $f^{-1}(p)$, we can find a k for which

$$S(f^{-1}(p), \mathcal{V}_k) \cap G = \emptyset.$$

For this k, there exist $\alpha_1^i, \ldots, \alpha_k^i \in A$, $i = 1, \ldots, l$, such that

$$f^{-1}(p) \subset \bigcup \{ U(\alpha_1^i, \ldots, \alpha_k^i) \mid i = 1, \ldots, l\} \cap Y \subset Y - G$$

and

$$f^{-1}(p) \cap U(\alpha_1, \ldots, \alpha_k) = \emptyset \quad \text{for } (\alpha_1, \ldots, \alpha_k) \neq (\alpha_1^i, \ldots, \alpha_k^i),$$

$$i = 1, \ldots, l \qquad (4)$$

(note that $\mathcal{V}_k$ is discrete). Now put

$$\mathcal{F} = \mathcal{F}_1 \wedge \mathcal{F}_2 \wedge \cdots \wedge \mathcal{F}_k ;$$

then $\mathcal{F}$ is a locally finite closed covering of X. Let $x \in G$. Then

$$x \notin U(\alpha_1^i, \ldots, \alpha_k^i) \cap Y \quad \text{for } i = 1, \ldots, l.$$

Hence $f(x) \in F$ for some $F \in \mathcal{F}$ with $F \neq F(\alpha_1^i, 1) \cap \cdots \cap F(\alpha_k^i, k)$, $i = 1, \ldots, l$. (Recall the definition (2) of $f(x)$.) Therefore

$$W = X - \{ F \mid F \in \mathcal{F}; F \neq F(\alpha_1^i, 1) \cap \cdots \cap F(\alpha_k^i, k), i = 1, \ldots, l\}$$

is a set which is disjoint from $f(G)$. On the other hand, if $p \in F = F(\alpha_1, 1) \cap \cdots \cap F(\alpha_k, k)$, where $(\alpha_1, \ldots, \alpha_k) \neq (\alpha_1^i, \ldots, \alpha_k^i)$, $i = 1, \ldots, l$, then it is easily seen that there is some $y \in U(\alpha_1, \ldots, \alpha_k) \cap Y$ such that $f(y) = p$. This implies

$$f^{-1}(p) \cap U(\alpha_1, \ldots, \alpha_k) \neq \emptyset,$$

contradicting (4). Hence $p \notin F$ for every member F of $\mathcal{F}$ with $F \neq F(\alpha_1^i, 1) \cap \cdots \cap F(\alpha_k^i, k)$, $i = 1, \ldots, l$. Therefore W is an open nbd of p which does not intersect $f(G)$. Thus $f(G)$ is closed, i.e., f is a closed mapping.

Sufficiency. Put

$$A_k = \{ (\alpha_1, \ldots, \alpha_k) \mid \alpha_i \in A, i = 1, \ldots, k \}.$$

For a given finite subset A_k' of A_k, we put

$$U(A_k') = X - \bigcup \{ f(U(\alpha_1, \ldots, \alpha_k)) \mid (\alpha_1, \ldots, \alpha_k) \in A_k - A_k' \}.$$

Then each $U(A_k')$ is an open set of X, because f is closed. Put

$$\mathcal{U}_k = \{U(A'_k) \mid A'_k \text{ is a finite subset of } A_k\}.$$

Then using the compactness of $f^{-1}(p)$ for each $p \in X$, we can easily show that $\bigcup_{k=1}^{\infty} \mathcal{U}_k$ is an open basis of X with a power $\leqslant \mu$.[1] On the other hand, the metrizability of X is a direct consequence of Theorem VI.14. Thus the proof of this corollary is complete.

5. Uniform space

There are various ways to generalize the concept of metric space. Among them, 'uniform space' is an attempt to generalize to non-metrizable spaces, concepts concerning the uniform topology of metric spaces. There are several mathematicians who reached the concept of uniform space. However, it seems today that this concept was first discovered by D. Kurepa [1], [2] and then given by A. Weil [1] a convenient definition.

In the present section we shall adopt the definition due to J. W. Tukey [1] which is essentially equivalent to that of Weil.

Definition VI.3. Let X be a set and $\{\mathcal{U}_\alpha \mid \alpha \in A\}$ a non-empty collection of coverings[2] of X which satisfies:

(i) if p, q are different elements of X, then $q \notin S(p, \mathcal{U}_\alpha)$ for some $\alpha \in A$,

(ii) if $\mathcal{U}$ is a covering of X such that $\mathcal{U}_\alpha < \mathcal{U}$ for some $\alpha \in A$, then $\mathcal{U} \in \{\mathcal{U}_\alpha \mid \alpha \in A\}$,

(iii) for every α, $\beta \in A$, there is $\gamma \in A$ such that $\mathcal{U}_\gamma^* < \mathcal{U}_\alpha \wedge \mathcal{U}_\beta$.

Then X is called a *uniform space* with the *uniformity* $\{\mathcal{U}_\alpha \mid \alpha \in A\}$. Each covering belonging to the uniformity is called a *uniform covering* of the uniform space. Let X be a uniform space with the uniformity $\{\mathcal{U}_\alpha \mid \alpha \in A\}$ and let $\{\mathcal{U}_\beta \mid \beta \in B\}$ be a subcollection of the uniformity. If for every $\alpha \in A$, there is $\beta \in B$ for which $\mathcal{U}_\beta < \mathcal{U}_\alpha$, then $\{\mathcal{U}_\beta \mid \beta \in B\}$ is called a *basis of the uniformity*. If $\{\mathcal{U}_\beta \mid \beta \in B\}$ and $\{\mathcal{U}_\gamma \mid \gamma \in \Gamma\}$ are bases of the same uniformity, then they are said to be *equivalent*.

---

[1] To prove $|\bigcup_{k=1}^{\infty} \mathcal{U}_k| \leqslant \mu$, we need, besides Corollary 3 to Theorem I.4, the fact that $\mu^2 = \mu$ for every infinite cardinal number μ. As for the proof, see, for example, F. Hausdorff [1].

[2] Although in II.5 we defined concepts concerning coverings of a topological space, some of those definitions are obviously valid for coverings of a set, and we may use them accordingly without further comment.

The proof of the following propositions, A), B) and C), is left to the reader.

A) *A collection* $\{\mathcal{U}_\beta \mid \beta \in B\}$ *of coverings of a set X is a basis for some uniformity if and only if it satisfies* (i) *and* (iii) *of Definition* VI.3.

B) *Every metric space X is a uniform space with the basis* $\{\mathcal{U}_n \mid n = 1, 2, \ldots\}$ *of the uniformity, where*

$$\mathcal{U}_n = \{S_{1/3^n}(p) \mid p \in X\}.$$

C) *Let X be a uniform space with the uniformity* (*or a basis of uniformity*) $\{\mathcal{U}_\alpha \mid \alpha \in A\}$. *We denote by $\mathcal{V}$ the collection of the subsets U of X such that for every point $p \in U$ and for some $\alpha \in A$, $S(p, \mathcal{U}_\alpha) \subset U$. Then $\mathcal{V}$ satisfies the condition of the topology* (*Definition* II.1). *Thus X is a topological space.*

D) *Under the same assumptions as in* C), *for each point p of X, $\{S(p, \mathcal{U}_\alpha) \mid \alpha \in A\}$ forms a nbd basis of p.*

Proof. Since for every open set U containing p, there is α for which $S(p, \mathcal{U}_\alpha) \subset U$, we should verify that every $S(p, \mathcal{U}_\alpha)$ contains an open set containing p. Put

$$V = \{q \mid S(q, \mathcal{U}_\beta) \subset S(p, \mathcal{U}_\alpha) \text{ for some } \beta \in A\}$$

and assume that q is a given point of V. Then

$$S(q, \mathcal{U}_\beta) \subset S(p, \mathcal{U}_\alpha) \quad \text{for some } \beta \in A.$$

Choose $\gamma \in A$ for which $\mathcal{U}_\gamma^* < \mathcal{U}_\beta$; then we note that every point q' of $S(q, \mathcal{U}_\gamma)$ satisfies

$$S(q', \mathcal{U}_\gamma) \subset S(q, \mathcal{U}_\beta).$$

Hence

$$S(q', \mathcal{U}_\gamma) \subset S(p, \mathcal{U}_\alpha),$$

which implies $q' \in V$. Since q' is a given point of $S(q, \mathcal{U}_\gamma)$, we obtain

$$S(q, \mathcal{U}_\gamma) \subset V.$$

This means that V is an open set of the topological space X.

Let X be a topological space and $\{\mathcal{U}_\alpha \mid \alpha \in A\}$ a uniformity of X such that $\{S(p, \mathcal{U}_\alpha) \mid \alpha \in A\}$ forms a nbd basis of each point p of X. Then we call the uniformity *compatible* with the topology of X. Suppose $\{\mathcal{U}_\alpha \mid \alpha \in A\}$ and $\{\mathcal{V}_\beta \mid \beta \in B\}$ are two uniformities compatible with the topology of a topological space X. If $\{\mathcal{U}_\alpha \mid \alpha \in A\} \subset \{\mathcal{V}_\beta \mid \beta \in B\}$, then the latter uniformity is called *stronger* than the former one.

Example VI.8. As seen in B), metric space is the most popular example of uniform space. Another significant example of uniform space is topological group. Let X be a T_1-space and a group at the same time. If the group operation of X is continuous, i.e., if for every $p, q \in X$ and nbd W of the element pq^{-1} of X, there are nbds U, V of p, q respectively such that

$$UV^{-1} = \{xy^{-1} \mid x \in U, y \in V\} \subset W,$$

then X is called a *topological group*.[1]
Let $\mathcal{V}$ be the collection of the nbds of the unit element e of X. Then for each $V \in \mathcal{V}$, we define a covering $\mathcal{U}(V)$ of X by

$$\mathcal{U}(V) = \{Vp \mid p \in X\},$$
where
$$Vp = \{xp \mid x \in V\}.$$

Then we can easily show that $\{\mathcal{U}(V) \mid V \in \mathcal{V}\}$ is a uniformity compatible with the topology of X.

E) *Let X be a topological space with a uniformity $\{\mathcal{U}_\alpha \mid \alpha \in A\}$ compatible with its topology. Then there is a basis of the uniformity consisting of open coverings only.*

Proof. For each $\mathcal{U}_\alpha$ we denote by $\mathcal{U}_\alpha^\circ$ the open collection defined by

$$\mathcal{U}_\alpha^\circ = \{U^\circ \mid U \in \mathcal{U}_\alpha\}.$$

[1] As for a detailed discussion on topological groups, see J. Pontrjagin [1].

Choose some $\beta \in A$ for which $\mathcal{U}_\beta^* < \mathcal{U}_\alpha$. Then we can assert that $\mathcal{U}_\beta <$ $\mathcal{U}_\alpha^\circ$. For, given a member U' of $\mathcal{U}_\beta$, we have

$$S(U', \mathcal{U}_\beta) \subset U \quad \text{for some } U \in \mathcal{U}_\alpha.$$

This means that every point $p \in U'$ satisfies

$$S(p, \mathcal{U}_\beta) \subset U,$$

i.e. $p \in U^\circ$. Therefore $U' \subset U^\circ$, proving $\mathcal{U}_\beta < \mathcal{U}_\alpha^\circ$. Thus we obtain by (ii) of Definition VI.3, that

$$\mathcal{U}_\alpha^\circ \in \{\mathcal{U}_\alpha \mid \alpha \in A\}$$

and hence $\{\mathcal{U}_\alpha^\circ \mid \alpha \in A\}$ is the desired basis of the uniformity $\{\mathcal{U}_\alpha \mid \alpha \in A\}$.

By C) every uniform space can be regarded as a topological space. But not every topological space has a uniformity compatible with its topology. As for uniformization of a topological space, we get the following answer which is much simpler than that for the metrization problem.

Theorem VI.15. *A topological space X has a uniformity compatible with its topology if and only if X is completely regular.*

Proof. Necessity. Suppose $\{\mathcal{U}_\alpha \mid \alpha \in A\}$ is the uniformity compatible with the topology of X. Then $\{S(p, \mathcal{U}_\alpha) \mid p \in X, \alpha \in A\}$ is a nbd basis of X. It follows from (i) of Definition VI.3 that X is T_1. Let p be a point of X and G a closed set which does not contain p. By use of E), we can choose open coverings $\mathcal{U}_0, \mathcal{U}_1, \mathcal{U}_2, \ldots$ from the uniformity of X such that

$$S(p, \mathcal{U}_0) \subset X - G$$

and

$$\mathcal{U}_0 > \mathcal{U}_1^* > \mathcal{U}_1 > \mathcal{U}_2^* > \mathcal{U}_2 > \cdots.$$

We define open sets $U(k/2^n)$, $k = 1, \ldots, 2^n - 1$; $n = 1, 2, \ldots$, as follows:

$$U(\tfrac{1}{2}) = S(p, \mathcal{U}_1),$$

$$U\left(\frac{1}{2^2}\right) = S(p, \mathcal{U}_2), \qquad U\left(\frac{3}{2^2}\right) = S(U(\tfrac{1}{2}), \mathcal{U}_2), \ldots.$$

Generally, assume we have defined $U(k/2^n)$, $k = 1, \ldots, 2^n - 1$; then we define $U(k'/2^{n+1})$ by

$$
U\left(\frac{k'}{2^{n+1}}\right) = \begin{cases} U\left(\dfrac{k}{2^n}\right) & \text{if } k' = 2k, \\[2mm] S(p, \mathcal{U}_{n+1}) & \text{if } k' = 1, \\[2mm] S\left(U\left(\dfrac{k}{2^n}\right), \mathcal{U}_{n+1}\right) & \text{if } k' = 2k + 1 \text{ and } k > 0. \end{cases}
$$

We can easily show that

$$
p \in U\left(\frac{k}{2^n}\right) \subset \overline{U\left(\frac{k}{2^n}\right)} \subset U\left(\frac{k+1}{2^n}\right) \subset \overline{U\left(\frac{k+1}{2^n}\right)} \subset X - F.
$$

Furthermore, we put

$$
U(1) = X.
$$

Now, let

$$
f(p) = \inf \{r \mid p \in U(r)\}, \quad p \in X.
$$

Using a similar argument to the one in the proof of Urysohn's lemma, we can show that f is a continuous function over X such that

$$
f(p) = 0, \qquad f(G) = 1 \quad \text{and} \quad 0 \leq f \leq 1.
$$

Therefore X is completely regular.

Sufficiency. Let X be a given completely regular space. For every pair p, U of a point p of X and its nbd U, we construct a continuous function f such that

$$
f(p) = 0, \qquad f(X - U) = 1 \quad \text{and} \quad 0 \leq f \leq 1.
$$

For each natural number n, we put

$$
\mathcal{U}(n, p, U) = \{f^{-1}(J_n(x)) \mid x \in [0, 1]\},
$$

where

$$
J_n(x) = \left\{y \mid y \in [0, 1] \cap \left(x - \frac{1}{3^n}, x + \frac{1}{3^n}\right)\right\}.
$$

Then it is easily seen that $\mathcal{U}(n, p, U)$, $n = 1, 2, \ldots$, are open coverings of X satisfying

$$\mathcal{U}(1, p, U) > \mathcal{U}^*(2, p, U) > \mathcal{U}(2, p, U) > \mathcal{U}^*(3, p, U) > \cdots$$

and

$$S(p, \mathcal{U}(1, p, U)) \subset U.$$

Let

$$\mu = \{\mathcal{U}(n, p, U) \mid p \in X, \ U \text{ is a nbd of } p, \ n = 1, 2, \ldots\},$$

$$\nu = \{\mathcal{U}_1 \wedge \cdots \wedge \mathcal{U}_k \mid \mathcal{U}_i \in \mu, \ i = 1, \ldots, k; \ k = 1, 2, \ldots\}.$$

Then ν is easily seen to be a basis of a uniformity compatible with the topology of X. (Note that generally $\mathcal{U}_i^* < \mathcal{V}_i$, $i = 1, \ldots, k$, for coverings $\mathcal{U}_i$ and $\mathcal{V}_i$ implies $(\wedge_{i=1}^{k} \mathcal{U}_i)^* < \wedge_{i=1}^{k} \mathcal{V}_i$.) Thus the proof of the theorem is complete.

Example VI.9. Let X be a topological space and $\mathcal{U}$ a covering of X. If there is a sequence $\mathcal{U}_1, \mathcal{U}_2, \ldots$ of open coverings of X such that

$$\mathcal{U} > \mathcal{U}_1^* > \mathcal{U}_1 > \mathcal{U}_2^* > \cdots,$$

then $\mathcal{U}$ is called a normal covering of X, as defined in II.5. In a completely regular space X, the collection of all normal coverings obviously forms a uniformity compatible with the topology. On the other hand, by (iii) of Definition VI.3 it is also clear that every uniform covering is a normal covering. Therefore the uniformity given in the above is the strongest uniformity compatible with the topology of X and it is called the *a-uniformity* of X.

If X is fully normal, then every open covering is a normal covering; hence the *a*-uniformity of X has the collection of all open coverings as a basis.

Let X be a uniform space with uniformity $\{\mathcal{U}_\alpha \mid \alpha \in A\}$. If we can introduce a metric ρ into X such that $\{\mathcal{S}_n \mid n = 1, 2, \ldots\}$, where $\mathcal{S}_n = \{S_{1/n}(p) \mid p \in X\}$, is a basis of the uniformity, then we call the uniform space *metrizable*, and the metric ρ is said to be *compatible* with the uniformity. As for metrization of uniform spaces we obtain the following proposition which is a direct consequence of Alexandroff–Urysohn's metrization theorem.

Theorem VI.16. *A uniform space X is metrizable if and only if there is a basis of the uniformity consisting of countably many coverings.*

Corollary (Kakutani's theorem). *A topological group is metrizable if and only if it is first-countable.*[1]

We should note that every metric space has a uniquely defined uniformity, and every uniform space has a uniquely defined topology while generally speaking neither the uniformity compatible with the topology of a given completely regular space nor the metric compatible with the uniformity of a given uniform space is uniquely determined.

Let X be a uniform space with uniformity $\{\mathscr{U}_\alpha \mid \alpha \in A\}$ and X' a subspace of the topological space X. Then $\{\mathscr{U}'_\alpha \mid \alpha \in A\}$ where $\mathscr{U}'_\alpha = \{U \cap X' \mid U \in \mathscr{U}_\alpha\}$ forms a basis of a uniformity compatible with the topology of the subspace X'. Thus we can define a *subspace X'* of the uniform space X. Now, we suppose $\mathscr{U}_\gamma$, $\gamma \in \Gamma$, are coverings of sets X_γ, $\gamma \in \Gamma$, respectively. Then we define their product $\prod\{\mathscr{U}_\gamma \mid \gamma \in \Gamma\}$ by

$$\prod\{\mathscr{U}_\gamma \mid \gamma \in \Gamma\} = \{\prod\{U_\gamma \mid \gamma \in \Gamma\} \mid U_\gamma \in \mathscr{U}_\gamma, \gamma \in \Gamma\}.$$

Then the product of the coverings is a covering of the Cartesian product $\prod\{X_\gamma \mid \gamma \in \Gamma\}$.

Let X_γ, $\gamma \in \Gamma$, be uniform spaces with uniformities $\{\mathscr{U}^\gamma_\alpha \mid \alpha \in A_\gamma\}$, respectively. We denote by $\mathscr{R}_\gamma$ the covering of X_γ consisting of X_γ only. Then

$$\{\mathscr{U}^{\gamma_1}_{\alpha_1} \times \cdots \times \mathscr{U}^{\gamma_k}_{\alpha_k} \times \prod\{\mathscr{R}_\alpha \mid \alpha \neq \alpha_i, i = 1, \ldots, k\} \mid \alpha_i \in A_{\gamma_i}, \gamma_i \in \Gamma,$$
$$i = 1, \ldots, k; k = 1, 2, \ldots\}$$

is easily seen to be a basis of a uniformity compatible with the topology of the product space $\prod\{X_\gamma \mid \gamma \in \Gamma\}$ of the topological spaces X_γ, $\gamma \in \Gamma$. Thus we can define the *product space $\prod\{X_\gamma \mid \gamma \in \Gamma\}$* of the uniform spaces X_γ, $\gamma \in \Gamma$.

Theorem VI.17. *A compact, completely regular space X has one and only one uniformity compatible with the topology.*[2]

Proof. The uniqueness of uniformity is the only thing to be proved. For that purpose it suffices to prove that the collection of all open coverings

---

[1] Due to S. Kakutani [1].

[2] The converse of this theorem is not true. Topological spaces which admit only one uniformity compatible with its topology were studied by R. Doss [1].

forms a basis of an arbitrary uniformity $\{\mathcal{U}_\alpha \mid \alpha \in A\}$. Suppose that $\mathcal{U}$ is a given open covering of X; then there is an $\alpha \in A$ for which $\mathcal{U}_\alpha < \mathcal{U}$. For, if we assume the contrary, then for every $\alpha \in A$, there is a point $\varphi(\alpha)$ of X such that $S(\varphi(\alpha), \mathcal{U}_\alpha)$ is contained in no member of $\mathcal{U}$. We can regard φ a net (of X) on A, where A is directed by the order: $\alpha > \beta$ if and only if $\mathcal{U}_\alpha < \mathcal{U}_\beta$. Since X is compact, there is a cluster point p of φ. Since $\{\mathcal{U}_\alpha \mid \alpha \in A\}$ is compatible with the topology of X, there is an $\alpha_0 \in A$ for which $S(p, \mathcal{U}_{\alpha_0})$ is contained in some member of $\mathcal{U}$. Take $\alpha \in A$ such that $\mathcal{U}_\alpha^* < \mathcal{U}_{\alpha_0}$. There is some $\beta > \alpha$ such that

$$\varphi(\beta) \in S(p, \mathcal{U}_\alpha).$$

Then

$$S(\varphi(\beta), \mathcal{U}_\beta) \subset S^2(p, \mathcal{U}_\alpha) \subset S(p, \mathcal{U}_{\alpha_0}) \subset U$$

for some $U \in \mathcal{U}$. But this contradicts the definition of $\varphi(\beta)$, and hence we conclude that $\mathcal{U}_\alpha < \mathcal{U}$ for some $\alpha \in A$. This conclusion combined with E) proves that the collection of all open coverings of X forms a basis of the uniformity $\{\mathcal{U}_\alpha \mid \alpha \in A\}$.[1]

Definition VI.4. A uniform space X is said to be *totally bounded* if every uniform covering of X has a finite subcovering. Let $\mathcal{F}$ be a filter of a uniform space X. If for each uniform covering $\mathcal{U}$ of X, there are $F \in \mathcal{F}$ and $U \in \mathcal{U}$ such that $F \subset U$, then $\mathcal{F}$ is called a *Cauchy filter*.[2] Let $\varphi(\Delta \mid >)$ be a net of X. If each uniform covering $\mathcal{U}$ of X has a member U in which φ is residual, then $\varphi(\Delta \mid >)$ is called a *Cauchy net*. If every Cauchy filter of a uniform space X converges, then X is called a *complete uniform space*.

The terminologies given in the above are extensions of those for metric spaces. The following propositions are extensions of theorems for metric spaces. Their proofs are left to the reader.

F) *A subspace Y of a complete uniform space X is complete if and only if Y is closed in X.*

G) *The product space of complete uniform spaces is complete.*

---

[1] We may also say that in a compact, completely regular space the collection of all finite open coverings forms a basis of the uniformity.

[2] Generally we can define *Cauchy filter basis* as a basis of a Cauchy filter. A filter $\mathcal{F}$ is Cauchy if and only if $\mathcal{U} \cap \mathcal{F} \neq \emptyset$ for every uniform covering $\mathcal{U}$.

Theorem VI.18. *A uniform space X is compact if and only if it is totally bounded and complete.*

Example VI.10. In the last part of the proof of Theorem VI.15, we dealt with the basis ν of uniformity. It is clear that ν gives a totally bounded uniformity. Therefore for every completely regular space, there is a totally bounded uniformity compatible with the topology. On the other hand, not every completely regular space admits a complete uniformity. For example, the normal space R_5 in Example II.1 can be seen to have no complete uniformity compatible with its topology, because even its a-uniformity is not complete.[1]

We can also extend the concept of uniformly continuous mapping to uniform spaces as follows.

Definition VI.5. Let f be a mapping of a uniform space X into a uniform space Y. If for every uniform covering $\mathcal{V}$ of Y, $f^{-1}(\mathcal{V}) = \{f^{-1}(V) \mid V \in \mathcal{V}\}$ is a uniform covering of X, then f is called a *uniformly continuous mapping*. If f is a one-to-one, uniformly continuous mapping of X onto Y and the inverse mapping f^{-1} is also uniformly continuous, then f is called a *unimorphic mapping* or a *unimorphism* of X onto Y. In this case X and Y are said to be *unimorphic*.

The following is an extension of a well-known theorem for metric spaces.

Theorem VI.19. *Let X be a uniform space with the a-uniformity and Y a uniform space. Then every continuous mapping of X into Y is uniformly continuous.*[2]

Proof. Let $\mathcal{V}$ be a uniform covering of Y. Then there is a sequence $\mathcal{V}_1, \mathcal{V}_2, \ldots$ of open coverings of Y such that

$$\mathcal{V} > \mathcal{V}_1^* > \mathcal{V}_1 > \mathcal{V}_2^* > \cdots .$$

Then $f^{-1}(\mathcal{V}_1), f^{-1}(\mathcal{V}_2), \ldots$ is a sequence of open coverings of X satisfying

---

[1] A topological space is called *Dieudonné complete* when it admits a complete uniformity compatible with the topology.

[2] Spaces for which every continuous function becomes uniformly continuous were studied by J. Nagata [1] and A. Atsuji [1] and others. Such a space is especially interesting in the metric case, when various equivalent conditions are given by J. Nagata [1], M. Atsuji [1], J. Rainwater [1], S. Willard [1]. A further development in this direction can be seen in A. Hohti [2]. See also S. Ginsburg–J. Isbell [1] and A. Hohti [1].

$$f^{-1}(\mathscr{V}) > f^{-1}(\mathscr{V}_1)^* > f^{-1}(\mathscr{V}_1) > f^{-1}(\mathscr{V}_2)^* > \cdots .$$

Hence $f^{-1}(\mathscr{V})$ is a normal open covering, i.e., it is a uniform covering of the uniform space X since X has the a-uniformity. This proves that f is uniformly continuous.

Combining this theorem with Theorem VI.17, we get:

Corollary. *Let X be a compact uniform space. Then every continuous mapping of X into a uniform space Y is uniformly continuous.*

Let (P) be a property of topological spaces or of uniform spaces. A uniform space X is called *uniformly locally* (P) if it has a uniform covering $\mathscr{U}$ each of whose elements has (P). In this connection, the following proposition may be of some interest.

H) *Let X be a connected, uniformly locally compact, uniform space. Then it is the union of countably many compact spaces, i.e., it is σ-compact.*

Proof. Since X is uniformly locally compact, there is a uniform covering $\mathscr{U}_\alpha$ consisting of compact sets. Take a uniform covering $\mathscr{U}_\beta$ with $\mathscr{U}_\beta^* < \mathscr{U}_\alpha$ and a non-empty member U of $\mathscr{U}_\beta$. Then we put

$$F_1 = \bar{U}$$

and

$$F_{n+1} = \overline{S(F_n, \mathscr{U}_\beta)}, \quad n = 1, 2, \ldots . \tag{1}$$

Then F_1 is compact because

$$F_1 = \bar{U} \subset U'$$

for some member U' of $\mathscr{U}_\alpha$ which is compact. We shall prove by induction on n that the F_n, $n = 1, 2, \ldots$, are compact. Assume that F_n is compact. Then

$$F_n \subset \bigcup \{U_i \mid i = 1, \ldots, k\}$$

for some $U_i \in \mathscr{U}_\beta$, $i = 1, \ldots, k$. Since $\mathscr{U}_\beta^* < \mathscr{U}_\alpha$, we can choose $U_i' \in \mathscr{U}_\alpha$, $i = 1, \ldots, k$, for which

$$S(U_i, U_\beta) \subset U_i', \quad i = 1, \ldots, k.$$

For these U_i', we can prove that

$$F_{n+1} \subset \bigcup \{U_i' \mid i = 1, \ldots, k\}.$$

Because, if $U \in \mathcal{U}_\beta$ and $U \cap F_n \neq \emptyset$, then $U \cap U_i \neq \emptyset$ for some i, which implies that

$$U \subset S(U_i, \mathcal{U}_\beta) \subset U_i' \subset \bigcup \{U_i' \mid i = 1, \ldots, k\}.$$

Therefore

$$S(F_n, \mathcal{U}_\beta) \subset \bigcup \{U_i' \mid i = 1, \ldots, k\}.$$

Since each U_i' is compact, it is closed and hence $\bigcup \{U_i' \mid i = 1, \ldots, k\}$ is a closed compact set. Therefore

$$F_{n+1} \subset \bigcup \{U_i' \mid i = 1, \ldots, k\} \quad \text{(see (1))}.$$

This means that F_{n+1} is compact since it is a closed subset of a compact set. Now, we can prove that $\bigcup_{n=1}^{\infty} F_n = X$ follows from the connectedness of X. For if $p \in \bigcup_{n=1}^{\infty} F_n$, then $p \in F_n$ for some n. Hence by (1)

$$S(p, \mathcal{U}_\beta) \subset F_{n+1} \subset \bigcup_{n=1}^{\infty} F_n.$$

Thus $\bigcup_{n=1}^{\infty} F_n$ is open. On the other hand, if $p \not\in \bigcup F_n$, then

$$S(p, \mathcal{U}_\beta) \cap (\bigcup F_n) = \emptyset.$$

For, if we assume the contrary, then

$$S(p, \mathcal{U}_\beta) \cap F_n \neq \emptyset \quad \text{for some } n.$$

This implies that $p \in F_n$ (see (1)), which is a contradiction. Thus $\bigcup_{n=1}^{\infty} F_n$ is open and closed. Therefore it follows from the connectedness of X that

$$\bigcup_{n=1}^{\infty} F_n = X,$$

and hence X is the countable sum of the compact sets F_n, $n = 1, 2, \ldots$.

A complete uniform space $\tilde{X}$ is called a *completion* of a uniform space X if X is unimorphic with a dense subspace of the uniform space $\tilde{X}$. The position which is occupied by completion in the theory of uniform spaces is somewhat like that of compactification in the theory of topological spaces.

Now, let us consider the problem of constructing a completion of a given uniform space X. If X is a metric space, then by Example VI.6 it is isometrically imbedded in a complete metric space $C^*(X)$. Thus the closure $\bar{X}$ of X in $C^*(X)$ is a completion of X (see F)). If X is a general uniform space, then by use of the following proposition we can uniformly imbed X in the product of metric spaces. Therefore X is unimorphic to a subspace of a complete uniform space because the product space of complete uniform spaces is complete (see G)). Thus $\bar{X}$ in the product space is a completion of X.

I) *Every uniform space X is unimorphic to a subset of the product space of metric spaces.*

Proof. Let us denote by $\{\rho_\alpha' \mid \alpha \in A\}$ the collection of all uniformly continuous functions ρ_α' over $X \times X$ which satisfy

$$\rho_\alpha'(p, q) = \rho_\alpha'(q, p) \geqslant 0, \qquad \rho_\alpha'(p, p) = 0$$

and

$$\rho_\alpha'(p, q) + \rho_\alpha'(q, r) \geqslant \rho_\alpha'(p, r).$$

For each $\alpha \in A$, the relation $\rho_\alpha'(p, q) = 0$ is an equivalence relation between two points p, q of X. Therefore classifying all points of X by use of this relation we obtain a decomposition $\mathcal{D}_\alpha$ of X. Define a function $\rho_\alpha(D, D')$ on pairs of members of $\mathcal{D}_\alpha$ by

$$\rho_\alpha(D, D') = \rho_\alpha'(p, p'),$$

where $p \in D$, $p' \in D'$. We can easily verify that ρ_α defined in this manner satisfies the conditions for a metric, and hence $\mathcal{D}_\alpha$ becomes a metric space which we denote by X_α. Moreover we denote by φ_α the natural mapping of X onto X_α, i.e.,

$$\varphi_\alpha(p) = D \in \mathcal{D}_\alpha \quad \text{for } p \in D.$$

Now define a mapping φ of X into the product space $P = \prod \{X_\alpha \mid \alpha \in A\}$

of the uniform spaces X_α as follows:

$$\varphi(p) = \{\varphi_\alpha(p) \mid \alpha \in A\}, \quad p \in X.$$

Since we can easily see that φ is a one-to-one uniformly continuous mapping, we shall prove only the uniform continuity of φ^{-1}.

Let $\mathcal{U}$ be a given uniform covering of X; then we choose a sequence $\{\mathcal{U}_n \mid n = 1, 2, \ldots\}$ of uniform coverings such that

$$\mathcal{U} > \mathcal{U}_1 > \mathcal{U}_2^* > \mathcal{U}_2 > \mathcal{U}_3^* \cdots.$$

For convenience of description, we put $\mathcal{U}_0 = \{X\}$. For $p, q \in X$, we define

$$\sigma(p, q) = \inf\{2^{-n} \mid q \in S(p, \mathcal{U}_n)\},$$

$$\rho'(p, q) = \inf\{\sigma(p_0, p_1) + \sigma(p_1, p_2) + \cdots + \sigma(p_{k-1}, p_k) \mid$$

$$p_i \in X, \, i = 1, \ldots, k-1; \, p_0 = p, \, p_k = q\}.$$

Then we can show that ρ' is a member of $\{\rho'_\alpha \mid \alpha \in A\}$. It is easy to prove $\rho'(p, q) = \rho'(q, p) \geq 0$, $\rho'(p, p) = 0$ and $\rho'(p, q) + \rho'(q, r) \geq \rho'(p, r)$ from the definition of ρ'. To see its uniform continuity, assume

$$\rho' \in S(p, \mathcal{U}_n), \qquad q' \in S(q, \mathcal{U}_n);$$

then $\sigma(p, p') \leq 2^{-n}$, $\sigma(q, q') \leq 2^{-n}$. This implies

$$\rho'(p', q') \leq \sigma(p', p) + \rho'(p, q) + \sigma(q, q') \leq 2^{1-n} + \rho'(p, q)$$

and

$$\rho'(p, q) \leq 2^{1-n} + \rho'(p', q'),$$

i.e.,

$$|\rho'(p, q) - \rho'(p', q')| \leq 2^{1-n},$$

proving that $\rho'(p, q)$ is uniformly continuous. Thus we assume

$$\rho' = \rho'_\alpha. \tag{1}$$

Now, let us prove that

$$\sigma(p_0, p_1) + \sigma(p_1, p_2) + \cdots + \sigma(p_{k-1}, p_k) \geq \tfrac{1}{2}\sigma(p_0, p_k) \tag{2}$$

for any choice of $p_0, p_1, \ldots, p_k \in X$.

We use induction on the number k. For $k = 1$, the inequality (2) is obviously true. Assume it is true for all $k < l$. Put

$$\sigma(p_0, p_1) + \cdots + \sigma(p_{l-1}, p_l) = s$$

and denote by m the largest number such that

$$\sigma(p_0, p_1) + \cdots + \sigma(p_{m-1}, p_m) \leqslant \tfrac{1}{2} s .$$

Then

$$\sigma(p_0, p_1) + \cdots + \sigma(p_m, p_{m+1}) > \tfrac{1}{2} s ,$$

which implies

$$\sigma(p_{m+1}, p_{m+2}) + \cdots + \sigma(p_{l-1}, p_l) \leqslant \tfrac{1}{2} s .$$

Therefore it follows from the induction hypothesis that

$$\tfrac{1}{2}\sigma(p_0, p_m) \leqslant \tfrac{1}{2} s , \qquad \tfrac{1}{2}\sigma(p_{m+1}, p_l) \leqslant \tfrac{1}{2} s .$$

On the other hand, $\sigma(p_m, p_{m+1}) \leqslant s$ is obvious. Denote by n the smallest number such that $2^{-n} \leqslant s$; then, in view of the definition of σ, we can assert

$$\sigma(p_0, p_m) \leqslant 2^{-n} , \qquad \sigma(p_m, p_{m+1}) \leqslant 2^{-n} , \qquad \sigma(p_{m+1}, p_l) \leqslant 2^{-n} .$$

Because, e.g. if $\sigma(p_0, p_m) > 2^{-n}$, then $\sigma(p_0, p_m) = 2^{-t}$ and $t < n$. Thus $2^{-t} \leqslant s$, which contradicts the definition of n. Hence from the definition of σ it follows that

$$p_m \in S(p_0, \mathcal{U}_n) , \qquad p_{m+1} \in S(p_m, \mathcal{U}_n) , \qquad p_l \in S(p_{m+1}, \mathcal{U}_n) .$$

Since $\mathcal{U}_n^* < \mathcal{U}_{n-1}$, these imply

$$p_l \in S(p_0, \mathcal{U}_{n-1}) ,$$

i.e.,

$$\sigma(p_0, p_l) \leqslant 2^{1-n} \leqslant 2s$$

proving (2) in the case $k = l$. Thus (2) is generally true.

Suppose $q \notin S(p, \mathcal{U})$; then by (2)

$$\sigma(p_0, p_1) + \cdots + \sigma(p_{k-1}, p_k) \geqslant \tfrac{1}{2}\sigma(p_0, p_k) = \tfrac{1}{2}$$

for $p_0 = p$, $p_k = q$ and for any choice of $p_1, \ldots, p_{k-1}$. Therefore, by the definition of ρ', $\rho'(p, q) = \rho'_\alpha(p, q) \geq \frac{1}{2}$ (see (1)). This means that if $\varphi^{-1}(p') = p$, $\varphi^{-1}(q') = q$ and $\rho_\alpha(p'_\alpha, q'_\alpha) < \frac{1}{2}$ in X_α, then $q \in S(p, \mathcal{U})$ in X. Thus φ^{-1} is uniformly continuous, and accordingly, φ is unimorphic.

Now, we shall give a direct method of constructing a completion of a given uniform space X without imbedding X in the product space of metric spaces. Let X be a uniform space with a basis $\{\mathcal{U}_\alpha \mid \alpha \in A\}$ of uniformity. We call two Cauchy filters $\mathcal{F}$, $\mathcal{G}$ of X *equivalent* if for every $\alpha \in A$, there is $U \in \mathcal{U}_\alpha$ satisfying $U \in \mathcal{F} \cap \mathcal{G}$. We denote this equivalence relation by $\mathcal{F} \sim \mathcal{G}$.

J) *If $\mathcal{F} \sim \mathcal{G}$ and $\mathcal{G} \sim \mathcal{H}$ for three Cauchy filters $\mathcal{F}$, $\mathcal{G}$ and $\mathcal{H}$, then $\mathcal{F} \sim \mathcal{H}$.*

Proof. Given $\alpha \in A$, we take $\beta \in A$ such that $\mathcal{U}_\beta^* < \mathcal{U}_\alpha$. Then there are $U \in \mathcal{U}_\beta$, $U' \in \mathcal{U}_\beta$ such that

$$U \in \mathcal{F} \cap \mathcal{G}, \qquad U' \in \mathcal{G} \cap \mathcal{H}.$$

Since

$$U, U' \in \mathcal{G},$$

we get

$$U \cap U' \neq \emptyset,$$

and hence

$$U \cup U' \subset U'' \quad \text{for some } U'' \in U_\alpha.$$

This means that

$$U'' \in \mathcal{F} \cap \mathcal{H},$$

i.e. $\mathcal{F} \sim \mathcal{H}$.

In view of J) we can classify all Cauchy filters of X and denote by $\tilde{X}$ the collection of all those classes. Suppose p is a point of X; then the filter $\mathcal{F}(p) = \{A \mid p \in A\}$ is clearly a Cauchy filter of X. We denote by $\{\mathcal{F}\}_p$ the class containing $\mathcal{F}(p)$. Then there is a one-to-one mapping between the collection X' of those special classes and X. For brevity of description we regard X as a subset of $\tilde{X}$ identifying X with X' (i.e. $\{\mathcal{F}\}_p = p$). Now, let us assume that the given basis $\{\mathcal{U}_\alpha \mid \alpha \in A\}$ of the uniformity of X consists of open coverings of X. Generally for a given open set U of X, we denote by $\tilde{U}$ the subset of $\tilde{X}$ defined by

$$\tilde{U} = \{q \mid q \in \tilde{X}, U \in \mathcal{F} \text{ for every } \mathcal{F} \in q\}.$$

Put

$$\tilde{\mathcal{U}}_\alpha = \{\tilde{U} \mid U \in \mathcal{U}_\alpha\}.$$

Then it is easily seen that the restriction of $\tilde{\mathcal{U}}_\alpha$ to X is $\mathcal{U}_\alpha$. Furthermore we can show:

K) $\tilde{\mathcal{U}}_\alpha$ *is a covering of* $\tilde{X}$ *for each* $\alpha \in A$.

Proof. Let q be a given point of $\tilde{X}$ and choose $\beta \in A$ for which $\mathcal{U}_\beta^* < \mathcal{U}_\alpha$. Suppose that $\mathcal{F}$ is a member of q regarded as a class of Cauchy filters; then since $\mathcal{F}$ is Cauchy, $F \subset U$ for some $F \in \mathcal{F}$ and $U \in \mathcal{U}_\beta$, i.e. $U \in \mathcal{F}$. We take a member U'' of $\mathcal{U}_\alpha$ for which

$$S(U, \mathcal{U}_\beta) \subset U''. \tag{1}$$

Let $\mathcal{G}$ be a given filter belonging to q; then it follows from $\mathcal{F} \sim \mathcal{G}$ that

$$U' \in \mathcal{F} \cap \mathcal{G} \quad \text{for some } U' \in \mathcal{U}_\beta.$$

Since $U, U' \in \mathcal{F}$, we obtain

$$U \cap U' \neq \emptyset,$$

and hence

$$U \cup U' \subset U'' \quad \text{(see (1))}.$$

Since $U' \in \mathcal{G}$, $U'' \in \mathcal{G}$.

Thus we have proved that there is a $U'' \in \mathcal{U}_\alpha$ such that $U'' \in \mathcal{G}$ for every $\mathcal{G} \in q$. Therefore $q \in \tilde{U}''$ follows from the definition of $\tilde{U}''$. Hence $\tilde{\mathcal{U}}_\alpha$ covers $\tilde{X}$.

L) *If* $\mathcal{U}_\alpha^* < \mathcal{U}_\beta$ *in* X *then* $\tilde{\mathcal{U}}_\alpha^* < \tilde{\mathcal{U}}_\beta$ *in* $\tilde{X}$.

Proof. We note first that if $U \subset V$ holds for open sets U, V of X, then $\tilde{U} \subset \tilde{V}$. Now, suppose that $U, U' \in \mathcal{U}_\alpha$ and $U \cap U' = \emptyset$. Then each filter cannot contain U and U' at the same time. This implies that $\tilde{U} \cap \tilde{U}' = \emptyset$. Therefore, for a given member $\tilde{U}_0$ of $\tilde{\mathcal{U}}_\alpha$, we obtain

$$S(\tilde{U}_0, \tilde{\mathcal{U}}_\alpha) = \bigcup \{ \tilde{U} \mid U \cap U_0 \neq \emptyset, U \in \mathcal{U}_\alpha \} \subset \tilde{U}'$$

for some $U' \in \mathcal{U}_\beta$ satisfying $S(U_0, \mathcal{U}_\alpha) \subset U'$ since $\mathcal{U}_\alpha^* < \mathcal{U}_\beta$. Hence $\tilde{\mathcal{U}}_\alpha^* < \tilde{\mathcal{U}}_\beta$.

M) *Suppose q, q' are two different points of $\tilde{X}$. Then $q' \notin S(q, \tilde{\mathcal{U}}_\alpha)$ for some $\alpha \in A$.*

Proof. Let $\mathcal{F} \in q$ and $\mathcal{F}' \in q'$. Then, since $\mathcal{F} \nsim \mathcal{F}'$, there is a $\mathcal{U}_\alpha$ each member of which does not belong to $\mathcal{F} \cap \mathcal{F}'$. Let U be a given member of $\mathcal{U}_\alpha$ such that $q \in \tilde{U}$. Then by the definition of $\tilde{U}$, $U \in \mathcal{F}$ and hence $U \notin \mathcal{F}'$, i.e. $q' \notin \tilde{U}$. Therefore $q' \notin S(q, \tilde{\mathcal{U}}_\alpha)$.

It follows from K) and M) that $\{ \tilde{\mathcal{U}}_\alpha \mid \alpha \in A \}$ is a basis of uniformity of $\tilde{X}$. Since the thus defined uniform space $\tilde{X}$ clearly contains X as a subspace, we shall show that $\tilde{X}$ is a complete uniform space satisfying $\bar{X} = \tilde{X}$. For that purpose we need the following proposition which is valid for any uniform space.

N) *Let p be a cluster point of a Cauchy filter basis $\mathcal{F}$. Then $\mathcal{F} \to p$.*

Proof. Given $\alpha \in A$, we choose $\beta \in A$ for which $\mathcal{U}_\beta^* < \mathcal{U}_\alpha$. Since $\mathcal{F}$ is Cauchy, $F \subset U$ for some $F \in \mathcal{F}$ and $U \in \mathcal{U}_\beta$. On the other hand, since $p \in \bar{F}$, there is $U' \in \mathcal{U}_\beta$ with $p \in U'$ and $U' \cap U \neq \emptyset$. Hence

$$\{p\} \cup F \subset U' \cup U \subset U'' \quad \text{for some } U'' \in \mathcal{U}_\alpha.$$

Thus $F \subset S(p, \mathcal{U}_\alpha)$. Since $\{ S(p, \mathcal{U}_\alpha) \mid \alpha \in A \}$ is a nbd basis of p, we can conclude that $\mathcal{F} \to p$.

O) $\bar{X} = \tilde{X}$.

Proof. Given $q_0 \in \tilde{X}$ and $\alpha \in A$. Suppose that

$$q_0 \in \tilde{U} \in \tilde{\mathcal{U}}_\alpha.$$

Then

$$\tilde{U} \cap X = U \neq \emptyset.$$

This implies that $q_0 \in \bar{X}$, proving the assertion.

P) $\bar{X}$ *is complete.*

Proof. Let $\mathcal{G}$ be a given Cauchy filter of $\bar{X}$. Then $\mathcal{F} = \{G \cap X \mid G \in \mathcal{G}$, and G is open in $\bar{X}\}$ is a Cauchy filter basis of X. We denote by q_0 the class which contains the Cauchy filter $\mathcal{F}'$ of X generated by $\mathcal{F}$. We regard q_0 as a point of $\bar{X}$, to prove that $\mathcal{G} \to q_0$. Given $\tilde{\mathcal{U}}_\alpha$, we suppose that $\tilde{U}$ is a member of $\tilde{\mathcal{U}}_\alpha$ such that $q_0 \in \tilde{U}$. Then $U \in \mathcal{F}'$, which implies that $U \cap F \neq \emptyset$ for every $F \in \mathcal{F}$. Therefore $\tilde{U} \cap G \neq \emptyset$ for every open $G \in \mathcal{G}$, which means that q_0 is a cluster point of the Cauchy filter basis $\mathcal{G}' = \{G \mid G \in \mathcal{G}$, and G is open in $\bar{X}\}$. It follows from N) that $\mathcal{G}' \to q_0$. Therefore $\mathcal{G} \to q_0$. This proves that $\bar{X}$ is complete.

Now pursuing the analogy between completion and compactification, we can assert that every real-valued uniformly continuous function over X can be extended to a uniformly continuous function over $\bar{X}$. In fact we can prove more generally the following proposition.

Q) *Let X' be a dense subspace of a uniform space X, i.e. $\overline{X'} = X$. Then every uniformly continuous mapping f of X' into a complete uniform space Y can be extended to a uniformly continuous mapping of X into Y.*[1]

Proof. Let p be a given point of X. Then

$$\mathcal{F} = \{U(p) \cap X' \mid U(p) \text{ is a nbd of } p \text{ in } X\}$$

is a Cauchy filter of X'. We can assert that

$$\mathcal{G} = \{G \mid G \supset f(F) \text{ for some } F \in \mathcal{F}\}$$

is a Cauchy filter of Y. To see it, suppose that $\mathcal{V}_\alpha$ is a given uniform covering of Y; then by the uniform continuity of f,

$$f^{-1}(\mathcal{V}_\alpha) = \{f^{-1}(V) \mid V \in \mathcal{V}_\alpha\}$$

---

[1] M. Atsuji [2] and H. H. Corson and J. R. Isbell [1] studied uniform spaces with the property that every uniformly continuous function over every subset can be extended to a uniformly continuous function over the whole space.

is a uniform covering of X. Therefore $F \subset f^{-1}(V)$ for some $F \in \mathcal{F}$ and $V \in \mathcal{V}_\alpha$. This implies that $f(F) \subset V$, i.e. $\mathcal{G}$ is a Gauchy filter of Y. Since Y is complete, $\mathcal{G}$ converges to a point $q \in Y$. Put $g(p) = q$. The mapping g defined in this manner is clearly an extension of f over X. The only thing to be proved is that g is uniformly continuous over X. Suppose that $\mathcal{V}_\alpha$ is a given uniform covering of Y. Then take an open uniform covering $\mathcal{V}_\beta$ with $\mathcal{V}_\beta^* < \mathcal{V}_\alpha$.

Now, we can prove that

$$\{X - \overline{X' - f^{-1}(V)} \mid V \in \mathcal{V}_\beta\} < g^{-1}(\mathcal{V}_\alpha) \quad \text{in } X. \tag{1}$$

For, if $p \in X - \overline{X' - f^{-1}(V)}$ for $V \in \mathcal{V}_\beta$, then

$$U(p) = X - \overline{X' - f^{-1}(V)}$$

is a nbd of p in X satisfying

$$U(p) \cap X \subset f^{-1}(V)$$

since $f^{-1}(V)$ is open in X'. Hence

$$g(p) \in \overline{f(U(p) \cap X)} \subset \bar{V} \subset V'$$

for a member V' of $\mathcal{V}_\alpha$ such that $S(V, \mathcal{V}_\beta) \subset V'$. Since p is a given point of $X - \overline{X' - f^{-1}(V)}$, we get $g(X - \overline{X' - f^{-1}(V)}) \subset V'$, i.e.,

$$X - \overline{X' - f^{-1}(V)} \subset g^{-1}(V'),$$

proving (1). Since f is uniformly continuous, $f^{-1}(\mathcal{V}_\beta)$ is a uniform covering of X'. This means that there is an open uniform covering $\mathcal{U}_\beta$ of X such that the restriction $\mathcal{U}_\beta'$ of $\mathcal{U}_\beta$ to X' is a refinement of $f^{-1}(\mathcal{V}_\beta)$. In other words, for a given member U of $\mathcal{U}_\beta$, there is a member V of $\mathcal{V}_\beta$ for which

$$U \cap X \subset f^{-1}(V).$$

Therefore

$$U \cap \overline{X' - f^{-1}(V)} = \emptyset$$

(note that U is open), i.e.,

$$U \subset X - \overline{X' - f^{-1}(V)},$$

proving

$$\mathcal{U}_\beta < \{X - \overline{X' - f^{-1}(V)} \mid V \in \mathcal{V}_\beta\}.$$

This relation combined with (1) implies that $g^{-1}(\mathcal{V}_\alpha)$ is a uniform covering of X, proving that g is a uniformly continuous mapping.

To end our discussion on completion, let us verify the uniqueness of completion as follows:

R) *Let $\tilde{X}$ and $\hat{X}$ be completions of a uniform space X. Then $\tilde{X}$ and $\hat{X}$ are unimorphic by a mapping which keeps X fixed.*

Proof. We consider the identity mapping f of X onto X. Then by Q) we can extend f to a uniformly continuous mapping g of $\tilde{X}$ into $\hat{X}$. In the same way, we can extend f^{-1} to a uniformly continuous mapping h of $\hat{X}$ into $\tilde{X}$. Suppose that $p \in \tilde{X}$, $q \in \hat{X}$ and $g(p) = q$. Let

$$\mathcal{G} = \{V(q) \cap X \mid V(q) \text{ is a nbd of } q \text{ in } \hat{X}\}.$$

Putting

$$\mathcal{F} = \{F \mid F \subset \tilde{X}, F \supset f^{-1}(G) \text{ for some } G \in \mathcal{G}\},$$

we get a Cauchy filter $\mathcal{F}$ of $\tilde{X}$. As seen in the proof of Q),

$$\mathcal{F} \to h(q). \tag{1}$$

On the other hand, we can assert that $\mathcal{F} \to p$ in $\tilde{X}$. To see it, let $U(p)$ be a given nbd of p in $\tilde{X}$. Then by the definition of g,

$$q \in \overline{f(U(p) \cap X)} \quad \text{in } \hat{X}.$$

Hence it follows from the definition of $\mathcal{G}$ that

$$G \cap f(U(p) \cap X) \neq \emptyset \quad \text{for every } G \in \mathcal{G},$$

which implies

$$f^{-1}(G) \cap U(p) \neq \emptyset \quad \text{for every } G \in \mathcal{G} \text{ in } \tilde{X}.$$

Thus $p \in \overline{f^{-1}(G)}$, which means that p is a cluster point of $\mathscr{F}$. Since $\mathscr{F}$ is a Cauchy filter, by N), $\mathscr{F} \to p$. Thus from (1) we can conclude that $h(q) = p$, i.e. $h = g^{-1}$. This proves that g is a unimorphic mapping of $\tilde{X}$ onto $\hat{X}$, proving the proposition.

Combining O), P), Q) and R) we obtain the following theorem.

Theorem VI.20. *Every uniform space X has a completion $\tilde{X}$ such that every uniformly continuous mapping f of X into a complete uniform space Y can be extended to a uniformly continuous mapping g of $\tilde{X}$ into Y. The completion $\tilde{X}$ is unique in the sense of* R).

In view of the process of constructing the completion, we can assert the following.

Corollary. *If X is a metric space, then its completion is a complete metric space. If X is a totally bounded uniform space, then its completion is a compact uniform space.*

Example VI.11. Let X be a set. To each element p of X, we assign a collection $\{U_\alpha(p) \,|\, \alpha \in A\}$ of subsets of X satisfying the following conditions (where we should note that the index set A is common for all elements p of X):

  (i) $\bigcap \{U_\alpha(p) \,|\, \alpha \in A\} = \{p\}$ for every $p \in X$,
  (ii) for every $\alpha, \beta \in A$, there is $\gamma \in A$ for which

$$U_\gamma(p) \subset U_\alpha(p) \cap U_\beta(p) \quad \text{at every } p \in X,$$

  (iii) for every $\alpha \in A$, there is $\beta \in A$ such that $p \in U_\alpha(q)$ whenever p, $q \in U_\beta(r)$ for some $r \in X$.

Then $\{U_\alpha(p) \,|\, p \in X, \, \alpha \in A\}$ is called a *uniform nbd basis* of X. It is easy to show that $\{\mathscr{U}_\alpha \,|\, \alpha \in A\}$ for $\mathscr{U}_\alpha = \{U_\alpha(p) \,|\, p \in X\}$ satisfies the conditions for a basis of uniformity. Thus we can define a uniform space by use of the uniform nbd basis. This method of defining a uniform space is due to A. Weil [1].

Conversely, let X be a uniform space with a basis $\{\mathscr{U}_\alpha \,|\, \alpha \in A\}$ of uniformity. Then we can easily show that $U_\alpha(p) = S(p, \mathscr{U}_\alpha)$, $\alpha \in A$, $p \in X$, satisfy the three conditions above. Thus the two treatments of uniform spaces, by uniformity and uniform nbd basis, are essentially equivalent and we can pursue discussions with the concept of uniform nbd basis in a quite parallel way as we did in the preceding part of this section.[1]

6. Proximity space

In the present section we shall give a brief account of proximity space which may be considered as another generalization of metric space. We owe this concept originally to V. A. Efremovič [1] and its development to Yu. M. Smirnov and other mathematicians.[2]

Definition VI.6. Let X be a set. A *proximity relation* (or merely *proximity*) in X is a binary relation $A\delta B$ between subsets A, B of X satisfying the following conditions:

(i) $A\delta B$ implies $B\delta A$,
(ii) $A\delta X$ if and only if $A \neq \emptyset$,
(iii) $A\delta B \cup C$ if and only if $A\delta B$ or $A\delta C$,
(iv) if for every subset C of X, either $A\delta C$ or $B\delta X - C$, then $A\delta B$,
(v) if $\{p\}\delta\{q\}$ for elements p, q of X, then $p = q$.

A set X with a proximity relation is called a *proximity space*.

Example VI.12. The most popular example of a proximity space is a metric space. For subsets A, B of a metric space X, we define that

$$A\delta B \text{ if and only if } \rho(A, B) = \inf \{\rho(p, q) \mid p \in A, q \in B\} = 0.$$

Then δ is easily seen to satisfy (i)–(v) of Definition VI.6, and therefore X is a proximity space with the proximity δ.

A) $A\delta A$ *for every non-empty set* A.

Proof. By (ii) of Definition VI.5, $A\delta X$. Since $X = (X - C) \cup C$ for every subset $C \subset X$,

$$A\delta X - C \quad \text{or} \quad A\delta C.$$

---

[1] Extensive investigations of uniform spaces were done by J. R. Isbell, Z. Frolik and his school, and other mathematicians. See, for example, J. R. Isbell [1].
[2] See Yu. M. Smirnov [3] and also E. M. Alfsen and J. E. Fernstad [1], E. M. Alfsen and O. Njåstad [1], S. Leader [1].

Thus from (iv) it follows that $A\delta A$.

B) $A \subset B$ *and* $A\delta C$ *imply* $B\delta C$.

Proof. By (i) and (iii),

$$C\delta A \cup B, \qquad A \cup B = B,$$

which implies $B\delta C$.

C) $A \cap B \neq \emptyset$ *implies that* $A\delta B$.

Proof. From A) it follows that $A \cap B\delta A \cap B$, which combined with B) implies that $B\delta A \cap B$.
Hence by (iii) $B\delta(A \cap B) \cup A$, i.e. $A\delta B$.

D) *Let* $A_i \bar{\delta} B_i$, $i = 1, \ldots, n$; *then*

$$\left(\bigcap_{i=1}^{n} A_i \right) \bar{\delta} \left(\bigcup_{i=1}^{n} B_i \right),$$

where we denote by $\bar{\delta}$ *the negation of* δ.

Proof. It suffices to prove the assertion for $n = 2$. Since $A_1 \cap A_2 \subset A_1$ and $A_1 \bar{\delta} B_1$, by B) we get

$$A_1 \cap A_2 \bar{\delta} B_1 .$$

Similarly we get

$$A_1 \cap A_2 \bar{\delta} B_2 .$$

Hence by (iii) of Definition VI.6,

$$A_1 \cap A_2 \bar{\delta} B_1 \cup B_2 .$$

E) *Let* X *be a proximity space and* A *a subset of* X. *Define the closure* $\bar{A}$ *of* A *by*

$$\bar{A} = \{ p \mid p \in X, \{p\}\delta A \} .$$

Then X *is a* T_1-*topological space.*

Proof. $\bar{\emptyset} = \emptyset$ is a direct consequence of (ii) of Definition VI.6 and B). $A \subset \bar{A}$ is a direct consequence of C). $\overline{A \cup B} = \bar{A} \cup \bar{B}$ is a direct consequence of (iii).

$\bar{A} \subset \bar{\bar{A}}$ follows from B) since $A \subset \bar{A}$ is proved. Thus all we have to show is that $\bar{A} \supset \bar{\bar{A}}$. For that purpose it suffices to show that $\{p\}\delta\bar{A}$ for a point p of X implies $\{p\}\delta A$. In fact, we can prove more generally that $B\delta\bar{A}$ if and only if $B\delta A$, where $B \subset X$.

Since it is clear that $B\delta A$ implies $B\delta\bar{A}$, let us assume that $B\delta\bar{A}$ and prove that $B\delta A$. It follows from B) that $B\delta C$ for every C with $C \supset \bar{A}$. Thus for a given $C \subset X$, either $B\delta C$ or $C \not\supset \bar{A}$ holds. In the latter case there is a point $p \in \bar{A} - C$, i.e.,

$$\{p\}\delta A \quad \text{and} \quad p \in X - C.$$

Hence it follows from B) that

$$A\delta X - C.$$

Thus we have proved that for every $C \subset X$, either $B\delta C$ or $A\delta X - C$ holds. Hence using (iv), we get $B\delta A$. Therefore all the conditions for closure are satisfied, i.e., X is a topological space. It follows from (v) that $\overline{\{p\}} = \{p\}$ for each point p of X, and hence X is a T_1-space.

F) *Let X be a uniform space with uniformity $\{\mathcal{U}_\gamma \mid \gamma \in \Gamma\}$. For two sets A and B of X, we define*

$$A\delta B \text{ if and only if } S(A, \mathcal{U}_\gamma) \cap B \neq \emptyset \text{ for every } \gamma \in \Gamma.$$

Then X is a proximity space.

Proof. The conditions (i), (ii) and (v) of Definition VI.6 are clearly satisfied. It is also clear that $A\delta B$ or $A\delta C$ implies $A\delta B \cup C$. Now, let

$$A\bar{\delta}B \quad \text{and} \quad A\bar{\delta}C$$

at the same time. Then

$$S(A, \mathcal{U}_\alpha) \cap B = \emptyset \quad \text{and} \quad S(A, \mathcal{U}_\beta) \cap C = \emptyset$$

for some $\alpha, \beta \in \Gamma$. Then

$$S(A, \mathcal{U}_\gamma) \cap (B \cup C) = \emptyset$$

for $\mathcal{U}_\gamma = \mathcal{U}_\alpha \wedge \mathcal{U}_\beta$. This means that

$$A\bar{\delta}B \cup C,$$

proving (iii).

Finally, suppose $A\bar{\delta}B$; then

$$S(A, \mathcal{U}_\gamma) \cap B = \emptyset$$

for some $\gamma \in \Gamma$. Take $\beta \in \Gamma$ for which $\mathcal{U}_\beta^* < \mathcal{U}_\gamma$. Then put

$$C = S(B, \mathcal{U}_\beta).$$

Now, it is easy to see that

$$S(A, \mathcal{U}_\beta) \cap C = \emptyset.$$

Therefore

$$A\bar{\delta}C \quad \text{and} \quad B\bar{\delta}X - C,$$

proving (iv) of Definition VI.6.

Let $\mathcal{U} = \{U_i \mid i = 1, \ldots, k\}$ be a finite covering of a proximity space X. If there is a covering $\mathcal{V} = \{V_i \mid i = 1, \ldots, k\}$ of X such that

$$V_i \bar{\delta} X - U_i, \quad i = 1, \ldots, k,$$

then $\mathcal{U}$ is called a *p-covering*. Let f be a mapping of a proximity space X into a proximity space Y. If $A\delta B$ in X implies $f(A)\delta f(B)$ in Y, then f is called a *p-mapping*. A one-to-one mapping of a proximity space X onto a proximity space Y is called *p-homeomorphic* if f and f^{-1} are *p*-mappings.

Let X be a topological space and δ a proximity relation in X. If $p \in \bar{A}$ is equivalent with $p\delta A$, then the proximity relation is called *compatible* with the topology of X. Let X be a uniform space with uniformity $\{\mathcal{U}_\gamma \mid \gamma \in \Gamma\}$ and δ a proximity relation in X. If $A\delta B$ if and only if $S(A, \mathcal{U}_\gamma) \cap B \neq \emptyset$ for every $\gamma \in \Gamma$, then the uniformity is called *compatible* with the proximity relation. Now, we shall study when we can introduce a compatible proximity relation into a given topological space and when we can introduce a compatible uniformity into a given proximity space.

G) *Let X be a given proximity space, then the collection of all p-coverings of X is a basis of a totally bounded uniformity which is compatible with the proximity of X.*

Proof. We denote by $\{\mathcal{U}_\gamma \mid \gamma \in \Gamma\}$ the collection of all p-coverings of X. Given two members $\mathcal{U}_\alpha$ and $\mathcal{U}_\beta$ of this collection, we suppose that

$$\mathcal{U}_\alpha = \{U_i \mid i = 1, \ldots, m\} \quad \text{and} \quad \mathcal{U}_\beta = \{V_j \mid j = 1, \ldots, n\}.$$

Then there exist coverings

$$\mathcal{U}'_\alpha = \{U'_i \mid i = 1, \ldots, m\} \quad \text{and} \quad \mathcal{U}'_\beta = \{V'_j \mid j = 1, \ldots, n\}$$

such that

$$U'_i \bar{\delta} X - U_i \quad \text{and} \quad V'_j \bar{\delta} X - V_j.$$

It follows from B) that

$$U'_i \cap V'_j \bar{\delta} X - U_i \quad \text{and} \quad U'_i \cap V'_j \bar{\delta} X - V_j$$

which combined with (iii) of Definition VI.6 implies that

$$U'_i \cap V'_j \bar{\delta} (X - U_i) \cup (X - V_j),$$

i.e.,

$$U'_i \cap V'_j \bar{\delta} (X - U_i \cap V_j), \quad i = 1, \ldots, m; j = 1, \ldots, n.$$

Since $\{U'_i \cap V'_j \mid i = 1, \ldots, m; \; j = 1, \ldots, n\}$ is a covering, $\mathcal{U}_\alpha \wedge \mathcal{U}_\beta$ is a p-covering.

To prove that $\{\mathcal{U}_\gamma \mid \gamma \in \Gamma\}$ satisfies (iii) of Definition VI.3, we suppose that $\mathcal{U}_\alpha = \{U_i \mid i = 1, \ldots, m\}$ is a p-covering of X. Then there is a covering $\mathcal{W} = \{W_i \mid i = 1, \ldots, m\}$ such that

$$W_i \bar{\delta} (X - U_i).$$

By (iv) of Definition VI.6, there is an M_i with

$$W_i \bar{\delta} (X - M_i) \quad \text{and} \quad X - U_i \bar{\delta} M_i. \tag{1}$$

For different integers $i_1, \ldots, i_k$ chosen from $\{1, \ldots, m\}$, we define

$$N(i_1, \ldots, i_k) = [\cap \{U_{i_j} \mid j = 1, \ldots, k\}] \cap [\cap \{X - W_i \mid i \neq i_1, \ldots, i_k\}]$$
(2)

and

$$M(i_1, \ldots, i_k) = [\cap \{M_{i_j} \mid j = 1, \ldots, k\}] \cap [\cap \{X - M_i \mid i \neq i_1, \ldots, i_k\}].$$
(3)

Then

$$\mathcal{U} = \{N(i_1, \ldots, i_k) \mid 1 \leq i_1, \ldots, i_k \leq m; k = 1, \ldots, m\}$$

is a covering of X, because $\mathcal{U}_\alpha$ is a covering, and $W_i \subset U_i$ as seen from $W_i \bar{\delta}(X - U_i)$ combined with C). For this covering $\mathcal{U}$ we can assert that $\mathcal{U}^\Delta < \mathcal{U}_\alpha$. For, let p be a given point of X. Then there is an i for which $p \in W_i$. Now, suppose that $p \in N(i_1, \ldots, i_k)$; then by the definition of $N(i_1, \ldots, i_k)$ it must be true that $i_j = i$ for some j. Therefore $N(i_1, \ldots, i_k) \subset U_i$. This means that

$$S(p, \mathcal{U}) \subset U_i \in \mathcal{U}_\alpha.$$

Thus $\mathcal{U}^\Delta < \mathcal{U}_\alpha$.

Furthermore, we can prove that $\mathcal{U}$ is a p-covering. To accomplish this, it suffices to show that

$$M(i_1, \ldots, i_k) \bar{\delta} X - N(i_1, \ldots, i_k),$$

because

$$\mathcal{M} = \{M(i_1, \ldots, i_k) \mid 1 \leq i_1, \ldots, i_k \leq m, k = 1, \ldots, m\}$$

is obviously a covering of X. Since it follows from (1) that

$$M_{i_j} \bar{\delta} X - U_{i_j}, \quad j = 1, \ldots, k \quad \text{and} \quad X - M_i \bar{\delta} W_i, \quad i \neq i_1, \ldots, i_k,$$

in view of D) and (3) we obtain

$$M(i_1, \ldots, i_k) \bar{\delta} [\cup \{X - U_{i_j} \mid j = 1, \ldots, k\}] \cup [\cup \{W_i \mid i \neq i_1, \ldots, i_k\}].$$

Thus by (2)

$$M(i_1, \ldots, i_k) \bar{\delta} X - N(i_1, \ldots, i_k),$$

proving our assertion that $\mathcal{U}$ is a p-covering.

Suppose p, q are different points of X. Then by (v) of Definition VI.6, $p \bar{\delta} q$. Hence by (iv), there is a set C for which

$p\bar{\delta}C$ and $q\bar{\delta}X - C$.

Since $\{C, X - C\}$ is a covering of X,

$$\mathcal{U}' = \{X - p, X - q\}$$

is a p-covering of X satisfying $p \notin S(p, \mathcal{U}')$. Thus the collection $\{\mathcal{U}_\gamma \mid \gamma \in \Gamma\}$ of all p-coverings satisfies the conditions for a basis of uniformity, i.e. (i) and (iii) of Definition VI.3. Since each covering $\mathcal{U}_\gamma$ is finite, the uniformity is totally bounded.

Finally, we shall prove that this uniformity is compatible with the proximity of X. Suppose that $A\bar{\delta}B$. Then by (iv) of Definition VI.6, there is a set C with

$$A\bar{\delta}X - C   \text{ and }   B\bar{\delta}C.$$

Hence the covering

$$\mathcal{U} = \{X - A, X - B\}$$

is a p-covering, i.e., a uniform covering satisfying

$$S(A, \mathcal{U}) \cap B = \emptyset.$$

Conversely, let

$$S(A, \mathcal{U}) \cap B = \emptyset$$

for sets A, B and a uniform covering $\mathcal{U}$. Take a p-covering $\mathcal{U}_\gamma = \{U_i \mid i = 1, \ldots, m\}$ with

$$\mathcal{U}_\gamma^\Delta < \mathcal{U}.$$

Then
$$S(A, \mathcal{U}_\gamma) \cap S(B, \mathcal{U}_\gamma) = \emptyset. \tag{4}$$

Since $\mathcal{U}_\gamma$ is a p-covering, there is a covering $\mathcal{V} = \{V_i \mid i = 1, \ldots, m\}$ with

$$V_i\bar{\delta}X - U_i.$$

We denote by $V_{i_1}, \ldots, V_{i_k}$ all the members of $\mathcal{V}$ which intersect A. Then by D)

$$\cup\{V_{i_j} \mid j = 1, \ldots, k\} \bar{\delta} \cap \{X - U_{i_j} \mid j = 1, \ldots, k\}. \tag{5}$$

Since

$$A \subset \cup\{V_{i_j} \mid j = 1, \ldots, k\}$$

and

$$X - S(A, \mathcal{U}_\gamma) \subset \cap\{X - U_{i_j} \mid j = 1, \ldots, k\},$$

we obtain from B) and (5) that

$$A \bar{\delta} X - S(A, \mathcal{U}_\gamma). \tag{6}$$

In the same way, we obtain

$$B \bar{\delta} X - S(B, \mathcal{U}_\gamma). \tag{7}$$

Since

$$S(A, \mathcal{U}_\gamma) \subset X - S(B, \mathcal{U}_\gamma)$$

follows from (4), by (7) and B) we obtain

$$B \bar{\delta} S(A, \mathcal{U}_\gamma).$$

Therefore, in view of (6), by (iv) of Definition VI.6 we can conclude that

$$A \bar{\delta} B.$$

This completes the proof of the proposition.

Theorem VI.21. *Let X be a proximity space. Then there is one and only one totally bounded uniformity compatible with the proximity of X.*

Proof. The existence of a totally bounded uniformity compatible with the proximity of X was proved in G). Let us consider a given totally bounded uniformity compatible with the proximity of X. Then, since it is totally bounded, as seen easily, the uniformity has a basis consisting of finite uniform coverings. Therefore we denote by $\{\mathcal{V}_\gamma \mid \gamma \in \Gamma\}$ the basis of the uniformity consisting of all finite uniform coverings. Then for each $\gamma \in \Gamma$, there is $\beta \in \Gamma$ such that $\mathcal{V}_\beta^* < \mathcal{V}_\gamma$. Thus we may assume that

$$\mathcal{V}_\gamma = \{V_i \mid i = 1, \ldots, n\}, \qquad \mathcal{V}_\beta = \{V_i' \mid i = 1, \ldots, n\}$$

and

$$S(V_i', \mathcal{V}_\beta) \subset V_i, \quad i = 1, \ldots, n,$$

where some V_i' might be empty, and V_i and V_j may coincide for different i and j. Since the uniformity is compatible with the proximity,

$$V_i' \bar{\delta} (X - V_i).$$

Therefore $\mathcal{V}_\gamma$ is a p-covering.

Conversely, let $\mathcal{V} = \{V_i \mid i = 1, \ldots, n\}$ be a p-covering of X. Then there is a covering $\mathcal{V}' = \{V_i' \mid i = 1, \ldots, n\}$ such that

$$V_i' \bar{\delta} (X - V_i), \quad i = 1, \ldots, n.$$

Hence there are $\gamma_i \in \Gamma$, $i = 1, \ldots, n$, for which

$$S(V_i', \mathcal{V}_{\gamma_i}) \subset V_i, \quad i = 1, \ldots, n. \tag{1}$$

Choose $\gamma \in \Gamma$ for which

$$\mathcal{V}_\gamma < \bigwedge_{i=1}^{n} \mathcal{V}_{\gamma_i}. \tag{2}$$

Then $\mathcal{V}_\gamma < \mathcal{V}$. For, let $V \in \mathcal{V}_\gamma$; then $V \cap V_i' \neq \emptyset$ for some member V_i' of $\mathcal{V}'$ since $\mathcal{V}'$ is a covering. This combined with (1) and (2) implies that

$$V \subset V_i \in \mathcal{V},$$

proving our assertion. Thus $\mathcal{V}$ is a uniform covering belonging to $\{\mathcal{V}_\gamma \mid \gamma \in \Gamma\}$. Therefore the only totally bounded uniformity compatible with the proximity of X is the one induced by the totality of p-coverings of X.[1]

H) *Let X and Y be proximity spaces and f a mapping of X into Y. In view of G) we can consider X and Y to be totally bounded uniform spaces. Then f is a p-mapping if and only if it is a uniformly continuous mapping of the uniform space X into the uniform space Y.*

Proof. It is left to the reader.

---

[1] Note that this is the weakest uniform topology compatible with the proximity. To the contrary, there is a proximity space which has no strongest compatible uniformity. See E. M. Alfsen and O. Njåstad [1].

I) *Let X be a proximity space and A and B subsets of X. Then $A\bar{\delta}B$ if and only if there is a p-mapping of X into $[0, 1]$ such that $f(A) = 0$ and $f(B) = 1$.*

Proof. Generally for subsets A, B of a uniform space X, we can easily show, in a way similar to the proof of Urysohn's lemma, that there is a uniformly continuous mapping f of X into $[0, 1]$ such that $f(A) = 0$ and $f(B) = 1$ if and only if $S(A, \mathscr{U}) \cap B = \emptyset$ for some uniform covering $\mathscr{U}$ of X. The present proposition follows directly from the above observation combined with H).

Theorem VI.22. *Let X be a topological space. Then we can introduce a proximity compatible with the topology of X if and only if X is completely regular.*

Proof. If X is a proximity space, then by G) there is a uniformity compatible with its proximity. This uniformity is easily seen to be also compatible with the topology of X. Therefore X is completely regular by Theorem VI.15. Conversely, if X is completely regular, then by Theorem VI.15, there is a uniformity $\{\mathscr{U}_\gamma \mid \gamma \in \Gamma\}$ compatible with the topology of X. If we define that $A\delta B$ if and only if $S(A, \mathscr{U}_\gamma) \cap B \neq \emptyset$ for every $\gamma \in \Gamma$, then this proximity is easily seen to be compatible with the topology of X.

Corollary. *Let X be a compact T_2-space. Then there is one and only one proximity compatible with the topology of X.*

Proof. Since the existence of a proximity is a direct consequence of Theorem VI.22, we shall prove only the uniqueness. Let δ be a proximity compatible with the topology of X. Then by Theorem VI.21, there is a totally bounded uniformity compatible with δ. Since the uniformity is also compatible with the topology of X, and X is compact, the collection of all finite open coverings of X forms a basis of the unique uniformity of X. Suppose A and B are two sets of X such that $\bar{A} \cap \bar{B} = \emptyset$. Then the finite open covering $\mathscr{U} = \{X - \bar{A}, X - \bar{B}\}$ satisfies $S(A, \mathscr{U}) \cap B = \emptyset$. Therefore $A\bar{\delta}B$.

Conversely, if $\bar{A} \cap \bar{B} \neq \emptyset$, then we take a point $p \in \bar{A} \cap \bar{B}$. Now, for every finite open covering $\mathscr{U}$ of X,

$$S(A, \mathscr{U}) \supset \bar{A} \ni p,$$

i.e. $S(A, \mathscr{U})$ is a nbd of p. Hence from $p \in \bar{B}$ it follows that

$$S(A, \mathscr{U}) \cap B \neq \emptyset.$$

Therefore $A\delta B$. Thus $A\delta B$ if and only if $\bar{A} \cap \bar{B} \neq \emptyset$. Namely the proximity δ is uniquely defined by the topology of X.

Let X be a proximity space and Y a subset of X. If for every pair of subsets A and B of Y we define that $A\delta B$ in Y if and only if $A\delta B$ in X, then Y also turns out to be a proximity space called a *subspace* of the proximity space X. We know that not every uniform space can be imbedded as a dense subspace of a compact uniform space. In fact, totally bounded uniform spaces, and only those, can be dense subsets of compact uniform spaces. But for proximity space the circumstance is different as seen in the following.

Theorem VI.23. *Every proximity space X is a dense subspace of a compact T_2 proximity space $\tilde{X}$ which is uniquely defined by X. Every p-mapping of X into a compact proximity space Y can be extended to a p-mapping of $\tilde{X}$ into Y.*

Proof. We regard X as a totally bounded uniform space with the uniformity given in G). We construct the completion $\tilde{X}$ of the uniform space X. Then, since X is totally bounded, $\tilde{X}$ is a compact T_2-space. We may consider $\tilde{X}$ to be a proximity space with the proximity defined by its uniformity. Thus $\tilde{X}$ is easily seen to be the desired proximity space.

Conversely, if X^* is a given compact proximity space which contains X as a dense subspace, then by the corollary to Theorem VI.22, we can regard X^* as a totally bounded uniform space with the uniformity given in G). We can assert that this uniform space coincides with the uniform space $\tilde{X}$. To prove this, it suffices to show that X^* is a completion of the uniform space X. Since X^* is a complete uniform space and contains X as a dense subset, all we have to prove is that X is a subspace of the uniform space X^*. Let $\{\mathscr{U}_\gamma \mid \gamma \in \Gamma\}$ be the uniformity of X^*; then we can prove that $\{\mathscr{U}'_\gamma \mid \gamma \in \Gamma\}$, where $\mathscr{U}'_\gamma = \{U \cap X \mid U \in \mathscr{U}_\gamma\}$, is a uniformity compatible with the proximity of X. Suppose that A and B are subsets of X; then $A\bar{\delta}B$ in X if and only if $A\bar{\delta}B$ in X^* which is true if and only if $S(A, \mathscr{U}_\gamma) \cap B = \emptyset$ for some $\gamma \in \Gamma$. The last statement is true if and only if $S(A, \mathscr{U}'_\gamma) \cap B = \emptyset$. Thus the uniformity $\{\mathscr{U}'_\gamma \mid \gamma \in \Gamma\}$ is compatible with the proximity of X. Since $\{\mathcal{U}_\gamma \mid \gamma \in \Gamma\}$ is a totally bounded uniformity, so is $\{\mathcal{U}'_\gamma \mid \gamma \in \Gamma\}$. Therefore by Theorem VI.21, it is the uniformity of X considered at the beginning of this proof. Therefore X is a subspace of the uniform space X^*, i.e., X^* is a completion of X. Thus by the uniqueness of completion (4.R)), $\tilde{X}$ and X^* are unimorphic by a unimorphism which keeps X fixed. Therefore by H), $\tilde{X}$ and X^* are p-homeomorphic by a mapping which keeps X fixed.

Finally, we can easily verify the last part of Theorem VI.23 by use of Theorem VI.20.[1]

7. P-space

In the present section we shall give an account of the theory of K. Morita [4], [5] which is aimed at characterizing spaces whose products with metric spaces are normal and leads us, as a result, to a new category of spaces which contains metric spaces and compact spaces as special cases.

Definition VI.7. A topological space X is called a *P-space*[2] if for every open collection $\{U(\alpha_1, \ldots, \alpha_i) \mid \alpha_1, \ldots, \alpha_i \in A; i = 1, 2, \ldots\}$ in X satisfying the condition

$$U(\alpha_1, \ldots, \alpha_i) \subset U(\alpha_1, \ldots, \alpha_i, \alpha_{i+1}), \quad \alpha_1, \ldots, \alpha_{i+1} \in A; i = 1, 2, \ldots,$$

there exists a closed collection

$$\{F(\alpha_1, \ldots, \alpha_i) \mid \alpha_1, \ldots, \alpha_i \in A; i = 1, 2, \ldots\}$$

in X satisfying:

---

[1] We can define terminologies like totally bounded proximity space and complete proximity space. See Yu. M. Smirnov [4], [5], S. Leader [1], E. M. Alfsen and O. Njåstad [1]. V. A. Efremovič and A. S. Švarc [1] considered a necessary and sufficient condition for a proximity space to be metrizable. Some mathematicians are studying theories unifying those of topological space, uniform space and proximity space. See A. Császár [1], D. Doičinov [1]. The concept of *category* is also useful for giving a systematic description of a formal part of the theory of topological spaces, uniform spaces and proximity spaces though the usefulness is restricted. See, e.g., J. R. Isbell [1], H. Herrlich [1], H. Herrlich–G. E. Strecker [1], O. Wyler [1].

[2] This terminology is sometimes used in a completely different sense. So, to be precise, we should call a space satisfying this condition a *Morita's P-space*.

(i) $F(\alpha_1, \ldots, \alpha_i) \subset U(\alpha_1, \ldots, \alpha_i)$,

(ii) if $\bigcup_{i=1}^{\infty} U(\alpha_1, \ldots, \alpha_i) = X$ for a sequence $\{\alpha_i \mid i = 1, 2, \ldots\}$, then $\bigcup_{i=1}^{\infty} F(\alpha_1, \ldots, \alpha_i) = X$.

A) *Every countably compact space is a P-space.*

Proof. Let $\{U(\alpha_1, \ldots, \alpha_i) \mid \alpha_1, \ldots, \alpha_i \in A; i = 1, 2, \ldots\}$ be a given open collection satisfying the condition of Definition VI.7. Then we define $F(\alpha_1, \ldots, \alpha_i)$ by

$$F(\alpha_1, \ldots, \alpha_i) = \begin{cases} U(\alpha_1, \ldots, \alpha_i) & \text{if } U(\alpha_1, \ldots, \alpha_i) = X, \\ \emptyset & \text{otherwise.} \end{cases}$$

Then $\{F(\alpha_1, \ldots, \alpha_i) \mid \alpha_1, \ldots, \alpha_i \in A; \ i = 1, 2, \ldots\}$ satisfies (i) and (ii) of the definition, because $\bigcup_{i=1}^{\infty} U(\alpha_1, \ldots, \alpha_i) = X$ implies that $U(\alpha_1, \ldots, \alpha_i) = X$ for some i, since X is countably compact.

B) *Every fully normal, Čech complete space is a P-space.*

Proof. Since the validity of this proposition is implied by A) if X is compact, we may assume that X is not compact. Then by 2.D), there is a sequence $\mathscr{F}_i$ of collections of zero sets with f.i.p. satisfying (i) and (ii) of the same proposition. We note here that (ii) holds for every non-convergent maximal closed filter $\mathscr{F}$, because X is normal. Putting

$$\mathscr{U}_i = \{X - F \mid F \in \mathscr{F}_i\},$$

we obtain open coverings $\mathscr{U}_i$ of X. Since X is fully normal, we can choose open coverings $\mathscr{V}_i$, $i = 1, 2, \ldots$, of X for which

$$\bar{\mathscr{V}}_i^{\Delta} < \mathscr{U}_i, \quad i = 1, 2, \ldots. \tag{1}$$

We may assume that $\mathscr{V}_{i+1} < \mathscr{V}_i$. Now, suppose that

$$\{U(\alpha_1, \ldots, \alpha_i) \mid \alpha_1, \ldots, \alpha_i \in A; i = 1, 2, \ldots\}$$

is a given open collection satisfying the condition of Definition VI.7. Then we define closed sets $F(\alpha_1, \ldots, \alpha_i)$ by

$$F(\alpha_1, \ldots, \alpha_i) = X - S(X - U(\alpha_1, \ldots, \alpha_i), \mathscr{V}_i). \tag{2}$$

It is clear that $F(\alpha_1, \ldots, \alpha_i)$ satisfies (i) of Definition VI.7. To see (ii), we suppose that

$$\bigcup_{i=1}^{\infty} U(\alpha_1, \ldots, \alpha_i) = X. \tag{3}$$

If we assume that $\bigcup_{i=1}^{\infty} F(\alpha_1, \ldots, \alpha_i) \neq X$, then there is some

$$p_0 \in \bigcap_{i=1}^{\infty} (X - F(\alpha_1, \ldots, \alpha_i)).$$

Then
$$\mathcal{S} = \{\overline{S(p_0, \mathcal{V}_i)} \mid i = 1, 2, \ldots\} \cup \{X - U(\alpha_1, \ldots, \alpha_i) \mid i = 1, 2, \ldots\}$$

has f.i.p., because if $i_1 \geq i_2 \geq \cdots \geq i_k$, then

$$[\cap\{\overline{S(p_0, \mathcal{V}_{i_j})} \mid j = 1, \ldots, k\}] \cap [\cap\{X - U(\alpha_1, \ldots, \alpha_{i_j}) \mid j = 1, \ldots, k\}]$$
$$= \overline{S(p_0, \mathcal{V}_{i_1})} \cap (X - U(\alpha_1, \ldots, \alpha_{i_1}))$$
$$\supset S(p_0, \mathcal{V}_{i_1}) \cap (X - U(\alpha_1, \ldots, \alpha_{i_1})) \neq \emptyset$$

since $p_0 \in X - F(\alpha_1, \ldots, \alpha_{i_1})$ (see (2)).

Hence there is a maximal closed filter $\mathcal{F}$ which contains $\mathcal{S}$ as a subcollection. It follows from (3) that $\mathcal{F}$ is non-convergent. On the other hand, for each i,

$$\overline{S(p_0, \mathcal{V}_i)} \in \mathcal{F},$$

which combined with (1) implies that $X - F \in \mathcal{F}$ for some $F \in \mathcal{F}_i$, i.e. $F \notin \mathcal{F}$; hence $\mathcal{F}_i \not\subset \mathcal{F}$. But this contradicts the condition (ii) of 2.D) (in a modified form as noted before). Thus we have proved

$$\bigcup_{i=1}^{\infty} F(\alpha_1, \ldots, \alpha_i) = X,$$

i.e. X is a P-space.

C) *Every normal P-space is countably paracompact.*

Proof. Let $\{U_i \mid i = 1, 2, \ldots\}$ be an open covering of a normal P-space X such that $U_i \subset U_{i+1}$. Then, since X is a P-space, there is a sequence $\{F_i \mid i = 1, 2, \ldots\}$ of closed sets such that

$$F_i \subset U_i \quad \text{and} \quad \bigcup_{i=1}^{\infty} F_i = X.$$

Therefore by Theorem V.5, X is countably paracompact.

D) *A covering $\mathcal{U}$ of a topological space X is normal if and only if there is a σ-locally finite open refinement $\mathcal{V}$ of $\mathcal{U}$ such that each member V of $\mathcal{V}$ is a cozero set, i.e.,*

$$V = \{p \mid f(p) > 0\}$$

for some real-valued continuous function f over X with $0 \le f \le 1$.

Proof. Let $\mathcal{U} = \{U_\alpha \mid \alpha < \tau\}$ be a normal covering. Then there exists a sequence $\mathcal{U}_1, \mathcal{U}_2, \ldots$ of open coverings such that

$$\mathcal{U} > \mathcal{U}_1^{\Delta} > \mathcal{U}_1 > \mathcal{U}_2^{\Delta} > \cdots.$$

Using the argument in the proof of Theorem V.2, we can define open sets W_α^n, $\alpha < \tau$, $n = 1, 2, \ldots$, such that

$$S(W_\alpha^n, \mathcal{U}_n) \subset U_\alpha$$

and

$$S(W_\alpha^n, \mathcal{U}_{n+1}) \cap W_\beta^n = \emptyset \quad \text{if } \alpha \ne \beta. \tag{1}$$

Put

$$S(W_\alpha^n, \mathcal{U}_{n+3}) = M_\alpha^n.$$

Then by a process similar to that in the proof of Urysohn's lemma, we can define a continuous function f_α^n such that

$$f_\alpha^n(W_\alpha^n) = 1, \qquad f_\alpha^n(X - M_\alpha^n) = 0 \quad \text{and} \quad 0 \le f_\alpha^n \le 1.$$

Put

$$N_\alpha^n = \{p \mid f_\alpha^n(p) > 0\};$$

then N_α^n is a cozero set satisfying

$$W_\alpha^n \subset N_\alpha^n \subset M_\alpha^n \subset U_\alpha.$$

It easily follows from $\mathcal{U}_{n+3}^{\Delta\Delta} < \mathcal{U}_{n+1}$ and (1) that

$$S(N_\alpha^n, \mathcal{U}_{n+3}) \cap N_\beta^n = \emptyset \quad \text{if } \alpha \ne \beta.$$

Thus

$$\mathcal{V} = \{N_\alpha^n \mid \alpha < \tau, n = 1, 2, \ldots\}$$

is the desired σ-locally finite open refinement of $\mathcal{U}$.

Conversely, we consider a covering $\mathcal{U}$ of X and a σ-locally finite open covering

$$\mathcal{V} = \bigcup_{n=1}^{\infty} \mathcal{V}_n < \mathcal{U},$$

where each $\mathcal{V}_n$ is a locally finite open collection whose elements are cozero sets. Let

$$\mathcal{V}_n = \{V_{\alpha n} \mid \alpha \in A_n\}, \quad V_{\alpha n} = \{p \mid f_{\alpha n}(p) > 0\}$$

for a real-valued continuous function $f_{\alpha n}$ over X satisfying $0 \le f_{\alpha n} \le 1$. Since $\mathcal{V}_n$ is locally finite,

$$f_n(p) = \sum_{\alpha \in A_n} f_{\alpha n}(p), \quad p \in X,$$

is a continuous function. Putting

$$f(p) = \sum_{n=1}^{\infty} \frac{f_n(p)}{2^n(1 + f_n(p))}, \quad p \in X,$$

we obtain a continuous function such that $f(p) > 0$ at every $p \in X$. Further, we put

$$g_{\alpha n}(p) = \frac{f_{\alpha n}(p)}{2^n f(p)(1 + f_n(p))}, \quad p \in X.$$

Then $g_{\alpha n}$, $\alpha \in A_n$, $n = 1, 2, \ldots$, are continuous functions satisfying

$$\sum_{n=1}^{\infty} \sum_{\alpha \in A_n} g_{\alpha n}(p) = 1, \quad p \in X, \tag{2}$$

and

$$V_{\alpha n} = \{p \mid g_{\alpha n}(p) > 0\}.$$

Let us consider a metric space Y whose points are the points $\{x_{\alpha n} \mid \alpha \in A_n, n = 1, 2, \ldots\}$ of the cartesian product of the copies $I_{\alpha n}$, $\alpha \in A_n$,

$n = 1, 2, \ldots$, of the unit segment $[0, 1]$, satisfying

$$\sum x_{an} = 1,$$

and whose metric is defined by

$$\rho(x, y) = \sum \{|x_{an} - y_{an}| \mid \alpha \in A_n, n = 1, 2, \ldots\}$$

for $x = \{x_{an}\}$ and $y = \{y_{an}\}$.

To every point p of X, we assign a point

$$g(p) = \{g_{an}(p) \mid \alpha \in A_n, n = 1, 2, \ldots\}$$

of Y (see (2)). Then g is seen to be a continuous mapping of X into Y. For, suppose p_0 is a given point of X, and ε is a positive number. Then we can choose $(\alpha_1, n_1), \ldots, (\alpha_k, n_k)$ and a nbd $U(p_0)$ of p_0 such that

$$\sum \{g_{an}(p_0) \mid (\alpha, n) \neq (\alpha_i, n_i), i = 1, \ldots, k\} < \varepsilon$$

and

$$\sum_{i=1}^{k} |g_{\alpha_i n_i}(p) - g_{\alpha_i n_i}(p_0)| < \varepsilon \quad \text{for every } p \in U(p_0).$$

Then, in view of the fact that

$$\sum g_{an}(p) = \sum g_{an}(p_0) = 1,$$

we obtain

$$\sum_{(\alpha, n) \neq (\alpha_i, n_i)} g_{an}(p) = \sum_{i=1}^{k} (g_{\alpha_i n_i}(p_0) - g_{\alpha_i n_i}(p)) + \sum_{(\alpha, n) \neq (\alpha_i, n_i)} g_{an}(p_0)$$

$$\leq \sum_{i=1}^{k} |g_{\alpha_i n_i}(p_0) - g_{\alpha_i n_i}(p)| + \sum_{(\alpha, n) \neq (\alpha_i, n_i)} g_{an}(p_0)$$

$$< 2\varepsilon \quad \text{for every } p \in U(p_0).$$

This implies that

$$\rho(g(p), g(p_0)) = \sum_{i=1}^{k} |g_{\alpha_i n_i}(p) - g_{\alpha_i n_i}(p_0)| + \sum_{(\alpha, n) \neq (\alpha_i, n_i)} |g_{an}(p) - g_{an}(p_0)|$$

$$< \varepsilon + \sum_{(\alpha, n) \neq (\alpha_i, n_i)} (g_{an}(p) + g_{an}(p_0)) < 4\varepsilon$$

for every $p \in U(p_0)$, which proves the continuity of g. Put

$$W_{\alpha n} = \{\{x_{\alpha n}\} \mid x_{\alpha n} > 0, \{x_{\alpha n}\} \in g(X)\},$$

then

$$\mathcal{W} = \{W_{\alpha n} \mid \alpha \in A_n, n = 1, 2, \ldots\}$$

is an open covering of $g(X)$ satisfying

$$g^{-1}(\mathcal{W}) = \{g^{-1}(W) \mid W \in \mathcal{W}\} < \mathcal{V},$$

because

$$g^{-1}(W_{\alpha n}) = V_{\alpha n}.$$

Since $g(X)$ is a metric space, we can construct open coverings $\mathcal{W}_1, \mathcal{W}_2, \ldots$ of $g(X)$ such that

$$\mathcal{W} > \mathcal{W}_1^\Delta > \mathcal{W}_1 > \mathcal{W}_2^\Delta > \cdots.$$

Now

$$\mathcal{V}_i = f^{-1}(\mathcal{W}_i), \quad i = 1, 2, \ldots,$$

are open coverings of X satisfying

$$\mathcal{U} > \mathcal{V} > \mathcal{V}_1^\Delta > \mathcal{V}_1 > \mathcal{V}_2^\Delta \cdots.$$

Hence $\mathcal{U}$ is a normal covering.

E) *Let $\mathcal{U}$ be an open covering of a topological space X. If there is a normal covering $\mathcal{V}$ of X such that for each $V \in \mathcal{V}$, the restriction $\{V \cap U \mid U \in \mathcal{U}\}$ of $\mathcal{U}$ to V is a normal covering of V, then $\mathcal{U}$, too, is a normal covering of X.*

Proof. Since $\mathcal{V}$ is normal, by D) there is a σ-locally finite open refinement $\bigcup_{i=1}^{\infty} \mathcal{V}_i$ of $\mathcal{V}$, where each $\mathcal{V}_i$ is a locally finite collection of cozero sets. We may assume that

$$\bigcup_{i=1}^{\infty} \bar{\mathcal{V}}_i < \mathcal{V}.$$

Let

$$\mathcal{V}_i = \{V_{\alpha i} \mid \alpha \in A_i\}, \quad V_{\alpha i} = \{p \mid f_{\alpha i}(p) > 0\},$$

where $f_{\alpha i}$ is a continuous function over X with $0 \leqslant f_{\alpha i} \leqslant 1$. To each $V_{\alpha i}$ we assign a member $V(V_{\alpha i})$ of $\mathcal{V}$ containing $\bar{V}_{\alpha i}$. Since $\{V(V_{\alpha i}) \cap U \mid U \in \mathcal{U}\}$

is a normal covering of $V(V_{\alpha i})$, we can construct a σ-locally finite open covering $\bigcup_{j=1}^{\infty} \mathcal{V}^j_{\alpha i}$ of the subspace $V(V_{\alpha i})$ such that

$$\bigcup_{j=1}^{\infty} \mathcal{V}^j_{\alpha i} < \{V(V_{\alpha i}) \cap U \mid U \in \mathcal{U}\},$$

where each $\mathcal{V}^j_{\alpha i}$ is locally finite in $V(V_{\alpha i})$ and consists of cozero sets of $V(V_{\alpha i})$. Now put

$$\mathcal{W}^j_{\alpha i} = \{V_{\alpha i} \cap V \mid V \in \mathcal{V}^j_{\alpha i}\}.$$

Then it easily follows from the local finiteness of $\mathcal{V}_i$ that

$$\mathcal{W}^j_i = \bigcup\{\mathcal{W}^j_{\alpha i} \mid \alpha \in A_i\}$$

is a locally finite open collection of X. Thus

$$\mathcal{W} = \bigcup_{i,j=1}^{\infty} \mathcal{W}^j_i$$

is a σ-locally finite open refinement of $\mathcal{U}$. On the other hand, for each $V \in \mathcal{V}^j_{\alpha i}$ there is a continuous function g, with $0 \leqslant g \leqslant 1$, over $V(V_{\alpha i})$ such that

$$V = \{p \mid g(p) > 0\}.$$

Put

$$h(p) = \begin{cases} g(p)f_{\alpha i}(p), & p \in V_{\alpha i}, \\ 0, & p \notin V_{\alpha i}. \end{cases}$$

Then h is a continuous function over X satisfying

$$V_{\alpha i} \cap V = \{p \mid h(p) > 0\}.$$

Thus $\mathcal{W}$ consists of cozero sets of X. Hence by D), $\mathcal{U}$ is a normal covering.

F) *Every σ-locally finite open covering $\mathcal{U}$ of a countably paracompact, normal space X is normal.*

Proof. Let $\mathcal{U} = \bigcup_{i=1}^{\infty} \mathcal{U}_i$ be a σ-locally finite open covering of X, where each $\mathcal{U}_i$ is locally finite. Putting

$$U_i = \bigcup \{U \mid U \in \mathcal{U}_i\},$$

we obtain an open covering $\{U_i \mid i = 1, 2, \ldots\}$ of X. Since X is countably paracompact, we can construct a locally finite open refinement $\{V_i \mid i = 1, 2, \ldots\}$ of $\{U_i \mid i = 1, 2, \ldots\}$ with $V_i \subset U_i$. Put

$$\mathcal{V}_i = \{V_i \cap U \mid U \in \mathcal{U}_i\};$$

then

$$\mathcal{V} = \bigcup_{i=1}^{\infty} \mathcal{V}_i$$

is a locally finite open refinement of $\mathcal{U}$. Let

$$\mathcal{V} = \{V_\alpha \mid \alpha \in A\};$$

then, since X is normal, there is a closed covering $\{F_\alpha \mid \alpha \in A\}$ of X such that $F_\alpha \subset V_\alpha$. For each $\alpha \in A$, we consider a continuous function f_α with

$$f_\alpha(F_\alpha) = 1, \qquad f_\alpha(X - V_\alpha) = 0 \quad \text{and} \quad 0 \leqslant f_\alpha \leqslant 1,$$

and put

$$W_\alpha = \{p \mid f_\alpha(p) > 0\}.$$

Then

$$\mathcal{W} = \{W_\alpha \mid \alpha \in A\}$$

is a locally finite open refinement of $\mathcal{V}$ consisting of cozero sets. Thus by V.5.D), $\mathcal{U}$ is normal.

Corollary. *Every σ-locally finite open cover $\mathcal{U}$ of a countably paracompact normal space X has a locally finite open refinement.*

Proof. The cover $\mathcal{V}$ in the proof of F) is the desired one.

G) *Let H be an F_σ-open set of a countably paracompact normal space X. Then H is also countably paracompact and normal.*

Proof. Since X is normal, by Urysohn's lemma we can construct cozero sets U_i, $i = 1, 2, \ldots$, such that

$$\bar{U}_i \subset U_{i+1}, \qquad \bigcup_{i=1}^{\infty} \bar{U}_i = H.$$

Let $\mathcal{V} = \{V_1, V_2, \ldots\}$ be a given countable open covering of H. It suffices to prove that $\mathcal{V}$ is normal. For, if it is so, then there is an open covering $\mathcal{V}_1$ of H with $\mathcal{V}_1^* < \mathcal{V}$. Put $W_i = \cup\{V \mid V \in \mathcal{V}_1, \ S(V, \mathcal{V}_1) \subset V_i\}$, $i = 1, 2, \ldots$. Then $\{W_i \mid i = 1, 2, \ldots\}$ is an open covering of H satisfying $\bar{W}_i \subset V_i$ in H. Thus by Theorem V.5, H is countably paracompact while it is normal by Exercise III.11. Since each $\bar{U}_i$ is countably paracompact and normal, there is a locally finite refinement $\mathcal{V}_i$ of $\mathcal{V}$ consisting of cozero sets of $\bar{U}_i$. Let

$$\mathcal{W}_i = \{V \cap U_i \mid V \in \mathcal{V}_i\}\,;$$

then it is easy to see that $\cup_{i=1}^{\infty} \mathcal{W}_i$ is a σ-locally finite open refinement of $\mathcal{V}$ consisting of cozero sets of H. Thus by V.5.D), $\mathcal{V}$ is a normal covering of H. Therefore H is normal and countably paracompact.

H) *Every open F_σ-set H of a normal space X is a cozero set.*

Proof. Let $H = \cup_{i=1}^{\infty} F_i$ for closed sets F_i, $i = 1, 2, \ldots$. Then for each i there is a continuous function f_i such that

$$f_i(F_i) = 1\,, \qquad f_i(X - H) = 0\,, \quad 0 \leq f_i \leq 1\,.$$

Then

$$f = \sum_{i=1}^{\infty} \frac{1}{2^i} f_i$$

is a continuous function such that $\{x \in X \mid f(x) \neq 0\} = H$.

I) *Let $\mathcal{U} = \{U_\alpha \mid \alpha \in A\}$ be a normal covering of a topological space X, and let Y be a given topological space. Then $\{U_\alpha \times Y \mid \alpha \in A\}$ is a normal covering of $X \times Y$.*

Proof. The easy proof is left to the reader.

J) *Let $\{U(\alpha_1 \ldots \alpha_i) \mid \alpha_1, \ldots, \alpha_i \in A; \ i = 1, 2, \ldots\}$ be an open collection in a normal space X such that*

$$U(\alpha_1 \ldots \alpha_i) \subset U(\alpha_1 \ldots \alpha_i \alpha_{i+1})\,.$$

Then the following conditions are equivalent:
  (i) *there exists a closed collection $\{F(\alpha_1 \ldots \alpha_i) \mid \alpha_1, \ldots, \alpha_i \in A; \ i = 1, 2, \ldots\}$ satisfying (i), (ii) of Definition VI.7,*

(ii) *there exists a collection* $\{G(\alpha_1 \ldots \alpha_i) \mid \alpha_1, \ldots, \alpha_i \in A; i = 1, 2, \ldots\}$ *of F_σ-sets of X satisfying* (i), (ii) *of Definition* VI.7,

(iii) *there exists a collection* $\{H(\alpha_1 \ldots \alpha_i) \mid \alpha_1, \ldots, \alpha_i \in A; i = 1, 2, \ldots\}$ *of open F_σ-sets of X satisfying* (i), (ii) *of Definition* VI.7.

Proof. (i) $\Rightarrow$ (iii). For each $(\alpha_1, \ldots, \alpha_i)$ we can define, by use of Urysohn's lemma, a continuous function f such that

$$f(F(\alpha_1 \ldots \alpha_i)) = 1, \qquad f(X - U(\alpha_1 \ldots \alpha_i)) = 0 \quad \text{and} \quad 0 \leq f \leq 1.$$

Then putting

$$H(\alpha_1 \ldots \alpha_i) = \{p \mid f(p) > 0\},$$

we obtain an open F_σ-subset satisfying

$$F(\alpha_1 \ldots \alpha_i) \subset H(\alpha_1 \ldots \alpha_i) \subset U(\alpha_1 \ldots \alpha_i).$$

Then $\{H(\alpha_1 \ldots \alpha_i) \mid \alpha_1, \ldots, \alpha_i \in A; i = 1, 2, \ldots\}$ is the desired collection.

(iii) $\Rightarrow$ (ii) is obvious.

(ii) $\Rightarrow$ (i). Put

$$G(\alpha_1 \ldots \alpha_i) = \bigcup_{k=1}^{\infty} K(\alpha_1 \ldots \alpha_i; k)$$

for closed sets $K(\alpha_1 \ldots \alpha_i; k)$, $k = 1, 2, \ldots$. Put

$$F(\alpha_1 \ldots \alpha_i) = \bigcup \{K(\alpha_1 \ldots \alpha_j; k) \mid j = 1, \ldots, i; k = 1, \ldots, i\};$$

then it is obvious that $F(\alpha_1 \ldots \alpha_i)$ is a closed set satisfying

$$F(\alpha_1 \ldots \alpha_i) \subset U(\alpha_1 \ldots \alpha_i).$$

On the other hand, if

$$\bigcup_{i=1}^{\infty} U(\alpha_1 \ldots \alpha_i) = X,$$

then

$$\bigcup_{i=1}^{\infty} G(\alpha_1 \ldots \alpha_i) = X.$$

Therefore every point $p \in X$ is contained in $G(\alpha_1 \ldots \alpha_i)$ for some i, and hence

$$p \in K(\alpha_1 \ldots \alpha_i; k) \quad \text{for some } k.$$

If $k \leq i$, then $p \in F(\alpha_1 \ldots \alpha_i)$. If $k \geq i$, then $p \in F(\alpha_1 \ldots \alpha_k)$. Therefore, in any case,

$$\bigcup \{F(\alpha_1 \ldots \alpha_i) \mid i = 1, 2, \ldots\} = X,$$

proving (i).

K) *Every perfectly normal space X is a P-space.*

Proof. Let $\{U(\alpha_1 \ldots \alpha_i) \mid \alpha_1, \ldots, \alpha_i \in A; i = 1, 2, \ldots\}$ be an open collection of a perfectly normal space X satisfying the condition of J). Then each $U(\alpha_1 \ldots \alpha_i)$ is F_σ because of the perfect normality of X. Thus putting

$$G(\alpha_1 \ldots \alpha_i) = U(\alpha_1 \ldots \alpha_i),$$

we get a collection satisfying (ii) of J). Therefore by the same proposition, there is a closed collection satisfying (i), which means that X is a P-space.

L) *Let X be a normal P-space and Y a subspace of a Baire's zero-dimensional space $N(A)$; then the product space $X \times Y$ is normal.*

Proof. Let $\mathcal{M} = \{M_k \mid k = 1, 2, \ldots\}$ be a given open covering of $X \times Y$. For every $\alpha_1, \ldots, \alpha_i \in A$, we put

$$V(\alpha_1 \ldots \alpha_i) = \{q \mid q = (\beta_1, \beta_2, \ldots) \in Y; \beta_j = \alpha_j, j = 1, \ldots, i\}.$$

For given $\alpha_1, \ldots, \alpha_i \in A$ and a positive integer k, we define an open set $L(\alpha_1 \ldots \alpha_i; k)$ of X by

$$L(\alpha_1 \ldots \alpha_i; k) = \bigcup \{L \mid L \times V(\alpha_1 \ldots \alpha_i) \subset M_k, L \text{ is an open set of } X\}.$$

Note that some of the $L(\alpha_1 \ldots \alpha_i; k)$ may be empty. Then $L(\alpha_1 \ldots \alpha_i; k)$ satisfies

$$L(\alpha_1 \ldots \alpha_i; k) \times V(\alpha_1 \ldots \alpha_i) \subset M_k$$

and

$$\{L(\alpha_1 \ldots \alpha_i; k) \times V(\alpha_1 \ldots \alpha_i) \mid \alpha_1, \ldots, \alpha_i \in A; i = 1, 2, \ldots;$$
$$k = 1, 2, \ldots\}$$

forms an open covering of $X \times Y$, because

$$\{V(\alpha_1 \ldots \alpha_i) \mid \alpha_1, \ldots, \alpha_i \in A; i = 1, 2, \ldots\}$$

is an open basis of Y. Note that for $j \leq i$,

$$L(\alpha_1 \ldots \alpha_j; k) \times V(\alpha_1 \ldots \alpha_j \ldots \alpha_i) \subset$$
$$\subset L(\alpha_1 \ldots \alpha_j; k) \times V(\alpha_1 \ldots \alpha_j) \subset M_k.$$

Hence putting

$$U(\alpha_1 \ldots \alpha_i; k) = \bigcup_{j=1}^{i} L(\alpha_1 \ldots \alpha_j; k), \tag{1}$$

we get open sets of X satisfying

$$U(\alpha_1 \ldots \alpha_i; k) \times V(\alpha_1 \ldots \alpha_i) \subset M_k \tag{2}$$

and

$$U(\alpha_1 \ldots \alpha_i; k) \subset U(\alpha_1 \ldots \alpha_i \alpha_{i+1}; k).$$

Furthermore, we put

$$U(\alpha_1 \ldots \alpha_i) = \bigcup_{k=1}^{\infty} U(\alpha_1 \ldots \alpha_i; k); \tag{3}$$

then the open sets $U(\alpha_1 \ldots \alpha_i)$ clearly satisfy

$$U(\alpha_1 \ldots \alpha_i) \subset U(\alpha_1 \ldots \alpha_i \alpha_{i+1}).$$

We can also prove that

$$\bigcup_{i=1}^{\infty} U(\alpha_1 \ldots \alpha_i) = X \quad \text{if } (\alpha_1, \alpha_2, \ldots) \in Y. \tag{4}$$

For, given a point p of X, then $(p, (\alpha_1, \alpha_2, \ldots))$ is a point of $X \times Y$. Since we have chosen $L(\alpha_1 \ldots \alpha_i; k)$ so that $\{L(\alpha_1 \ldots \alpha_i; k) \times V(\alpha_1 \ldots \alpha_i)\}$ forms a covering of $X \times Y$,

$$(p, (\alpha_1, \alpha_2, \ldots)) \in L(\alpha_1 \ldots \alpha_i ; k) \times V(\alpha_1 \ldots \alpha_i)$$

for some i and k. Therefore it follows from (1) and (3) that

$$p \in L(\alpha_1 \ldots \alpha_i ; k) \subset U(\alpha_1 \ldots \alpha_i ; k) \subset U(\alpha_1 \ldots \alpha_i),$$

proving (4).

Since X is a normal P-space, by J) there is a collection

$$\{H(\alpha_1 \ldots \alpha_i) \mid \alpha_1, \ldots, \alpha_i \in A ; i = 1, 2, \ldots\}$$

of open F_σ-sets of X such that

$$H(\alpha_1 \ldots \alpha_i) \subset U(\alpha_1 \ldots \alpha_i), \tag{5}$$

$$\bigcup_{i=1}^{\infty} H(\alpha_1 \ldots \alpha_i) = X \quad \text{if } (\alpha_1, \alpha_2, \ldots) \in Y. \tag{6}$$

Note that (6) follows from (4). Since X is countably paracompact and normal by C), it follows from G) that $H(\alpha_1 \ldots \alpha_i)$ is countably paracompact and normal. Hence, by F), the open covering $\{U(\alpha_1 \ldots \alpha_i ; k) \cap H(\alpha_1 \ldots \alpha_i) \mid k = 1, 2, \ldots\}$ of $H(\alpha_1 \ldots \alpha_i)$ is normal. (See (3) and (5) for the reason why it is a covering of $H(\alpha_1 \ldots \alpha_i)$.) Hence by I),

$$\mathcal{U}(\alpha_1 \ldots \alpha_i) = \{[U(\alpha_1 \ldots \alpha_i ; k) \cap H(\alpha_1 \ldots \alpha_i)]$$
$$\times V(\alpha_1 \ldots \alpha_i) \mid k = 1, 2, \ldots\} \tag{7}$$

is a normal covering of the subspace $H(\alpha_1 \ldots \alpha_i) \times V(\alpha_1 \ldots \alpha_i)$ of $X \times Y$. Since $H(\alpha_1 \ldots \alpha_i)$ and $V(\alpha_1 \ldots \alpha_i)$ are open F_σ-sets of normal spaces X and Y respectively, by H) they are cozero sets. Hence

$$H(\alpha_1 \ldots \alpha_i) = \{p \in X \mid f(p) > 0\}$$

and

$$V(\alpha_1 \ldots \alpha_i) = \{q \in Y \mid g(q) > 0\}$$

for some continuous functions f and g over X and Y, respectively, with $0 \leqslant f \leqslant 1$ and $0 \leqslant g \leqslant 1$. Thus

$$H(\alpha_1 \ldots \alpha_i) \times V(\alpha_1 \ldots \alpha_i) = \{(p, q) \mid f(p) g(q) > 0\}$$

is also a cozero set of $X \times Y$. Since $\{V(\alpha_1 \ldots \alpha_i) \,|\, \alpha_1, \ldots, \alpha_i \in A; \; i = 1, 2, \ldots\}$ is a σ-locally finite (actually σ-discrete) open collection of Y,

$$\mathcal{N} = \{H(\alpha_1 \ldots \alpha_i) \times V(\alpha_1 \ldots \alpha_i) \,|\, \alpha_1, \ldots, \alpha_i \in A; i = 1, 2, \ldots\}$$

is a σ-locally finite open collection in $X \times Y$. As a matter of fact, by virtue of (6), $\mathcal{N}$ is easily seen to be a σ-locally finite open covering of $X \times Y$ whose members are cozero sets. Hence it follows from V.5.D) that $\mathcal{N}$ is normal.

Now, we consider the restriction

$$\mathcal{M}(\alpha_1 \ldots \alpha_i) = \{M_k \cap [H(\alpha_1 \ldots \alpha_i) \times V(\alpha_1 \ldots \alpha_i)] \,|\, k = 1, 2, \ldots\}$$

of the original open covering

$$\mathcal{M} = \{M_k \,|\, k = 1, 2, \ldots\}$$

to each member $H(\alpha_1 \ldots \alpha_i) \times V(\alpha_1 \ldots \alpha_i)$ of $\mathcal{N}$. Then it follows from (2) that

$$[U(\alpha_1 \ldots \alpha_i; k) \cap H(\alpha_1 \ldots \alpha_i)] \times V(\alpha_1 \ldots \alpha_i) \subset$$
$$\subset M_k \cap [H(\alpha_1 \ldots \alpha_i) \times V(\alpha_1 \ldots \alpha_i)],$$

which implies that

$$\mathcal{U}(\alpha_1 \ldots \alpha_i) < \mathcal{M}(\alpha_1 \ldots \alpha_i) \quad \text{(see (7))}.$$

Since by (7) $\mathcal{U}(\alpha_1 \ldots \alpha_i)$ is a normal covering of $H(\alpha_1 \ldots \alpha_i) \times V(\alpha_1 \ldots \alpha_i)$, so is $\mathcal{M}(\alpha_1 \ldots \alpha_i)$.

Thus by virtue of V.5.E), the given open covering $\mathcal{M}$ is a normal covering of X because $\mathcal{N}$ is normal. Therefore $X \times Y$ is normal and actually countably paracompact, too.[1]

M) *Let X be a topological space such that the product space $X \times Y$ of X with every subset Y of $N(A)$ for every A is normal. Then X is a normal P-space.*

---

[1] Actually K. Morita proved that if the product $X \times Y$ of a normal space X and a metric space Y is countably paracompact, then it is normal.

Proof. First we can prove that $X \times Y$ is countably paracompact for every subset Y of $N(A)$. Let us denote by D the discrete space of two points 0, 1.[1]

Then we define a mapping f of $Y \times N(D)$ into $N(A \cup D)$ by

$$f(\alpha, \varepsilon) = (\alpha_1, \varepsilon_1, \alpha_2, \varepsilon_2, \ldots)$$

for

$$\alpha = (\alpha_1, \alpha_2, \ldots) \in Y \quad \text{and} \quad \varepsilon = (\varepsilon_1, \varepsilon_2, \ldots) \in N(D).$$

We can easily see that f is a topological mapping. Therefore $Y \times N(D)$ is homeomorphic with a subset of $N(A \cup D)$. Thus $X \times (Y \times N(D))$ is normal by the hypothesis of this proposition. Since $X \times (Y \times N(D))$ is homeomorphic with $(X \times Y) \times N(D)$, the latter is also normal. Thus from Theorem V.8 it follows that $X \times Y$ is countably paracompact and normal.

We suppose that $\{U(\alpha_1 \ldots \alpha_i) \mid \alpha_1, \ldots, \alpha_i \in A; \ i = 1, 2, \ldots\}$ is a given open collection of X such that

$$U(\alpha_1, \ldots \sigma_i) \subset U(\alpha_1 \ldots \alpha_i \alpha_{i+1}).$$

Putting

$$Y = \left\{ (\alpha_1, \alpha_2, \ldots) \mid (\alpha_1, \alpha_2, \ldots) \in N(A), \bigcup_{i=1}^{\infty} U(\alpha_1 \ldots \alpha_i) = X \right\},$$

(1)

we get a subset Y of the Baire's zero-dimensional space $N(A)$. For each $(\alpha_1, \ldots, \alpha_i)$ with $\alpha_j \in A$, $j = 1, \ldots, i$, we define an open covering $V(\alpha_1 \ldots \alpha_i)$ of Y by

$$V(\alpha_1 \ldots \alpha_i) = \{(\beta_1, \beta_2, \ldots) \mid (\beta_1, \beta_2, \ldots) \in Y; \beta_j = \alpha_j, j = 1, \ldots, i\}.$$

Then, by (1),

$$\mathcal{W} = \{U(\alpha_1, \ldots \alpha_i) \times V(\alpha_1 \ldots \alpha_i) \mid \alpha_1, \ldots, \alpha_i \in A; i = 1, 2, \ldots\}$$

is an open covering of $X \times Y$.

Furthermore, for each i $\{V(\alpha_1 \ldots \alpha_i) \mid \alpha_1, \ldots, \alpha_i \in A\}$ is locally finite in Y, and hence $\mathcal{W}$ is a σ-locally finite open covering of $X \times Y$. Since $X \times Y$ is countably paracompact and normal, by the corollary of F), we can

---

[1] Suppose $A \cap D = \emptyset$. Note that $N(A)$ is homeomorphic with $A^{\aleph_0}$, the product of countably many copies of the discrete space A.

construct a locally finite open covering

$$\mathscr{L} = \{L_\lambda \mid \lambda \in \Lambda\}$$

of $X \times Y$ such that

$$\bar{\mathscr{L}} < \mathscr{W}. \tag{2}$$

Putting

$$L(\alpha_1 \ldots \alpha_i; \lambda) = \bigcup\{U \mid U \times V(\alpha_1 \ldots \alpha_i) \subset L_\lambda,$$
$$U \text{ is an open set of } X\},$$

we can construct an open covering

$$\mathscr{L}' = \{L(\alpha_1 \ldots \alpha_i; \lambda) \times V(\alpha_1 \ldots \alpha_i) \mid \alpha_1 \ldots \alpha_i \in A, \lambda \in \Lambda, i = 1, 2, \ldots\}$$

of $X \times Y$ such that

$$L(\alpha_1 \ldots \alpha_i; \lambda) \times V(\alpha_1 \ldots \alpha_i) \subset L_\lambda, \tag{3}$$

where we note that some of the $L(\alpha_1 \ldots \alpha_i; \lambda)$ may be empty. On the other hand, it follows from (2) that for each $\lambda \in \Lambda$, there is a finite sequence $(\beta_1, \ldots, \beta_j)$ of elements of A for which

$$\bar{L}_\lambda \subset U(\beta_1 \ldots \beta_j) \times V(\beta_1 \ldots \beta_j). \tag{4}$$

For every $\alpha_1, \ldots, \alpha_i \in A$, $\lambda \in \Lambda$, we put

$$M(\alpha_1 \ldots \alpha_i; \lambda) = \bigcup\{L(\alpha_1 \ldots \alpha_j; \lambda) \mid j \leq i,$$

$$\overline{L(\alpha_1 \ldots \alpha_j; \lambda)} \subset U(\alpha_1 \ldots \alpha_j \ldots \alpha_i)\}. \tag{5}$$

Now we can prove that

$$\mathscr{M} = \{M(\alpha_1 \ldots \alpha_i; \lambda) \times V(\alpha_1 \ldots \alpha_i) \mid \alpha_1, \ldots, \alpha_i \in A,$$
$$\lambda \in \Lambda, i = 1, 2, \ldots\}$$

is an open covering of $X \times Y$. To see it, suppose that (p, α) is a given point of $X \times Y$. Then since $\mathscr{L}'$ is a covering of $X \times Y$,

$$(p, \alpha) \in L(\alpha_1 \ldots \alpha_i; \lambda) \times V(\alpha_1 \ldots \alpha_i)$$

for some $\alpha_1, \ldots, \alpha_i \in A$ and $\lambda \in \Lambda$.

Then it follows from (3) and (4) that

$$(p, \alpha) \in L(\alpha_1 \ldots \alpha_i; \lambda) \times V(\alpha_1 \ldots \alpha_i)$$
$$\subset \overline{L(\alpha_1 \ldots \alpha_i; \lambda)} \times V(\alpha_1 \ldots \alpha_i) \subset \bar{L}_\lambda$$
$$\subset U(\beta_1 \ldots \beta_j) \times V(\beta_1 \ldots \beta_j) \tag{6}$$

for some $\beta_1, \ldots, \beta_j \in A$. If $j \le i$, then

$$\beta_1 = \alpha_1, \quad \ldots, \quad \beta_j = \alpha_j,$$

and hence

$$U(\beta_1 \ldots \beta_j) = U(\alpha_1 \ldots \alpha_j) \subset U(\alpha_1 \ldots \alpha_j \ldots \alpha_i),$$

which combined with (6) implies

$$\overline{L(\alpha_1 \ldots \alpha_i; \lambda)} \subset U(\alpha_1 \ldots \alpha_i).$$

Therefore it follows from (5) that

$$L(\alpha_1 \ldots \alpha_i; \lambda) \subset M(\alpha_1 \ldots \alpha_i; \lambda).$$

Since
$$p \in L(\alpha_1 \ldots \alpha_i; \lambda) \quad (\text{see (6)}),$$

we obtain

$$(p, \alpha) \in M(\alpha_1 \ldots \alpha_i; \lambda) \times V(\alpha_1 \ldots \alpha_i).$$

If $j > i$, then

$$\beta_1 = \alpha_1, \quad \ldots, \quad \beta_i = \alpha_i$$

(note that $\alpha = (\alpha_1 \ldots, \alpha_i \ldots) \in V(\beta_1 \ldots \beta_j)$).
 Thus by (6) we obtain

$$\overline{L(\alpha_1 \ldots \alpha_i; \lambda)} \subset U(\alpha_1 \ldots \alpha_i \beta_{i+1} \ldots \beta_j).$$

Therefore by (5)

$$L(\alpha_1 \ldots \alpha_i; \lambda) \subset M(\alpha_1 \ldots \alpha_i \beta_{i+1} \ldots \beta_j; \lambda),$$

which combined with (6) implies

$$(p, \alpha) \in M(\alpha_1 \ldots \alpha_i \beta_{i+1} \ldots \beta_j ; \lambda) \times V(\alpha_1 \ldots \alpha_i \beta_{i+1} \ldots \beta_j).$$

In any case (p, α) belongs to a member of $\mathcal{M}$, and hence $\mathcal{M}$ is an open covering of $X \times Y$. Furthermore, we note that

$$M(\alpha_1 \ldots \alpha_i ; \lambda) \times V(\alpha_1 \ldots \alpha_i) \subset L_\lambda \tag{7}$$

follows from (3), (5) and that

$$\overline{M(\alpha_1 \ldots \alpha_i ; \lambda)} \subset U(\alpha_1 \ldots \alpha_i) \tag{8}$$

follows from (5).

Finally, we put

$$F(\alpha_1 \ldots \alpha_i) = \bigcup \{ \overline{M(\alpha_1 \ldots \alpha_i ; \lambda)} \mid \lambda \in \Lambda \}.$$

Then, since $\mathcal{L} = \{ L_\lambda \mid \lambda \in \Lambda \}$ is locally finite, it follows from (7) that $\{ M(\alpha_1 \ldots \alpha_i ; \lambda) \mid \lambda \in \Lambda \}$ is locally finite, and hence $F(\alpha_1 \ldots \alpha_i)$ is a closed set of X. On the other hand, (8) implies

$$F(\alpha_1 \ldots \alpha_i) \subset U(\alpha_1 \ldots \alpha_i).$$

Since $\mathcal{M} = \{ M(\alpha_1 \ldots \alpha_i ; \lambda) \times V(\alpha_1 \ldots \alpha_i) \mid \alpha_1, \ldots, \alpha_i \in A, \lambda \in \Lambda, i = 1, 2, \ldots \}$ is a covering of $X \times Y$, $\{ F(\alpha_1 \ldots \alpha_i) \times V(\alpha_1 \ldots \alpha_i) \mid \alpha_1, \ldots, \alpha_i \in A; i = 1, 2, \ldots \}$ is also a covering of $X \times Y$. Therefore, if

$$\bigcup_{i=1}^{\infty} U(\alpha_1 \ldots \alpha_i) = X,$$

then by (1)

$$\alpha = (\alpha_1, \alpha_2, \ldots) \in Y,$$

and hence for each point p of X, we obtain

$$(p, \alpha) \in F(\alpha_1 \ldots \alpha_i) \times V(\alpha_1 \ldots \alpha_i) \quad \text{for some } i.$$

This implies $p \in F(\alpha_1 \ldots \alpha_i)$, and hence

$$\bigcup_{i=1}^{\infty} F(\alpha_1 \ldots \alpha_i) = X.$$

Thus we have proved that X is a P-space. Since X is normal as a closed set of the normal space $X \times Y$, the proof of this proposition is complete.

Now, we can prove the following theorem.

Theorem VI.24. *Let X be a topological space. Then the product space $X \times Y$ of X with every metric space Y is normal if and only if X is a normal P-space.*

Proof. The necessity of the condition follows directly from M).

To prove the sufficiency, let X be a given normal P-space and Y a metric space. Then by Corollary 2 to Theorem VI.14, there is a subspace Z of a Baire's zero-dimensional space $N(A)$ and a perfect mapping f of Z onto Y. Put

$$g(p, t) = (p, f(t)), (p, t) \in X \times Z.$$

Then we can assert that g is a closed continuous mapping of $X \times Z$ onto $X \times Y$. Since g is clearly continuous and onto, we shall prove only that g is closed. Given a closed set F of $X \times Z$, we consider an arbitrary point (p, q) of $X \times Y$ with

$$(p, q) \notin g(F).$$

Then

$$g^{-1}(p, q) = \{p\} \times f^{-1}(q) \subset X \times Z - F.$$

Since $f^{-1}(q)$ is compact, there is an open set V of Z containing $f^{-1}(q)$ and an open nbd U of p such that

$$F \subset X \times Z - U \times V. \tag{1}$$

By the definition of g,

$$g(X \times Z - U \times V) = [(X - U) \times Y] \cup [U \times f(Z - V)],$$

which is easily seen to be closed because $f(Z - V)$ is a closed set of Y. On the other hand, it is clear that

$$(p, q) \notin g(X \times Z - U \times V)$$

and hence

$$W = X \times Y - g(X \times Z - U \times V)$$

is an open nbd of (p, q) in $X \times Y$. It is also clear that (1) implies

$$W \cap g(F) = \emptyset .$$

Therefore $g(F)$ is a closed set of $X \times Y$ proving that g is a closed mapping. Thus there is a closed continuous mapping of $X \times Z$ onto $X \times Y$. On the other hand, it follows from L) that $X \times Z$ is a normal space, and hence $X \times Y$ is normal (Exercise III.33).[1]

Corollary 1. *If a normal space X is either countably compact, fully normal and Čech complete or perfectly normal, then the product space $X \times Y$ of X with every metric space Y is normal.*

Proof. This is a direct consequence of Theorem VI.24 combined with A), B) and K).

Corollary 2. *Let X be a topological space. Then the product space $X \times Y$ of X with every separable metric space Y is normal if and only if X is a normal space satisfying the condition of Definition VI.7 for an arbitrary countable set A.*

Proof. We can apply the proof of Theorem VI.24 to this case, but considering that A denotes a countable set. The details are left to the reader.

8. Various generalized metric spaces

We have already learned quite a few generalizations of a metric space,

---

[1] It is known that if X is not only normal P but also paracompact, then the product space $X \times Y$ is paracompact. See K. Morita [5].

M. E. Rudin–M. Starbird [1] proved the following important results: (i) Suppose X is metric and Y is μ-paracompact. If $X \times Y$ is normal, then $X \times Y$ is μ-paracompact. (ii) Suppose X is metric and $X \times Y$ is normal. If there is a closed continuous map from Y onto Z, then $X \times Z$ is normal.

e.g. developable space, uniform space, proximity space and P-space. In the present section we are going to discuss more generalizations which are becoming increasingly important in recent research of general topology. A direct method to obtain a generalization of a metric space is to relax the conditions for the metric function. We defined 'pseudo-metric' in such a way. Another example is 'semi-metric' defined in the following.

Definition VI.8. Let X be a topological space and ρ a real-valued function defined on $X \times X$. If ρ satisfies the following conditions, then it is called a *semi-metric*:

(i) $\rho(x, y) \geqslant 0$,

(ii) $\rho(x, y) = 0$ if and only if $x = y$,

(iii) $\rho(x, y) = \rho(y, x)$.

(Namely, a semi-metric is not required to satisfy the triangle axiom among the conditions for a metric.) If $\{S_\varepsilon(x) \mid \varepsilon > 0\}$, where $S_\varepsilon(x) = \{y \in X \mid \rho(x, y) < \varepsilon\}$, is a nbd base at each point $x \in X$, then X with ρ is called a *semi-metric space*. If there is such a semi-metric of X, then X is called *semi-metrizable*.

Every semi-metric space is obviously a T_1-space.[1] Another method of generalization is to consider a metric function which takes on values in an abstracter range, abstracter than the non-negative real numbers, e.g. in a totally ordered commutative semi-group. In fact various interesting results are being obtained in this way.[2]

---

[1] A similar but somewhat weaker condition is 'symmetrizability'. A topological space X is called *symmetrizable* if X has a semi-metric ρ such that a subset F of X is closed if and only if $\rho(x, F) > 0$ for every $x \in X - F$. See A. Arhangelskii [6] for results surrounding this concept.

[2] See G. Kurepa [1], Z. Mamuzić [2], M. Antonovskii [1], M. Antonovskii–V. Boltyanskii–T. Sarymsakov [1], F. W. Stevenson–W. J. Thron [1], H.-C. Reichel [1], [2], [3], H.-C. Reichel–W. Ruppert [1], Y. Yasui [2] for generalizations in this direction.

'Statistical metric space' is an interesting generalized metric space initiated by K. Menger [1], where the metric function takes on values in a set of functions. Let X be a set and $\mathscr{F}$ a mapping defined on $X \times X$ such that for every $x, y \in X$, $\mathscr{F}(x, y) = F_{xy}$ is a real-valued left-continuous (i.e. for each $r \in R$, $F_{xy}|_{(-\infty, r]}$ is continuous at r) non-decreasing function defined on the real line R taking values between 0 and 1 satisfying:

(i) $F_{xy}(0) = 0$,

(ii) $F_{xy} = F_{yx}$ for every $x, y \in X$,

(iii) $F_{xy}(r) = 1$ for all $r > 0$ if and only if $x = y$,

(iv) $F_{xy}(r) = 1$ and $F_{yz}(s) = 1$ imply $F_{xz}(r + s) = 1$.

Then $\langle X, \mathscr{F} \rangle$ is called a *statistical metric space*. For intuitive understanding we may regard $F_{xy}(r)$ as the probability for the distance between x and y to be smaller than r. Namely, the distance between two points is given only in a probability. Now, let X be a given statistical

Another group of generalized metric spaces was obtained by modifying various metrizability conditions. In the following are those obtained by modifying Nagata–Smirnov's metrization theorem and also related spaces.

Definition VI.9. A regular space X is called an M_1-*space* if it has a σ-closure-preserving base; a regular space X is called an M_2-*space* if it has a σ-closure-preserving *quasi-base* $\mathcal{U}$, i.e., for each $x \in X$ and every nbd W of x, there is $U \in \mathcal{U}$ satisfying $x \in U^\circ \subset U \subset W$; a T_1-space X is called an M_3-*space* (or a *stratifiable space*) if it has a σ-cushioned pair base, where a collection $\mathcal{P}$ of ordered pairs (P_1, P_2) of an open set P_1 and a subset P_2 of X is called a *pair base* if for each $x \in X$ and every nbd U of x there is $(P_1, P_2) \in \mathcal{P}$ such that $x \in P_1 \subset P_2 \subset U$, and a pair base $\mathcal{P}$ is called σ-*cushioned* if it is the sum of countably many subcollections $\mathcal{P}_i$, $i = 1, 2, \ldots$, such that for each i $\{P_1 \mid (P_1, P_2) \in \mathcal{P}_i\}$ is cushioned in $\{P_2 \mid (P_1, P_2) \in \mathcal{P}_i\}$.

Let X be a topological space and $\mathcal{O}$ its topology (the collection of all open sets) and $\mathcal{C}$ the collection of all closed sets of X. Furthermore, we denote by N the set of all natural numbers in the rest of this section. Then X is called *semi-stratifiable* if there is a map G from $\mathcal{O} \times N$ into $\mathcal{C}$ such that:

(i) $U = \bigcup_{n=1}^{\infty} G(U, n)$ for each $U \in \mathcal{O}$,

(ii) $G(U, n) \subset G(V, n)$, $n = 1, 2, \ldots$, whenever $U \subset V$.

The map G is called a *semi-stratification* of X.

Let $\mathcal{G}$ be a collection of subsets of a topological space X. If each open set of X is a sum of members of $\mathcal{G}$, then $\mathcal{G}$ is called a *network* of X. (Thus every base is a network. On the other hand, members of a network are not required to be open.) A topological space X is called a σ-*space* if it has a σ-locally finite network.[1]

(*Footnote continued from p.* 337)

metric space, $x \in X$ and $\varepsilon, \delta > 0$; then we put

$$N_x(\varepsilon, \delta) = \{y \in X \mid F_{xy}(\varepsilon) > 1 - \delta\}.$$

Then $\{N_x(\varepsilon, \delta) \mid \varepsilon, \delta > 0\}$ does not necessarily satisfy the conditions for a nbd basis at x. But if it does for every $x \in X$ and accordingly induces a topology τ, then $\langle X, \tau \rangle$ is known to be semi-metrizable. Conversely every semi-metric space is homeomorphic to a statistical metric space with the topology induced by $\{N_x(\varepsilon, \delta)\}$. In this sense topological statistical metric spaces are equivalent with semi-metrizable spaces. See e.g. B. Schweizer–A. Sklar [1], B. Schweizer–A. Sklar–E. Thorp [1], E. Thorp [1], J. Brown [1] and B. Morrel–J. Nagata [1] for results in this aspect.

[1] We owe the definitions of M_i-spaces for $i = 1, 2, 3$ to J. Ceder [1], semi-stratifiable space to G. Creede [1], network to A. V. Arhangelskii and σ-space to A. Okuyama [3].

A) *A T_1-space X is stratifiable if and only if there is a map G from $\mathcal{O} \times N$ into $\mathcal{C}$ such that*:

   (i) *$U = \bigcup_{n=1}^{\infty} G(U, n) = \bigcup_{n=1}^{\infty} (G(U, n))^{\circ}$ for each $U \in \mathcal{O}$,*

   (ii) *$G(U, n) \subset G(V, n)$, $n = 1, 2, \ldots$, whenever $U \subset V$,*

where $\mathcal{O}$ and $\mathcal{C}$ denote the collection of all open sets and of closed sets, respectively. Such a map G is called a stratification *of X.*

Proof. Let X be a stratifiable space with a σ-cushioned pair base $\mathcal{P} = \bigcup_{n=1}^{\infty} \mathcal{P}_n$. Then for each $U \in \mathcal{O}$ and $n \in N$, we define

$$G(U, n) = \overline{\bigcup \{P_1 \mid (P_1, P_2) \in \mathcal{P}_n, P_2 \subset U\}}.$$

Then it is obvious that G satisfies the said conditions.

Conversely, let G be a stratification of X. Then it is obvious that X is regular. For each $n \in N$ we consider

$$\mathcal{P}_n = \{(G(U, n)^{\circ}, U) \mid U \in \mathcal{O}\}.$$

Then it is obvious that $\bigcup_{n=1}^{\infty} \mathcal{P}_n$ is a σ-cushioned pair base of X. Thus X is stratifiable.

Theorem VI.25. *Let X be a T_1-space and suppose that a sequence $\{U(n, x) \mid n = 1, 2, \ldots\}$ of open nbds of x is defined at each point x of X. Then we consider the following conditions for the sequence*:

   (i) *$\{U(n, x) \mid n = 1, 2, \ldots\}$ is a nbd base of x,*

   (ii) *if $y \in U(n, x)$, then $U(n, y) \subset U(n, x)$,*

   (iii) *if $x \notin F$ for a closed set F, then $x \notin \bigcup \{U(n, y) \mid y \in F\}$ for some n,*

   (iv) *if $x \notin F$ for a closed set F, then $x \notin \overline{\bigcup \{U(n, y) \mid y \in F\}}$ for some n,*

   (v) *if $\{x, x_n\} \subset U(n, y_n)$, $n = 1, 2, \ldots$, then x is a cluster point of $\{x_n\}$.*

Now, we can characterize various generalized metric spaces in terms of $\{U(n, x)\}$ as follows: Semi-stratifiable = (iii), $\sigma = $ (ii) and (iii), stratifiable = (iv), $M_2 = $ (ii) and (iv), semi-metrizable = (i) and (iii), developable = (v), where, e.g., the first equality means that X is a semi-stratifiable space if and only if there is a sequence $\{U(n, x) \mid n = 1, 2, \ldots\}$ of open nbds of each $x \in X$ satisfying the condition (iii).[1]

Proof. The methods of proofs are somewhat similar in all cases. So we shall prove only the first case and the last two cases to leave the rest to

---

[1] These characterizations are due to G. Creede [1], R. Heath–R. Hodel [1], R. Heath [2], J. Nagata [7], respectively, and the last two cases to R. Heath [1].

the reader. Assume that X is a semi-stratifiable space with a semi-stratification G. Then put

$$U(n, x) = X - G(X - \{x\}, n).$$

To see that (iii) is satisfied by the sequence $\{U(n, x) \mid n = 1, 2, \ldots\}$, let $x \not\in F$ for a closed set F. Then for some n we have

$$x \in G(X - F, n) \subset X - F.$$

Then for each $y \in F$ it holds that $G(X - F, n) \subset G(X - \{y\}, n)$, because $X - F \subset X - \{y\}$. This implies that

$$U(n, y) \subset X - G(X - F, n) \not\ni x,$$

proving (iii).

Conversely, assume that $\{U(n, x) \mid n = 1, 2, \ldots\}$ is given to satisfy the condition (iii). Then define a map G from $\mathcal{O} \times N$ into $\mathscr{C}$ by

$$G(U, n) = X - \cup\{U(n, x) \mid x \in X - U\}.$$

It is obvious that G is a semi-stratification of X.

Suppose $\langle X, \rho \rangle$ is a given semi-metric space. Then put

$$U(n, x) = (S_{1/n}(x))^\circ, \quad S_{1/n}(x) = \{y \in X \mid \rho(x, y) < 1/n\}.$$

Now it is obvious that $\{U(n, x) \mid n = 1, 2, \ldots\}$ satisfies (i) and (iii).

Conversely, assume that $\{U(n, x)\}$ satisfies (i) and (iii). Note that we may assume $U(n, x) \supset U(n + 1, x)$, $n = 1, 2, \ldots$, at each $x \in X$. Then define a function ρ on $X \times X$ by

$$\rho(x, y) = \inf\{1/n \mid x \in U(n, y) \text{ or } y \in U(n, x)\}.$$

It is obvious that $\rho(x, x) = 0$ and $\rho(x, y) = \rho(y, x)$. Now, to prove that $\{S_{1/n}(x) \mid n = 1, 2, \ldots\}$ is a nbd base at each $x \in X$, let U be a nbd of x. Then there is n for which $U(n, x) \subset U$ and $x \not\in \cup\{U(n, y) \mid y \in X - U\}$ hold at the same time. Then for each $y \in X - U$, $\rho(x, y) \geq 1/n$, i.e. $S_{1/n}(x) \subset U$. Thus X is a semi-metrizable space.

Let X be a developable space and $\{\mathcal{U}_n \mid n = 1, 2, \ldots\}$ a development of X. Then for each $n \in N$ and $x \in X$ we fix an element $U(n, x)$ of $\mathcal{U}_n$ such that $x \in U(n, x)$. Then the condition (v) is obviously satisfied by $\{U(n, x)\}$.

Conversely, assume that $\{U(n, x)\}$ satisfies (v). Then put

$$\mathcal{U}_n = \{U(n, x) \mid x \in X\}.$$

To prove that $\{\mathcal{U}_n\}$ is a development, suppose U is a given nbd of $x \in X$. If $S(x, \mathcal{U}_n) \not\subset U$ for $n = 1, 2, \ldots$, then select $x_n \in S(x, \mathcal{U}_n) - U$, $n = 1, 2, \ldots$. Then there are y_n, $n = 1, 2, \ldots$, such that $\{x, x_n\} \subset U(n, y_n)$. Since $x_n \not\in U$, $n = 1, 2, \ldots$, x cannot be a cluster point of $\{x_n\}$ contradicting (v). Thus $S(x, \mathcal{U}_n) \subset U$ for some n, proving that $\{\mathcal{U}_n\}$ is a development of X. Hence X is a developable space.

Corollary. *A T_1-space X is a semi-metrizable space if and only if it is first countable and semi-stratifiable.*

Proof. The 'only if' part is obvious because of the above theorem.

Assume that X is first countable and semi-stratifiable. Then it has a sequence $\{U(n, x)\}$ satisfying the condition (i) of the theorem and a sequence $\{U'(n, x)\}$ satisfying the condition (iii). Put

$$U''(n, x) = U(n, x) \cap U'(n, x).$$

Then $\{U''(n, x)\}$ satisfies (i) and (iii) at the same time, and hence X is semi-metrizable.

B) *Let X be a T_1-space. Then relations between various generalized metric spaces are given in the following diagram:*

$$metrizable \Rightarrow M_1 \Rightarrow M_2 \Rightarrow M_3 \ (stratifiable) \Rightarrow \sigma \Rightarrow semi\text{-}stratifiable$$

$$developable \Rightarrow semi\text{-}metrizable$$

where, e.g., $M_1 \Rightarrow M_2$ means that every M_1-space is M_2.

Proof. The implications metrizable $\Rightarrow M_1 \Rightarrow M_2 \Rightarrow M_3$ are obvious. The implications $\sigma \Rightarrow$ semi-metrizable and semi-metrizable $\Rightarrow$ semi-stratifiable follow from Theorem VI.25. Let X be a developable space with a development $\{\mathcal{U}_n\}$. Then define ρ by

$$\rho(x, y) = \inf\{1/n \mid y \in S(x, \mathcal{U}_n)\}.$$

Now it is obvious that ρ is a semi-metric of X such that $\{S_\varepsilon(x) \mid \varepsilon > 0\}$ is a nbd base at each $x \in X$. Hence X is semi-metrizable. Suppose

Put
$$\mathcal{U}_n = \{U_\alpha^n \mid 0 \leq \alpha < \tau_n\}, \quad n = 1, 2, \ldots.$$

$$F_{\alpha m}^n = X - S(X - U_\alpha^n, \mathcal{U}_m), \quad m = 1, 2, \ldots,$$

$$G_{\alpha m}^n = F_{\alpha m}^n - \bigcup \{U_\beta^n \mid 0 \leq \beta < \alpha\},$$

$$\mathcal{G}_{nm} = \{G_{\alpha m}^n \mid 0 \leq \alpha < \tau_n\}.$$

Then each $\mathcal{G}_{nm}$ is a discrete closed collection. To prove that $\mathcal{G} = \bigcup_{n, m=1}^{\infty} \mathcal{G}_{nm}$ is a network of X, suppose that U is a nbd of $x \in X$. Then $S(x, \mathcal{U}_n) \subset U$ for some n. Suppose $x \in U_\alpha \in \mathcal{U}_n$ and $x \notin U_\beta$ for all $U_\beta \in \mathcal{U}_n$ with $\beta < \alpha$. Select m for which $S(x, \mathcal{U}_m) \subset U_\alpha$. Then it is obvious that

$$x \in G_{\alpha m}^n \subset U_\alpha \subset U.$$

Thus $\mathcal{G}$ is a network of X, proving that X is a σ-space.

The most difficult part is the proof of the implication $M_3 \Rightarrow \sigma$, which was first proved by R. W. Heath [3] as follows. Let X be an M_3-space; then by Theorem VI.25 there is a sequence $\{U(n, x) \mid n = 1, 2, \ldots\}$ of open nbds of each $x \in X$ satisfying the condition (iv). We may assume that $U(n, x) \supset U(n+1, x)$ for every n and x. Well-order the points of X to put

$$X = \{x_\alpha \mid 0 \leq \alpha < \tau\}.$$

Now we define

$$D(\alpha, i, n) = X - (\bigcup \{U(n, y) \mid y \notin U(i, x_\alpha)\})$$
$$\cup (\bigcup \{U(i, x_\beta) \mid \beta < \alpha\}), \quad (1)$$
$$0 \leq \alpha < \tau, \quad (i, n) \in N \times N.$$

Then it is obvious that $D(\alpha, i, n) \subset U(i, x_\alpha)$. Let

$$\mathcal{D}(i, n) = \{D(\alpha, i, n) \mid 0 \leq \alpha < \tau\};$$

then each $\mathcal{D}(i, n)$ is obviously a discrete closed collection. Further define

$$E(\alpha, i, n, m) = \{z \in D(\alpha, i, n) \mid x_\alpha \in U(m, z)\},$$

$$0 \leq \alpha < \tau, \ (i, n, m) \in N \times N \times N, \tag{2}$$

$$\mathscr{E}(i, n, m) = \{E(\alpha, i, n, m) \mid 0 \leq \alpha < \tau\},$$

$$\mathscr{E} = \cup \{\mathscr{E}(i, n, m) \mid (i, n, m) \in N \times N \times N\}.$$

Since $\mathscr{D}(i, n)$ is discrete, so is $\mathscr{E}(i, n, m)$. Thus $\mathscr{E}$ is a σ-discrete collection. We claim that $\mathscr{E}$ is a network of X. To prove it, let W be an open nbd of $x \in X$. For each $i \in N$, let α_i be the least ordinal number such that $x \in U(i, x_{\alpha_i})$. Then it follows from the condition (iv) of Theorem VI.25 that $x_{\alpha_i} \to x$. By the same condition there is $m \in N$ such that

$$x \notin \overline{\cup \{U(m, y) \mid y \in X - W\}}. \tag{3}$$

Thus there is $i \geq m$ such that

$$x_{\alpha_i} \notin \overline{\cup \{U(m, y) \mid y \in X - W\}} \tag{4}$$

and

$$x_{\alpha_i} \in U(m, x). \tag{5}$$

Since $x \in U(i, x_{\alpha_i})$, there is $n \in N$ satisfying

$$x \notin \cup \{U(n, z) \mid z \in X - U(i, x_{\alpha_i})\}. \tag{6}$$

Now, we can prove that

$$x \in E(\alpha_i, i, n, m) \subset W.$$

First $x \in E(\alpha_i, i, n, m)$ follows from (5), $x \in D(\alpha_i, i, n)$ and (2). (Note that $x \in D(\alpha_i, i, n)$ follows from (6), the definition of α_i and (1).) $E(\alpha_i, i, n, m) \subset W$ follows from the fact that $y \notin W$ implies $U(m, y) \not\ni x_{\alpha_i}$ (because of (4)), i.e. $y \notin E(\alpha_i, i, n, m)$ (because of (2)). Hence $\mathscr{E}$ is a σ-discrete network of X, and hence X is a σ-space.

C) *Every M_3-space is hereditarily paracompact.*

Proof. It directly follows from Theorem V.4.

D) *Every semi-stratifiable space X is subparacompact. Thus compactness* *and countable compactness coincide for semi-stratifiable spaces, and every collectionwise normal semi-stratifiable space is paracompact.*

Proof. Let $\mathcal{U} = \{U_\alpha \mid 0 \le \alpha < \tau\}$ be an open cover of X and G a semi-stratification of X. Put

$$F_{\alpha n} = G(U_\alpha, n) - \bigcup \{U_\beta \mid \beta < \alpha\}.$$

Then $\mathcal{F}_n = \{F_{\alpha n} \mid 0 \le \alpha < \tau\}$ is discrete, and $\bigcup_{n=1}^{\infty} \mathcal{F}_n$ is a σ-discrete closed refinement of $\mathcal{U}$. Hence X is subparacompact. See V.4 for the rest of the claim.

Theorem VI.26. *For a regular space X the following conditions are equivalent:*
  (i) *X has a σ-closure-preserving network,*
  (ii) *X is a σ-space,*
  (iii) *X has a σ-discrete network.*[1]

Proof. The implication (iii) $\Rightarrow$ (ii) $\Rightarrow$ (i) is obviously true. To prove (i) $\Rightarrow$ (iii), assume that X has a σ-closure-preserving network $\mathcal{G} = \bigcup_{n=1}^{\infty} \mathcal{G}_n$, where we may assume that $\mathcal{G}_n$ is a discrete closed collection because X is regular. Let (n, m) be a fixed element of $N \times N$. Then for each $G \in \mathcal{G}_n$ we define

$$H_m(G) = \bigcup \{G' \in \mathcal{G}_m \mid G' \cap G = \emptyset\},$$

and also
$$\mathcal{H}_m(G) = \{G, H_m(G)\}.$$

Note that $H_m(G)$ is a closed set. Further we define a closed collection $\mathcal{H}_{mn}$ by

$$\mathcal{H}_{mn} = \wedge \{\mathcal{H}_m(G) \mid G \in \mathcal{G}_n\}.$$

Then we claim that each $\mathcal{H}_{mn}$ is discrete. To prove the claim, it suffices to show that $\mathcal{H}_{mn}$ is disjoint and closure-preserving. Since each $\mathcal{H}_m(G)$ is disjoint, $\mathcal{H}_{mn}$ is obviously disjoint. Assume that

$$x \notin \bigcup \{H_\gamma \mid \gamma \in \Gamma\} = H,$$

---
[1] Due to F. Siwiec–J. Nagata [1].

where $\{H_\gamma \mid \gamma \in \Gamma\} \subset \mathcal{H}_{mn}$. Then from the definition of $\mathcal{H}_{mn}$ it follows that $U = X - \bigcup\{G \mid x \notin G \in \mathcal{G}_n \cup \mathcal{G}_m\}$ is an open nbd of x such that $U \cap H = \emptyset$. This proves that H is closed. Hence $\mathcal{H}_{mn}$ is closure-preserving and accordingly discrete. To show that $\mathcal{H} = \bigcup_{n, m=1}^{\infty} \mathcal{H}_{mn}$ is a network of X, let W be a nbd of $x \in X$. Then there is $n \in N$ and $G_0 \in \mathcal{G}_n$ such that $x \in G_0 \subset W$. Select $m \in N$ such that

$$x \in G_1 \subset X - \bigcup\{G \mid x \notin G \in \mathcal{G}_n\} \quad \text{for some } G_1 \in \mathcal{G}_m. \tag{1}$$

Define $H \in \mathcal{H}_{mn}$ by

$$H = \bigcap\{G \mid x \in G \in \mathcal{G}_n\} \cap (\bigcap\{H_m(G) \mid x \notin G \in \mathcal{G}_n\}). \tag{2}$$

Then $x \in H$ follows from (1) and (2), because $x \notin G \in \mathcal{G}_n$ implies that $x \in G_1 \subset H_m(G)$. On the other hand, $H \subset G_0 \subset W$ follows from (2). Thus $\mathcal{H}$ is a discrete network of X, proving (iii).

Corollary. *Let f be a closed continuous map from a regular σ-space X onto a topological space Y; then Y is a σ-space.*

Proof. Consider a σ-closure-preserving closed network of X. Then its image by f is a σ-closure-preserving network of Y. Hence by the theorem Y is a σ-space.

Definition VI.10. Generally, let $\mathcal{H}$ be a class of topological spaces. Then we consider various conditions to be satisfied by $\mathcal{H}$, as follows:

(a) If $X' \subset X \in \mathcal{H}$, then $X' \in \mathcal{H}$.

(b) If $X_i \in \mathcal{H}$, $i = 1, 2, \ldots$, then $\prod_{i=1}^{\infty} X_i \in \mathcal{H}$.

(c) If $X \in \mathcal{H}$, and there is a closed continuous map f from X onto a topological space Y, then $Y \in \mathcal{H}$.

(d) If $X = \bigcup_{i=1}^{\infty} X_i$ for closed subsets $X_i \in \mathcal{H}$, $i = 1, 2, \ldots$, of X, then $X \in \mathcal{H}$.

(e) If X is dominated by a closed cover $\{X_\alpha \mid \alpha \in A\}$ such that $X_\alpha \in \mathcal{H}$ for all α, then $X \in \mathcal{H}$, where the closed cover $\{X_\alpha \mid \alpha \in A\}$ is said to *dominate* X if the following holds: A subset F of X is closed if there is a subcollection $\{X_\alpha \mid \alpha \in A'\}$ of the cover such that $\bigcup\{X_\alpha \mid \alpha \in A'\} \supset F$ and such that $X_\alpha \cap F$ is closed for every $\alpha \in A'$.[1]

---

[1] It is obvious that every locally finite closed cover of X dominates X, and every dominating cover is closure-preserving.

(a') If X' is a closed subset of $X \in \mathcal{H}$, then $X' \in \mathcal{H}$.

(c') If $X \in \mathcal{H}$, and there is a perfect map f from X onto Y, then $Y \in \mathcal{H}$.

Recent developments of the theory of generalized metric spaces are motivated by several factors; one of them is that some generalized metric spaces behave, as a class, better than the metric spaces. The class $\mathcal{M}$ of metrizable spaces satisfies only (a), (a'), (b) and (c'). The non-metrizable hedgehog $S'(N)$ gives an example to show that $\mathcal{M}$ satisfies no other condition in Definition VI.10. On the other hand, e.g. the class of semi-stratifiable spaces satisfies all conditions there. Thus $S'(N)$ is semi-stratifiable. There are many other such spaces which are non-metrizable but 'near-metrizable' in the sense that they are constructed from metric spaces through a simpler process. Thus it is desirable to systematically study those non-metrizable spaces.

E) M_1-spaces satisfy (b). M_3-spaces satisfy all conditions in Definition VI.10 except (d). σ-spaces satisfy all conditions if X is regular in (c) and (e). Semi-stratifiable spaces satisfy all conditions. Semi-metrizable spaces satisfy (a), (a') and (b), and so do developable spaces.

Proof. We have already proved that (c) is satisfied by σ-spaces. We shall prove here only some of the non-trivial claims and leave the rest to the reader. To prove (e) for the class of semi-stratifiable spaces, let $\{X_\alpha \mid \alpha \in A\}$ be a closed cover dominating X, where each X_α is semi-stratifiable. Then we denote by G_α a semi-stratification of X. Well-order all members of the cover to put

$$\{X_\alpha \mid \alpha \in A\} = \{X_\alpha \mid 0 \leq \alpha < \tau\}.$$

Suppose U is an open set of X and $n \in N$. Then we define $G(U, n)$ by

$$G(U, n) = \bigcup \{F_\alpha \mid \alpha \in A\},$$

where

$$F_\alpha = G_\alpha(U \cap X_\alpha - \bigcup \{X_\beta \mid \beta < \alpha\}, n).$$

Then it is easy to see that G is a semi-stratification of X. To see that $G(U, n)$ is a closed set of X, let $x \in X - G(U, n)$. Suppose $x \in X_\alpha$, and $x \notin X_\beta$ for all $\beta < \alpha$. Then $V = X - \bigcup \{X_\beta \mid \beta < \alpha\}$ is an open nbd of X, which is disjoint from $\bigcup \{F_\beta \mid \beta < \alpha\}$. (Note that $\{X_\alpha\}$ is closure-preserving.) On the other hand, x has an open nbd W_α in X_α such that

$W_\alpha \cap F_\alpha = \emptyset$. Now, for each γ with $\alpha \leq \gamma < \tau$, we can construct an open set W_γ in $Y_\gamma = \bigcup\{X_{\alpha'} \mid \alpha \leq \alpha' \leq \gamma\}$ such that $W_\gamma \cap Y_{\gamma'} = W_{\gamma'}$ whenever $\alpha \leq \gamma' < \gamma < \tau$, and such that $W_\gamma \cap F_\gamma = \emptyset$. Then put

$$W = \bigcup\{W_\gamma \mid \alpha \leq \gamma < \tau\}.$$

Since $\{X_\alpha \mid \alpha \in A\}$ dominates X, W is an open nbd of x in X, because $W \subset \bigcup\{X_\gamma \mid \alpha \leq \gamma < \tau\}$, and $W \cap X_\gamma$ is open in X_γ. Also observe that $W \cap F_\gamma = \emptyset$ for all $\gamma \geq \alpha$. Thus $V \cap W$ is a nbd of x which is disjoint from $G(U, n)$. This proves that $G(U, n)$ is closed. Hence X is semi-stratifiable.

To prove (e) for σ-spaces, let $\{X_\alpha \mid \alpha \in A\}$ be a closed cover dominating X, where each X_α is a regular σ-space. Let $\mathcal{G}_\alpha = \bigcup_{i=1}^\infty \mathcal{G}_{\alpha i}$ be a σ-closure-preserving closed network of X_α, where each $\mathcal{G}_{\alpha i}$ is closure-preserving. Then it is easy to prove that $\mathcal{H}_i = \bigcup\{\mathcal{G}_{\alpha i} \mid \alpha \in A\}$ is closure-preserving for each i. Thus $\bigcup_{i=1}^\infty \mathcal{H}_i$ is a σ-closure-preserving network of X, proving that X is a σ-space.

Finally, let us prove that M_3-spaces satisfy (c). Let X be an M_3-space with a stratification G. Then to every pair (F, U) of a closed set F and an open set U such that $F \subset U$, we can assign a closed set $H(F, U)$ satisfying

$$X - U \subset (H(F, U))^\circ \subset H(F, U) \subset X - F, \tag{1}$$

$$H(F, U) \supset H(F', U') \quad \text{whenever } F \subset F' \text{ and } U \subset U'.[1]$$

To do so, we put

$$H(F, U) = X - \bigcup_{n=1}^\infty [(G(F^c, n))^c \cap (G(U, n))^\circ] \,;$$

then it is easy to prove that the desired condition is satisfied. (The detail is left to the reader.) Now, assume that f is a closed continuous map from X onto Y. Then we can define a stratification G' of Y as follows. Let V be an open set of Y. Then $F(V, n) = f(G(f^{-1}(V), n))$ is a closed set of Y satisfying

$$\bigcup_{n=1}^\infty F(V, n) = V. \tag{2}$$

---

[1] Generally a T_1-space X is called *monotonically normal* if for every pair (F, U) of a closed set F and an open set $U \supset F$, there is a closed set $H(F, U)$ satisfying this condition.

Then $f^{-1}(F(V, n))$ is a closed set of X contained in $f^{-1}(V)$. Put

$$H'(V, n) = Y - f(H(f^{-1}(F(V, n)), f^{-1}(V))) \, .$$

Then $H'(V, n)$ is an open set of Y contained in V such that

$$F(V, n) \subset H'(V, n) \subset \overline{H'(V, n)} \subset V, \tag{3}$$

which follows from (1). Further we put

$$G'(V, n) = \overline{H'(V, n)} \, .$$

Thus it follows from (2) and (3) that

$$\bigcup_{n=1}^{\infty} H'(V, n) = \bigcup_{n=1}^{\infty} (G'(V, n))^{\circ} = \bigcup_{n=1}^{\infty} G'(V, n) = V \, .$$

It is obvious that $G'(V, n) \subset G'(W, n)$, whenever $V \subset W$ for open sets V and W. Thus G' is a stratification of Y, proving that Y is M_3. The condition (e) for M_3-spaces will be proved later in VII.5.

In the following is a generalization of a proposition on metrizable spaces.

F) *Separability and Lindelöf property coincide for every collectionwise normal semi-stratifiable space and accordingly for every M_3-space as well.*

Proof. Let X be Lindelöf and $\{U(x, n) \mid n = 1, 2, \ldots\}$ a sequence of open nbds of x satisfying the condition (iii) of Theorem VI.25. Then for each n there is a countable subcover of $\{U(n, x) \mid x \in X\}$ which we denote by $\{U(n, x_i^n) \mid i = 1, 2, \ldots\}$. Now, we can show that $\{x_i^n \mid i, n = 1, 2, \ldots\}$ is dense in X. To this end, let $x \in X$ and consider a given nbd U of x. Then there is n for which

$$x \notin \bigcup \{U(n, y) \mid y \in X - U\} \, .$$

Thus $x_i^n \in U$ for some i, because otherwise $x \notin \bigcup_{i=1}^{\infty} U(n, x_i^n)$, which is impossible. This proves that $\{x_i^n\}$ is dense, and hence X is separable. (Actually we used only semi-stratifiability of X to prove this part of the proposition.)

Conversely, let X be separable and $\{x_i \mid i = 1, 2, \ldots\}$ a countable dense subset of X. Suppose $\mathcal{U}$ is a given open cover of X. Then, since X is paracompact by D), there is a locally finite open refinement $\mathcal{V}$ of $\mathcal{U}$. It suffices to show that $\mathcal{V}$ is countable. To each x_i we assign $\mathcal{V}_i = \{V \in \mathcal{V} \mid x_i \in V\}$. Then $\mathcal{V}_i$ is a finite subcollection of $\mathcal{V}$. Since $\mathcal{V} = \bigcup_{i=1}^{\infty} \mathcal{V}_i$ is obvious, $\mathcal{V}$ is countable. This proves that X is Lindelöf.

Example VI.13. It follows from E) that the non-metrizable hedgehog $S'(N)$ is stratifiable. In fact one can prove that $S'(N)$ is an M_1-space. (The easy proof is left to the reader.) The non-metrizable space X given in Example VI.3 is a first countable M_1-space. It is known that M_2 and M_3 coincide, but it is an open question if M_1 and M_2 coincide or not; we shall discuss this later. Niemytzki space is developable and accordingly a σ-space but not stratifiable, because it is not paracompact. Also observe that it is separable but not Lindelöf. See E. S. Berney [1] for an example of a Tychonoff semi-metric space which is not a σ-space (and accordingly not developable). The above example $S'(N)$ shows that developable spaces do not satisfy (c). On the other hand, J. M. Worrell [1] proved that developable spaces satisfy (c'). The condition (d) is not satisfied by M_3-spaces. In fact R. Heath [4] gave an example of a countable space which is not M_3. We can show that (c') is not satisfied by semi-stratifiable spaces. Consider $X = \{(x, y) \mid -\infty < x, y < +\infty\}$ with the following special topology: Let $p = (x, y) \in X$ and $n \in N$; then put

$$U(n, p) = \{p\} \cup \left\{(x', y') \in X \mid \left(y' - y - \frac{1}{n}(x' - x)\right)\right.$$

$$\left. \cdot \left(y' - y + \frac{1}{n}(x' - x)\right) < 0, \ |x' - x| < \frac{1}{n}\right\}.$$

Define that $\{U(n, p) \mid n = 1, 2, \ldots\}$ is a nbd base of each point p of X. Then it is easy to see that X is a semi-metrizable space, because $\{U(n, p)\}$ satisfies the conditions (i), (iii) of Theorem VI.25. Identify all points of $I = \{(x, 0) \in X \mid 0 \leqslant x \leqslant 1\}$ and leave the other points of X as singletons to define the quotient space X' of X. Then the natural (quotient) map f is obviously a perfect map. But X' is not semi-metrizable, because it is not first countable. To prove it, let $\{V_i \mid i = 1, 2, \ldots\}$ be a given sequence of open nbds of $y = f(I)$. For each i select an open set U_i which is a finite sum of $U(n, p)$, $p \in I$, and satisfies $I \subset U_i \subset f^{-1}(V_i)$. Denote by P_i the set of points $p = (x, y) \in I$ such that

$$U_i \cap \{(x, y') \mid -\infty < y' < +\infty\} = \{p\}\,.$$

Then P_i is finite. Select

$$q = (u, v) \in I - \bigcup_{i=1}^{\infty} P_i\,.$$

Consider an open nbd W of I in X such that

$$W \cap \{(u, y') \mid -\infty < y' < +\infty\} = \{q\}\,.$$

Then $W \not\supset U_i$, $i = 1, 2, \ldots$. Hence $W \not\supset f^{-1}(V_i)$. Thus $X' - f(X - W)$ is an open nbd of y in X' which contains none of V_i, $i = 1, 2, \ldots$. Hence $\{V_i \mid i = 1, 2, \ldots\}$ is no nbd base of y, proving that X' is not first countable.

Theorem VI.27 in the following indicates a merit of the concept of σ-space.

G) *Let X and Y be paracompact T_2-σ-spaces; then so is the product space $X \times Y$.*

Proof. It is obvious that $X \times Y$ is T_2 and σ. Let $\bigcup_{n=1}^{\infty} \mathcal{V}_n$ be a σ-discrete closed network of X, where $\mathcal{V}_n = \{V(n, \beta) \mid \beta \in B_n\}$ is discrete. Suppose $\mathcal{U} = \{U_\alpha \mid \alpha \in A\}$ is a given open cover of $X \times Y$. Put

$$\begin{aligned}
P(n, \beta, \alpha) &= \bigcup \{P \mid P \text{ is an open set of } Y \text{ such} \\
&\qquad \text{that } V(n, \beta) \times P \subset U_\alpha\}\,. \\
P(n, \beta) &= \bigcup \{P(n, \beta, \alpha) \mid \alpha \in A\}\,, \quad \beta \in B_n,\ n = 1, 2, \ldots\,.
\end{aligned}$$

Note that $P(n, \beta)$ is an open and accordingly F_σ-set of Y, because Y is perfectly normal. Thus $P(n, \beta)$ is paracompact by V.2.A). By use of the paracompactness and perfect normality of $P(n, \beta)$, we can find its open cover $\{P(n, \beta, \alpha, i) \mid \alpha \in A_n,\ i = 1, 2, \ldots\}$ such that $P(n, \beta, \alpha, i) \subset P(n, \beta, \alpha)$, and for each i $\{P(n, \beta, \alpha, i) \mid \alpha \in A\}$ is locally finite in Y. Now we claim that

$$\mathcal{W} = \{V(n, \beta) \times P(n, \beta, \alpha, i) \mid n = 1, 2, \ldots,\ \beta \in B_n,\ \alpha \in A,\ i = 1, 2, \ldots\}$$

covers $X \times Y$. To see it, let $(x, y) \in X \times Y$. Then $V(n, \beta) \times P \subset U_\alpha$ for some n, $\beta \in B_n$ with $x \in V(n, \beta)$, $\alpha \in A$, and an open nbd P of y. Hence

$$(x, y) \in V(n, \beta) \times P(n, \beta, \alpha) \subset V(n, \beta) \times P(n, \beta),$$

which implies

$$(x, y) \in V(n, \beta) \times P(n, \beta, \alpha', i) \quad \text{for some } \alpha' \text{ and } i.$$

Thus $\mathcal{W}$ covers $X \times Y$. Since X is collectionwise normal, for each n there is a discrete open collection $\{W(n, \beta) \mid \beta \in B_n\}$ in X such that $V(n, \beta) \subset W(n, \beta)$. Then $\{W(n, \beta) \times P(n, \beta, \alpha, i) \mid \beta \in B_n, \alpha \in A\}$ is locally finite for fixed n and i. Hence $\{W(n, \beta) \times P(n, \beta, \alpha, i) \cap U_\alpha \mid n = 1, 2, \ldots, \beta \in B_n, \alpha \in A, i = 1, 2, \ldots\}$ is a σ-locally finite open refinement of $\mathcal{U}$, proving that $X \times Y$ is paracompact.

H) *If $X_1 \times \cdots \times X_n$ is perfectly normal and paracompact for every $n \in N$, then $\prod_{i=1}^{\infty} X_i$ is perfectly normal and paracompact.*[1]

Proof. Let U be an open set of $\prod_{i=1}^{\infty} X_i$. Put

$$U_n = \Big\{ (x_1, x_2, \ldots) \in U \mid U(x_1) \times \cdots \times U(x_n) \times \prod_{j>n} X_j \subset U$$

$$\text{for some open nbds } U(x_i) \text{ of } x_i \text{ in } X_i, i = 1, 2, \ldots, n \Big\}.$$

Then $U = \bigcup_{n=1}^{\infty} U_n$ is obvious. By the assumption each U_n is F_σ. Hence U is also F_σ. Let $\mathcal{V}$ be a given open cover of $\prod_{i=1}^{\infty} X_i$. Then there is an open refinement $\bigcup_{n=1}^{\infty} \mathcal{V}_n$ of $\mathcal{V}$ such that each $V \in \mathcal{V}_n$ is of the form $V_i \times \cdots \times V_n \times \prod_{j>n} X_j$, where V_i is an open set of X_i for $i \leq n$. Then by a method similar to the one used in the previous proof, we can construct a σ-locally finite open cover $\mathcal{W}$ of $\prod_{i=1}^{\infty} X_i$ such that $\mathcal{W} < \bigcup_{n=1}^{\infty} \mathcal{V}_n$. Hence $\prod_{i=1}^{\infty} X_i$ is paracompact.

Theorem VI.27. *Let X_i, $i = 1, 2, \ldots$, be paracompact, T_2 and σ; then so is the product space $\prod_{i=1}^{\infty} X_i$.*

Proof. By G), $\prod_{i=1}^{n} X_i$ is paracompact T_2 and σ for each n. Hence paracompactness of $\prod_{i=1}^{\infty} X_i$ follows from H), while $\prod_{i=1}^{\infty} X_i$ is σ by E).

----

[1] Due to K. Morita.

I) *Let X be a paracompact T_2-σ-space. If each closed set F of X has a σ-closure-preserving outer nbd base $\mathcal{G}(F)$ (i.e., for every open set U containing F, there is a member G of $\mathcal{G}(F)$ such that $F \subset G° \subset G \subset U$), then X is an M_2-space. If moreover we can select $\mathcal{G}(F)$ consisting of open sets, then X is an M_1-space.*

Proof. Let $\mathcal{F} = \bigcup_{n=1}^{\infty} \mathcal{F}_n$ be a σ-discrete closed network of X, where $\mathcal{F}_n = \{F_\alpha \mid \alpha \in A_n\}$ is discrete. Then select a discrete open collection $\{W_\alpha \mid \alpha \in A_n\}$ for each n such that $F_\alpha \subset W_\alpha$. We consider $\mathcal{G}'(F_\alpha) = \{G \in \mathcal{G}(F_\alpha) \mid G \subset W_\alpha\}$. Then $\bigcup \{\mathcal{G}'(F_\alpha) \mid \alpha \in \bigcup_{n=1}^{\infty} A_n\}$ is a σ-closure-preserving quasi-base of X. Hence X is M_2.

J) *The following statements are equivalent*:
　(i) *Any M_3-space is M_2 (M_1).*
　(ii) *Any point of any M_3-space has a σ-closure-preserving (open) nbd base.*

Proof. It is obvious that (i) implies (ii). Assume (ii). Let X be a given M_3-space, and F a closed set of X. Then identify the points of F while leaving the other points as singletons. Denote by Y thus obtained quotient space and by f the quotient map. Note that Y is an M_3-space. Let $f(F) = y \in Y$. Then by the hypothesis, y has a σ-closure-preserving nbd base $\mathcal{V}$. $f^{-1}(\mathcal{V}) = \{f^{-1}(V) \mid V \in \mathcal{V}\}$ is a σ-closure-preserving nbd base of F in X. Thus by I) X is an M_2-space. If $\mathcal{V}$ consists of open sets, then so is $f^{-1}(\mathcal{V})$. Hence in this case X is an M_1-space.

　The following interesting theorem was first proved by H. Junnila [1] and G. Gruenhage [1] independently. Here we give a sketch of Gruenhage's proof.

Theorem VI.28. *Every M_3-space is M_2. Thus M_2 and M_3 are equivalent.*

Proof. We denote by $H(F, U)$ the open set in the proof of E), assigned to the pair of a closed set F and an open set U with $F \subset U$. Let p be a fixed point of X. Then for each closed set F with $F \not\ni p$, we put

$$E(F) = H(F, X - \{p\}), \qquad E^2(F) = X - H(E(F), X - \{p\}).$$

In a similar way we can define $E^n(F)$ for all $n \in N$. Let $\mathcal{F}'$ be a σ-discrete network of X. Put

$$\{F \mid p \notin F \in \mathcal{F}'\} = \mathcal{F} = \bigcup_{n=1}^{\infty} \mathcal{F}_n,$$

where each $\mathcal{F}_n$ is discrete. Then, since X is hereditarily paracompact, there are locally finite open covers $\mathcal{U}_n$, $n = 1, 2, \ldots$, of $X - \{p\} = X'$ such that

$$S(x, \mathcal{U}_n) \cap E(F) = \emptyset \quad \text{for each } x \in F \in \mathcal{F}_n, \; \mathcal{U}_{n+1}^{*} < \mathcal{U}_n,$$

$$p \notin \overline{S(x, U_1)} \quad \text{for every } x \in X'. \tag{1}$$

To each $x \in X'$ we assign an open nbd W_x of p defined by

$$W_x = \bigcup_{n=1}^{\infty} E^2(\overline{S(x, \mathcal{U}_n)}).$$

Then it is easy to show that $S(x, \mathcal{U}_n) \cap W_x = \emptyset$ for some n follows from (1) and the fact that $\mathcal{F}$ is a network. Now, put

$$n(x) = \min \{n \mid S(x, \mathcal{U}_n) \cap W_x = \emptyset\},$$

$$X_n = \{x \in X' \mid n(x) \le n\}, \quad n = 1, 2, \ldots.$$

Then one can prove that X_n is closed in X'. For each $x \in X'$ we define an open nbd $P(x)$ by

$$P(x) = \cap \left\{ U \mid x \in U \in \bigcup_{k=1}^{n(x)} U_k \right\} - X_{n(x)-1}.$$

Then it can be proved that

$$y \in P(x) \text{ implies } P(y) \subset P(x), \tag{2}$$

$$P(x) \cap E^4(\{x\}) = \emptyset. \tag{3}$$

We denote by $\mathscr{C}$ the collection of all closed sets of X which do not contain p. Then for each $C \in \mathscr{C}$ we put

$$P(C) = \cup \{P(x) \mid x \in C\}.$$

From (3) it follows that $p \notin \overline{P(C)}$. Thus $\mathscr{P} = \{X - P(C) \mid C \in \mathscr{C}\}$ is a nbd base of p. To prove that $\mathscr{P}$ is closure-preserving, let

$$y \notin \cup \{X - P(C) \mid C \in \mathscr{D}\}, \quad \mathscr{D} \subset \mathscr{C}.$$

For each $C \in \mathscr{D}$, $y \notin P(C)$. Thus there is $x(C) \in C$ such that $y \in P(x(C))$. Then $P(y) \subset P(x(C))$ follows from (2). Hence

$$P(y) \cap (X - P(C)) = \emptyset \quad \text{for all } C \in \mathscr{D},$$

proving that $y \notin \overline{\cup \{X - P(C) \mid C \in \mathscr{D}\}}$. Thus $\mathscr{P}$ is closure-preserving. Hence this theorem follows from J).

It is still an open question whether or not M_1 and M_3 coincide. However, recently there have been remarkable developments on this question. R. Heath–H. Junnila [1] proved that every M_3-space is a retract of an M_1-space by a perfect map. Thus the following statements are equivalent:

(i) Every M_3-space is M_1.

(ii) Every closed set of an M_1-space is M_1 (i.e., M_1 satisfies (a') of Definition VI.10).

(iii) The image of an M_1-space by a perfect map is M_1 (i.e., M_1 satisfies (c') of Definition VI.10).

Naturally the following two conditions are also equivalent with the above.

(iv) M_1 satisfies (a).

(v) M_1 satisfies (c).

G. Gruenhage [3] proved that every M_3-space which is the sum of countably many closed metrizable sets is an M_1-space. M. Ito [1] proved that the image of an hereditarily M_1-space by a closed continuous map is hereditarily M_1. M. Ito [2] proved that an M_3-space X is M_1 if every point of X has a closure-preserving open nbd base, and especially that every first countable M_3-space is M_1.[1]

In the following we define generalized metric spaces, which are somewhat different in their nature from those defined in Definition VI.9.

Definition VI.11. A Tychonoff space X is called a *p-space* if there is a sequence $\mathscr{U}_1, \mathscr{U}_2, \ldots$ of open collections in $\beta(X)$ such that

---

[1] See also G. Gruenhage [2], T. Mizokami [1], [2] for other results in this aspect. A first countable M_3-space is sometimes called a *Nagata space*.

(i) each $\mathcal{U}_i$ covers X,

(ii) for each $x \in X$, $\bigcap_{i=1}^{\infty} S(x, \mathcal{U}_i) \subset X$.[1]

A topological space X is called an *M-space* if it has a sequence $\mathcal{U}_1, \mathcal{U}_2, \ldots$ of open covers of X such that

(i) $\mathcal{U}_1 > \mathcal{U}_2^* > \mathcal{U}_2 > \mathcal{U}_3^* > \cdots$,

(ii) if $x_i \in S(x, \mathcal{U}_i)$, $i = 1, 2, \ldots$, for a fixed point x, then the point sequence $\{x_i\}$ has a cluster point.[2]

We call (ii) *wΔ-condition*.

The condition for *p*-space is obviously weaker than that for Čech complete space. The condition (ii) for *M*-space is weaker than

(ii′) if $x_i \in S(x, U_i)$, $i = 1, 2, \ldots$, for a fixed point x, then $x_i \to x$. The condition (ii′) is equivalent to the second condition of Alexandroff–Urysohn's metrization theorem. Thus we obtain:

K) *Every metrizable space as well as every Čech complete space is a p-space. Every metrizable space as well as every countably compact space is an M-space.*

Proof. To prove that every metrizable space X is a *p*-space, let

$$U_i = \{\beta(X) - \overline{(X - S_{1/i}(x))} \mid x \in X\},$$

where $S_{1/i}(x)$ denotes the $1/i$-nbd of x in X and the closure is taken in $\beta(X)$.

L) *M-spaces satisfy* (a′) *of Definition VI.10, and p-spaces satisfy* (a′) *and* (b).[3]

Proof. The easy proof is left to the reader.

Theorem VI.29. *A Tychonoff space X is a p-space if and only if it has a sequence $\mathcal{V}_1, \mathcal{V}_2, \ldots$ of open covers satisfying the following conditions: If $x \in V_i \in \mathcal{V}_i$, $i = 1, 2, \ldots$, then*

(i) $\bigcap_{i=1}^{\infty} \bar{V}_i$ *is compact,*

(ii) *if $x_n \in \bigcap_{i=1}^{n} \bar{V}_i$, $n = 1, 2, \ldots$, then $\{x_n\}$ has a cluster point.*[4]

[1] Due to A. V. Arhangelskii [4].

[2] Due to K. Morita [5]. Generally a topological space X is called a *wΔ-space* if it has a sequence $\{U_i\}$ of open covers satisfying (ii).

[3] See T. Isiwata [2] for an example of two *M*-spaces whose product is not *M*.

[4] Due to D. Burke [1].

Proof. We give here a sketchy proof and leave the detail to the reader. Let X be a p-space with open collections $\mathcal{U}_1, \mathcal{U}_2, \ldots$ in $\beta(X)$ satisfying (i), (ii) of Definition VI.11. For each i select an open collection $\mathcal{U}'_i$ in $\beta(X)$ such that $\mathcal{U}'_i$ covers X, and $\bar{\mathcal{U}}'_i < \mathcal{U}_i$, where here and in the rest of the proof closure is taken in $\beta(X)$ unless it is denoted by $^{-X}$ meaning the closure is taken in X. Then we have

$$\bigcap_{i=1}^{\infty} S(x, \bar{\mathcal{U}}'_i) \subset X \quad \text{for each } x \in X. \tag{1}$$

Let $\mathcal{V}_i = \{U \cap X \mid U \in \mathcal{U}'_i\}$. Assume

$$x \in V_i \in \mathcal{V}_i, \quad i = 1, 2, \ldots,$$

and $V_i = U_i \cap X$, $U_i \in \mathcal{U}'_i$, $\bar{U}_i \subset U^i \in \mathcal{U}_i$. Then

$$\bigcap_{i=1}^{\infty} \bar{V}_1^X = \bigcap_{i=1}^{\infty} \bar{U}_i \cap X = \bigcap_{i=1}^{\infty} \bar{U}_i .$$

Since the last set is closed in $\beta(X)$, the condition (i) of the theorem is satisfied by $\{V_i\}$.

Assume that $x_n \in \bigcap_{i=1}^{n} \bar{V}_i$, $n = 1, 2, \ldots$, while $\{x_n\}$ has no cluster point in X. Put $F_m = \{x_n \mid n \geqslant m\}$. Then for each m F_m is closed, and $F_m \cap (\bigcap_{i=1}^{n} \bar{V}_i^X) \neq \emptyset$ for all n. Since $\bigcap_{m=1}^{\infty} F_m = \emptyset$, we obtain

$$\left(\bigcap_{i=1}^{\infty} \bar{V}_i^X \right) \cap F_m = \emptyset \quad \text{for some } m . \tag{2}$$

On the other hand, $\{\overline{\bigcap_{i=1}^{n} \bar{V}_i^X \cap F_m} \mid n = 1, 2, \ldots\}$ is a decreasing sequence of non-empty closed sets of $\beta(X)$. Hence there is y in their intersection. Observe that $y \in \beta(X) - X$ follows from (2). Note that $x \in U_i \in \mathcal{U}'_i$ and $y \in \bar{V}_i^X \subset \bar{U}_i$ for $i = 1, 2, \ldots$. Thus

$$y \in S(x, \bar{\mathcal{U}}'_i) \cap (\beta(X) - X),$$

contradicting (1). Hence $\{x_n\}$ has a cluster point, proving the 'if' part.

Conversely, let $\mathcal{V}_1, \mathcal{V}_2, \ldots$ satisfy (i), (ii) of the theorem. Put

$$\mathcal{U}_i = \{\beta(X) - \overline{(X - V)} \mid V \in \mathcal{V}_i\}.$$

Then $\mathcal{U}_i$ is an open collection in $\beta(X)$ covering X. Assume $x \in X$ and

$y \in \beta(X) - X$. Then y may be regarded as a non-convergent maximal zero-filter of X, which we denote by $\mathcal{F}$. Suppose

$$x \in U_i = \beta(X) - \overline{(X - V_i)} \in \mathcal{U}_i, \quad i = 1, 2, \ldots, \quad V_i \in \mathcal{V}_i. \tag{3}$$

Select $F \in \mathcal{F}$ such that $F \cap (\cap_{i=1}^{\infty} \bar{V}_i^X) = \emptyset$. Then $F \cap (\cap_{i=1}^{n} \bar{V}_i^X) = \emptyset$ for some n, because of (ii) of the theorem. Hence for every nbd W of y in $\beta(X)$, we have

$$W \cap \left(X - \bigcap_{i=1}^{n} \bar{V}_i^X \right) \neq \emptyset.$$

Thus for some $i \leqslant n$, $W \cap (X - \bar{V}_i^X) \neq \emptyset$ holds for every nbd W of y. Hence $y \in X - \bar{V}_i^X \subset \overline{X - V_i}$. This combined with (3) implies $y \notin U_i$. Thus $\cap_{i=1}^{\infty} S(x, \mathcal{U}_i) \subset X$ is proved. Hence X is a p-space.

Corollary 1. *Every Tychonoff developable space is a p-space.*

Corollary 2. *A paracompact T_2-space is a p-space if and only if it is an M-space.*

Proof. Suppose X is an M-space with a sequence $\{\mathcal{U}_i \,|\, i = 1, 2, \ldots\}$ of open covers satisfying the conditions in Definition VI.11. Then we can prove that $\{\mathcal{U}_i\}$ satisfies the conditions (i), (ii) of the theorem. Suppose $x \in U_i \in \mathcal{U}_i$, $i = 1, 2, \ldots$. Then it is obvious that $\cap_{i=1}^{\infty} \bar{U}_i$ is countably compact and accordingly compact. The condition (ii) follows from (ii) of the definition of M-space. Hence X is a p-space.

Assume X is a p-space. Then we can select a normal sequence $\{\mathcal{U}_i\}$ of open covers satisfying (i), (ii) of the theorem. Now it is easy to see that $\{\mathcal{U}_i\}$ satisfies the second condition of M-space, too. Hence X is an M-space.

Example VI.14. As seen in the above corollary, M and p coincide for a paracompact space. However, they are distinct in general. For example, Niemytzki space is a p-space by Corollary 1 but not M (as will be seen later). M and p are a generalization of metric space into a different direction from those spaces defined in Definition VI.9. Those two directions are complementary in the sense that they combined together produce metrizability, which will be seen in the next chapter. The space R_5 in Example II.1 is countably compact and Čech complete, and hence it is M and p. However, R_5 is no σ-space, because it is collectionwise normal and non-paracompact. Actually it is neither semi-stratifiable. We shall see in the next chapter that any compact T_2 non-metrizable space is not semi-stratifiable and that every M_3-space is neither M nor p unless it is metrizable.[1]

We finish this section with another diagram to supplement the one given in B).

$$\text{compact} \Longrightarrow \begin{array}{c} \text{countably compact} \Rightarrow M \Leftarrow \text{metric} \\[2mm] \Downarrow \\ \text{locally compact} \Rightarrow \text{Čech complete} \Rightarrow p \Leftarrow \text{developable} \end{array}$$

We assume T_2 and Tychonoff for the implications locally compact $\Rightarrow$ Čech complete and developable $\Rightarrow p$, respectively.

Exercise VI

1. Let X be a T_1-space and $\varphi(p, q)$ a real-valued function defined over $X \times X$ such that:

(i) $\varphi(p, q) = \varphi(q, p) \geq 0$,

(ii) for every $\varepsilon > 0$, $\varphi(p, q) < \varepsilon$ and $\varphi(q, r) < \varepsilon$ imply $\varphi(p, r) < 2\varepsilon$,

(iii) $\{S_{1/n}(p) \mid n = 1, 2, \ldots\}$, where $S_{1/n}(p) = \{q \mid \varphi(p, q) < 1/n\}$,

forms a nbd basis of each point p of X.

Then X is metrizable (E. W. Chittenden's theorem). (Apply Theorem VI.1.)

2. A T_1-space X is metrizable if and only if there exists a sequence $\{\mathcal{U}_n \mid n = 1, 2, \ldots\}$ of open coverings such that for every nbd U of each point p of X there is a nbd V of p, such that $S(V, \mathcal{U}_n) \subset U$ for some n (A. Arhangelskii's theorem).

3. The fully normal space M in Example V.2 has a σ-point-finite open basis but is not metrizable.

4. Alexandroff's one-point compactification $\alpha(X)$ is metrizable if and only if X is metrizable, and $\alpha(X)$ is T_2 and first countable.

---

[1] D. Burke–D. Lutzer's survey article [1] is recommended for further information on generalized metric spaces.

5. Let f be a closed, open continuous mapping of a metric space X onto a topological space Y, then Y is also metrizable (due to W. K. Balachandran).

6. A Tychonoff space X is Čech complete if and only if it is homeomorphic to a closed set of the product of countably many locally compact spaces.

7. Let A be a Čech complete subset of a metric space X. Then A is a G_δ-set of X.

8. Any G_δ-set of a Čech complete space is Čech complete. Thus X is Čech complete if and only if it is homeomorphic to a G_δ-set of a compact T_2-space.

9. The countable product of Čech complete spaces is Čech complete.

10. A topological space of weight $\leqslant \mu$ is completely metrizable if and only if it is homeomorphic to a closed set of $H(A)$, where $|A| = \mu$.

11. Every metrizable space has a first countable T_2-compactification. (Use Theorem VI.11.)

12. Every paracompact T_2-space is Dieudonné complete.

13. A uniform space X is complete if and only if every Cauchy net of X converges.

14. A uniform space is totally bounded if and only if its uniformity has a basis consisting of finite uniform coverings.

15. The product space of totally bounded uniform spaces is totally bounded.

16. Let A, B be subsets of a uniform space X with uniformity $\{\mathscr{U}_\gamma \mid \gamma \in \Gamma\}$. Then there is a uniformly continuous function f such that

$$f(A) = 0, \qquad f(B) = 1 \quad \text{and} \quad 0 \leqslant f \leqslant 1$$

if and only if $S(A, \mathscr{U}_\gamma) \cap B = \emptyset$ for some $\gamma \in \Gamma$.

17. Give an example of a uniformly continuous mapping which maps a complete uniform space onto a non-complete uniform space.

18. If X is a uniform space satisfying the conditions in 4.G), then it is strongly paracompact. How about if connectedness is dropped from the condition?

19. Let f be a mapping of a proximity space X onto a proximity space Y. Then f is a p-mapping if and only if $A\bar{\delta}(Y - B)$ for subsets A and B of Y always implies $f^{-1}(A)\bar{\delta}(X - f^{-1}(B))$ in X.

20. Suppose X and Y are proximity spaces.

For two subsets A, B of the cartesian product $X \times Y$, we define that $A\delta B$ if and only if $S(A, \mathcal{U} \times \mathcal{V}) \cap B \neq \emptyset$ for every pair of p-coverings $\mathcal{U}$ of X and $\mathcal{V}$ of Y. Then δ is a proximity compatible with the topology of the product space $X \times Y$.

21. Every p-mapping of a metric space X into a metric space Y is uniformly continuous.

22. Let f be a closed continuous map from X onto Y such that $f^{-1}(y)$ is countably compact for each $y \in Y$. If Y is a P-space, then so is X.

23. Let f be a closed continuous mapping of a P-normal space X onto a topological space Y. Then Y is also normal P.

24. Every semi-stratifiable, monotonically normal space is M_3.

25. For an M_3-space the following conditions are equivalent: Lindelöf, separability and hereditary separability.

26. Every closed set of an M_2-space has a closure-preserving outer nbd base.

27. Every closed set of the product of a metric space and a countably compact space is an M-space.

28. Any non-metrizable hedgehog is neither M- nor p-space.

29. Every G_δ-set of the product of a metric space and a compact T_2-space is a p-space.

30. The product of countably many p-spaces is a p-space.

31. Every separable stratifiable space has a countable network (due to E. Michael).

CHAPTER VII

TOPICS RELATED TO MAPPINGS

This chapter is a combination of various topics related to the concept of mapping, rather than a systematic description of a fixed subject.

In the first section we shall learn various methods of introducing topologies into a set of mappings and of approximating a given continuous function over a T_2-space by special functions. The second section is concerned with to what extent the properties of the range space and domain space of a continuous mapping will affect each other, and especially with characterizations of spaces as images and inverse images of nice spaces. For example, M-spaces are characterized as the inverse images of metric spaces by continuous maps satisfying certain conditions, and metrization of M-spaces is studied in the following section. In the fourth section we shall try to characterize some spaces as the inverse limit spaces of simpler spaces. Sections 5 and 6 are devoted to extensions of Tietze's extension theorem into two different directions. In the last section we shall discuss rather sporadic results concerning characterizations of topological properties of X in terms of $C(X)$ and $C^*(X)$.

1. Mapping space

Let X be a set and Y a topological space. We denote by $F(X, Y)$ a set of mappings of X into Y. There are several different methods to introduce a topology into $F(X, Y)$, which is the subject of study in the present section.

Definition VII.1. Suppose f is a given element of $F(X, Y)$. Let $p_1, \ldots, p_n$ be points of X and $U_1, \ldots, U_n$ nbds in Y of $f(p_1), \ldots, f(p_n)$, respectively. Then we define a subset $U(p_1 \ldots p_n; U_1 \ldots U_n; f)$ of $F(X, Y)$ by

$$U(p_1 \ldots p_n; U_1 \ldots U_n; f) = \{g \mid g \in F(X, Y), g(p_i) \in U_i$$
$$\text{for } i = 1, \ldots n\}.$$

Then $\mathcal{U}(f) = \{U(p_1 \ldots p_n ; U_1 \ldots U_n ; f) \,|\, p_i \in X,\ U_i$ is a nbd of $f(p_i)$ in Y, $i = 1, \ldots n;\ n = 1, 2, \ldots\}$ is easily seen to satisfy the conditions for a nbd basis of f in $F(X, Y)$. Thus $F(X, Y)$ turns out to be a topological space called a mapping space with the *weak topology* or the *topology of pointwise convergence*.

A) *A mapping space $F(X, Y)$ with the weak topology is homeomorphic to a subspace of the product space $\prod \{Y_p \,|\, p \in X\}$ where each Y_p is a copy of the topological space Y.*

Proof. Mapping each $f \in F(X, Y)$ to $\{f(p) \,|\, p \in X\} \in \prod \{Y_p \,|\, p \in X\}$, we obtain a topological mapping of $F(X, Y)$ onto a subspace of the product space.

B) *If Y is completely regular (regular, T_2 or T_1), then $F(X, Y)$ with the weak topology is also completely regular (regular, T_2 or T_1).*

Proof. This is an immediate consequence of A).

C) *Let Y be a compact T_2-space. We denote by Y^X the mapping space consisting of all mappings of X into Y with the weak topology. Then a mapping space $F(X, Y)$ with the weak topology is compact if and only if it is closed in Y^X.*

Proof. Since Y^X is the product space of the copies of Y, it is compact T_2 by Tychonoff's product theorem. From this, the present proposition is derived.

D) *Let $F(X, Y)$ be a mapping space with a topology stronger than the weak topology, where we assume that Y is T_2. If $F(X, Y)$ is compact, then its topology coincides with the weak topology.*

Proof. We denote by $F_0(X, Y)$ the mapping space with the elements of $F(X, Y)$ and the weak topology. Then the identity mapping of $F(X, Y)$ onto $F_0(X, Y)$ is continuous and one-to-one. Since $F(X, Y)$ is compact and since by B), $F_0(X, Y)$ is T_2, this mapping is a homeomorphism by III.3.D).

Example VII.1. Suppose f is a member of $F(X, Y)$, a collection of mappings of X into Y. To every point $p \in X$ we assign a nbd U_p of $f(p)$ in Y. Then define a subset $U(\{U_p\}, f)$ of $F(X, Y)$ by $U(\{U_p\}, f) =$

$\{g \mid g \in F(X, Y),\ g(p) \in U_p$ for every $p \in X\}$. It is easily seen that $\{U(\{U_p\}, f) \mid U_p$ is a nbd of $p,\ p \in X\}$ satisfies the condition for nbd basis of f and gives a stronger topology of $F(X, Y)$ than the weak topology. This topology is called the *strong topology* of $F(X, Y)$.

Example VII.2. Suppose Y is a uniform space with uniformity $\{\mathcal{U}_\alpha \mid \alpha \in A\}$; then we define the *weak uniformity* of $F(X, Y)$ as follows. Let $\mathcal{U}_\alpha = \{U_\gamma \mid \gamma \in \Gamma_\alpha\}$ and $p \in X$. Then we put

$$U(p; U_\gamma) = \{g \mid g \in F(X, Y),\ g(p) \in U_\gamma\}$$

and

$$\mathcal{U}(p; \alpha) = \{U(p; U_\gamma) \mid \gamma \in \Gamma_\alpha\}.$$

For $p_i \in X,\ \alpha_i \in A,\ i = 1, \ldots, n$, we put

$$\mathcal{U}(p_1 \ldots p_n; \alpha_1 \ldots \alpha_n) = \wedge \{\mathcal{U}(p_i; \alpha_i) \mid i = 1, 2, \ldots, n\}.$$

Then

$$\{\mathcal{U}(p_1 \ldots p_n; \alpha_1 \ldots \alpha_n) \mid p_i \in X, \alpha_i \in A, i = 1, \ldots, n; n = 1, 2, \ldots\}$$

forms a basis for a uniformity compatible with the weak topology. The thus defined uniformity is called the weak uniformity of $F(X, Y)$.

Definition VII.2. Let us suppose that X and Y are topological spaces, and $F(X, Y)$ is a set of mappings of X into Y. For a finite number of compact sets $C_1, \ldots, C_n$ of X and open sets $U_1, \ldots, U_n$ of Y, we define a subset $U(C_1 \ldots C_n; U_1 \ldots U_n)$ of $F(X, Y)$ by

$$U(C_1 \ldots C_n; U_1 \ldots U_n) = \{f \mid f \in F(X, Y), f(C_i) \subset U_i, i = 1, \ldots, n\}.$$

Then the collection $\{U(C_1 \ldots C_n; U_1 \ldots U_n) \mid C_1, \ldots, C_n$ are compact sets of X; $U_1, \ldots, U_n$ are open sets of X; $n = 1, 2, \ldots\}$ is easily seen to satisfy the condition for an open basis. The topology induced by this open basis is called the *compact open topology* of $F(X, Y)$.

E) *The compact open topology is stronger than the weak topology.*

Proof. It is left to the reader.

F) *If Y is T_2, then so is the mapping space $F(X, Y)$ with the compact open topology. If Y is completely regular (or regular) and $F(X, Y)$ consists of continuous mappings, then $F(X, Y)$ is also completely regular (or regular).*

Proof. In case that Y is T_2, this proposition is directly derived from B) and E). If Y is completely regular and $F(X, Y)$ consists of continuous mappings, then we suppose that G is a given closed set and f is a given point of $F(X, Y)$ such that $f \notin G$. There are compact sets $C_1, \ldots, C_n$ of X and open sets $U_1, \ldots, U_n$ of Y such that

$$f \in U(C_1 \ldots C_n; U_1 \ldots U_n) \subset F(X, Y) - G.$$

Since $f(C_i) \subset U_i$, for each point p of $f(C_i)$ there is a continuous function φ_p over Y such that

$$\varphi_p(p) = -1, \qquad \varphi_p(Y - U_i) = 1 \quad \text{and} \quad -1 \leq \varphi_p \leq 1.$$

Put

$$U(p) = \{y \mid y \in Y, \varphi_p(y) < 0\}$$

for every $p \in f(C_i)$; then

$$\bigcup \{U(p) \mid p \in f(C_i)\} \supset f(C_i).$$

Since $f(C_i)$ is compact, we can cover $f(C_i)$ with a finite number of $U(p)$, say $U(p_1), \ldots, U(p_k)$.

Put

$$\varphi^i = \max\{\min\{\varphi_{p_1}, \ldots, \varphi_{p_k}\}, 0\};$$

then φ^i is a continuous function over Y such that

$$\varphi^i(f(C_i)) = 0, \qquad \varphi^i(Y - U_i) = 1 \quad \text{and} \quad 0 \leq \varphi^i \leq 1.$$

We define a real-valued function Ψ^i over $F(X, Y)$ by

$$\Psi^i(g) = \max\{\varphi^i(p) \mid p \in g(C_i)\}, \quad g \in F(X, Y). \tag{1}$$

Then it is clear that

$$\Psi^i(f) = 0, \qquad \Psi^i(F(X, Y) - U(C_i; U_i)) = 1$$

and

$$0 \leq \Psi^i \leq 1.$$

To prove the continuity of Ψ^i we consider a given point g_0 of $F(X, Y)$.

Since $g_0(C_i)$ is compact, we may suppose there is a point $p_0 \in g_0(C_i)$ for which

$$\varphi^i(p_0) = \max\{\varphi^i(p) \mid p \in g_0(C_i)\} . \tag{2}$$

For a given $\varepsilon > 0$, we choose an open nbd $V(p_0)$ of p_0 in Y such that

$$|\varphi^i(p) - \varphi^i(p_0)| < \varepsilon \tag{3}$$

for every $p \in V(p_0)$. Take a point

$$x_0 \in g_0^{-1}(p_0) \cap C_i . \tag{4}$$

Put

$$W = \{p \mid p \in Y, \varphi^i(p) < \varphi^i(p_0) + \varepsilon\} . \tag{5}$$

Now, we consider the open set $U(x_0, C_i; V(p_0), W)$ of $F(X, Y)$. It is easily seen to be an open nbd of g_0 because $g_0(x_0) = p_0 \in V(p_0)$ follows from (4), while $g_0(C_i) \subset W$ follows from (2) and (5). (If $x \in C_i$, then $g_0(x) \in g_0(C_i)$ which implies $\varphi^i(g_0(x)) \leqslant \varphi^i(p_0) < \varphi^i(p_0) + \varepsilon$.) On the other hand, if

$$g \in U(x_0, C_i; V(p_0), W) ,$$

then $g(x_0) \in V(p_0)$, which implies, by property (3) of $V(p_0)$, that

$$\varphi^i(g(x_0)) > \varphi^i(p_0) - \varepsilon .$$

Since $x_0 \in C_i$, $g(x_0) \in g(C_i)$; hence by (1) we get

$$\Psi^i(g) = \max\{\varphi^i(p) \mid p \in g(C_i)\} > \varphi^i(p_0) - \varepsilon . \tag{6}$$

Since $g(C_i) \subset W$, it follows from (5) that

$$\Psi^i(g) < \varphi^i(p_0) + \varepsilon . \tag{7}$$

Note that it follows from (1) and (2) that

$$\varphi^i(p_0) = \Psi^i(g_0) ;$$

hence, combining (6) and (7), we obtain

$$|\Psi^i(g) - \Psi^i(g_0)| < \varepsilon$$

for every point $g \in U(x_0, C_i; V(p_0), W)$.

Therefore Ψ^i is a continuous function over $F(X, Y)$. We put

$$\Psi = \max\{\Psi^1, \dots, \Psi^n\}.$$

Then Ψ is a continuous function over $F(X, Y)$ satisfying

$$\Psi(f) = 0, \qquad \Psi(G) = 1 \quad \text{and} \quad 0 \leqslant \Psi \leqslant 1,$$

which follows from the properties of Ψ^i. Thus $F(X, Y)$ is completely regular. It is not difficult to prove the regular case in a somewhat analogous way and so it is left to the reader.[1]

Definition VII.3. Let us suppose X is a set and Y is a uniform space with uniformity $\{\mathcal{U}_\alpha \mid \alpha \in A\}$, where $\mathcal{U}_\alpha = \{U_\gamma \mid \gamma \in \Gamma_\alpha\}$. For each $f \in F(X, Y)$ and $\alpha \in A$, we define a subset $U(\alpha, f)$ of $F(X, Y)$ by $U(\alpha, f) = \{g \mid g \in F(X, Y),\ g(p) \in S(f(p), \mathcal{U}_\alpha)$ for every $p \in X\}$, and $\mathcal{U}(\alpha) = \{U(\alpha, f) \mid f \in F(X, Y)\}$.

Then we can easily see that $\{\mathcal{U}(\alpha) \mid \alpha \in A\}$ satisfies the condition for a basis of a uniformity. Thus $F(X, Y)$ turns out to be a uniform space called a mapping space with the *uniformity of uniform convergence*. The topology induced by this uniformity is called the *topology of uniform convergence*.

Every mapping space with the above topology is at least completely regular, but the most important is the following metric case.

G) *Let $F(X, Y)$ be a collection of bounded mappings of a set X into a metric space Y.[2] Then the uniformity of uniform convergence of $F(X, Y)$ coincides with the uniformity induced by the metric of $F(X, Y)$ defined by*

---

[1] J. L. Kelley [1] and A. H. Stone [4] investigated conditions insuring that $F(X, Y)$ with the compact open topology be compact. The latter also considered conditions for normality. Generally speaking, considerably strong conditions are required for normality or compactness of $F(X, Y)$. For example, denoting by I the unit segment, A. H. Stone proved that Y^I is not normal if Y is the product of uncountably many unit segments and that Y^I is compact if and only if Y is compact and contains no arc, and that the only space Y for which Y^X is compact for every compact metric space X, is the space with at most one point.

[2] A mapping f of X into Y is called bounded if $\rho'(f(p), f(p_0))$ is bounded for some definite point $p_0 \in R$.

$$\rho(f, g) = \sup\{\rho'(f(p), g(p)) \mid p \in X\},$$

where ρ' denotes the metric of Y.

The following proposition is of extensive use.

H) *Let $C(X, Y)$ be the uniform space with the uniformity of uniform convergence consisting of all continuous mappings of a topological space X into a uniform space Y. If Y is complete, then $C(X, Y)$ is also complete.*

Proof. Let $\mathscr{F}$ be a Cauchy filter of $C(X, Y)$. Suppose p is a given point of X. For each member $F \in \mathscr{F}$, we define a subset F_p of Y by

$$F_p = \{f(p) \mid f \in F\}. \tag{1}$$

Then by the definition of uniformity of uniform convergence,

$$\mathscr{F}_p = \{A \mid A \subset Y, A \supset F_p \text{ for some } F \in \mathscr{F}\} \tag{2}$$

forms a Cauchy filter of Y. Hence, by virtue of the completeness of Y, $\mathscr{F}_p \to f(p)$ for some point $f(p)$ of Y. Now we can assert that f is a continuous mapping of X into Y. Given $p_0 \in X$ and a nbd $V(f(p_0))$ of $f(p_0)$ in X, we choose $\alpha \in A$ for which

$$S(f(p_0), \mathscr{U}_\alpha) \subset V(f(p_0)).$$

Take β such that $\mathscr{U}_\beta^{**} < \mathscr{U}_\alpha$.
Since $\mathscr{F}$ is a Cauchy filter of $C(X, Y)$, there is some $U \in \mathscr{U}(\beta)$ such that

$$U \in \mathscr{F} \quad \text{(see Definition VII.3)}. \tag{3}$$

Since $\mathscr{F}_{p_0} \to f(p_0)$ and $U_{p_0} \in \mathscr{F}_{p_0}$ (see (1), (2)),

$$f(p_0) \in \bar{U}_{p_0},$$

and hence

$$f(p_0) \in S(U_{p_0}, \mathscr{U}_\beta) \quad \text{in } Y.$$

Thus by (1) there is some $h \in U$ such that

$$f(p_0) \in S(h(p_0), \mathcal{U}_\beta). \tag{4}$$

Since h is continuous over X, there is a nbd $U(p_0)$ of p_0 such that $p \in U(p_0)$ implies

$$h(p) \in S(h(p_0), \mathcal{U}_\beta). \tag{5}$$

Suppose p is a given point of $U(p_0)$. Then since $\mathcal{F}_p \to f(p)$,

$$f(p) \in \bar{U}_p \quad \text{in } Y \text{ (see (2), (3))}.$$

Hence there is some $g \in U$ such that

$$f(p) \in S(g(p), \mathcal{U}_\beta) \quad \text{(see (1))}. \tag{6}$$

Note that it follows from $g, h \in U \in \mathcal{U}(\beta)$ that

$$g(p) \in S^2(h(p), \mathcal{U}_\beta) \quad \text{(see Definition VII.3)}. \tag{7}$$

Thus combining (6), (7), (5) and (4), we get

$$f(p) \in S^5(f(p_0), \mathcal{U}_\beta) \subset S(f(p_0), \mathcal{U}_\beta^{**}) \subset S(f(p_0), \mathcal{U}_\alpha) \subset V(f(p_0)),$$

i.e.,

$$f(U(p_0)) \subset V(f(p_0)).$$

This means that f is continuous over X, i.e.,

$$f \in C(X, Y).$$

To see that $\mathcal{F} \to f$, we consider a given nbd W of f in $C(X, Y)$. Then

$$S(f, \mathcal{U}(\alpha)) \subset W$$

for some α (see Definition VII.3). Select β such that $\mathcal{U}_\beta^* < \mathcal{U}_\alpha$. Since $\mathcal{F}$ is Cauchy, there is some $U \in \mathcal{F} \cap \mathcal{U}(\beta)$. Suppose

$$U = U(\beta, f_0);$$

then for each point p of X,

$U_p \subset S(f_0(p), \mathcal{U}_\beta)$ in Y (see (1)) .

On the other hand, since $\mathcal{F}_p \to f(p)$ by the definition of f,

$$S(f(p), \mathcal{U}_\beta) \cap U_p \neq \emptyset \quad \text{in } Y$$

(see (1) and (2)). Therefore

$$U_p \subset S^3(f(p), \mathcal{U}_\beta) .$$

This implies that

$$U_p \subset S(f(p), \mathcal{U}_\alpha) \quad \text{for every } p \in X,$$

because $\mathcal{U}_\beta^* < \mathcal{U}_\alpha$. Therefore for every $f' \in U$, we obtain

$$f'(p) \in S(f(p), \mathcal{U}_\alpha), \quad \text{for every } p \in X,$$

i.e. $f' \in U(\alpha, f)$. Namely, $U \subset U(\alpha, f) \in \mathcal{U}(\alpha)$; hence

$$U \subset S(f, \mathcal{U}(\alpha)) \subset W .$$

This proves that $\mathcal{F} \to f$ which completes the proof of the proposition.

I) *Let $F(X, Y)$ be a mapping space of continuous mappings of a topological space X into a uniform space Y. Then in $F(X, Y)$ the topology of uniform convergence is stronger than the compact open topology. If, moreover, X is compact, then they coincide.*

Proof. Let C be a compact set of X and U an open set of Y. To prove the first half of the proposition, it suffices to show that $U(C; U)$ is open with respect to the topology of the uniform convergence. Suppose

$$f \in U(C; U) \subset F(X, Y),$$

i.e. $f(C) \subset U$. We can assert that

$$S(f(C), \mathcal{U}_\alpha) \subset U \tag{1}$$

for some uniform covering $\mathcal{U}_\alpha$ of U. (We denote by $\{\mathcal{U}_\alpha \mid \alpha \in A\}$ the uniformity of Y.) For, if not, then for every $\alpha \in A$ we can choose points $\varphi(\alpha)$ and $\psi(\alpha)$ such that

$$\varphi(\alpha) \in f(C), \qquad \psi(\alpha) \in Y - U \quad \text{and} \quad \psi(\alpha) \in S(\varphi(\alpha), \mathcal{U}_\alpha).$$

Since $f(C)$ is compact, the net $\varphi(A \mid >)$ has a cluster point $q \in f(C)$. Then we can easily show that q is also a cluster point of $\psi(A \mid >)$. Hence $q \in \overline{Y - U} = Y - U$, which is a contradiction. Thus (1) is proved. Now, suppose that g is a member of $F(X, Y)$ such that

$$g \in S(f, \mathcal{U}(\beta)) \quad \text{(see Definition VII.3)},$$

where $\beta \in A$ is chosen to satisfy $\mathcal{U}_\beta^* < \mathcal{U}_\alpha$. Then

$$g(p) \in S(f(p), \mathcal{U}_\alpha)$$

for every point p, especially for every point of C. Therefore from (1) it follows that

$$g(C) \subset S(f(C), \mathcal{U}_\alpha) \subset U,$$

i.e. $g \in U(C; U)$. Thus we have proved that

$$S(f, \mathcal{U}(\beta)) \subset U(C; U).$$

This means that $U(C; U)$ is an open set for the topology of uniform convergence. Hence every open set of $F(X, Y)$ for the compact open topology is also open for the topology of uniform convergence, which means that the latter topology is stronger than the former.

Now, let us suppose that X is compact, and f is a member of $F(X, Y)$. To complete our proof it suffices to show that for every $\alpha \in A$, $S(f, \mathcal{U}(\alpha))$ is also a nbd of f for the compact open topology. Since $f(X)$ is compact, we can cover it with a finite number of members of $\mathcal{U}_\alpha$, say $U_1, \ldots, U_k$. Since $f(X)$ is normal, we can construct a closed covering $\{G_1, \ldots, G_k\}$ of $f(X)$ such that $G_i \subset U_i$, $i = 1, \ldots, k$. Put $f^{-1}(G_i) = C_i$; then C_i, $i = 1, \ldots, k$, are closed, and therefore are compact sets of X satisfying

$$f(C_i) \subset U_i \quad \text{and} \quad \bigcup_{i=1}^{k} C_i = X.$$

Therefore $U(C_1 \ldots C_k; U_1 \ldots U_k)$ is an open nbd of f. To prove that this nbd is contained in $S(f, \mathcal{U}(\alpha))$, we take a given point $g \in U(C_1 \ldots C_k; U_1 \ldots U_k)$. Suppose that p is an arbitrary point of X; then $p \in C_i$ for some i. Hence $g(p) \in U_i$; thus it follows from $f(p) \in U_i \in \mathcal{U}_\alpha$ that

$$g(p) \in S(f(p), \mathcal{U}_\alpha).$$

Thus

$$g \in S(f, \mathcal{U}(\alpha)),$$

i.e.,

$$U(C_1 \ldots C_k; U_1 \ldots U_k) \subset S(f, \mathcal{U}(\alpha)),$$

proving that $S(f, \mathcal{U}(\alpha))$ is a nbd of f. This ends the proof of the proposition.[1]

Example VII.3. We can modify the definition of the topology of uniform convergence for $F(X, Y)$ as follows:

Let $\mathcal{C}$ be a collection of subsets of X. Suppose $C \in \mathcal{C}$ and $f \in F(X, Y)$. Then for each uniform covering $\mathcal{U}_\alpha$ of Y, we put

$$U(C, \alpha, f) = \{g \mid g \in F(X, Y), g(p) \in S(f(p), \mathcal{U}_\alpha) \text{ for every } p \in C\},$$

$$\mathcal{U}(C, \alpha) = \{U(C, \alpha, f) \mid f \in F(X, Y)\}.$$

Then $\{\mathcal{U}(C, \alpha) \mid C \in \mathcal{C}, \alpha \in A\}$ defines for $F(X, Y)$ the *uniformity* (and the *topology*) *of uniform convergence with respect to* $\mathcal{C}$.

J) *If $\mathcal{C}$ is the collection of all compact sets of X and $F(X, Y)$ is a collection of continuous mappings, then the topology of uniform convergence with respect to $\mathcal{C}$ coincides with the compact open topology.*

Let $F(X, Y)$ be a collection of continuous mappings of a topological space X into a uniform space Y with uniformity $\{\mathcal{U}_\alpha \mid \alpha \in A\}$. If for each $\alpha \in A$ and for each $p \in X$, there is a nbd $N(p)$ of p such that

---

[1] Generally, a topology of $F(X, Y)$ is called *joint continuous* if the mapping P of $F(X, Y) \times X$ into Y defined by

$$P(f, p) = f(p)$$

is continuous. The topology of uniform convergence is an example of joint continuous topology. R. H. Fox [1] studied relations between joint continuity and the compact open topology. See J. L. Kelley [1], too.

$$f(N(p)) \subset S(f(p), \mathcal{U}_\alpha)$$

for every $f \in F(X, Y)$, then $F(X, Y)$ is called *equicontinuous*.

K) *Let $F(X, Y)$ be a collection of continuous mappings of a compact space X into a uniform space Y. If $F(X, Y)$ is equicontinuous, then the weak topology coincides with the topology of uniform convergence.*

Proof. Since by E) and I) the topology of uniform convergence is stronger than the weak topology, we need only prove that the converse is also true. Suppose $\{\mathcal{U}_\alpha \mid \alpha \in A\}$ is the uniformity of Y. Let $f \in F(X, Y)$ and V be a given nbd of f with respect to the topology of uniform convergence. Then we can assert that V is also a nbd of f with respect to the weak topology. First we take $\alpha \in A$ for which

$$S(f, \mathcal{U}(\alpha)) \subset V.$$

Take $\beta \in A$ such that $\mathcal{U}_\beta^* < \mathcal{U}_\alpha$. Since $F(X, Y)$ is equicontinuous, we can assign an open nbd $U(p)$ to every $p \in X$ such that

$$g(U(p)) \subset S(g(p), \mathcal{U}_\beta) \quad \text{for all } g \in F(X, Y). \tag{1}$$

Since X is compact, we can cover X with a finite number of the $U(p)$, say $U(p_1), \ldots, U(p_k)$. We consider an open nbd

$$U_0(f) = U(p_1 \ldots p_k ; S(f(p_1), \mathcal{U}_\beta) \ldots S(f(p_k), \mathcal{U}_\beta); f) \tag{2}$$

of f with respect to the weak topology (see Definition VII.1). Suppose $g \in U_0(f)$; then for each $p \in X$, there is i for which $p \in U(p_i)$. It follows from the definition (1) of $U(p_i)$ that

$$g(p) \in S(g(p_i), \mathcal{U}_\beta) \tag{3}$$

and

$$f(p) \in S(f(p_i), \mathcal{U}_\beta). \tag{4}$$

On the other hand, the definition (2) of $U_0(f)$ implies

$$g(p_i) \in S(f(p_i), \mathcal{U}_\beta). \tag{5}$$

Combining (3), (4) and (5) we get

$$g(p) \in S(f(p), \mathcal{U}_\beta^*) \, .$$

This implies

$$g(p) \in S(f(p), \mathcal{U}_\alpha)$$

because $\mathcal{U}_\beta^* < \mathcal{U}_\alpha$. Since this is valid for all $p \in X$, we conclude

$$g \in S(f, \mathcal{U}(\alpha)) \, ,$$

which means that

$$U_0(f) \subset S(f, \mathcal{U}(\alpha)) \subset V \, .$$

Therefore V is also a nbd of f with respect to the weak topology. Thus the weak topology is stronger than the topology of uniform convergence, and consequently they coincide.

Theorem VII.1. *Let $C(X, Y)$ be the space of all continuous mappings of a compact space X into a compact uniform space Y. We assume that $C(X, Y)$ has the topology of uniform convergence. Then a subset C' of $C(X, Y)$ is compact if and only if C' is equicontinuous and closed in $C(X, Y)$.*

Proof. Assume that C' is equicontinuous and closed in $C(X, Y)$ with the topology of uniform convergence. First we note that by K) the topology of C' coincides with the weak topology. Let us show that with respect to the weak topology, C' is closed in the space Y^X of all mappings of X into Y. Suppose $f \in Y^X - C'$. If $f \in C(X, Y)$, then by the hypothesis $f \notin \bar{C}'$ in $C(X, Y)$ for the topology of uniform convergence. To prove that $f \notin \bar{C}'$ is true for the weak topology of $C(X, Y)$, We assume the contrary: $f \in \bar{C}'$ in $C(X, Y)$. Then for a given $\alpha \in A$, using the equicontinuity of C' we take $\beta \in A$ with $\mathcal{U}_\beta^* < \mathcal{U}_\alpha$ and assign an open nbd $U(p)$ to each $p \in X$ such that

$$g(U(p)) \subset S(g(p), \mathcal{U}_\beta) \quad \text{for all } g \in C' \, .$$

We cover the compact space X with a finite number of the $U(p)$, say $U(p_1), \ldots, U(p_k)$. Putting

$$U_0(f) = U(p_1 \ldots p_k ; S(f(p_1), \mathcal{U}_\beta) \ldots S(f(p_k), \mathcal{U}_\beta); f)$$

we obtain a nbd of f with respect to the weak topology. It follows from our assumption that

$$U_0(f) \cap C' \neq \emptyset.$$

Take $g \in U_0(f) \cap C'$; then for each $p \in X$ we choose i with $p \in U(p_i)$. From the definitions of $U_0(f)$ and $U(p_i)$ we obtain

$$g(p_i) \in S(f(p_i), \mathcal{U}_\beta), \qquad g(p) \in S(g(p_i), \mathcal{U}_\beta)$$

and

$$f(p) \in S(f(p_i), \mathcal{U}_\beta).$$

Therefore

$$g(p) \in S(f(p), \mathcal{U}_\alpha)$$

follows from $\mathcal{U}_\beta^* < \mathcal{U}_\alpha$. Thus

$$g \in S(f, \mathcal{U}(\alpha)) \cap C',$$

i.e.,

$$S(f, \mathcal{U}(\alpha)) \cap C' \neq \emptyset,$$

where $\alpha \in A$ is arbitrary. This contradicts the fact that $f \notin \bar{C}'$ for the topology of uniform convergence. Thus $f \notin \bar{C}'$ for the weak topology in $C(X, Y)$ and therefore in Y^X, too. (Remember that we assumed $f \in C(X, Y)$.)

If $f \notin C(X, Y)$, then there exists $p \in X$ and a nbd V of $f(p)$ in Y such that $f^{-1}(V)$ is not a nbd of p in X. We take a nbd W of $f(p)$ such that

$$S^2(W, \mathcal{U}_\alpha) \subset V \tag{1}$$

for some uniform covering $\mathcal{U}_\alpha$ of Y. Then since C' is equicontinuous, there is a nbd U of p such that

$$g(U) \subset S(g(p), \mathcal{U}_\alpha) \quad \text{for every } g \in C'. \tag{2}$$

Since $U \not\subset f^{-1}(V)$, we can choose a point

$$p' \in U - f^{-1}(V).$$

Then

$$f(p') \in Y - V \subset S(Y - V, \mathcal{U}_\alpha),$$

and hence $U(p, p'; W, V'; f)$, where $V' = S(Y - V, \mathcal{U}_\alpha)$, is a nbd of f in Y^X for the weak topology. Given $g \in U(p, p'; W, V'; f)$, then $g(p) \in W$ and $g(p') \in V'$. Note that

$$S(g(p), \mathcal{U}_\alpha) \cap V' = \emptyset$$

follows from (1). Therefore,

$$g(p') \notin S(g(p), \mathcal{U}_\alpha),$$

which combined with (2) implies $g \notin C'$. Thus

$$U(p, p'; W, V'; f) \cap C' = \emptyset,$$

which means $f \notin \bar{C}'$ in Y^X.

Thus C' is closed in Y^X with respect to the weak topology. Therefore it follows from C) that C' is compact for the weak topology and, consequently, for the topology of uniform convergence, too.

Conversely, if C' is compact, then it must be closed in $C(X, Y)$. To prove that C' is equicontinuous, we take an arbitrary $\alpha \in A$. Choose $\beta \in A$ for which $\mathcal{U}_\beta^* < \mathcal{U}_\alpha$. Since C' is compact, we can cover it with finitely many of the open sets

$$U(\beta, f) = \{g \mid g \in C', g(p) \in S(f(p), \mathcal{U}_\beta) \text{ for all } p \in X\}, \qquad (3)$$

say $U(\beta, f_1), \ldots, U(\beta, f_k)$, where $f_i \in C'$, $i = 1, \ldots, k$. To each $p \in X$, using the continuity of f_i, we assign a nbd $U(p)$ such that

$$f_i(U(p)) \subset S(f_i(p), \mathcal{U}_\beta), \quad i = 1, \ldots, k. \qquad (4)$$

Suppose $g \in C'$ and $p' \in U(p)$; then $g \in U(\beta, f_i)$ for some i. By (3) and (4) this implies that

$$g(p') \in S(f_i(p'), \mathcal{U}_\beta), \qquad f_i(p') \in S(f_i(p), \mathcal{U}_\beta)$$

and

$$g(p) \in S(f_i(p), \mathcal{U}_\beta).$$

Thus it follows from $\mathcal{U}_\beta^* < \mathcal{U}_\alpha$ that

$$g(p') \in S(g(p), \mathcal{U}_\alpha),$$

which means that

$$g(U(p)) \subset S(g(p), \mathcal{U}_\alpha) \quad \text{for every } g \in C'.$$

Therefore C' is equicontinuous.

Corollary (Ascoli–Arzela's theorem). *Let $C(X, Y)$ be the collection of all continuous mappings of a compact space X into a compact metric space Y. Suppose $C(X, Y)$ has the topology of uniform convergence. If a subset F of $C(X, Y)$ is equicontinuous, then every point sequence of F contains a subsequence which converges in $C(X, Y)$.*

Proof. It is easily seen that $\bar{F}$ is equicontinuous and closed in $C(X, Y)$. Therefore, by Theorem VII.1, $\bar{F}$ is compact. Since $C(X, Y)$ is a metric space, $\bar{F}$ is also sequentially compact. Hence every point sequence of F contains a subsequence which converges in $\bar{F}$.

Finally, we shall deal with the problem of approximating a given continuous function over a T_2-space by functions belonging to a ring of continuous functions.

L) *Let X be a T_2-space and $C(X)$ the ring of all real-valued continuous functions over X. We consider $C(X)$ to be a ring with real scalar multiplication. Suppose D is a subring of $C(X)$, i.e., $f + g$, fg, $\alpha f \in D$ for every f, $g \in D$ and every real number α. Moreover, we assume that D satisfies:*
   (i) *$1 \in D$,*
   (ii) *for every pair of distinct points p, q of X, there is some $g \in D$ for which $g(p) \neq g(q)$.*
   We let $C(X)$ have the compact open topology. Then for every disjoint compact sets F, G of X, there is some $\varphi \in \bar{D}$ such that

$$\varphi(F) = 0, \quad \varphi(G) = 1 \quad \text{and} \quad 0 \leq \varphi \leq 1.$$

Proof. For each pair of points $p \in F$, $q \in G$, we choose $g \in D$ satisfying $g(p) \neq g(q)$. Define $h \in C(X)$ by

$$h(p) = \left(\frac{g(x) - g(q)}{g(p) - g(q)}\right)^2, \quad x \in X.$$

Then $h \in D$ and h satisfies

$$h(p) = 1, \qquad h(q) = 0 \quad \text{and} \quad h \geq 0.$$

Hence, putting

$$U(p) = \{x \in X \mid h(x) > \tfrac{1}{2}\},$$

we get an open nbd of p. We let p run through F while fixing q. Then we obtain an open covering $\{U(p) \mid p \in F\}$ of F. Since F is a compact set, we can cover it with a finite number of $U(p)$, say $U(p_1), \ldots, U(p_n)$. We denote by h_i, $i = 1, \ldots, n$, the continuous functions defined in the above for $p_1, \ldots, p_n$, respectively, i.e.,

$$h_i(p_i) = 1, \qquad h_i(q) = 0 \quad \text{and} \quad h_i \geq 0, \quad i = 1, \ldots, n.$$

Put

$$h'_q(x) = \sum_{i=1}^{n} h_i(x), \quad x \in X;$$

then $h'_q \in D$ and it satisfies

$$h'_q(q) = 0, \qquad h'_q(F) > \tfrac{1}{2} \quad \text{and} \quad h'_q \geq 0.$$

We define a real-valued continuous function $u(y)$ of a real variable y by

$$u(y) = 2 \min\{y, \tfrac{1}{2}\}.$$

Then we put

$$u_q(x) = u(h'_q(x)), \quad x \in X.$$

Then u_q is a continuous function over X satisfying

$$u_q(q) = 0, \qquad u_q(F) = 1 \quad \text{and} \quad 0 \leq u_q \leq 1.$$

On the other hand, by the well-known Weierstrass' approximation theorem, for every $\varepsilon > 0$ and every compact set K of X, there is a polynomial $P(y)$ such that

$$|P(y) - u(y)| < \varepsilon \quad \text{whenever } 0 \le y \le k,$$

where $k = \max\{h'_q(x) \mid \in K\}$. Therefore

$$|P(h'_q(x)) - u_q(x)| < \varepsilon, \quad x \in K.$$

Since $P(h'_q) \in D$ by J) we obtain $u_q \in \bar{D}$. Now, we note that $\bar{D}$ is also a subring of $C(X)$ because, as easily seen, the operations $f + g$, fg, αf are continuous. Putting $u'_q = 1 - u_q$, we obtain an element u'_q of $\bar{D}$ satisfying

$$u'_q(q) = 1, \qquad u'_q(F) = 0 \quad \text{and} \quad 0 \le u'_q \le 1.$$

Furthermore, we put

$$V(q) = \{x \in X \mid u'_q(x) > \tfrac{1}{2}\}.$$

Then $V(q)$ is an open nbd of q. Since

$$\bigcup \{V(q) \mid q \in G\} \supset G$$

and G is compact, we can cover G with a finite number of $V(q)$, say $V(q_1), \ldots, V(q_m)$. Put

$$u'(x) = \sum_{i=1}^{m} u'_{q_i}(x), \quad x \in X.$$

Then

$$u' \in \bar{D}, \qquad u'(F) = 0, \qquad u'(G) > \tfrac{1}{2}, \qquad 0 \le u' \le m.$$

We again use the real-valued continuous function u in the above argument and put

$$\varphi(x) = u(u'(x)), \quad x \in X.$$

Then

$$\varphi(F) = 0, \qquad \varphi(G) = 1 \quad \text{and} \quad 0 \le \varphi \le 1.$$

Using Weierstrass' approximation theorem, for every $\varepsilon > 0$ we can find a polynomial P' such that

$$|P'(u'(x)) - \varphi(x)| < \varepsilon, \quad x \in X.$$

Since $P'(u') \in \bar{D}$, by J) this implies that $\varphi \in \bar{\bar{D}} = \bar{D}$. Thus the proposition is proved.

From proposition L) we can derive the following useful theorem which is an extension of Weierstrass' approximation theorem in analysis.

Theorem VII.2 (M. H. Stone–Weierstrass' approximation theorem).[1] *Let X be a T_2-space and D a ring of continuous functions with real scalar multiplication satisfying the conditions in L). Then D is dense in the mapping space $C(X)$ of all continuous functions over X with the compact open topology.*

Proof. By virtue of J) it suffices to prove that for a given $f \in C(X)$ and for every compact set of X and $\varepsilon > 0$ there is $\varphi \in \bar{D}$ satisfying $|\varphi(x) - f(x)| < \varepsilon$ for all $x \in K$. Since f is bounded on K, we assume $|f(x)| \leq \alpha$ for all $x \in K$. Then we choose $\alpha_1, \ldots, \alpha_n$ such that

$$-\alpha = \alpha_1 < \alpha_2 < \cdots < \alpha_n = \alpha$$

and

$$\alpha_i - \alpha_{i-1} < \varepsilon, \quad i = 2, \ldots, n.$$

Put

$$F_i = \{x \mid x \in K \text{ and } f(x) \geq \alpha_i\}$$

and

$$G_i = \{x \mid x \in K \text{ and } f(x) \leq \alpha_{i-1}\}.$$

Using L) we construct continuous functions $\varphi_i \in \bar{D}$ such that

$$\varphi_i(F_i) = 1, \quad \varphi_i(G_i) = 0 \quad \text{and} \quad 0 \leq \varphi_i \leq 1.$$

Put

$$\varphi(x) = \sum_{i=2}^{n} (\alpha_i - \alpha_{i-1})\varphi_i(x) + \alpha_1, \quad x \in X.$$

Then φ is the desired function. Because, if $p \in K$, and

$$\alpha_{j-1} \leq f(p) \leq \alpha_j,$$

then

$$p \in F_i, \quad i = 2, \ldots, j-1,$$

$$p \in G_i, \quad i = j+1, \ldots, n.$$

---

[1] The present version of the proof is essentially due to H. Nakano's proof of the following corollary, which was first proved by M. H. Stone [1]. There are various other generalizations of Weierstrass' theorem; see for example M. Krein and S. Krein [1] and E. Hewitt [2].

Therefore

$$\varphi(p) \geqslant \sum_{i=2}^{j-1} (\alpha_i - \alpha_{i-1})\varphi_i(p) + \alpha_1 = \sum_{i=2}^{j-1} (\alpha_i - \alpha_{i-1}) + \alpha_1 = \alpha_{j-1},$$

$$\varphi(p) = \sum_{i=2}^{j} (\alpha_i - \alpha_{i-1})\varphi_i(p) + \sum_{i=j+1}^{n} (\alpha_i - \alpha_{i-1})\varphi_i(p) + \alpha_1$$

$$\leqslant \sum_{i=2}^{j} (\alpha_i - \alpha_{i-1}) + \alpha_1 = \alpha_j.$$

Thus

$$|\varphi(p) - f(p)| \leqslant \alpha_j - \alpha_{j-1} < \varepsilon.$$

Since $\varphi \in \bar{D}$ is obvious, φ satisfies our requirement. Thus the theorem is proved.

Corollary. *Let X be a compact T_2-space and D a ring of continuous functions with real scalar multiplication satisfying the conditions in* L). *Then for every continuous function f over X and for every $\varepsilon > 0$, there is a member ψ of D for which*

$$|\psi(p) - f(p)| < \varepsilon \quad \text{for all } p \in X.$$

Proof. Combine the theorem with G) and I).

Example VII.4. Let F be a zero set of $X = I^T = \prod \{I_\alpha \mid \alpha \in T\}$, the product of closed segments I_α, $\alpha \in T$. For each finite subset A of T we put $I^A = \prod \{I_\alpha \mid \alpha \in A\}$. Denote by $C_A(X)$ the subset of $C(X)$ consisting of functions which depend only on A (namely, functions which can be expressed as $\varphi \circ \pi_A^T$ for the projection π_A^T from I^T onto I^A and a function φ defined on I^A). Now put

$$D = \bigcup \{C_A(X) \mid A \text{ is a finite subset of } T\}.$$

Then D satisfies the conditions in L), and hence by the above corollary D is dense in $C(X)$ with the topology of uniform convergence. Suppose $F = \{x \in X \mid f(x) = 0\}$, where $f \in C(X)$. For each $n \in N$ there is $f_n \in D$ such that

$$\sup\{|f(x) - f_n(x)| \mid x \in X\} < 1/n.$$

Assume $f_n \in C_{A_n}(X)$, $n = 1, 2, \ldots$. Then it is obvious that f depends only on $A' = \bigcup_{n=1}^{\infty} A_n$. Hence F depends on at most countably many coordinates for $\alpha \in A'$.

2. Metric space, paracompact space and continuous mapping

The purpose of the present section is to learn to what extent a continuous mapping satisfying certain conditions transfers properties of the domain space to the range space (or properties of the range space to the domain space). From a little different point of view we may ask, following P. S. Alexandroff:[1]

(a) which spaces can be represented as images of 'nice' spaces under 'nice' continuous mappings? and

(b) which spaces can be mapped onto 'nice' spaces under 'nice' continuous mappings?

We have already dealt with such problems in the previous chapters and applied a result of the study to prove normality of a product space. In fact, in the final section of Chapter VI, we used Corollary 2 to Theorem VI.14, which asserts that every metric space is the image, by a perfect map, of a subspace of Baire's zero-dimensional metric space. To accomplish a more detailed study of this subject, we shall adopt, in this section, chiefly metric spaces and paracompact spaces as our 'nice' spaces. To begin with, we give here some additional terminologies for mappings.

Let f be a mapping of a topological space X into a topological space Y. Suppose that $\mathcal{U}$ is an open covering of X. If there is an open covering $\mathcal{V}$ of Y such that $f^{-1}(\mathcal{V}) = \{f^{-1}(V) \mid V \in \mathcal{V}\}$ is a refinement of $\mathcal{U}$ in X, then f is called a $\mathcal{U}$-*mapping*. We often assume certain property of the inverse image $f^{-1}(q)$ of each point q of Y. Thus if $f^{-1}(q)$, for each point q of Y, is compact, countably compact, etc., then we call the mapping f *compact*, *countably compact*, etc., respectively. As we defined before, a compact, closed and continuous map is called a *perfect map*. We call a countably compact, closed and continuous map a *quasi-perfect map*.

---

[1] P. S. Alexandroff [4]. This paper also contains various results in this aspect of investigation. We should also point out that there are many conditions of maps and related results which we could not discuss in this book. A. V. Arhangelskii's extensive survey article [6] is recommended to readers who are interested in more details of this aspect. Also see E. Michael [9], F. Siwiec [1], and J. Nagata [7]. Some mathematicians studied multivalued mappings. For example, see V. Ponomarev [3], A. Okuyama [1], C. J. R. Borges [2], J. Nagata [8].

First, we shall deal with metric spaces and their images and inverse images by continuous mappings. We showed in Section 4 of Chapter VI that the image of a metric space by a certain type of closed continuous mapping is also metric. As for open continuous mappings, we obtain the following.

A) *A T_1-space X is the image of a metric space by an open continuous mapping if and only if X satisfies the 1st axiom of countability.*[1]

[1] This proposition is due to V. Ponomarev [1] and S. Hanai [1]. Ponomarev also proved that a topological space is the image of a metric space X by an open continuous mapping such that each $f^{-1}(q)$ satisfies the 2nd axiom of countability, if and only if X has a point-countable open basis. S. Hanai (loc. cit) and A. Arhangelskii [2] proved that if a collection-wise normal space Y is the image of a metric space X by a compact open continuous mapping, then Y is metrizable. Actually, the latter proved that the T_1-spaces with a uniform open basis, and only they, are open compact images of metric spaces. H. Wicke [2] characterized the regular spaces which are open continuous images of complete metric spaces. E. Michael [8], A. Arhangelskii [5] and S. Franklin [1] characterized the images of metric spaces by (continuous) bi-quotient, (continuous) pseudo-open, and quotient mappings, respectively. A mapping from X onto Y is *bi-quotient* if for each $y \in Y$ and each open collection $\mathcal{U}$ in X which covers $f^{-1}(y)$, there is a finite subcollection $\mathcal{U}'$ of $\mathcal{U}$ such that $\bigcup \{f(U) \mid U \in \mathcal{U}'\}$ is a nbd of y. Such mappings (invented by O. Hájek and extensively studied by E. Michael [8]) have some interesting properties; e.g., the product of bi-quotient mappings is bi-quotient while the same is not true for quotient mappings (where for a collection $\{f_\alpha \mid \alpha \in A\}$ of mappings from X_α into Y_α, the mapping $f(x) = \{f_\alpha(x_\alpha) \mid \alpha \in A\}$, $x = \{x_\alpha\}$ from $\prod_{\alpha \in A} X_\alpha$ into $\prod_{\alpha \in A} Y_\alpha$ is the product of the mappings), and a continuous mapping f from a space X onto a T_2-space Y is bi-quotient if and only if the product $f \times i_Z$ is a quotient mapping for every space Z, where i_Z is the identity mapping from Z onto Z. A mapping f from X onto Y is *pseudo-open* (due to A. Arhangelskii and Yu. Smirnov) if for each $y \in Y$ and each open set $U \supset f^{-1}(y)$, $f(U)$ is a nbd of y. Let f be a quotient mapping from X onto Y; then it is known that the restriction of f to $f^{-1}(Y')$, where Y' is a subset of Y, is not necessarily a quotient mapping. Namely, 'quotient' is not hereditary. But pseudo-open continuous mappings are known to be precisely hereditarily quotient mappings. N. S. Laśnev [2] characterized the images of metric spaces by closed continuous mappings. F. G. Slaughter [1] proved that such images are M_1-spaces. Interrelations between various (continuous) mappings are shown in the following diagram:

$$\text{open} \Rightarrow \text{bi-quotient} \Rightarrow \text{pseudo-open} \Rightarrow \text{quotient}$$
$$\Uparrow \qquad\qquad\qquad \Uparrow$$
$$\text{perfect} \quad \Rightarrow \quad \text{closed}$$

A collection $\mathcal{P}$ of subsets of a topological space X is called a *pseudo-base* of X if whenever $C \subset U$ with C compact and U open in X, then $C \subset P \subset U$ for some $P \in \mathcal{P}$. A regular space X is called an $\aleph_0$-*space* if it has a countable pseudo-base. E. Michael [7] defined $\aleph_0$-space and studied its properties. All separable metric spaces are $\aleph_0$. Every $\aleph_0$-space is Lindelöf separable and M_1. Michael characterized $\aleph_0$-spaces as the images of separable metric spaces by *compact-covering mappings*. A continuous mapping f from X into Y is called compact-covering if every compact set of Y is the image of a compact set of X. The same author also studied properties of the continuous images of separable metric spaces.

Proof. Since the necessity is clear, we shall prove only the sufficiency. Let $\{\mathcal{U}_\alpha \mid \alpha \in A\}$ be the collection of all open sets of a topological space X satisfying the 1st axiom of countability. Then using its index set A, we construct Baire's zero-dimensional space $N(A)$. We define a subset S of $N(A)$ by

$$S = \{(\alpha_1, \alpha_2, \ldots) \mid U_{\alpha_1}, U_{\alpha_2}, \ldots \text{ form a nbd basis of a point } p \text{ of } X\}.$$

Now we put

$$f(\alpha) = p \quad \text{if } \alpha = (\alpha_1, \alpha_2, \ldots) \text{ and } \{U_{\alpha_1}, U_{\alpha_2}, \ldots\} \text{ is a nbd basis of } p.$$

Since X is T_1, it is clear that this uniquely defines a mapping of S onto X.

To prove the continuity of f, we suppose that U is a given nbd of $f(\alpha) = p$, where $\alpha = (\alpha_1, \alpha_2, \ldots)$. Then

$$p \in U_{\alpha_i} \subset U \quad \text{for some } i.$$

If $\rho(\alpha, \alpha') < 1/i$ for α and a point $\alpha' = (\alpha_1', \alpha_2', \ldots)$ of S, then $\alpha_1' = \alpha_1, \ldots, \alpha_i' = \alpha_i$. Therefore

$$f(\alpha') \in \bigcap_{k=1}^{i} U_{\alpha_k} \subset U,$$

which proves that f is continuous.

Finally, to prove that f is open, it suffices to show that $f(S_{1/k}(\alpha))$ is open for every natural number k and $\alpha \in S$, because $\{S_{1/k}(\alpha) \mid k = 1, 2, \ldots;$ $\alpha \in S\}$ is an open basis of S. To this end we suppose $\alpha = (\alpha_1, \alpha_2, \ldots)$; then we can show that

$$f(S_{1/k}(\alpha)) = \bigcap_{i=1}^{k} U_{\alpha_i}.$$

Since

$$f(S_{1/k}(\alpha)) \subset \bigcap_{i=1}^{k} U_{\alpha_i}$$

is clear, to verify the inverse relation, we take a given point $p \in \bigcap_{i=1}^{k} U_{\alpha_i}$. We choose a sequence $U_{\beta_{k+1}}, U_{\beta_{k+2}}, \ldots$ of open sets such that $\{U_{\beta_j} \mid j = k+1, k+2, \ldots\}$ is a nbd basis of p. Then $\alpha' = (\alpha_1, \ldots, \alpha_k, \beta_{k+1}, \ldots)$ is a point of S for which $f(\alpha') = p$. Since $\rho(\alpha, \alpha') < 1/k$ by the definition of the metric of $N(A)$, $p \in f(S_{1/k}(\alpha))$. Thus $f(S_{1/k}(\alpha)) = \bigcap_{i=1}^{k} U_{\alpha_i}$ is proved. Since $\bigcap_{i=1}^{k} U_{\alpha_i}$ is open, we have proved that f is an open mapping.

In the following is a result of the problem (b).

B) *A T_2-space X is paracompact if and only if for every open covering $\mathcal{U}$ of X, X can be mapped onto a metric space X by a continuous $\mathcal{U}$-mapping.*[1]

Proof. Let f be a continuous $\mathcal{U}$-mapping of a topological space X onto a metric space Y. Then by the definition of $\mathcal{U}$-mapping there is an open covering $\mathcal{V}$ of Y such that

$$f^{-1}(\mathcal{V}) = \{f^{-1}(V) \mid V \in \mathcal{V}\} < \mathcal{U}.$$

Since Y is metric and therefore paracompact, there is a locally finite open refinement $\mathcal{V}'$ of $\mathcal{V}$. Then $f^{-1}(\mathcal{V}')$ is obviously a locally finite open refinement of $\mathcal{U}$. Therefore X is paracompact.

To prove the converse, let $\mathcal{U}$ be a given open covering of a paracompact T_2-space X. Then by III.2.C) there are locally finite open coverings $\mathcal{V}$ and $\mathcal{W}$ of X such that

$$\mathcal{V} < \mathcal{U}, \qquad \mathcal{V} = \{V_\alpha \mid \alpha \in A\},$$
$$\mathcal{W} = \{W_\alpha \mid \alpha \in A\} \quad \text{and} \quad \bar{W}_\alpha \subset V_\alpha.$$

For the index set A, we consider the generalized Hilbert space $H(A)$. For each $\alpha \in A$, we define a continuous function f_α over X such that

$$f_\alpha(W_\alpha) = 1, \qquad f_\alpha(X - V_\alpha) = 0 \quad \text{and} \quad 0 \leq f_\alpha \leq 1. \tag{1}$$

We put

$$f(p) = \{f_\alpha(p) \mid \alpha \in A\}, \quad p \in X.$$

Then it is clear that f is a continuous mapping of X onto a subspace Y of $H(A)$, because $\mathcal{V}$ is locally finite. To prove that f is a $\mathcal{U}$-mapping, we suppose that p is a given point of X; then $f(p) = \{f_\alpha(p) \mid \alpha \in A\}$. Now, $p \in W_{\alpha_0}$ for some member W_{α_0} of the covering $\mathcal{W}$; then for this index α_0 we can prove that

---

[1] Essentially proved by C. H. Dowker [2]. V. Ponomarev [4] obtained interesting results in this aspect.

$$f^{-1}(S_1(f(p))) \subset V_{\alpha_0}. \tag{2}$$

For if $p' \in f^{-1}(S_1(f(p)))$, then

$$\rho(f(p'), f(p)) < 1 \quad \text{in } Y.$$

Since $f_{\alpha_0}(p) = 1$ by the definition (1) of f_{α_0}, $f_{\alpha_0}(p') > 0$ follows from the above inequality. Therefore $p' \in V_{\alpha_0}$ by (1), i.e. (2) is proved. Since V_{α_0} is contained in some member of $\mathcal{U}$ (remember that $\mathcal{V} < \mathcal{U}$), we have proved that f is a $\mathcal{U}$-mapping.

We can extend Corollary 2 of Theorem VI.14 to paracompact spaces as follows.

C) *Let X be a T_2-space. Then there is a zero-dimensional paracompact T_2-space S and a perfect mapping f of S onto X if and only if X is paracompact.*[1]

Proof. Since the necessity of the condition is a direct consequence of Corollary 2 of Theorem V.3, we shall prove only the sufficiency. We denote by $\{\mathcal{F}_\lambda \mid \lambda \in \Lambda\}$ the totality of the locally finite closed coverings of the paracompact T_2-space X. Suppose

$$\mathcal{F}_\lambda = \{F_{\alpha'} \mid \alpha' \in A_\lambda\}, \quad \lambda \in \Lambda \ ;$$

then we construct the product space

$$P = \prod \{A_\lambda \mid \lambda \in \Lambda\}$$

of the discrete spaces A_λ. We define a subset S of P by

$$S = \{\alpha \mid \alpha = \{\alpha_\lambda \mid \lambda \in \Lambda\} \in P, \ \cap \{F_{\alpha_\lambda} \mid \lambda \in \Lambda\} \neq \emptyset\}.$$

It is clear that if $\alpha \in S$, then $\cap \{F_{\alpha_\lambda} \mid \lambda \in \Lambda\}$ is a point of X. Put, for every point $\alpha = \{\alpha_\lambda \mid \lambda \in \Lambda\}$ of S,

$$f(\alpha) = \cap \{F_{\alpha_\lambda} \mid \lambda \in \Lambda\} = p. \tag{1}$$

Then f is a mapping of S onto X. Let $\alpha = \{\alpha_\lambda \mid \lambda \in \Lambda\}$ be an arbitrary

---

[1] Due to K. Nagami [1]. V. Ponomarev [2] proved a similar theorem for normal spaces.

point of S and V a nbd of $f(\alpha) = p$. Then there is a binary (and accordingly locally finite) closed covering $\mathscr{F}_{\lambda_0}$ such that

$$S(p, \mathscr{F}_{\lambda_0}) = F_{\alpha_{\lambda_0}} \in \mathscr{F}_{\lambda_0} \quad \text{and} \quad F_{\alpha_{\lambda_0}} \subset V.$$
$$U(\alpha_{\lambda_0}) = \{\beta \mid \beta = \{\beta_\lambda \mid \lambda \in \Lambda\} \in S, \beta_{\lambda_0} = \alpha_{\lambda_0}\}$$

is a nbd of α such that $f(U(\alpha_{\lambda_0})) \subset V$. Hence f is continuous.

Let p be a given point of X. For each $\lambda \in \Lambda$,

$$B_\lambda = \{\alpha' \mid \alpha' \in A_\lambda, p \in F_{\alpha'}\}$$

is finite, i.e., B_λ is a compact subspace of A_λ. Since

$$f^{-1}(p) = \prod \{B_\lambda \mid \lambda \in \Lambda\},$$

$f^{-1}(p)$ is compact.

To see that f is closed, we consider a closed set G of S and a point p of X such that

$$p \notin f(G).$$

Then

$$f^{-1}(p) \cap G = \emptyset \quad \text{in } S.$$

We can assert that there are $\lambda_1, \ldots, \lambda_k \in \Lambda$ such that for every $\alpha = \{\alpha_\lambda \mid \lambda \in \Lambda\} \in f^{-1}(p)$ and $\beta = \{\beta_\lambda \mid \lambda \in \Lambda\} \in G$, $(\alpha_{\lambda_1}, \ldots, \alpha_{\lambda_k}) \neq (\beta_{\lambda_1}, \ldots, \beta_{\lambda_k})$. For, if not, then we denote by Δ the set of all finite subsets of Λ. We consider Δ as a directed set with respect to the usual inclusion relation. For each $\delta = (\lambda_1, \ldots, \lambda_k) \in \Delta$ we choose $\varphi(\delta) \in f^{-1}(p)$ and $\psi(\delta) \in G$ such that the coordinates for $\lambda_1, \ldots, \lambda_k$ of $\varphi(\delta)$ coincide with those of $\psi(\delta)$. Since $f^{-1}(p)$ is compact, the net $\varphi(\Delta \mid >)$ has a cluster point $\alpha \in f^{-1}(p)$. It is easily seen that α is also a cluster point of $\psi(\Delta \mid >)$. Thus $\alpha \in \bar{G}$, contradicting the fact that G is closed.

Now we choose $\lambda_1, \ldots, \lambda_k \in \Lambda$ satisfying the above condition. For each point $\beta = \{\beta_\lambda \mid \lambda \in \Lambda\}$ of G, we consider $F_{\beta_{\lambda_1}} \cap \cdots \cap F_{\beta_{\lambda_k}}$; then this closed set obviously contains $f(\beta)$ but not p. For if

$$p \in F_{\beta_{\lambda_1}} \cap \cdots \cap F_{\beta_{\lambda_k}},$$

then there is

$$\alpha = \{\alpha_\lambda | \lambda \in \Lambda\} \in f^{-1}(p)$$

such that

$$\alpha_{\lambda_i} = \beta_{\lambda_i}, \quad i = 1, \ldots, k,$$

which is impossible. Thus

$$\mathscr{F} = \wedge \{\mathscr{F}_{\lambda_i} \mid i = 1, \ldots, k\}$$

is a locally finite closed covering of X satisfying

$$f(G) \subset \cup \{F \mid p \notin F \in \mathscr{F}\}.$$

Hence

$$V = X - \cup \{F \mid p \notin F \in \mathscr{F}\}$$

is an open nbd of p which does not intersect $f(G)$. Thus $f(G)$ is a closed set of X, i.e. f is a closed mapping.

It follows from V.2.D) that S is paracompact T_2. Finally we must prove that S is a zero-dimensional space. Let $\mathscr{U}$ be a given finite open covering of S. Note that for each point $\alpha = \{\alpha_\lambda \mid \lambda \in \Lambda\}$ of S,

$$\{U(\alpha_{\lambda_1}) \cap \cdots \cap U(\alpha_{\lambda_k}) \mid \lambda_1, \ldots, \lambda_k \in \Delta, k = 1, 2, \ldots\}$$

forms a nbd basis consisting of open closed sets of S, where we put

$$U(\alpha_\lambda) = \{\beta \mid \beta = \{\beta_\lambda \mid \lambda \in \Lambda\} \in S, \beta_\lambda = \alpha_\lambda\}. \tag{2}$$

Therefore there is an open closed covering $\mathscr{V}$ with $\mathscr{V} < \mathscr{U}$. Since for each $p \in X$, $f^{-1}(p)$ is compact,

$$f^{-1}(p) \subset V_{p1} \cup \cdots \cup V_{pm(p)} = V_p \tag{3}$$

for a finite number of elements $V_{p1}, \ldots, V_{pm(p)}$ of $\mathscr{V}$. Putting

$$W(p) = X - f(S - V_p),$$

we get an open nbd $W(p)$ of p since f is a closed mapping. Since

$\{W(p) \mid p \in X\}$ is an open covering of the paracompact T_2-space X, there is an index $\lambda_0 \in \Lambda$ for which

$$\mathscr{F}_{\lambda_0} < \{W(p) \mid p \in X\}.$$

Then $\mathscr{U}' = \{U(\alpha_{\lambda_0}) \mid \alpha_{\lambda_0} \in A_{\lambda_0}\}$ is an open covering of S (see (1), (2)) such that

$$\mathscr{U}' < f^{-1}(\mathscr{F}_{\lambda_0}) = \{f^{-1}(F) \mid F \in \mathscr{F}_{\lambda_0}\}$$

and

$$\operatorname{ord} \mathscr{U}' \leq 1.$$

Thus

$$\mathscr{U}' < \{f^{-1}(W(p)) \mid p \in X\} < \{V_p \mid p \in X\}.$$

We well-order all the points of X and put

$$\{V_p \mid p \in X\} = \{V_p \mid 0 \leq p < \tau\},$$

where p is regarded as a variable ordinal number. Then we put

$$U_p = \bigcup\{U \mid U \in \mathscr{U}', U \subset V_p, U \not\subset V_q \text{ for every } q > p\}.$$

Then $\{U_p \mid p \in X\}$ is an open covering of X with order ≤ 1 satisfying $U_p \subset V_p$ for all p. Further we put

$$U_{pi} = U_p \cap \left(V_{pi} - \bigcup_{j=1}^{i-1} V_{pj}\right), \quad i = 1, \ldots, m(p) \text{ (see (3))}.$$

Then, since each V_{pj} is open and closed, U_{pi} are open and mutually disjoint satisfying

$$U_{pi} \subset V_{pi} \cap U_p,$$

$$U_p \subset \bigcup_{i=1}^{m(p)} U_{pi}.$$

Thus

$$\mathscr{U}'' = \{U_{pi} \mid i = 1, \ldots, m(p); p \in X\}$$

is an open refinement of $\mathscr{U}$ with order ≤ 1, because $V_{pi} \in \mathscr{V} < \mathscr{U}$. This proves that S is zero-dimensional.

In the following is a classical theorem of P. S. Alexandroff, who pioneered to represent spaces as images of zero-dimensional spaces.[1] The proof is similar to the previous proof.

D) *A T_2-space X with weight μ is compact if and only if there is a closed subset C of D^μ (see Theorem V.8), and a continuous mapping f of C onto X. (Note that such a closed set C is a zero-dimensional compact T_2-space.)*

Proof. The sufficiency of the condition is clear.

To prove the necessity, we assume that X is a compact T_2-space with weight μ and denote by $\{U_\lambda \mid \lambda \in \Lambda\}$ an open basis of X with $|\Lambda| = \mu$. We repeat the process in the proof of C), replacing the collection of all locally finite closed coverings there with the collection $\{\mathscr{F}_\lambda \mid \lambda \in \Lambda\}$ of binary closed coverings, where $\mathscr{F}_\lambda = \{\bar{U}_\lambda, X - U_\lambda\}$. Put $\mathscr{F}_\lambda = \{F_\alpha \mid \alpha \in A_\lambda\}$. Then we can prove in a similar way that there is a continuous mapping f of a subset C of $D^\mu = \prod\{A_\lambda \mid \lambda \in \Lambda\}$ onto X, where

$$C = \{\alpha \mid \alpha = \{\alpha_\lambda \mid \lambda \in \Lambda\} \in D^\mu, \; \cap\{F_{\alpha_\lambda} \mid \lambda \in \Lambda\} \neq \emptyset\}.$$

Now we have to prove that C is closed. To do so, suppose $\gamma = \{\gamma_\lambda \mid \lambda \in \Lambda\} \not\in C$ where $\gamma_\lambda \in A_\lambda, \lambda \in \Lambda$. Then by the definition of C,

$$\cap\{F_{\gamma_\lambda} \mid \lambda \in \Lambda\} = \emptyset \quad \text{in } X.$$

Since X is compact,

$$\cap\{F_{\gamma_{\lambda_i}} \mid i = 1, \ldots, k\} = \emptyset$$

for some $\lambda_1, \ldots, \lambda_k \in \Lambda$. Therefore, $U(\gamma_{\lambda_1}) \cap \cdots \cap U(\gamma_{\lambda_k})$ is a nbd of γ in D^μ which does not intersect C, where

$$U(\gamma_\lambda) = \{\gamma' \mid \gamma' = \{\gamma'_\lambda \mid \lambda \in \Lambda\} \in D^\mu, \; \gamma'_\lambda = \gamma_\lambda\}.$$

Hence C is closed.

---

[1] There is an interesting result concerning inverse image of a compact space. A. Gleason [1] proved that for any compact T_2-space X, there is an extremely disconnected, compact, T_2-space S and an irreducible continuous mapping ϕ from S onto X and moreover that such (S, ϕ) is unique in the sense that if there is another (S', ϕ') satisfying the same condition, then there is a topological mapping ψ from S' onto S satisfying $\phi' = \phi \circ \psi$. A continuous mapping ϕ from S onto X is *irreducible* if for any closed set $C \subsetneqq S$, $\phi(C) \neq X$. This uniquely determined space S is called the *absolute* of X. Theory of absolute was extended by V. I. Ponomarev [5], S. Iliadis [1], J. Flachsmeyer [2] and others to more general spaces. See also P. Alexandroff–V. Ponomarev [2], V. I. Ponomarev [6].

In the following is an answer by K. Morita [5] and A. V. Arhangelskii [4] to Alexandroff's problem in a very important case.

Theorem VII.3. *A T_1-space X is an M-space if and only if there is a quasi-perfect map f from X onto a metric space Y.*

Proof. Let f be a quasi-perfect map from X onto a metric space Y with a sequence $\{\mathcal{V}_i \mid i = 1, 2, \ldots\}$ of open covers satisfying the conditions of Axexandroff–Urysohn's metrization theorem. Then $\mathcal{U}_i = f^{-1}(\mathcal{V}_i)$, $i = 1, 2, \ldots$, form a normal sequence of open covers of X. It also satisfies the $w\Delta$-condition in Definition VI.11. To see it, assume $x_i \in S(x, \mathcal{U}_i)$, $i = 1, 2, \ldots$, and $\{x_i\}$ has no cluster point. If $x_i \in f^{-1}(f(x))$ for infinitely many i's, then $\{x_i\}$ has a cluster point, because f is countably compact. Otherwise $x_i \notin f^{-1}(f(x))$ for $i \geq i_0$. Then $F = \{x_i \mid i \geq i_0\}$ is closed in X while $f(F)$ is not, because $f(x_i) \to f(x)$ in Y and $f(x) \notin f(F)$. This is a contradiction. Hence $\{x_i\}$ must have a cluster point, proving that X is an M-space.

Conversely, assume that X is an M-space with a normal sequence $\{\mathcal{U}_i \mid i = 1, 2, \ldots\}$ of open covers satisfying $w\Delta$-condition. Define equivalence $\sim$ in X by

$$x \sim y \text{ if and only if } y \in \bigcap_{i=1}^{\infty} S(x, \mathcal{U}_i).$$

Then we denote by Y the quotient space $X/\sim$ and by f the quotient map from X onto Y. It is obvious that Y is a T_1-space and f is a countably compact continuous map. To see that f is closed, suppose F is a closed set of X and $x \notin f^{-1}(f(F))$. Put $f(x) = y$; then $x \in f^{-1}(y) \subset X - F$. Thus for some n $S(x, \mathcal{U}_n) \cap F = \emptyset$. Because otherwise there are $x_n \in S(x, \mathcal{U}_n) \cap F$, $n = 1, 2, \ldots$, and thus a cluster point x' of $\{x_n\}$, which satisfies $x' \in F \cap f^{-1}(y)$, a contradiction. Hence

$$S(x, \mathcal{U}_{n+1}) \cap f^{-1}(f(F)) = \emptyset.$$

Therefore $x \notin \overline{f^{-1}(f(F))}$. This proves that $f^{-1}(f(F))$ is a closed set of X. Hence $f(F)$ is a closed set of Y proving our claim. Hence f is a quasi-perfect map. Put

$$\mathcal{V}_n = \{Y - f(X - U) \mid U \in \mathcal{U}_n\}.$$

Then, as easily seen, $\{\mathcal{V}_n \mid n = 1, 2, \ldots\}$ is a sequence of open covers of Y

satisfying the conditions of Alexandroff–Urysohn's metrization theorem. Thus Y is a metrizable space.

Corollary 1. *A T_1-space X is paracompact and M if and only if there is a perfect map from X onto a metric space.*

Proof. If there is such a perfect map, then X is M by the theorem and also paracompact by V.2.D). Conversely, if X is paracompact and M, then there is a quasi-perfect map f from X onto a metric space Y. Since $f^{-1}(y)$ is paracompact and countably compact for each $y \in Y$, it is compact. Namely, f is a perfect map.

Corollary 2. *Let X_i, $i = 1, 2, \ldots$, be paracompact T_1 and M; then so is the product space $\prod_{i=1}^{\infty} X_i$.*

Proof. Note that the product of perfect maps is a perfect map; then this corollary follows from the previous one.

Corollary 3. *A T_2-space X is paracompact and M if and only if it is homeomorphic to a closed set of the product of a compact T_2-space and a metric space.[1]*

Proof. The sufficiency of the condition is obvious. Assume that X is paracompact T_2 and M. Then by Corollary 1 there is a perfect map f from X onto a metric space Y. Then we define a map φ from X into $\beta(X) \times Y$ by

$$\varphi(x) = (i(x), f(x)), \quad x \in X,$$

where i denotes the imbedding of X into $\beta(X)$. Then it is obvious that φ is a topological imbedding of X into $\beta(X) \times Y$. To prove that $\varphi(X)$ is closed in $\beta(X) \times Y$, let $(u, v) \in \beta(X) \times Y - \varphi(X)$. Then $u \notin f^{-1}(v)$ holds in $\beta(X)$. Since $f^{-1}(v)$ is compact, it is a closed set of $\beta(X)$. Hence there are open sets U and V of $\beta(X)$ such that

$$u \in U, \quad f^{-1}(v) \subset V, \quad U \cap V = \emptyset.$$

---

[1] Due to J. Nagata [11]. A. Kato [1] and D. Burke–E. v. Douwen [1] gave an example of an M-space which is not homeomorphic to a closed set of any product of a countably compact space and a metric space. See also A. Kato [2].

Put $W = Y - f(X - V)$. Then W is an open nbd of v in Y. Thus $U \times V$ is a nbd of (u, v) in $\beta(X) \times Y$ satisfying $U \times V \cap \varphi(X) = \emptyset$. This proves that $\varphi(X)$ is closed in $\beta(X) \times Y$.

Corollary 4. *Every paracompact T_2-M-space X is a P-space.*

Proof. By Corollary 2 the product of X with every metric space is normal. Hence this corollary follows from Theorem VI.24.

Corollary 5. *A Tychonoff space X is paracompact and Čech complete if and only if there is a perfect map f from X onto a complete metric space Y.*[1]

Proof. Assume that such f and Y exist. Then, by use of the method of the proof of Corollary 3 to Theorem VII.3, we can show that X is homeomorphic to a closed set of the product space $\beta(X) \times Y$. Since the last space is Čech complete and paracompact, so is X.

Conversely, assume that X is paracompact and Čech complete. Then by Theorem VII.3 there is a perfect map f from X onto a metric space Y. By Exercise VI.30 Y is Čech complete. Hence by Čech's theorem Y is completely metrizable.

Example VII.5. Sorgenfrey line S is a P-space. For, let $\{U(\alpha_1, \ldots, \alpha_i) \mid \alpha_1, \ldots, \alpha_i \in A; \ i = 1, 2, \ldots\}$ be an open collection of S satisfying $U(\alpha_1, \ldots, \alpha_i) \subset U(\alpha_1, \ldots, \alpha_{i+1})$. For each i we put $F(\alpha_1, \ldots, \alpha_i) =$ the closure of the set $\{x \mid x \in S, \ (x - 1/i, \ x + 1/i) \subset U(\alpha_1, \ldots, \alpha_i)$ or $(x - \varepsilon, x + \varepsilon) \not\subset U(\alpha_1, \ldots, \alpha_i)$ for every $\varepsilon > 0$ and $[x, x + 1/i) \subset U(\alpha_1, \ldots, \alpha_i)\}$. Then it is easy to see that $F(\alpha_1, \ldots, \alpha_i)$ satisfies (i), (ii) of Definition VI.7. But S is not an M-space. Because if it were, then $S \times S$ would be paracompact T_2 and M, which is not true.

Example VII.6. Let X be a Tychonoff space. Then we denote by $\mu(X)$ the completion of X with respect to its strongest uniformity. Suppose T is an arbitrary metric space and f a continuous map from X into T. Then, since $\mu(T) = T$, f can be extended to a continuous map $\mu(f): \mu(X) \to T$. Since X is dense and C-embedded in $\mu(X)$, $\mu(X) \subset \gamma(X) \subset \beta(X)$ follows. Also observe that $\mu(X)$ is the smallest Dieudonné complete space in which X is dense and C-embedded (C^*-embedded). Because if X is dense and C^*-embedded in a Dieudonné complete space Y, then $X \subset$

---

[1] Due to Z. Frolik [3].

$Y \subset \beta(Y) = \beta(X)$. Let φ be the topological imbedding of X into Y. Then $\mu(\varphi)$ maps $\mu(X)$ into $\mu(Y) = Y$. Since $\mu(\varphi)$ is easily seen to be a topological imbedding, $\mu(X) \subset Y$ follows. Thus $\mu(X)$ can be expressed as

$$\mu(X) = \{x \in \beta(X) \mid \text{every continuous map from } X \text{ into any}$$
$$\text{metric space } T \text{ can be continuously}$$
$$\text{extended to } x\}$$
$$= \bigcap \{\beta(f)^{-1}(T) \mid f \text{ is a continuous map from } X \text{ into a}$$
$$\text{metric space } T\},$$

where $\beta(f): \beta(X) \to \beta(T)$ denotes the continuous extension of f.
    Now assume that X is an M-space. Then it is easy to see that

$$\mu(X) = \bigcap \{\beta(f)^{-1}(T) \mid f \text{ is a quasi-perfect map from } X \text{ onto a}$$
$$\text{metric space } T\}.$$

Because, let f be a continuous map from X into a metric space T. Then there is a metric space T' and a quasi-perfect map g from X onto T' and a continuous map h from T' into T such that $f = h \circ g$. Then $\beta(g)^{-1}(T') \subset \beta(f)^{-1}(T)$ follows, and so does the above expression of $\mu(X)$.
    Suppose $f : X \to T$ and $g : X \to T'$ are onto quasi-perfect maps, where T and T' are metric spaces. If $f = h \circ g$ for a continuous map h from T' into T, then h is a perfect map. Hence it is easy to see that $\beta(h)^{-1}(T) = T'$, and accordingly

$$\beta(g)^{-1}(T') = \beta(f)^{-1}(T).$$

Thus $\mu(X) = \beta(f)^{-1}(T)$ for any metric space T and a quasi-perfect map f from X onto T. This implies that the restriction $\mu(f)$ of $\beta(f)$ to $\mu(X)$ is a perfect map. Thus X is a paracompact M-space.[1]

E) *Every normal M-space X is countably paracompact and collectionwise normal.*

Proof. By the theorem there is a quasi-perfect map f from X onto a metric space Y. Suppose $\{F_\alpha \mid \alpha \in A\}$ is a given locally finite closed collection in X. For each $y \in Y f^{-1}(y)$ is countably compact, and hence it

---

[1] We owe the concept and properties of $\mu(X)$ in the above to K. Morita [6]. See the same paper for further results on $\mu(X)$. $\mu(X)$ may be called the *paracompactification* of X when it is paracompact.

hits at most finitely many of F_α, $\alpha \in A$. Thus $\{f(F_\alpha) \mid \alpha \in A\}$ is point-finite. Since f is closed, $\{f(F_\alpha) \mid \alpha \in A\}$ is a closure-preserving and accordingly locally finite closed collection in Y. Hence there is a locally finite open collection $\{V_\alpha \mid \alpha \in A\}$ in Y such that $f(F_\alpha) \subset V_\alpha$. Now $\{f^{-1}(V_\alpha) \mid \alpha \in A\}$ is a locally finite open collection in X satisfying $F_\alpha \subset f^{-1}(V_\alpha)$. Thus the proposition follows from V.3.D).

F) *Let f be a quasi-perfect map from a normal M-space X onto Y. Then Y is an M-space.*[1]

Proof. Let $\{\mathcal{U}_i \mid 1, 2, \ldots\}$ be a normal sequence of open covers of X satisfying $w\Delta$-condition. By virtue of V.5.C) and VI.7.D) we may assume that each $\mathcal{U}_i$ is locally finite. Then by III.2.C) we can shrink $\mathcal{U}_i$ to a locally finite closed cover $\mathcal{F}_i$. Since $f^{-1}(y)$ is countably compact for each $y \in Y$, it intersects at most finitely many members of $\mathcal{F}_i$. Thus $\mathcal{G}_i = f(\mathcal{F}_i)$ is a point-finite closed cover of Y. Since $\mathcal{G}_i$ is closure-preserving because of the closedness of f, it is locally finite. Put

$$\mathcal{G}_i = \{G_\alpha \mid \alpha \in A_i\},$$
$$W_\alpha = Y - \cup \{G_\beta \mid \beta \in A_i, \; G_\beta \cap G_\alpha = \emptyset\}.$$

Note that Y is collectionwise normal and countably paracompact because of E) and Exercises V.5 and V.14. Thus by V.3.D) each G_α can be expanded to an open set P_α such that $P_\alpha \subset W_\alpha$ and such that $\mathcal{W}_i = \{W_\alpha \mid \alpha \in A_i\}$ is locally finite. Hence $\mathcal{W}_i$ is a normal open cover of Y such that $\mathcal{W}_i < \mathcal{G}_i^*$. We can show that $\{\mathcal{W}_i\}$ satisfies $w\Delta$-condition. Suppose $y_i \in S(y, \mathcal{W}_i)$, $i = 1, 2, \ldots$, where y is a fixed point of Y. Then there are $G_1^i, G_2^i, G_3^i \in \mathcal{G}_i$ and $y_1^i, y_2^i \in X$ such that

$$y, y_1^i \in G_1^i, \qquad y_1^i, y_2^i \in G_2^i, \qquad y_2^i, y_i \in G_3^i.$$

Thus there are $F_1^i, F_2^i, F_3^i \in \mathcal{F}_i$ such that

$$f^{-1}(y) \cap F_1^i \neq \emptyset, \qquad F_1^i \cap f^{-1}(y_1^i) \neq \emptyset, \qquad f^{-1}(y_1^i) \cap F_2^i \neq \emptyset,$$
$$F_2^i \cap f^{-1}(y_2^i) \neq \emptyset, \qquad f^{-1}(y_2^i) \cap F_3^i \neq \emptyset, \qquad F_3^i \cap f^{-1}(y_i) \neq \emptyset.$$

---

[1] Due to T. Ishii [3] and K. Morita [8]. V. Filippov [2] proved it in case that X is paracompact.

Pick $x_i \in f^{-1}(y_i)$, $i = 1, 2, \ldots$. Then we can prove, by use of quasi-perfectness of f and $w\Delta$-condition and normality of $\{\mathcal{U}_i\}$, that $\{x_i\}$ has a cluster point. Thus it follows that $\{y_i\}$ has a cluster point. The detail is left to the reader. Select a normal sequence $\{\mathcal{W}'_i\}$ of open covers in Y such that $\mathcal{W}'_i < \mathcal{W}_i$. Then this sequence also satisfies $w\Delta$-condition. Hence Y is an M-space.[1]

The following interesting theorem was first proved by N. S. Lašnev [1] for metric spaces and extended by V. Fillipov [1] to paracompact M-spaces and the present form is due to T. Ishii [4].[2]

Theorem VII.4. *Let f be a closed continuous map from a normal M-space X onto Y. Then Y can be decomposed as $Y = \bigcup_{n=0}^{\infty} Y_n$, where Y_n is a discrete (closed) subset of Y for each $n \geqslant 1$, and $f^{-1}(y)$ is countably compact for every $y \in Y_0$.*

Proof. Let $\{\mathcal{U}_n\}$ be a normal sequence of open covers of X satisfying $w\Delta$-condition. Defining equivalence $\sim$ in X by

$$x \sim y \text{ if and only if } y \in \bigcap_{n=1}^{\infty} S(x, \mathcal{U}_n),$$

we decompose X into the sum of disjoint equivalence classes, which are countably compact closed sets of X and denoted by F_α, $\alpha \in A$. Put $U_{\alpha n} = S(F_\alpha, \mathcal{U}_n)$, $n = 1, 2, \ldots$. Then $\{U_{\alpha n} \mid n = 1, 2, \ldots\}$ is an outer base of F_α. In the rest of the proof we mean by a sequence a point sequence, say $\{y_i\}$, such that $y_i \neq y_j$ whenever $i \neq j$. For each $n \in N$ we put

$$Y_n = \{y \in Y \mid \text{for each sequence } \{y_i\} \text{ in } Y, \text{ there is } \alpha \in A \text{ for}$$
$$\text{which } F_\alpha \cap f^{-1}(y) \neq \emptyset \text{ and } U_{\alpha n} \cap (\bigcup \{f^{-1}(y_i) \mid i \in$$
$$N'\}) = \emptyset \text{ for some infinite subset } N' \text{ of } N\}.$$

Also put

---

[1] The image of a general M-space by a perfect map is not necessarily an M-space. The images of M-spaces by perfect maps and by closed continuous maps were characterized by J. Nagata [8] and K. Morita–T. Rishel [1], respectively. H. Wicke [1] characterized T_2-spaces which are open continuous images of paracompact T_2-M-spaces. See K. Morita [7] for further results on M-spaces.

[2] Ishii proved this theorem under a more general condition. A. Okuyama [5] and R. A. Stoltenberg [1] proved the same theorem for a normal σ-space and for a normal semi-stratifiable space, respectively.

$$Y_0 = Y - \bigcup_{n=1}^{\infty} Y_n .$$

Let us prove that Y_n is discrete for each $n \geq 1$. Assume the contrary; then $\{f^{-1}(y) \mid y \in Y_n\}$ has a cluster point $x \in X$, i.e., each nbd of x intersects $f^{-1}(y)$ for infinitely many y's, because f is closed. Suppose $x \in F_\alpha$. Then there is a sequence $\{y_i\}$ in Y_n such that

$$f^{-1}(y_i) \cap U_{\alpha i} \neq \emptyset, \quad i = 1, 2, \ldots . \tag{1}$$

We claim that

$$\{f^{-1}(y_i) \mid i = 1, 2, \ldots\} \text{ has no discrete subsequence.} \tag{2}$$

Let N' be an arbitrary infinite subset of N. If $f^{-1}(y_i) \cap F_\alpha \neq \emptyset$ for infinitely many $i \in N'$, then, since F_α is countably compact, $\{f^{-1}(y_i) \mid i \in N'\}$ is not discrete. Assume $f^{-1}(y_i) \cap F_\alpha = \emptyset$ for all $i \in N'$ with $i \geq i_0$. Then $F = \bigcup \{f^{-1}(y_i) \mid i \in N', i \geq i_0\}$ is disjoint from F_α. Since $F \cap U_{\alpha i} \neq \emptyset$ for all i follows from (1), F is not closed. Hence $\{f^{-1}(y_i) \mid i \in N'\}$ is not discrete, proving our claim (2).

Since $y_1 \in Y_n$, there is $\alpha(1) \in A$ such that

$$F_{\alpha(1)} \cap f^{-1}(y_1) \neq \emptyset$$

and a subsequence N_2 of $N_1 = N$ such that

$$U_{\alpha(1)n} \cap (\bigcup \{f^{-1}(y_i) \mid i \in N_2\}) = \emptyset .$$

Let $i(1) = 1$ and choose $i(2) \in N_2$ such that $i(2) > i(1)$. Since $y_{i(2)} \in Y_n$, there is $\alpha(2) \in A$ such that

$$F_{\alpha(2)} \cap f^{-1}(y_{i(2)}) \neq \emptyset$$

and a subsequence N_3 of N_2 such that

$$U_{\alpha(2)n} \cap (\bigcup \{f^{-1}(y_i) \mid i \in N_3\}) = \emptyset .$$

Choose $i(3) \in N_3$ such that $i(3) > i(2)$. Repeating the same process we obtain a subsequence $i(1) < i(2) < \cdots$ of N, a sequence $\{\alpha(k) \mid k = 1, 2, \ldots\}$ of elements of A and a sequence $N_1 \supset N_2 \supset N_3 \supset \cdots$ of infinite subsets of N such that

$$i(k) \in N_k, \qquad F_{\alpha(k)} \cap f^{-1}(y_{i(k)}) \neq \emptyset, \quad k = 1, 2, \ldots,$$

$$U_{\alpha(k)n} \cap (\cup \{f^{-1}(y_i) \mid i \in N_{k+1}\}) = \emptyset. \tag{3}$$

Select

$$x_k \in F_{\alpha(k)} \cap f^{-1}(y_{i(k)}), \quad k = 1, 2, \ldots.$$

Then the sequence $\{x_k\}$ does not cluster. Because, if x is a cluster point of $\{x_k\}$, then

$$x \in \overline{\left(\bigcup_{k=1}^{\infty} F_{\alpha(k)} \right)},$$

and also by (3)

$$x \notin \bigcup_{k=1}^{\infty} U_{\alpha(k)}.$$

This is a contradiction, because

$$S\left(\bigcup_{k=1}^{\infty} F_{\alpha(k)}, \mathcal{U}_n \right) = \bigcup_{k=1}^{\infty} U_{\alpha(k)n}.$$

Hence $\{x_k\}$ is a discrete sequence. Hence $\{f^{-1}(y_{i(k)}) \mid k = 1, 2, \ldots\}$ is also discrete, which contradicts (2). Thus Y_n is discrete

Now, let y be a fixed point of Y_0. Then we claim that $f^{-1}(y)$ is countably compact. Suppose

$$x_i \in f^{-1}(y) \cap F_{\alpha(i)}, \quad i = 1, 2, \ldots.$$

If $F_{\alpha(i)}$ coincide for infinitely many distinct i's, then $\{x_i\}$ has a cluster point, because F_α is countably compact. So we assume that $F_{\alpha(i)} \neq F_{\alpha(j)}$ whenever $i \neq j$. Now we assume that $\{F_{\alpha(i)} \mid i = 1, 2, \ldots\}$ is discrete. Then, since X is normal, there is a discrete open collection $\{V_i \mid i = 1, 2, \ldots\}$ such that $V_i \supset F_{\alpha(i)}$. Select a subsequence $\{l(i) \mid i = 1, 2, \ldots\}$ of N such that $U_{\alpha(i) l(i)} \subset U_i$. For each $n \geq 1$ there is a sequence $\{y_i^n \mid i = 1, 2, \ldots\}$ in Y such that for every α with $F_\alpha \cap f^{-1}(y) \neq \emptyset$ and for every subsequence N' of N,

$$U_{\alpha n} \cap (\cup \{f^{-1}(y_i^n) \mid i \in N'\}) \neq \emptyset,$$

because $y \notin Y_n$. Then there is $j(n) \in N$ such that

$$U_{\alpha(n)\,l(n)} \cap f^{-1}(y_j^{l(n)}) \neq \emptyset \quad \text{for all } j \geq j(n) \tag{4}$$

and

$$U_{\alpha(1)\,l(n)} \cap f^{-1}(y_j^{l(n)}) \neq \emptyset \quad \text{for all } j \geq j(n). \tag{5}$$

We can select $\{j(n)\}$ such that

$$j(1) < j(2) < \cdots,$$

and $\{y_{j(n)}^{l(n)} \mid n = 1, 2, \ldots\}$ is a sequence. Pick

$$z_n \in U_{\alpha(n)\,l(n)} \cap f^{-1}(y_{j(n)}^{l(n)}), \quad n = 1, 2, \ldots \text{ (by use of (4))}.$$

Then $\{z_n\}$ is discrete, because $\{U_{\alpha(n)\,l(n)}\}$ is discrete. Since f is closed, this means that $\{f^{-1}(y_{j(n)}^{l(n)}) \mid n = 1, 2, \ldots\}$ is also discrete. Thus there is n_0 such that

$$U_{\alpha(1)\,l(n)} \cap f^{-1}(y_{j(n)}^{l(n)}) = \emptyset \quad \text{for all } n \geq n_0.$$

But this contradicts (5). Thus $\{F_{\alpha(i)} \mid i = 1, 2, \ldots\}$ has a cluster point. From the property of $\{\mathscr{U}_i\}$ it follows that $\{x_i\}$ has a cluster point. Hence $f^{-1}(y)$ is countably compact, and the theorem is proved.

Definition VII.4. Let f be a mapping from a topological space X onto a topological space Y. If no closed proper subset of X is mapped onto Y by f, then f is called an *irreducible mapping*. A topological space X is called a *Fréchet–Urysohn space* if for every $A \subset X$ and every $x \in \bar{A}$, there is a point sequence $\{x_n\} \subset A$ such that $x_n \to x$.[1] A topological space X is called a *Lašnev space* if there is a metric space M and a closed continuous map f from M onto X.[2]

G) *Every Lašnev space X is Fréchet–Urysohn.*

Proof. Let f be a closed continuous map from a metric space M onto X. Suppose $x \in \bar{A}$ in X. Then there is $p \in f^{-1}(x) \cap \overline{f^{-1}(A)}$, because otherwise

---

[1] A. V. Arhangelskii [10] proved that a T_2-space is Fréchet–Urysohn if and only if it is hereditarily k-space.

[2] N. Lašnev [2] gave an internal characterization of Lašnev spaces.

$X - f(\overline{f^{-1}(A)})$ is a nbd of x disjoint from A, which is impossible. Thus there is $\{p_n\} \subset f^{-1}(A)$ such that $p_n \to p$. Hence $\{f(p_n)\} \subset A$ and $f(p_n) \to x$ in X, proving that X is Fréchet–Urysohn.

H) *Let f be a closed continuous map from a metric space M onto X. Then there is a closed set F of M such that $f(F) = X$ and the map f is irreducible on F.*[1]

Proof. Denote by X_0 the set of all isolated points of X. For each $x \in X_0$ select $p(x) \in f^{-1}(x)$ to put

$$M_0 = \{p(x) \mid x \in X_0\}, \qquad M_1 = f^{-1}(X - X_0).$$

Then $M_0 \cup M_1$ is a closed set of M. Put

$$\mathscr{G} = \{G \mid G \text{ is a closed subset of } M_0 \cup M_1 \text{ such that } f(G) = X\}.$$

Regard $\mathscr{G}$ as a partially ordered set with the inclusion order. Suppose $\mathscr{G}'$ is a totally ordered subset of $\mathscr{G}$. Let

$$G' = \cap \{G \mid G \in \mathscr{G}'\}.$$

Then G' is a closed subset of $M_0 \cup M_1$. We claim that $G' \in \mathscr{G}$. Assume the contrary; then there is $x \in X - X_0$ such that

$$f^{-1}(x) \subset M_1 - G' = M - G'.$$

Since X is Fréchet–Urysohn by G), there is a sequence $\{x_n\} \subset X - \{x\}$ such that $x_n \to x$. Now, observe that

$$B = f^{-1}(x) \cap \overline{\bigcup_{n=1}^{\infty} f^{-1}(x_n)} \neq \emptyset$$

is compact. Because otherwise we can select a point sequence $\{p_n\} \subset B$ with no cluster point and also a point sequence $\{q_n\}$ such that

$$q_n \in f^{-1}(x_{i(n)}) \cap S_{1/n}(p_n), \quad n = 1, 2, \ldots,$$

[1]Due to N. Lašnev [1].

where $i(1) < i(2) < \cdots$. Then $\{q_n\}$ has no cluster point, and thus $\{f(q_n) \mid n = 1, 2, \ldots\} = \{x_{i(n)} \mid n = 1, 2, \ldots\}$ is a closed set. But this contradicts $x \in \overline{\{x_{i(n)}\}} - \{x_{i(n)}\}$. Thus B is compact.

Since $B \subset M - \cap\{G \mid G \in \mathcal{G}'\}$, $B \subset M - G$ for some $G \in \mathcal{G}'$. Since $x_n \to x$, there is $n_0 \in N$ such that $f^{-1}(x_n) \subset M - G$ for $n \geq n_0$. Because otherwise $f^{-1}(x_n) \cap G \neq \emptyset$ for infinitely many n's, which contradicts that G and f are closed. Thus $x_n \notin f(G)$, which contradicts $G \in \mathcal{G}$. Therefore $G' \in \mathcal{G}$. Hence by Zorn's lemma $\mathcal{G}$ has a minimum element F, which obviously has the desired property.

I) *Let f be a closed irreducible map from X onto Y. Suppose $\mathcal{U}$ is a closure-preserving open collection in X. Then $\mathcal{V} = \{Y - f(X - U) \mid U \in \mathcal{U}\}$ is a closure-preserving open collection in Y.*

Proof. It is obvious that each element of $\mathcal{V}$ is open. Suppose $\mathcal{U}' \subset \mathcal{U}$, and

$$y \in \overline{\cup\{Y - f(X - U) \mid U \in \mathcal{U}'\}} \quad \text{in } Y.$$

Then

$$f^{-1}(y) \cap \overline{\cup \mathcal{U}'} \neq \emptyset \quad \text{in } X,$$

because otherwise $Y - f(\overline{\cup \mathcal{U}'})$ is a nbd of y disjoint from $\cup\{X - f(X - U) \mid U \in \mathcal{U}'\}$, which is impossible. Since $\mathcal{U}$ is closure-preserving,

$$f^{-1}(y) \cap \bar{U} \neq \emptyset \quad \text{for some } U \in \mathcal{U}'.$$

Pick $x \in f^{-1}(y) \cap \bar{U}$. Then

$$y = f(x) \in f(\bar{U}) \subset \overline{f(U)}.$$

Let V be a given open nbd of y in Y. Then $V \cap f(U) \neq \emptyset$, and thus $f^{-1}(V) \cap U \neq \emptyset$. Since f is irreducible, there is $y' \in Y$ such that

$$f^{-1}(y') \subset f^{-1}(V) \cap U.$$

Then

$$y' \in V \cap (Y - f(X - U)) \neq \emptyset$$

follows. Hence $y \in \overline{Y - f(X - U)}$, proving that

$$\overline{\cup\{Y - f(X - U) \mid U \in \mathcal{U}'\}} = \cup\{\overline{Y - f(X - U)} \mid U \in \mathcal{U}'\}.$$

Namely, $\mathcal{V}$ is closure-preserving.

J) *Let G be a closed subset of a metric space X. Then G has a closure-preserving outer base consisting of open sets.*

Proof. Let $\mathcal{U}_n$ be a locally finite open cover of X such that mesh $\mathcal{U}_n <$ $1/n$.[1] Then, let Σ be the family of all subcollections $\mathcal{U}'$ of $\bigcup_{n=1}^{\infty} \mathcal{U}_n$ satisfying $\bigcup \mathcal{U}' \supset G$ and $U \cap G \neq \emptyset$ for all $U \in \mathcal{U}'$. Define

$$\mathcal{W} = \{\bigcup \mathcal{U}' \mid \mathcal{U}' \in \Sigma\} .$$

Then $\mathcal{W}$ is obviously an outer base of G consisting of open sets. We can prove, as follows, that $\mathcal{W}$ is closure-preserving. Suppose $\mathcal{W}' \subset \mathcal{W}$, $x \notin \bigcup \{\bar{W} \mid W \in \mathcal{W}'\}$. Let $d(x, G) > 1/i$ for $i \in N$. Put

$$P(x) = X - \bigcup \left\{ \bar{U} \mid x \notin \bar{U}, U \in \bigcup_{n=1}^{2i} \mathcal{U}_n \right\} .$$

Then $P(x)$ is an open nbd of x. $P(x) \cap S_{1/2i}(x)$ is an open nbd of x disjoint from $\bigcup \mathcal{W}'$. Hence $x \notin \overline{\bigcup \mathcal{W}'}$, proving that $\mathcal{W}$ is closure-preserving.

K) *Every Lašnev space X is an M_1-space. Actually X is hereditarily M_1.*[2]

Proof. Let f be a closed continuous map from a metric space M onto X. Then by H) there is a closed set F of M such that $f(F) = X$, and f is irreducible on F. Let G be an arbitrary closed set of X. Then by J) $f^{-1}(G) \cap F$ has a closure-preserving open outer base $\mathcal{U}$ in F. Hence by I) $\{X - f(F - U) \mid U \in \mathcal{U}\}$ is a closure-preserving open outer base of G in X. Thus by VI.8.I) X is M_1. It is also easy to see that every subspace of X is M_1.

3. Metrization of M-spaces

A motivation of the study of generalized metric spaces is to factor metrizability into several simpler conditions. There are various combinations of conditions which induce metrizability. In the following we shall discuss just a few of them centering M-spaces.

---

[1] Generally, for a collection $\mathcal{V}$ of subsets of a metric space we define that mesh $\mathcal{V} =$ sup{diameter $V \mid V \in \mathcal{V}$}.

[2] First proved by F. G. Slaughter [1], and we owe this simpler proof to C. Borges–D. Lutzer [1].

A) *A T_2-space X is metrizable if and only if it is a paracompact M-space with a G_δ-diagonal.*[1]

Proof. Necessity of the condition is obvious. To prove sufficiency, let X be a paracompact M-space with a sequence $\{\mathcal{U}_i\}$ of open covers such that

$$\bigcap_{n=1}^{\infty} S(x, U_n) = \{x\} \quad \text{for each } x \in X \text{ (by use of VI.1.B))}. \tag{1}$$

Select a sequence $\{\mathcal{V}_n\}$ of open covers satisfying $w\Delta$-condition. Further select locally finite open covers $\mathcal{W}_n$, $n = 1, 2, \ldots$, such that

$$\bar{\mathcal{W}}_n < \mathcal{U}_n \wedge \mathcal{V}_n.$$

Let V be a given open nbd of a given point x of X. Then we claim that $S(x, \mathcal{W}_n) \subset V$ for some n. Because otherwise we can choose $x_n \in S(x, \mathcal{W}_n) - V$, $n = 1, 2, \ldots$. Then $\{x_n\}$ has a cluster point $y \in X - V$, because $\{\mathcal{W}_n\}$ satisfies $w\Delta$-condition. Observe that

$$y \in \bigcap_{n=1}^{\infty} S(x, \bar{\mathcal{W}}_n) = \{x\}$$

follows from the local finiteness of $\mathcal{W}_n$ and (1). Hence $\{\mathcal{W}_n\}$ is a development of X, and hence the proposition follows from the corollary to Theorem VI.1.

Theorem VII.5. *Every countably compact T_2-space X with a G_δ-diagonal is metrizable.*[2]

Proof. It suffices to show that X is Lindelöf. Let $\{\mathcal{U}_i\}$ be a sequence of open covers of X such that $\bigcap_{n=1}^{\infty} S(x, \mathcal{U}_n) = \{x\}$ for each $x \in X$. Assume that X is not Lindelöf. Then there is an open cover $\mathcal{V}$ which has no countable subcover. Observe that if Y is a subset of X and no countable subcollection of $\mathcal{V}$ covers Y, then for each $y \in Y$ and for some $n \in N$, $Y - S(x, \mathcal{U}_n)$ is not covered by any countable subcollection of $\mathcal{V}$. Now we select transfinite sequences $\{x_\beta \mid 0 \leq \beta < \omega_1\}$ of points of X and $\{n_\beta \mid 0 \leq \beta < \omega_1\}$ of natural numbers satisfying

---

[1] Due to C. Borges [1] and A. Okuyama [2].
[2] Due to J. Chaber [1].

(1) $x_\beta \notin S(x_\alpha, \mathcal{U}_{n_\alpha})$ if $\alpha < \beta$,

(2) $X - \bigcup\{S(x_\alpha, \mathcal{U}_{n_\alpha}) \mid 0 \le \alpha \le \beta\}$ is covered by no countable sub-collection of $\mathcal{V}$.

The selection is done by use of the induction on β. Select arbitrary $x_0 \in X$ and $n_0 \in N$ such that $X - S(x_0, \mathcal{U}_{n_0})$ is covered by no countable sub-collection of $\mathcal{V}$.

Assume that x_β and n_β have been selected for all $\beta < \gamma$. Then we claim that $X - \bigcup_{\beta < \gamma} S(x_\beta, \mathcal{U}_{n_\beta})$ is covered by no countable subcollection of $\mathcal{V}$. Assume the contrary; then it is covered by $\mathcal{V}' \subset \mathcal{V}$ with $|\mathcal{V}'| \le \aleph_0$. Then $\mathcal{V}' \cup \{S(x_\beta, \mathcal{U}_{n_\beta}) \mid \beta < \gamma\}$ is a countable open cover of X. Since X is countably compact, this cover has a finite subcover. Hence for some $\beta < \gamma$, $\mathcal{V}' \cup \{S(x_\alpha, \mathcal{U}_{n_\alpha}) \mid \alpha \le \beta\}$ covers X. Therefore $\mathcal{V}'$ covers $X - \bigcup\{S(x_\alpha, \mathcal{U}_{n_\alpha}) \mid \alpha \le \beta\}$ contradicting the induction hypothesis (2). Thus our claim is proved. Hence we can pick

$$x_\gamma \in X - \bigcup_{\beta < \gamma} S(x_\beta, \mathcal{U}_{n_\beta}).$$

Then recall the previous observation to select $n_\gamma \in N$ for which $X - \bigcup\{S(x_\beta, \mathcal{U}_{n_\beta}) \mid \beta \le \gamma\}$ is covered by no countable subcollection of $\mathcal{V}$. Thus the induction is complete.

Note that there is $n \in N$ such that $B = \{\beta \mid 0 \le \beta < \omega_1, n_\beta = n\}$ is infinite. Choose $\beta_1, \beta_2, \ldots \in B$ with $\beta_1 < \beta_2 < \cdots$. It follows from (1) that $\{x_{\beta_i} \mid i = 1, 2, \ldots\}$ is a discrete point sequence of X, which contradicts that X is countably compact. Hence X is Lindelöf and accordingly metrizable by A).

Corollary 1. *A T_2-space X is metrizable if and only if it is an M-space with a G_δ-diagonal.*

Proof. Assume that X is an M-space with a G_δ-diagonal. Then by Theorem VII.3 there is a quasi-perfect map f from X onto a metric space Y. Since $f^{-1}(y)$ is countably compact for each $y \in Y$, it is metrizable by Theorem VII.5. Hence $f^{-1}(y)$ is compact, i.e. f is perfect. Thus X is paracompact by V.2.D). Hence X is metrizable by A).

Corollary 2. *A T_2-space X is metrizable if and only if it is M and semi-stratifiable.*

Proof. If X is semi-stratifiable, then so is $X \times X$. Hence X has a G_δ-diagonal. Thus this corollary follows from the previous one.

Example VII.7. Niemytzki space R_4 is not M. Because it is developable and accordingly semi-stratifiable but non-metrizable, and hence by Corollary 2 it cannot be M. For a similar reason a non-metrizable hedgehog is not M either.

B) *Let $\mathcal{U}$ be a point-countable collection of subsets of X. Then there are at most countably many minimum finite covers of X by members of $\mathcal{U}$, where we mean by a* minimum cover *one of which no proper subcollection covers* X.[1]

Proof. Assume the contrary and put

$$\Sigma_n = \{\mathcal{V} \mid \mathcal{V} \subset \mathcal{U}, \mathcal{V} \text{ is a minimum cover of } X \text{ such that } |\mathcal{V}| = n\},$$
$$n = 1, 2, \ldots.$$

Then $|\Sigma_n| > \aleph_0$ for some n. For distinct elements $U_1, \ldots, U_k$ of $\mathcal{U}$ we define

$$\Sigma(U_1, \ldots, U_k) = \{\mathcal{V} \in \Sigma_n \mid U_1, \ldots, U_k \in \mathcal{V}\}.$$

Now, fix a point $x_1 \in X$. Then for some $U_1 \in \mathcal{U}$ with $x_1 \in U_1$ we have $|\Sigma(U_1)| > \aleph_0$, because $\Sigma_n = \bigcup \{\Sigma(U) \mid x \in U \in \mathcal{U}\}$. Hence $U_1 \neq X$ and hence we can select $x_2 \in X - U_1$. There is $U_2 \in \mathcal{U}$ with $x_2 \in U_2$ and

$$|\Sigma(U_1, U_2)| > \aleph_0,$$

because
$$\Sigma(U_1) = \bigcup \{\Sigma(U_1, U_2) \mid x \in U_2 \in \mathcal{U}\}.$$

Continue the same process until getting $\Sigma(U_1, \ldots, U_n)$ such that

$$|\Sigma(U_1, \ldots, U_n)| > \aleph_0.$$

However, $\Sigma(U_1, \ldots, U_n)$ consists of the only element $\mathcal{U}_0 = \{U_1, \ldots, U_n\}$. This contradiction proves the proposition.

Theorem VII.6. *Let $\mathcal{U}$ be a point-countable p-base of a countably compact T_2-space X. Then X is metrizable.*[2]

---

[1] Due to A. Miščenko [1]. In this proposition X may be just a set, because no topology of X is used in the following proof.

[2] Due to A. Miščenko [1]. An open collection $\mathcal{U}$ in X is called a *p-base* if for each $x \in X$
$\bigcap \{U \mid x \in U \in \mathcal{U}\} = \{x\}$.

Proof. First we claim that every cover $\mathcal{V}$ of X by elements of $\mathcal{U}$ has a countable subcover and accordingly a finite subcover, too.

To prove it, assume the contrary. Pick $x_1 \in X$; then $S(x_1, \mathcal{V}) \neq X$, because $\mathcal{V}$ is countable at x_1. Select

$$x_2 \in X - S(x_1, \mathcal{V}).$$

Then

$$S(x_1, \mathcal{V}) \cup S(x_2, \mathcal{V}) \neq X.$$

Select

$$x_3 \in X - S(x_1, \mathcal{V}) \cup S(x_2, \mathcal{V}).$$

Continue the same process to get a sequence $\{x_i\}$ such that

$$x_n \in X - \bigcup_{i=1}^{n-1} S(x_i, \mathcal{V}).$$

Then $\{x_i\}$ has no cluster point, which is impossible. Thus $\mathcal{V}$ must have a countable subcover.

By B) there are only countably many minimum covers by members of $\mathcal{U}$, which we denote by $\mathcal{V}_1, \mathcal{V}_2, \ldots$. Let x and y be distinct points of X. Then we can show that $y \notin S(x, \mathcal{V}_n)$ for some n. Since $\mathcal{U}$ is a p-base, there is $U \in \mathcal{U}$ such that

$$x \in U, \qquad y \notin U.$$

For each $z \in X - U$, choose $U(z) \in \mathcal{U}$ such that

$$z \in U(z), \qquad x \notin U(z).$$

Then, as observed before, $\{U, U(z) \mid z \in X - U\}$ contains a minimum subcover, say $\mathcal{V}_n$. It is obvious that $S(x, \mathcal{V}_n) = U \not\ni y$. Thus the sequence $\{\mathcal{V}_n \mid n = 1, 2, \ldots\}$ satisfies

$$\bigcap_{n=1}^{\infty} S(x, \mathcal{V}_n) = \{x\} \quad \text{for each } x \in X.$$

Namely, X has a G_δ-diagonal. Hence by Theorem VII.5 X is metrizable.

Theorem VII.7. *A T_2-space X is metrizable if and only if it is M and has a point-countable p-base.*[1]

Proof. We shall prove only the 'if' part. It follows from Theorem VII.3 and Theorem VII.6 that X is the inverse image of a metric space by a perfect map, and hence it is paracompact. Hence we can select a sequence $\{\mathcal{U}_i\}$ of locally finite open covers of X satisfying $w\Delta$-condition. We put $\mathcal{U}_n = \{U_\alpha \mid \alpha \in A_n\}$ and denote by $\mathcal{V}$ a point-countable p-base of X. Then for a fixed α there are at most countably many (finite) minimum covers of U_α by members of $\mathcal{V}$ (B)), which we denote by $\mathcal{V}_{\alpha m}$, $m = 1, 2, \ldots$. We can express that

$$\bigcup_{m=1}^{\infty} \mathcal{V}_{\alpha m} = \{V_{\alpha h} \mid h = 1, 2, \ldots\},$$

because $\bigcup_{m=1}^{\infty} \mathcal{V}_{\alpha m}$ is a countable collection. For each $(n, l) \in N \times N$, we define

$$\mathcal{W}_{nl} = \{U(\alpha, h_1, \ldots, h_k) \mid \alpha \in A_n ; 1 \leq h_1, \ldots, h_k \leq l ;$$
$$k = 1, \ldots, l\}, \quad (1)$$

where

$$U(\alpha, h_1, \ldots, h_k) = U_\alpha - V_{\alpha h_1} \cup \cdots \cup V_{\alpha h_k}. \quad (2)$$

Then it is obvious that $\mathcal{W}_{nl}$ is locally finite. Now we claim that $\bigcup_{n,l=1}^{\infty} \mathcal{W}_{nl}$ is a network of X. Suppose P is a given open nbd of a given point x of X. Put

$$C = \bigcap_{n=1}^{\infty} S(x, \mathcal{U}_n).$$

Then C and accordingly $C - P$, too, are compact. Thus there are $V_1, \ldots, V_k \in \mathcal{V}$ such that

$$C - P \subset V_1 \cup \cdots \cup V_k,$$
$$V_i \cap (C - P) \neq \emptyset, \quad i = 1, \ldots, k, \quad (3)$$
$$x \notin V_1 \cup \cdots \cup V_k.$$

[1] This theorem was first proved by V. Filippov [1] in the form that a paracompact T_2-space is metrizable if and only if it is M and has a point-countable base, and then improved by several people including J. Nagata [12], T. Shiraki [1], F. G. Slaughter and others.

For each i with $1 \le i \le k$, we can construct a minimum cover $\mathscr{V}_i$ of C by members of $\mathscr{V}$ such that $V_i \in \mathscr{V}_i$. Then

$$C \subset \left(\bigcap_{i=1}^{k} V_i' \right) \cap (P \cup V_1 \cup \cdots \cup V_k),$$

where

$$V_i' = \bigcup \{ V \mid V \in \mathscr{V}_i \}. \tag{4}$$

Now, we can find $n \in N$ and $\alpha \in A_n$ such that

$$C \subset U_\alpha \subset \left(\bigcap_{i=1}^{k} V_i' \right) \cap (P \cup V_1 \cup \cdots \cup V_k). \tag{5}$$

Then we have

$$U_\alpha - P \subset V_1 \cup \cdots \cup V_k, \tag{6}$$

because $x' \in U_\alpha - P$ implies

$$x' \in P \cup V_1 \cup \cdots \cup V_k - P \subset V_1 \cup \cdots \cup V_k.$$

Note that $V_1, \ldots, V_k \in \bigcup_{m=1}^{\infty} V_{\alpha m}$, because by (4) and (5) $\mathscr{V}_i$ is a minimum cover of U_α, i.e. $V_i \in \mathscr{V}_i = \mathscr{V}_{\alpha m_i}$ for some m_i. Now, assume that

$$V_i = V_{\alpha h_i}, \quad i = 1, \ldots, k,$$

and $h = \max\{h_1, \ldots, h_k\}$. Then from (1), (2), (3) and (6) it follows that

$$x \in U(\alpha, h_1, \ldots, h_k) \subset P \quad \text{and} \quad U(\alpha, h_1, \ldots, h_k) \in \mathscr{W}_{nh}.$$

This proves that $\bigcup_{n,h=1}^{\infty} W_{nh}$ is a network of X. Hence X is a σ-space and accordingly semi-stratifiable. Thus X is metrizable by Corollary 2 of Theorem VII.5.

There are many other remarkable results obtained on metrizability of generalized metric spaces. For example, A. V. Arhangelskii [12] proved that every regular space which is hereditarily Lindelof M is second-countable. A T_2-space which is hereditarily M is called an F_{pp}-*space*. Z. Balogh and A. V. Arhangelskii proved that every F_{pp}-space contains a dense metrizable subset.[1] The latter also gave a characterization of

---

[1] See Z. Balogh [1], [2] and A. V. Arhangelskii [11].

non-metrizable F_{pp}-space and showed that every perfectly normal F_{pp}-space is metrizable.

Recently, subconditions of metrizability like developability, p and m are further factored into weaker conditions. R. Hodel and others obtained many interesting results in this aspect.[1] Some examples of such factorization are: A regular space is developable if and only if it is semi-stratifiable and $w\Delta$.[2] A T_2-space is a σ-space if and only if it is a Σ-space with a G_δ-diagonal,[3] where a topological space X is called a Σ-space if it has a sequence $\mathscr{F}_1, \mathscr{F}_2, \ldots$ of locally finite closed covers such that $x_n \in \cap \{F \mid x \in F \in \mathscr{F}_n\}$, $n = 1, 2, \ldots$, for a fixed point x implies that $\{x_n\}$ has a cluster point. Every M-space and every regular σ-space are Σ-space, and Theorem VI.27 can be extended to Σ-spaces. In the following is an interesting metrization theorem of Σ-space due to E. Michael, T. Shiraki and F. Slaughter: A topological space is metrizable if and only if it is a collectionwise normal Σ-space with a point-countable base[4]; another interesting theorem[5] in this respect is that a semi-metric space with a point-countable base is developable.

4. Theory of inverse limit space

In preceding sections we have learned to characterize certain types of spaces as the continuous images or inverse images of concreter spaces. In the present section we shall try to characterize spaces in terms of the concept of inverse limit space, which was defined in Definition II.18. Namely, some types of compact spaces will be characterized as the limit spaces of concreter spaces.

Let E be a finite set of elements which are called *vertices* and K a collection of subsets of E such that every subset of a set belonging to K also belongs to K; then we call K an *(abstract) complex*. We call a set of $n + 1$ vertices $a_0, \ldots, a_n$ which belongs to K an *(abstract) n-simplex* and denote it by $|a_0 \ldots a_n|$. An m-simplex $|a_{i_0} \ldots a_{i_m}|$ whose vertices are chosen from $\{a_0, \ldots, a_n\}$ is called an *m-face* of $|a_0 \ldots a_n|$.

[1] See R. Hodel [1], [2], [5].

[2] Due to C. Creede [1]. H. Wicke–J. Worrell [1] and H. Brandenburg [1], [2] also gave interesting characterizations of developability.

[3] K. Nagami [2] invented the concept of Σ-space and proved this theorem for paracompact spaces, which was then improved by several other people.

[4] See E. Michael [11] and T. Shiraki [1].

[5] Due to R. Heath [5].

Now, let us consider $r + 1$ points

$$a_i = (a_1^{(i)}, \ldots, a_n^{(i)}), \quad i = 0, \ldots, r,$$

in an n-dimensional Euclidean space E^n. If the rank of the matrix $(b_j^{(i)})$, where $b_j^{(i)} = a_j^{(i)}$ for $i = 0, \ldots, r$, $j = 1, \ldots, n$, and $b_{n+1}^{(i)} = 1$ for $i = 0, \ldots, r$ is equal to $r + 1$, then the points $\{a_0, \ldots, a_r\}$ are called *linearly independent*. For linearly independent points $a_0, \ldots, a_r$, we define a subset

$$[a_0 \ldots a_r] = \{(x_1, \ldots, x_n) \mid x_i = \lambda_0 a_i^{(0)} + \cdots + \lambda_r a_i^{(r)}, i = 1, \ldots, n,$$

$$\lambda_0 + \cdots + \lambda_r = 1, 0 \leqslant \lambda_0, \ldots, \lambda_r \leqslant 1\}$$

of E^n and call it the *geometrical r-simplex* spanned by $a_0, \ldots, a_r$. The latter are called the vertices of the simplex. The point $(x_1, \ldots, x_n)$ of the simplex with $\lambda_0 = \lambda_1 = \cdots = \lambda_n = 1/(r + 1)$ is called the *barycenter* of the simplex. Each geometrical s-simplex spanned by $s + 1$ points chosen from $a_0, \ldots, a_r$ is called an *s-face* of $[a_0 \ldots a_r]$. Geometrical 1-, 2- and 3-simplices are a segment, triangle and tetrahedron, respectively.

Suppose K is a finite collection of geometrical simplices in E^n. If it satisfies:

(i) every face of a simplex belonging to K also belongs to K,

(ii) the intersection of two simplices belonging to K is either empty or a face of each of those simplices,

then K is called a *geometrical complex*. The union of the simplices (regarded as a subset of E^n) belonging to a geometrical complex forms a closed subset of E^n and is called a *polyhedron* in E^n.

Example VII.8. Suppose K is a geometrical complex. We consider all vertices of the simplices belonging to K and consider the vertices $a_0, \ldots, a_r$ to form an abstract simplex if and only if they are the vertices of a simplex belonging to K. In this way every geometrical complex determines an abstract complex. Another significant example of abstract complex is a *nerve*. Suppose X is a topological space and $\mathcal{U} = \{U_i \mid i = 1, \ldots, n\}$ is a covering of X. Define a subset $\{U_{i_k} \mid k = 0, \ldots, r\}$ of $\mathcal{U}$ to be a simplex if and only if

$$\bigcap_{k=0}^{r} U_{i_k} \neq \emptyset.$$

Then we obtain an abstract complex having $\{U_i \mid i = 1, \ldots, n\}$ as its vertices. This complex is called the nerve of the covering $\mathcal{U}$. The concept of nerve is a key which connects general topology with algebraic topology.

Let K be a geometrical complex. Then we define the *barycentric subdivision* K_1 of K as follows: For each n-simplex x^n of K, we take the barycenter $a(x^n)$ of x^n. Then K_1 is the collection of all the simplices $[a(x_{\alpha_0}^{n_0}) a(x_{\alpha_1}^{n_1}) \ldots a(x_{a_i}^{n_i})]$, where

$$ n_0 > n_1 > \cdots > n_i, $$

and $x_{\alpha_k}^{n_k}$ is a face of $x_{\alpha_{k-1}}^{n_{k-1}}$. As easily seen, K_1 is also a complex. It is called the barycentric subdivision of K. We often consider successive barycentric subdivisions $K_1, K_2, K_3, \ldots$. Then it is easy to see that mesh $K_n \to 0$ as $n \to \infty$.

First we shall characterize every compact T_2-space as the limit space of polyhedra.

A) *Let $\{X_\alpha, \pi_\beta^\alpha \mid \alpha, \beta \in A, \alpha > \beta\}$ be an inverse system of compact T_2-spaces. Then the inverse limit space*

$$ X = \varprojlim \{X_\alpha, \pi_\beta^\alpha\} $$

is a compact T_2-space.

Proof. Since X is a subspace of the product space P of the compact T_2-spaces X_α, $\alpha \in A$, all we have to prove is that X is closed in P. Let

$$ p = \{p_\alpha \mid \alpha \in A\} \in X - P. $$

Then there are α, $\beta \in A$ such that $\beta < \alpha$ and $\pi_\beta^\alpha(p_\alpha) \neq p_\beta$ in X_β. Since X_β is T_2, we can choose disjoint nbds U and V of $\pi_\beta^\alpha(p_\alpha)$ and p_β, respectively. Since π_β^α is continuous, there is a nbd W of p_α in X_α such that $\pi_\beta^\alpha(W) \subset U$. Then the nbd

$$ N(p) = \{q \mid q = \{q_\alpha \mid \alpha \in A\} \in P, q_\alpha \in W, q_\beta \in V\} $$

of p in P does not intersect X, because $\pi_\beta^\alpha(q_\alpha) \neq q_\beta$ holds for every $q \in N(p)$. Therefore X is a closed set of P, i.e. X is a compact T_2-space.

B) *Let F be a closed set of a finite-dimensional cube* I^{n} [1] *and U an open set of* I^{n} *such that* $F \subset U$*. Then there exists a polyhedron P such that*

$$F \subset \operatorname{Int} P \subset P \subset U.$$

Proof. Since I^{n} is a polyhedron, we can construct its successive barycentric subdivisions $K_1, K_2, \ldots$. We denote by K'_n the complex which consists of the simplices of K_n intersecting F and of their faces. Let P_n be the polyhedron determined by K'_n; then since mesh $K'_n \to 0$,

$$\bigcap_{n=1}^{\infty} \operatorname{Int} P_n = \bigcap_{n=1}^{\infty} P_n = F.$$

Hence by the compactness of I^{n}, $P_n \subset U$ for some n. (Note that $P_n \supset P_{n+1}$.) Thus $P_n = P$ is the desired polyhedron.

Theorem VII.8. *A topological space X is compact T_2 if and only if it is homeomorphic with the inverse limit space of an inverse system of polyhedra.*

Proof. The sufficiency is implied by A). To prove the necessity we assume that X is a given compact T_2-space. Then by Tychonoff's imbedding theorem it is homeomorphic with a closed set X' of the product space $I^{A} = \prod \{I_{\alpha} \mid \alpha \in A\}$ of closed unit segments $I_{\alpha} = [0, 1]$, $\alpha \in A$. Let B be a subset of A; then we denote by I^{B} the product space $I^{B} = \{I_{\alpha} \mid \alpha \in B\}$. Suppose B and C are subsets of A with $B \supset C$; then we denote by f_C^B the projection of I^{B} onto I^{C}, i.e.,

$$f_C^B(\{x_{\alpha} \mid \alpha \in B\}) = \{x_{\alpha} \mid \alpha \in C\}$$

for each point $\{x_{\alpha} \mid \alpha \in B\}$ of I^{B}. Therefore f_B^B is the identity mapping of I^{B} onto itself. Furthermore we denote by

$$\mathscr{P} = \{P_{\gamma} \mid \gamma \in \Gamma\}$$

the collection of all polyhedra P_{γ} contained in I^{B} for some finite subset B of A satisfying

---

[1] Namely I^{n} is the product space of n copies of the unit segment $[0, 1]$.

$$f^A_B(X') \subset \text{Int } P_\gamma \quad \text{in } I^B.$$

Incidentally, we denote by $B(\gamma)$ the smallest finite subset of A such that P_γ is in $I^{B(\gamma)}$. We define an order between two members γ and δ of Γ as follows:

$$\gamma < \delta \quad \text{if and only if} \quad B(\gamma) \subset B(\delta) \text{ and } f^{B(\delta)}_{B(\gamma)}(P_\delta) \subset P_\gamma.$$

To prove that Γ is a directed set, we suppose $\gamma, \delta \in \Gamma$. Put

$$C = B(\gamma) \cup B(\delta) \, ;$$

then

$$P_\varepsilon = (f^C_{B(\gamma)})^{-1}(P_\gamma) \cap (f^C_{B(\delta)})^{-1}(P_\delta)$$

is a polyhedron in I^C satisfying

$$f^A_C(X') \subset \text{Int } P_\varepsilon,$$

$$f^C_{B(\gamma)}(P_\varepsilon) \subset P_\gamma$$

and

$$f^C_{B(\delta)}(P_\varepsilon) \subset P_\delta.$$

Thus $P_\varepsilon \in \mathscr{P}$, and ε is a member of Γ such that $\varepsilon > \gamma$ and $\varepsilon > \delta$. Therefore Γ is a directed set.

Now, we consider an inverse system

$$\{P_\gamma, \pi^\gamma_\delta \mid \gamma, \delta \in \Gamma, \gamma > \delta\},$$

where π^γ_δ is the restriction of $f^{B(\gamma)}_{B(\delta)}$ to P_γ. Let

$$X_\infty = \lim_{\leftarrow} \{P_\gamma, \pi^\gamma_\delta\} \, ;$$

then we can prove that X' and X_∞ are homeomorphic. To this end we consider a mapping f of X' into X_∞ defined by

$$f(x) = \{f^A_{B(\gamma)}(x) \mid \gamma \in \Gamma\}, \quad x \in X'.$$

From the property of projection, it is clear that $f(x)$ is a continuous mapping of X' into X_∞. To prove that f is onto, take a point $y = \{y_\gamma \mid \gamma \in \Gamma\}$ of X_∞. For each $\alpha \in A$, $I_\alpha \in \mathscr{P}$; hence we put $I_\alpha = P_{\gamma(\alpha)}$, where

$\gamma(\alpha) \in \Gamma$. Now $x = \{x_\alpha \mid \alpha \in A\}$ for $x_\alpha = y_{\gamma(\alpha)}$ is a point of I^A. It is easily seen that $f(x) = y$ if $x \in X'$. Assume $x \in I^A - X'$; then since X' is closed,

$$f^A_B(x) \notin f^A_B(X')$$

for some finite subset B of A. By B) there is a polyhedron P_γ in I^B such that

$$f^A_B(X') \subset \text{Int } P_\gamma \subset P_\gamma \not\ni f^A_B(x). \tag{1}$$

We may assume without loss of generality that $B(\gamma) = B$. On the other hand, from the definition of x it follows that

$$f^A_B(x) = \{x_\alpha \mid \alpha \in B\} = y_{\gamma'}, \tag{2}$$

where we put $I^B = P_{\gamma'}$. However, since

$$B(\gamma) = B(\gamma') = B, \qquad P_\gamma \subset P_\gamma',$$

we have $\gamma > \gamma'$. Hence

$$f^B_B(y_\gamma) = y_\gamma = y_{\gamma'}.$$

This combined with (1) and (2) implies $y_\gamma \notin P_\gamma$, which contradicts that $y = \{y_\gamma \mid \gamma \in \Gamma\}$ is a point of X_∞. Therefore $x \in X'$, i.e. f is a continuous mapping of X' onto X_∞.

Finally, to prove that f is one-to-one, we assume that x and x' are different points of X'. Then[1]

$$f^A_\alpha(x) \neq f^A_\alpha(x') \quad \text{for some } \alpha \in A.$$

Therefore, assuming that $P_\gamma = I_\alpha$, we obtain

$$f^A_{B(\gamma)}(x) \neq f^A_{B(\gamma)}(x'),$$

i.e.,

$$f(x) \neq f(x').$$

---

[1] Actually the subindex α means the one-element set $\{\alpha\}$.

Applying III.3.D) we can conclude that f is a homeomorphic mapping of X' onto X_∞. Thus X and X_∞ are homeomorphic.

Revising the proof of the preceding theorem slightly we get the following theorem originally due to H. Freudenthal [1].

Corollary.[1] *A topological space X is a compact metric space if and only if it is homeomorphic with the limit space of an inverse sequence $\{P_i, \pi_j^i \mid i, j = 1, 2, \ldots ; i > j\}$ of polyhedra P_i.*

Proof. To adjust the preceding proof to the present case, we imbed X into the product I^ω of countably many unit segments and consider a sequence of finite dimensional cubes I^n, $n = 1, 2, \ldots$, and a sequence of polyhedra P_n in I^n, $n = 1, 2, \ldots$, such that

$$\pi_n^\omega(X') \subset \text{Int } P_n \subset S_{1/n}(\pi_n^\omega(X'))$$

and

$$\pi_n^{n+1}(P_{n+1}) \subset P_n \quad \text{in } I^n,$$

where π_n^k denotes the projection of I^k onto I^n. Then we can show that X' is homeomorphic with the limit space $\varprojlim \{P_i, \pi_j^i\}$. The detailed proof is left to the reader.

Next, let us turn to the theory, which was originated by P. S. Alexandroff, of representing a given compact space as the limit space of finite spaces. We shall describe the theory along the line of J. Flachsmeyer [1] who used decomposition spaces while the original method of Alexandroff used the concept of nerve.

Definition VII.5. A topological space X is called a T_0-*space* if for each pair p, q points of X, there is either a nbd U of p with $q \notin U$ or a nbd V of q with $p \notin V$. T_0 is a weaker condition than T_1.

Let us temporarily denote by $p \leqslant q$ the relation between two points p, q of a T_0-space X that $p \in \overline{\{q\}}$. Suppose $\{X_\alpha, \pi_\beta^\alpha \mid \alpha, \beta \in A, \alpha > \beta\}$ is an inverse system of T_0-spaces X_α; then we define order between two points

---

[1] This type of theorem is especially interesting in connection with dimension. As a matter of fact, Freudenthal proved that every compact metric space of dimension $\leqslant n$ is homeomorphic with the limit space of an inverse sequence of polyhedra of dimension $\leqslant n$. S. Mardešić [1] proved that every compact T_2-space of dim $\leqslant n$ is an inverse limit of compact metric spaces of dim $\leqslant n$.

$p = \{p_\alpha \mid \alpha \in A\}$ and $q = \{q_\alpha \mid \alpha \in A\}$ of the product space $\prod \{X_\alpha \mid \alpha \in A\}$ of X_α as follows:

$$p \leqslant q \text{ if and only if } p_\alpha \leqslant q_\alpha \text{ for all } \alpha \in A .$$

Now in view of the fact that X_α is T_0, $\prod X_\alpha$, and consequently the limit space X_∞ of the inverse system, turns out to be a partially ordered set with respect to this order. Then we mean by a *minimal point of* X_∞ a minimal element of this partially ordered set.

C) *Let* $\{X_\alpha, \pi_\beta^\alpha \mid \alpha, \beta \in A, \alpha > \beta\}$ *be an inverse system of finite* T_0-*spaces,* X_α. *Then the subspace* $\tilde{X}$ *of the minimal points of the limit space* X_∞ *is a compact* T_1-*space.*

Proof. Suppose p, q are different points of $\tilde{X}$. Then since p, q are minimal points,

$$p \not\leqslant q \quad \text{and} \quad q \not\leqslant p ,$$

i.e.,

$$p_\alpha \notin \overline{\{q_\alpha\}} \quad \text{and} \quad q_\beta \notin \overline{\{p_\beta\}}$$

for some $\alpha, \beta \in A$. This implies that $p \notin \overline{\{q\}}$ and $q \notin \overline{\{p\}}$, i.e., X satisfies the condition for T_1.

To prove the compactness of $\tilde{X}$, we suppose that $\varphi(\Delta \mid >)$ is a maximal net of $\tilde{X}$. We denote by P the set of the convergence points of φ in X_∞, i.e.,

$$P = \{p \mid \varphi(\Delta \mid >) \rightarrow p \in X_\infty\} .$$

To prove $P \neq \emptyset$, denoting by $\varphi_\alpha(\delta)$ the α-coordinate of $\varphi(\delta)$, we note that for each $\alpha \in A$, $\varphi_\alpha(\Delta \mid >)$ is a maximal net of X_α. Since X_α is finite, $\varphi_\alpha(\Delta \mid >)$ is residual in some point $p_\alpha \in X_\alpha$. We take such a point p_α for each $\alpha \in A$ to construct a point $p = \{p_\alpha \mid \alpha \in A\}$ of $\prod X_\alpha$. Then clearly $\varphi(\Delta \mid >) \rightarrow p$. Let $\alpha > \beta$, $\alpha, \beta \in A$; then

$$\varphi_\alpha(\delta) = p_\alpha \quad \text{and} \quad \varphi_\beta(\delta) = p_\beta \quad \text{for some } \delta \in \Delta ,$$

which implies

$$\pi_\beta^\alpha(p_\alpha) = p_\beta$$

because $\varphi(\delta) \in X_\infty$. Thus $p \in X_\infty$, i.e. $p \in P$, which implies $P = \emptyset$.

Suppose P' is a totally ordered subset of P. Then for each $\alpha \in A$, we put

$$p'_\alpha = \min\{p_\alpha \mid \{p_\beta \mid \beta \in A\} \in P'\}\,;$$

this is possible because X_α is finite. Then $p' = \{p'_\alpha \mid \alpha \in A\}$ satisfies $p' \leq p$ for every point $p \in P'$. Let $\alpha, \beta \in A$, $\alpha > \beta$; then there is $p \in P'$ such that $p_\beta = p'_\beta$. Then $\pi^\alpha_\beta(p_\alpha) = p_\beta$ since $p \in X_\infty$. Since $p'_\alpha \leq p_\alpha$, $p'_\alpha \in \overline{\{p_\alpha\}}$. Therefore from the continuity of π^α_β it follows that $\pi^\alpha_\beta(p'_\alpha) \in \overline{\{p'_\beta\}}$, i.e. $\pi^\alpha_\beta(p'_\alpha) \leq p'_\beta$. Thus by the definition of p'_β we obtain $\pi^\alpha_\beta(p'_\alpha) = p'_\beta$. This means that $p' \in X_\infty$. It is also easily seen that $\varphi(\Delta \mid >) \to p'$. Thus p' is a greatest lower bound of P' in P. Hence by Zorn's lemma there is a minimal element q of P.

To prove that $q \in \tilde{X}$, we assume

$$r \leq q, \quad r \in X_\infty\,.$$

If $r \neq q$, then from the definition of q it follows that $r \notin P$, i.e. $\varphi(\Delta \mid >) \not\to r$. Hence there is an open nbd $U(r)$ of r in X_∞ such that $\varphi(\Delta \mid >)$ is residual in $\tilde{X} - U(r)$. Therefore from $\varphi(\Delta \mid >) \to q$, we obtain that $q \notin U(r)$, i.e., $q_\alpha \notin U_\alpha(r_\alpha)$ for some $\alpha \in A$ and some nbd $U_\alpha(r_\alpha)$ of r_α in X_α. Therefore $r_\alpha \not\leq q_\alpha$, which contradicts the assumption $r \leq q$. Thus we obtain $r = q$, i.e., q is a minimal point of X_∞ and hence $q \in \tilde{X}$. Since $q \in P$, $\varphi(\Delta \mid >) \to q$ in $\tilde{X}$. This proves that $\tilde{X}$ is compact.

Suppose X is a T_1-space and $\{\mathcal{U}_\alpha \mid \alpha \in A\}$ the collection of all finite open coverings of X. Then for each α, we define an equivalence relation between two points p, q of X as follows: $p \sim_\alpha q$ means that for each $U \in \mathcal{U}_\alpha$, $p \in U$ if and only if $q \in U$. Then we denote by $\mathcal{D}_\alpha$ the decomposition of X with respect to the relation $\sim_\alpha$.

From now on, for brevity, we shall denote by X_α the decomposition space $X(\mathcal{D}_\alpha)$ of X with respect to $\mathcal{D}_\alpha$. The following proposition is easy to prove, so its proof is left to the reader.

D) *Every X_α is a finite T_0-space, where X_α is the space defined in the above.*

We define order between two elements α, β of A by

$$\beta < \alpha \quad \text{if and only if} \quad \mathcal{D}_\alpha < \mathcal{D}_\beta\,.$$

Then A turns out to be a directed set. For, let $\alpha, \beta \in A$; then $\mathcal{U}_\gamma =$

$\mathcal{U}_\alpha \wedge \mathcal{U}_\beta$ is a finite open covering and it is easily seen that $p \sim_\gamma q$ implies $p \sim_\alpha q$ and $p \sim_\beta q$, i.e. $\gamma > \alpha, \beta$. Suppose $\alpha > \beta$; then to each $D \in \mathcal{D}_\alpha$ we assign $D' \in \mathcal{D}_\beta$ such that $D' \supset D$. It is easily seen that this defines a continuous mapping π_β^α of X_α onto X_β. Namely $\{X_\alpha, \pi_\beta^\alpha \mid \alpha, \beta \in A, \alpha > \beta\}$ forms an inverse system. It follows from C) that $\check{X} \subset X_\infty = \varprojlim\{X_\alpha, \pi_\beta^\alpha\}$ is a compact T_1-space.

Let us denote by f_α the natural mapping of X onto the decomposition space X_α, i.e., $f_\alpha(p)$ is the member of $\mathcal{D}_\alpha$ which contains p. Now, we define a mapping f of X into X_∞ by

$$f(p) = \{f_\alpha(p) \mid \alpha \in A\} \in X_\infty, \quad p \in X.$$

Since $\pi_\beta^\alpha(f_\alpha(p)) = f_\beta(p)$ obviously holds for every $\alpha, \beta \in A$ with $\alpha > \beta$, this definition determines a mapping of X into X_∞. Since each natural mapping f_α is continuous, f is a continuous mapping. We can say more about f.

E) *f is a topological mapping of X into X_∞.*

Proof. Suppose $U(p)$ is a given open nbd of a point p of X. Let

$$\mathcal{U}_\alpha = \{X - \{p\}, U(p)\};$$

then $\mathcal{U}_\alpha$ is a finite open covering of X, i.e., it is a member of $\{\mathcal{U}_\alpha \mid \alpha \in A\}$. Then $\mathcal{D}_\alpha$, and accordingly X_α, consists of three elements $\{p\}$, $U(p) - \{p\}$ and $X - U(p)$. We denote them, regarded as the points of X_α, by q_α, q_α' and q_α'' respectively. Then $V_\alpha = \{q_\alpha, q_\alpha'\}$ is a nbd of q_α in X_α, and therefore

$$V = (V_\alpha \times \prod\{X_{\alpha'} \mid \alpha' \neq \alpha, \alpha' \in A\}) \cap X_\infty$$

is a nbd of $f(p)$ such that

$$f^{-1}(V) \subset U(p).$$

Thus f maps every open set of X to an open set of $f(X)$. On the other hand, if p, p' are different points of X, then, since X is T_1, $X - \{p'\}$ is an open nbd of p. Hence there is a nbd V of $f(p)$ in X_∞ such that

$$f^{-1}(V) \subset X - \{p'\}.$$

Therefore $f(p') \neq f(p)$, i.e. f is one-to-one. Thus f is a topological mapping of X into X_∞.

F) $f(X) \subset \tilde{X}$.

Proof. Suppose p is a given point of X and q a point of X_∞ such that $q \neq f(p)$. Then for some $\alpha \in A$, $q_\alpha \neq f_\alpha(p)$, where f_α denotes the natural mapping of X onto X_α. Hence q_α, considered as a member of $\mathcal{D}_\alpha$, is a set $D \in \mathcal{D}_\alpha$ with $p \notin D$. Put

$$\mathcal{U}_\beta = \{X - \{p\}, X\}$$

and

$$\mathcal{U}_\gamma = \mathcal{U}_\alpha \wedge \mathcal{U}_\beta .$$

Then

$$X - \{p\} = \bigcup \{D' \mid D' \in \mathcal{D}_\gamma, D' \not\ni p\} ,$$

because every point p' different from p is contained in a member of $\mathcal{U}_\gamma$ which does not contain p, and hence p' is contained in a member of $\mathcal{D}_\gamma$, which is contained in $X - \{p\}$. On the other hand,

$$q_\gamma \in f_\gamma(X - \{p\})$$

easily follows from the fact that $q_\alpha \neq f_\alpha(p)$ and accordingly $q_\gamma \neq f_\gamma(p)$.

Therefore $f_\gamma(X - \{p\})$ is an open nbd of q_γ in X_γ which does not contain $f_\gamma(p)$. This implies that $f(p)$ is a closed set in X_∞. Thus $f(p)$ is a minimal point of X_∞, i.e. $f(p) \in \tilde{X}$. This proves our assertion.

Identifying X with $f(X)$, we regard X as a subspace of $\tilde{X}$. Generally, for a subset F of X we put

$$F^* = \{p \mid p = \{p_\alpha\} \in \tilde{X}, p_\alpha \subset F^1 \text{ for some } \alpha \in A\} .$$

G) *If F is a closed set of X, then its closure $\bar{F}$ in $\tilde{X}$ coincides with F^*.*

Proof. If $F = \emptyset$ or X, then the proposition is obviously true. So we assume $F \neq \emptyset$, $F \neq X$. Let

$$\mathcal{D}_\beta = \{F, X - F\} .$$

---

[1] Remember that p_α is a member of the decomposition $\mathcal{D}_\alpha$ at the same time.

(Namely $\mathcal{U}_\beta = \{X, X - F\}$.) Then we regard F as a point of X_β and denote it by x_0. Then it is obvious that

$$F^* = \tilde{f}_\beta^{-1}(x_0),$$

where $\tilde{f}_\beta$ denotes the projection of $\tilde{X}$ into X_β. Since $\{x_0\}$ is a closed set of X_β, F^* is closed in $\tilde{X}$. Observe that $F \subset F^*$; then we obtain $\bar{F} \subset F^*$.

Conversely, let $p = \{p_\alpha\}$ be a given point of F^*. Then $p_{\alpha_0} \subset F$ in X for some $\alpha_0 \in A$. Let $\beta_1, \ldots, \beta_k \in A$ be arbitrary. Choose $\gamma \in A$ such that $\gamma > \alpha_0, \beta_1, \ldots, \beta_k$. Then, since $p \in X_\infty$,

$$p_\gamma \subset p_{\alpha_0} \cap p_{\beta_1} \cap \cdots \cap p_{\beta_k} = F \cap p_{\alpha_0} \cap p_{\beta_1} \cap \cdots \cap p_{\beta_k} \quad \text{in } X,$$

where p_γ, p_{α_0}, etc. are regarded as subsets of X. Pick up a point x from p_γ. Then x can be expressed in $\tilde{X}$ as $x = \{x_\alpha \mid \alpha \in A\}$ satisfying $x_{\beta_i} = p_{\beta_i}$, $i = 1, \ldots, k$. Namely, the β_i-coordinates of x are equal to the β_i-coordinates of p for $i = 1, \ldots, k$. Since $x \in F$, this implies that every nbd of p in $\tilde{X}$ intersects F. Thus $p \in \bar{F}$, i.e. $F^* \subset \bar{F}$. Therefore our assertion is proved.

Theorem VII.9. *$\tilde{X}$ is the Wallman's compactification of X.*

Proof. Identifying X with $f(X)$, we regard X as a subspace of $\tilde{X}$. Then by virtue of IV.2.B), all we have to prove is that the conditions of Theorem IV.3 are satisfied by X, $\tilde{X}$ and the collection $\mathcal{G}$ of all closed sets of X. It is clear that $\tilde{X}$ is a T_1-compactification of X. To prove that $\{\bar{F} \mid F$ is a closed set of $X\}$ forms a closed basis of $\tilde{X}$, we suppose G is a given closed set of $\tilde{X}$ and $p = \{p_\alpha\}$ is a point of $\tilde{X}$ with $p \notin G$. Then we can find $\beta_1, \ldots, \beta_k \in A$ and open nbds $U_i(p_{\beta_i})$ of p_{β_i} in X_{β_i}, $i = 1, \ldots, k$, such that

$$\left(\prod_{i=1}^k U_i(p_{\beta_i}) \times \prod \{X_\alpha \mid \alpha \in A, \alpha \neq \beta_i\} \right) \cap \tilde{X} \subset \tilde{X} - G \quad \text{in } \tilde{X}, \quad (1)$$

where we may assume that

$$U_i(p_{\beta_i}) = f_{\beta_i}(U_i) \quad (2)$$

for an open set U_i of X, which contains $p_{\beta_i} \in \mathcal{D}_{\beta_i}$ as a subset and is a sum of elements of $\mathcal{D}_{\beta_i}$.

Then from (1) it follows that

$$\bigcap_{i=1}^{k} U_i \subset X - G \quad \text{in } X.$$

Put

$$U = \bigcap_{i=1}^{k} U_i \,;$$

then $F = X - U$ is a closed set of X satisfying $G \subset F^*$. For, if $q = \{q_\alpha\} \notin F^*$ in $\tilde{X}$, then

$$q_\alpha \cap \left(\bigcap_{i=1}^{k} U_i \right) \neq \emptyset \quad \text{in } X \text{ for all } \alpha \in A,$$

and especially

$$q_{\beta_i} \cap U_i \neq \emptyset, \quad i = 1, \dots, k.$$

Therefore

$$q_{\beta_i} \subset U_i, \quad i = 1, \dots, k \quad (\text{see (2)}),$$

proving

$$q \in \prod_{i=1}^{k} U_i(p_{\beta_i}) \times \prod \{X_\alpha \mid \alpha \in A, \alpha \neq \beta_i\},$$

i.e. $q \notin G$ (see (1)). This proves $G \subset F^*$.

On the other hand, we can easily see that U^* is a nbd of p in $\tilde{X}$ such that

$$U^* \cap F = \emptyset.$$

(Note that U is a union of members of $\mathscr{D}_\gamma$ for each $\gamma > \beta_i, i = 1, \dots, k$.) Thus by G), F is a closed set of X, satisfying

$$G \subset \bar{F} \not\ni p,$$

proving our assertion.

Finally, to prove

$$\overline{\bigcap_{k=1}^{n} G_k} = \bigcap_{k=1}^{n} \bar{G}_k$$

in $\tilde{X}$ for closed sets $G_k, k = 1, \dots, n$, of X, we take an arbitrary point $p = \{p_\alpha\}$ of $\bigcap_{k=1}^{n} \bar{G}_k$. Then by G),

$$p \in \bar{G}_k = G_k^*, \quad k = 1, \ldots, n,$$

which implies

$$p_{\alpha_k} \subset G_k, \quad k = 1, \ldots, n, \text{ for some } \alpha_k \in A.$$

Take α such that $\alpha > \alpha_k, k = 1, \ldots, n$; then

$$p_\alpha \subset p_{\alpha_k}, \quad k = 1, \ldots, n, \text{ in } X,$$

$$p_\alpha \subset \bigcap_{k=1}^{n} G_k.$$

Therefore

$$p \in \left(\bigcap_{k=1}^{n} G_k \right)^* = \overline{\bigcap_{k=1}^{n} G_k},$$

proving

$$\overline{\bigcap_{k=1}^{n} G_k} \supset \bigcap_{k=1}^{n} \bar{G}_k.$$

Conversely,

$$\overline{\bigcap_{k=1}^{n} G_k} \subset \bigcap_{k=1}^{n} \bar{G}_k$$

holds in every topological space. Therefore $\overline{\bigcap_{k=1}^{n} G_k} = \bigcap_{k=1}^{n} \bar{G}_k$ is concluded. Thus by IV.2.E) $\tilde{X}$ is the Wallman's compactification of X.

Corollary. *A topological space X is a compact T_1-space if and only if it is homeomorphic with the set of all minimal points of the limit space of an inverse system of finite T_0-spaces.*

Proof. This proposition is a direct consequence of Theorem VII.4 combined with C).[1]

---

[1] J. Flachsmeyer [1] also constructed Čech–Stone compactification by use of inverse limit space. As for non-compact spaces, B. Pasynkov [1] obtained interesting results with respect to inverse systems. See also B. Pasynkov [3] and V. Kljušin [1].

Theorem VII.10. *A T_2-space X is paracompact and M if and only if it is the inverse limit space of an inverse system of metric spaces with perfect bonding maps.*[1]

Proof. Suppose $X = \varprojlim\{X_\alpha, \pi_\beta^\alpha \mid \alpha, \beta \in A, \alpha > \beta\}$, where each X_α is metric and each π_β^α perfect. Fix $\beta \in A$ and denote by f_β the projection from X into X_β. We shall prove that f_β is a perfect map. Let $y \in X_\beta$ be a given point and $x \in f_\beta^{-1}(y)$. Then for any $\alpha \in A$ choose $\gamma = \gamma(\alpha) \in A$ such that $\gamma > \alpha, \beta$. Then $f_\alpha(x) \in f_\alpha^\gamma(f_\beta^{\gamma^{-1}}(y))$. Thus $f_\beta^{-1}(y)$ is a closed subset of $\prod\{f_\alpha^\gamma(f_\beta^{\gamma^{-1}}(y)) \mid \alpha \in A\}$ and hence it is compact.

Let F be a closed set of X and assume $y^i \in f_\beta(F)$, $i = 1, 2, \ldots$, satisfying $y^i \to y$ in X_β. Suppose

$$x^i \in F, \qquad x^i = \{x_\alpha^i \mid \alpha \in A\}, \qquad f_\beta(x^i) = x_\beta^i = y^i, \quad i = 1, 2, \ldots.$$

We claim that for each $\alpha \in A$ $\{x^i\}$ has a cluster point in X_α. Because if not, then select $\gamma \in A$ such that $\gamma > \alpha, \beta$. Then $\{f_\alpha^{\gamma^{-1}}(x_\alpha^i) \mid i = 1, 2, \ldots\}$ is discrete in X_γ. Thus $\{x_\gamma^i \mid i = 1, 2, \ldots\}$ and accordingly $\{f_\beta^\gamma(x_\gamma^i) \mid i = 1, 2, \ldots\} = \{x_\beta^i \mid i = 1, 2, \ldots\}$, too, is discrete, which is not true. Now, fix a cluster point x_α of $\{x_\alpha^i\}$ for each $\alpha \neq \beta$. Then we claim that $\{x_\alpha \mid \alpha \in A\}$ is a cluster point of $\{x^i \mid i = 1, 2, \ldots\}$ in $\prod_\alpha X_\alpha$. To see it, let $\alpha_1, \ldots, \alpha_k \in A$ and let U_{α_j} be a nbd of x_{α_j} in X_{α_j}, $j = 1, \ldots, k$. Select $\alpha \in A$ such that $\alpha > \alpha_j$, $j = 1, \ldots, k$, and a nbd U_α of x_α in X_α such that $f_{\alpha_j}^\alpha(U_\alpha) \subset U_{\alpha_j}$, $j = 1, \ldots, k$. For any $i_0 \in N$ there is $i \geq i_0$ such that $x_\alpha^i \in U_\alpha$. Then

$$f_{\alpha_j}^\alpha(x_\alpha^i) = x_{\alpha_j}^i \in U_{\alpha_j}, \quad j = 1, \ldots, k.$$

Thus $\{x_\alpha \mid \alpha \in A\}$ is a cluster point of $\{x^i\}$ in $\prod_\alpha X_\alpha$.

As implied by the proof of A), X is closed in $\prod_\alpha X_\alpha$. Hence $\{x_\alpha\} \in X$, i.e. $\{x_\alpha\} \in \bar{F}$ in X. Hence $\{x_\alpha\} \in F$ follows because F is closed. Hence $y = x_\beta \in f_\beta(F)$, proving that $f_\beta(F)$ is closed in X_β. Thus f_β is a perfect map from X onto a metric space. Namely, X is paracompact and M by Corollary 1 of Theorem VII.3.

Conversely, let X be paracompact T_2 and M. Denote by A the set of all normal covers of X satisfying $w\Delta$-condition. Let $\mu = \{\mathcal{U}_i \mid i = 1, 2, \ldots\} \in A$. Then define equivalence relation $\sim_\mu$ in X by

$$x \sim_\mu y \text{ if and only if } y \in \bigcap_{i=1}^{\infty} S(x, \mathcal{U}_i).$$

[1] The 'only if' part is due to V. Kljusin [1] and the 'if' part to K. Morita [6].

We define direction between two elements μ, ν of A by

$$\mu > \nu \text{ if and only if } x \sim_\mu y \text{ implies } x \sim_\nu y.$$

Then $\langle A, > \rangle$ is a directed set. As we saw in the proof of Theorem VII.3, the quotient space $X_\mu = X/\sim_\mu$ is a metrizable space, and the quotient map f_μ from X onto X_μ is a perfect map. For each μ, $\nu \in A$ with $\mu > \nu$, we define a map f^μ_ν from X_μ into X_ν by

$$f^\mu_\nu(x_\mu) = f_\nu(f_\mu^{-1}(x_\mu)), \quad x_\mu \in X_\mu.$$

Then it is easy to see that f^μ_ν is a perfect map from X_μ onto X_ν. (The detail is left to the reader.) It is also obvious that $f^\mu_\nu \circ f^\lambda_\mu = f^\lambda_\nu$ whenever $\lambda > \mu > \nu$. Let

$$\tilde{X} = \varprojlim \{X_\mu, f^\mu_\nu \mid \mu, \nu \in A, \mu > \nu\}.$$

We define a map φ from $\tilde{X}$ to X as follows. Consider a given point $\tilde{x} = \{x_\mu \mid \mu \in A\}$ of $\tilde{X}$. Then $\{f_\mu^{-1}(x_\mu) \mid \mu \in A\}$ is a closed filter base of X consisting of compact sets, and hence its intersection is non-empty. If x and y are two distinct points of X, then there is $\{\mathcal{U}_i\} = \mu \in A$ such that $y \notin S(x, \mathcal{U}_i)$ for some i. Hence $\cap \{f_\mu^{-1}(x_\mu) \mid \mu \in A\}$ is a singleton $\{x\}$. We define that $\varphi(\tilde{x}) = x$. It is easy to see that φ is a homeomorphism from $\tilde{X}$ onto X. Thus the theorem is proved.

5. Theory of selection

In the present section we are going to generalize Tietze's extension theorem by use of a new point of view due to E. Michael [3]. Tietze-type extension theory is treated as a special case of Michael's selection theory, which gives new characterizations of normal spaces, collectionwise normal spaces and fully normal spaces.

Definition VII.6. Let X and Y be topological spaces and 2^Y the collection of all non-empty subsets of Y. In this section we call a mapping φ of X into 2^Y a *carrier*. Suppose F is a subset of X; then we shall denote by φ_F the restriction of φ to F. φ_F is then a carrier of F into 2^Y. A carrier φ of X into 2^Y is called *lower semi-continuous* (*l.s.c.*) if for each open set V of Y,

$$\varphi^{-1}\{V\} = \{p \mid p \in X, \varphi(p) \cap V \neq \emptyset\}$$

is an open set of X.

Suppose φ is a carrier of X into 2^Y. If a continuous mapping f of X into Y satisfies

$$f(p) \in \varphi(p) \quad \text{for every } p \in X,$$

then f is called a *selection* for φ.

Example VII.9. A mapping φ of X into Y may be considered a special carrier of X into 2^Y such that each $\varphi(p)$ is a one-point subset of Y. Then this carrier is l.s.c. if and only if the mapping φ is continuous, because $\varphi^{-1}\{V\} = \varphi^{-1}(V)$ for every subset V of Y.

Let F be a closed set of X and f a continuous mapping defined over F. Then the carrier φ defined by

$$\varphi(p) = \begin{cases} f(p), & p \in F, \\ X, & p \in X - F, \end{cases}$$

is also l.s.c. If there is a selection g for φ, then it is clearly a continuous extension of f over X.

A) *A carrier φ of X into 2^Y is l.s.c. if and only if for every $p \in X$, $q \in \varphi(p)$ and nbd V of q, there is a nbd U of p such that for every $p' \in U$, $\varphi(p') \cap V \neq \emptyset$.*

Proof. Suppose φ is a l.s.c. carrier and $p \in X$, $q \in \varphi(p)$ and V is a nbd of q. We may assume that V is an open set of Y. Hence by the definition of l.s.c., $U = \varphi^{-1}\{V\}$ is an open nbd of p. If $p' \in U$, then $\varphi(p') \cap V \neq \emptyset$, and hence the condition is necessary.

Conversely, assume the condition is satisfied by a carrier φ of X into 2^Y. Suppose V is a given open set of X, and $p \in \varphi^{-1}\{V\}$. Then

$$\varphi(p) \cap V \neq \emptyset.$$

Take $q \in \varphi(p) \cap V$; then V is a nbd of q, and hence by the hypothesis, there is a nbd U of p such that $p' \in U$ implies

$$\varphi(p') \cap V \neq \emptyset.$$

Hence

$$U \subset \varphi^{-1}\{V\},$$

which means that $\varphi^{-1}\{V\}$ is an open set of X, i.e. φ is a l.s.c. carrier.

B) *Let X and Y be topological spaces and assume that to each point q of Y, an open nbd $V(q)$ is assigned so that if $q \in V(q')$, then there are nbds W of q and W' of q' such that $q'' \in W'$ implies $W \subset V(q'')$. Suppose φ and ψ are l.s.c. carriers of X into 2^Y satisfying*

$$\varphi(p) \cap V(\psi(p)) \neq \emptyset \quad \text{for every } p \in X,$$

where $V(\psi(p)) = \bigcup \{V(q) \mid q \in \psi(p)\}$. Then $\theta(p) = \varphi(p) \cap V(\psi(p))$ is a l.s.c. carrier.

Proof. Let $q \in \theta(p)$ and M a given nbd of q in Y. Since $q \in V(\psi(p))$, there is $q' \in \psi(p)$ such that $q \in V(q')$. We take a nbd W of q and a nbd W' of q' satisfying the condition mentioned in the proposition. Then, since ψ is l.s.c., there is a nbd U of p such that $p' \in U$ implies

$$\psi(p') \cap W' \neq \emptyset.$$

On the other hand, since $q \in \varphi(p)$ and φ is l.s.c., there is a nbd U' of p such that $p' \in U'$ implies

$$\varphi(p') \cap W \cap M \neq \emptyset.$$

Thus $U'' = U \cap U'$ is a nbd of p such that $p' \in U''$ implies

$$\theta(p') \cap M = \varphi(p') \cap V(\psi(p')) \cap M \supset \varphi(p') \cap W \cap M \neq \emptyset.$$

(Note that $\psi(p') \cap W' \neq \emptyset$ and thus we can select $q'' \in \psi(p') \cap W'$ for which $V(q'') \supset W$ and hence $V(\psi(p')) \supset W$.) Therefore $\theta(p') \cap M \neq \emptyset$. Hence by A), θ is l.s.c.

C) *Let X and Y be topological spaces and $\mathcal{A}$ a subcollection of 2^Y containing all one-point subsets as members. Then the following two conditions are equivalent:*

   (i) *If φ is a l.s.c. carrier of X into $\mathcal{A}$, then for every closed set F of X, each selection for φ_F can be extended to a selection for φ.*

   (ii) *Every l.s.c. carrier φ of X into $\mathcal{A}$ has a selection.*

Proof. Since (i) clearly implies (ii), we shall prove only that (ii) implies (i).

We suppose g is a selection for φ_F, i.e., g is a continuous mapping of F into Y satisfying

$$g(p) \in \varphi(p) \quad \text{for all } p \in F.$$

We define a carrier ψ of X into $\mathscr{A}$ by

$$\psi(p) = \begin{cases} \{g(p)\} & \text{for } p \in F, \\ \varphi(p) & \text{for } p \notin F. \end{cases}$$

We can assert that ψ is l.s.c. For, let V be an open set of Y. Then

$$\psi^{-1}\{V\} = \{p \mid \psi(p) \cap V \neq \emptyset\} = (\varphi^{-1}\{V\} - F) \cup g^{-1}(V),$$

and hence in view of the fact that φ is l.s.c. we can verify that $\psi^{-1}\{V\}$ is open. This means that ψ is l.s.c. Thus it follows from (ii) that there is a selection h for ψ. Now it is obvious that h is an extension of g over X, and a selection for φ.

Let us recall Definition IV.2, to observe some simple properties of normed linear spaces.

D) *Let Y be a normed linear space. We consider mappings $f(x, y) = x + y$ and $g(\alpha, x) = \alpha x$ of $Y \times Y$ into Y and of $E^1 \times Y$ into Y. Then f and g are both continuous mappings.*[1]

E) *Let α, A and V be a real number, a set and an open set of a normed linear space Y, respectively. Then $A + V = \{x + v \mid x \in A, v \in V\}$ and $\alpha V = \{\alpha v \mid v \in V\}$ are open sets of Y.*

F) *Let Y be a normed linear space. Then for every nbd U of 0 there is a convex symmetric open nbd V of 0 such that $V \subset U$.*

Proof. Take a spherical nbd $S_\varepsilon(0) \subset U$. Then it is easy to see that $V = S_\varepsilon(0)$ is a desired nbd of 0.

---

[1] If a linear (not necessarily normed) space Y is a T_2-space at the same time, and f and g are continuous, then Y is called a *linear topological space*.

G) *Let X be a paracompact T_2-space. Then for every open covering $\mathcal{U}$ of X, there is a collection $\{f_\alpha \mid \alpha \in A\}$ of non-negative, real-valued continuous functions over X such that:*

(i) $\sum \{f_\alpha \mid \alpha \in A\} = 1$,

(ii) *every point p of X has a nbd in which all but finitely many of the f_α vanish,*

(iii) *for every α, there is $U \in \mathcal{U}$ for which $f_\alpha(X - U) = 0$.*

Conversely, if X is a T_1-space, and for every open covering $\mathcal{U}$ there is a collection $\{f_\alpha \mid \alpha \in A\}$ of non-negative, real-valued continuous functions over X satisfying (i) *and* (iii), *then X is paracompact T_2.*[1]

Proof. Suppose X is paracompact T_2. We take a locally finite open refinement $\mathcal{V} = \{V_\alpha \mid \alpha \in A\}$ of the given open covering $\mathcal{U}$. Since X is normal, there is an open covering $\mathcal{W} = \{W_\alpha \mid \alpha \in A\}$ with $\bar{W}_\alpha \subset V_\alpha$. Using Urysohn's lemma, for each α we construct a continuous function g_α over X such that

$$g_\alpha(\bar{W}_\alpha) = 1, \qquad g_\alpha(R - V_\alpha) = 0 \quad \text{and} \quad 0 \leq g_\alpha \leq 1.$$

Putting

$$f_\alpha = g_\alpha \bigg/ \sum_{\alpha \in A} g_\alpha,$$

we get continuous functions f_α, $\alpha \in A$, satisfying the desired conditions.

Conversely, suppose X satisfies the condition. Let p and F be a point and a closed set of X respectively, such that $p \notin F$. Since $\{X - \{p\}, X - F\}$ is an open covering of X, we can construct a collection $\{f_\alpha \mid \alpha \in A\}$ of continuous functions satisfying (i) and (iii). By virtue of (i) there is an α for which $f_\alpha(p) > 0$. Then $f_\alpha(F) = 0$ follows from (iii). Hence

$$U = \{x \in X \mid f_\alpha(x) > \tfrac{1}{2} f_\alpha(p)\}$$

is an open nbd of p, satisfying

$$\bar{U} \cap F = \emptyset,$$

which implies that X is regular.

Now, let $\mathcal{U}$ be a given open covering and $\{f_\alpha \mid \alpha \in A\}$ a collection of continuous functions satisfying (i) and (iii). Then for each pair $\alpha \in A$ and

[1] Such a collection $\{f_\alpha \mid \alpha \in A\}$ is called a *partition of unity* subordinated to $\mathcal{U}$.

a natural number n, we put

$$U_{\alpha n} = \{x \in X \mid f_\alpha(x) > 1/n\}.$$

It is easy to see that $\{U_{\alpha n} \mid \alpha \in A\}$ is locally finite for each n. For, let p be a given point of X, then by (i) $\sum f_\alpha(p) = 1$, and hence there are α_i, $i = 1, \ldots, k$, such that

$$\sum_{i=1}^{k} f_{\alpha_i}(p) > 1 - \frac{1}{n}.$$

Putting

$$V = \left\{x \in X \mid \sum_{i=1}^{k} f_{\alpha_i}(x) > 1 - \frac{1}{n}\right\},$$

we get an open nbd V of p. Then it follows from (i) that V does not intersect $U_{\alpha n}$ if $\alpha \neq \alpha_i$, $i = 1, \ldots, k$. Thus $\{U_{\alpha n} \mid \alpha \in A\}$ is locally finite. Hence by (iii) $\{U_{\alpha n} \mid \alpha \in A, n = 1, 2, \ldots\}$ is a σ-locally finite open refinement of $\mathscr{U}$. Thus by Theorem V.1, X is paracompact.

In the rest of this section we denote by $\mathscr{K}(Y)$, $\mathscr{F}(Y)$ and $\mathscr{C}(Y)$ the collections of all non-empty convex sets, all non-empty convex closed sets and all non-empty convex compact sets plus Y, respectively, of a linear space Y.

H) *Let X be a paracompact T_2-space, Y a normed linear space, and ψ a l.s.c. carrier of X into $\mathscr{K}(Y)$. Suppose V is a convex open nbd of $0 \in Y$. Then there is a continuous mapping f of X into Y such that*

$$f(p) \in \psi(p) + V \quad \text{for every } p \in X.$$

Proof. For each point q of Y, we put

$$U_q = \{p \in X \mid q \in \psi(p) + V\}. \tag{1}$$

Then

$$U_q = \{p \mid \psi(p) \cap (q - V) \neq \emptyset\} = \psi^{-1}\{q - V\},$$

where $q - V = \{q - v \mid v \in V\}$. Since $q - V$ is an open set of Y and ψ is l.s.c., U_q is an open set of X. Let p be a given point of X. Then for an arbitrary point $q \in \psi(p)$,

$$q \in \psi(p) \cap (q - V) \neq \emptyset,$$

and hence $p \in U_q$. Therefore $\mathcal{U} = \{U_q \mid q \in Y\}$ is an open covering of X. Since X is paracompact T_2, we can construct functions f_α, $\alpha \in A$, which satisfy the conditions of G).

For each $\alpha \in A$, we choose $q(\alpha)$ such that

$$f_\alpha(X - U_{q(\alpha)}) = 0 . \tag{2}$$

Putting

$$f(p) = \sum \{f_\alpha(p)q(\alpha) \mid \alpha \in A\}, \quad p \in X,$$

we get the desired mapping. (Note that all $f_\alpha(p)$ vanish except at most finitely many.) The continuity of f follows from (ii) of G). On the other hand, by (2), $f_\alpha(p) \neq 0$ implies $p \in U_{q(\alpha)}$, and hence

$$q(\alpha) \in \psi(p) + V$$

by the definition (1) of $U_{q(\alpha)}$. Since $f_\alpha(p) \neq 0$ holds only for finitely many α, it follows from (i) of G) that

$$f(p) = \sum f_\alpha(p)q(\alpha) \in \psi(p) + V,$$

because both of $\psi(p)$ and V are convex sets of Y and therefore $\psi(p) + V$ is convex.

Theorem VII.11. *A T_1-space X is paracompact T_2 if and only if for every Banach space Y and every l.s.c. carrier φ of X into $\mathcal{F}(Y)$, there is a selection.*

Proof. Necessity. By F) there is a nbd basis $\{V_i \mid i = 1, 2, \ldots\}$ of 0 in X consisting of symmetric, convex, open sets V_i. In view of E) we may assume that V_i satisfies

$$V_{i+1} \subset \tfrac{1}{2}V_i = \{\tfrac{1}{2}v \mid v \in V_i\}, \quad i = 1, 2, \ldots, \tag{1}$$

and

$$V_1 \subset S_1(0) .$$

Now, we can construct continuous mappings f_i, $i = 1, 2, \ldots$, of X into Y such that

$$f_1(p) \in \varphi(p) + V_1 \quad \text{for every } p \in X$$

and

$$f_i(p) \in (f_{i-1}(p) + 2V_{i-1}) \cap (\varphi(p) + V_i)$$

$$\text{for every } p \in X, \ i = 2, 3, \ldots .^{[1]}$$

For $i = 1$, the existence of such a mapping f_1 is a direct consequence of H). To define f_i by induction on i, we assume that $f_1, \ldots, f_{i-1}$ have been defined. Since

$$f_{i-1}(p) \in \varphi(p) + V_{i-1},$$

and V_{i-1} is symmetric,

$$\varphi(p) \cap (f_{i-1}(p) + V_{i-1}) \neq \emptyset, \quad p \in X.$$

Since we can easily show that $\{q + V_{i-1} \mid q \in Y\}$ satisfies the condition of $\{V(q) \mid q \in Y\}$ in B), by the same proposition we can conclude that

$$\varphi_i(p) = \varphi(p) \cap (f_{i-1}(p) + V_{i-1}), \quad p \in X, \tag{2}$$

defines a l.s.c. carrier of X into $\mathcal{K}(Y)$. (We regard $f_{i-1}(p)$ as a l.s.c. carrier.) It follows from H) that there is a continuous mapping f_i of X into Y such that

$$f_i(p) \in \varphi_i(p) + V_i \quad \text{for every } p \in X.$$

This combined with (2) implies

$$f_i(p) \in f_{i-1}(p) + V_{i-1} + V_i \subset f_{i-1}(p) + 2V_{i-1} \quad \text{for every } p \in X$$

and

$$f_i(p) \in \varphi(p) + V_i \quad \text{for every } p \in X. \tag{3}$$

(Note that (1) combined with the convexity of V_{i-1} implies $V_i \subset \frac{1}{2}V_{i-1} \subset V_{i-1}$.)

Thus $f_i(p)$ satisfies the desired condition. Hence it follows from (1) that

$$f_i(p) \in f_{i-1}(p) + 2V_{i-1} \subset f_{i-1}(p) + (\tfrac{1}{2})^{i-3}V_1 \subset f_{i-1}(p) + (\tfrac{1}{2})^{i-3}S_1(0),$$

---

[1] As shown in E), we define that $2V_{i-1} = \{2v \mid v \in V_{i-1}\}$, which is, by the convexity of V_{i-1}, equal to $V_{i-1} + V_{i-1} = \{v + v' \mid v, v' \in V_{i-1}\}$.

which implies

$$\|f_i(p) - f_{i-1}(p)\| < (\tfrac{1}{2})^{i-3}.$$

Therefore $\{f_i(p)\}$ is a Cauchy point sequence of Y, and hence $\{f_i\}$ uniformly converges to a continuous mapping f of X into (see 1.H)). If we assume $f(p) \not\in \varphi(p)$ for some $p \in X$, then

$$\varphi(p) \cap (f(p) + V_i) = \emptyset \quad \text{for some } i,$$

because $\varphi(p)$ is closed, and $\{f(p) + V_i \mid i = 1, 2, \ldots\}$ is a nbd basis of $f(p)$. Hence

$$(\varphi(p) + V_{i+1}) \cap (f(p) + V_{i+1}) = \emptyset \tag{4}$$

follows from (1). Since $f_i(p) \to f(p)$, there is some $k \geq i + 1$ such that

$$f_k(p) \in f(p) + V_{i+1}.$$

On the other hand, from (3) it follows that

$$f_k(p) \in \varphi(p) + V_k \subset \varphi(p) + V_{i+1},$$

which contradicts (4). Therefore $f(p) \in \varphi(p)$, i.e. f is the desired selection.

Sufficiency. Let $\mathcal{U}$ be a given open covering of X. We define a normed linear space $X = l_1(\mathcal{U})$ as follows. The points of $l_1(\mathcal{U})$ are the real-valued functions y defined over $\mathcal{U}$ such that

$$\sum \{|y(U)| \mid U \in \mathcal{U}\} < +\infty,$$

and the norm of y is defined by

$$\|y\| = \sum \{|y(U)| \mid U \in \mathcal{U}\}.$$

We can easily verify that $Y = l_1(\mathcal{U})$ is a Banach space. Put

$$C = \{y \mid y \in Y, y(U) \geq 0 \text{ for all } U \in \mathcal{U}, \text{ and } \|y\| = 1\}. \tag{5}$$

Then it is almost obvious that $C \in \mathcal{F}(Y)$. It is also clear that, for each $p \in X$,

$$C'(p) = \{y \mid y \in Y, \ y(U) = 0 \text{ if } U \in \mathscr{U} \text{ and } p \notin U\} \in \mathscr{F}(Y). \quad (6)$$

Therefore,

$$C \cap C'(p) \in \mathscr{F}(Y) \quad \text{for every } p \in X.$$

Put

$$\varphi(p) = C \cap C'(p), \quad p \in X ; \quad\quad (7)$$

then we can assert that φ is a l.s.c. carrier of X into $\mathscr{F}(Y)$.

To prove that φ is l.s.c., we show that given $y \in C$ and $\varepsilon > 0$, then there are $y' \in C$ and $U_1, \ldots, U_k \in \mathscr{U}$ such that

$$\|y - y'\| < \varepsilon, \quad\quad (8)$$

$$y(U_i) > 0, \quad i = 1, \ldots, k, \quad\quad (9)$$

and

$$y'(U) = 0 \quad \text{if } U \neq U_i, \ i = 1, \ldots, k. \quad\quad (10)$$

For, since it follows from (5) that

$$\sum \{|y(U)| \mid U \in \mathscr{U}\} = 1$$

and

$$y(U) \geqslant 0, \quad U \in \mathscr{U},$$

we can choose $U_1, \ldots, U_k \in \mathscr{U}$ such that

$$y(U_1) + \cdots + y(U_k) = \delta > 1 - \tfrac{1}{2}\varepsilon$$

and

$$y(U_i) > 0, \quad i = 1, \ldots, k.$$

Now, we define a function y' over $\mathscr{U}$ by

$$y'(U) = 0 \quad \text{if } U \neq U_i, \ i = 1, \ldots, k,$$

$$y'(U_1) = y(U_1) + 1 - \delta$$

and

$$y'(U_i) = y(U_i), \quad i = 2, \ldots, k.$$

Then it is clear that $y' \in C$ (see (5)). Furthermore,

$$\|y - y'\| = \sum \{|y(U) - y'(U)| \mid U \in \mathcal{U}\}$$

$$= 1 - \delta + \sum \{y(U) \mid U \neq U_i, i = 1, \ldots, k\}$$

$$= 2(1 - \delta) < \varepsilon$$

holds. Thus y' is the desired function.

To show that φ defined by (7) is l.s.c., we take a given point $y \in \varphi(p)$ and an ε-nbd $S_\varepsilon(y)$ of y for a given $\varepsilon > 0$. Then we choose $U_1, \ldots, U_k \in \mathcal{U}$ and $y' \in C$ which satisfy (8)–(10). Since by (9)

$$y(U_i) > 0, \quad i = 1, \ldots, k,$$

and

$$y \in \varphi(p) \subset C'(p),$$

it follows from the definition (6) of $C'(p)$ that

$$p \in U_i, \quad i = 1, \ldots, k.$$

Therefore $U_0 = U_1 \cap \cdots \cap U_k$ is a nbd of p. Suppose p' is an arbitrary point of U_0; then it follows from (10) that

$$y'(U) = 0 \quad \text{for every } U \in \mathcal{U} \text{ with } p' \notin U.$$

Hence $y' \in C'(p')$ follows from (6). This combined with $y' \in C$ implies that $y' \in \varphi(p')$ (see (7)). Therefore by (8)

$$y' \in \varphi(p') \cap S_\varepsilon(y),$$

i.e., $\varphi(p') \cap S_\varepsilon(y) \neq \emptyset$ for each point p' of the nbd U_0 of p. Thus by A), φ is l.s.c.

Now, using the hypothesis we can choose a selection f for φ. For each $U \in \mathcal{U}$, we define a real-valued function f_U over X by

$$f_U(p) = [f(p)](U), \quad p \in X.$$

Since f is continuous, f_U is also continuous by the definition of the topology of $l_1(\mathcal{U})$. Furthermore, we note that it follows from $f(p) \in \varphi(p) \subset C$ and (5) that $f_U(p) \geq 0$ and $\sum \{f_U(p) \mid U \in \mathcal{U}\} = 1$, $p \in X$. On the other hand, if $p \notin U$, then it follows from $f(p) \in \varphi(p) \subset C'(p)$ and (6) that

$$f_U(p) = [f(p)](U) = 0\,.$$

Therefore $\{f_U \mid U \in \mathcal{U}\}$ satisfies the conditions (i) and (iii) of G), and hence X is paracompact T_2 by virtue of the last part of $G)$.

We need the following properties of normal spaces and collectionwise normal spaces to extend selection theory to those spaces.

I) *Every point-finite, countable, open covering of a normal space has a locally finite open refinement.*

Proof. The proof is quite analogous to the proof (iv) $\Rightarrow$ (i) of Theorem V.5, so it is left to the reader.

J) *Let $\mathcal{U}$ be a point-finite open covering of a collectionwise normal space X. Then there is a locally finite open refinement of $\mathcal{U}$.*

Proof. Suppose $\mathcal{U} = \{U_\alpha \mid \alpha \in A\}$; then since X is normal, there is an open covering $\mathcal{V} = \{V_\alpha \mid \alpha \in A\}$ such that $\bar{V}_\alpha \subset U_\alpha$. For each finite subset $\{\alpha_1, \ldots, \alpha_k\}$ of A we define

$$U(\alpha_1 \ldots \alpha_k) = \{p \mid p \in U_{\alpha_i}, i = 1, \ldots, k;\ p \notin U_\alpha$$
$$\text{for } \alpha \neq \alpha_i, i = 1, \ldots, k\}\,.$$

Now, we shall define open sets $V(\alpha_1 \ldots \alpha_k)$ and V'_k which satisfy the following conditions (1), (2) and (3):

$$U_{\alpha_1} \cap \cdots \cap U_{\alpha_k} \supset V(\alpha_1 \ldots \alpha_k) \supset U(\alpha_1 \ldots \alpha_k) - \bigcup_{i=1}^{k-1} V'_i, \qquad (1)$$

where

$$V'_i = \bigcup \{V(\alpha_1 \ldots \alpha_i) \mid \alpha_1, \ldots, \alpha_i \in A\}, \qquad (2)$$

$$\{V(\alpha_1 \ldots, \alpha_k) \mid \alpha_1, \ldots, \alpha_k \in A\} \text{ is discrete for each } k\,. \qquad (3)$$

Note that (1) and (2) imply

$$\bigcup_{i=1}^{k} V'_i \supset \bigcup \{U(\alpha_1 \ldots \alpha_i) \mid \alpha_1, \ldots, \alpha_i \in A, i = 1, \ldots, k\}\,. \qquad (4)$$

We shall define such open sets $V(\alpha_1 \ldots \alpha_k)$ and V'_k by induction on the number k.

For $k = 1$, $\{U(\alpha_1) \mid \alpha_1 \in A\}$ is a discrete closed collection, and hence by the collectionwise normality of X there is a discrete open collection $\{V'(\alpha_1) \mid \alpha_1 \in A\}$ satisfying $U(\alpha_1) \subset V'(\alpha_1)$ (see V.3.B)). Since $U(\alpha_1) \subset U_{\alpha_1}$, putting

$$V(\alpha_1) = V'(\alpha_1) \cap U_{\alpha_1},$$

we get $V(\alpha_1)$, $\alpha_1 \in A$, which satisfy (1) and (3).

Assume we have constructed $V(\alpha_1 \ldots \alpha_i)$ and accordingly V'_i (by (2)) for $i = 1, \ldots, k - 1$. Now we can assert that $\{U(\alpha_1 \ldots \alpha_k) - \bigcup_{i=1}^{k-1} V'_i \mid \alpha_1, \ldots, \alpha_k \in A\}$ is a discrete closed collection of X. To prove that $U(\alpha_1 \ldots \alpha_k) - \bigcup_{i=1}^{k-1} V'_i$ is closed, take a point

$$p \notin U(\alpha_1 \ldots \alpha_k) - \bigcup_{i=1}^{k-1} V'_i. \tag{5}$$

If

$$p \in \bigcup_{i=1}^{k-1} V'_i,$$

then $\bigcup_{i=1}^{k-1} V'_i$ is an open nbd of p which does not intersect $U(\alpha_1 \ldots \alpha_k) - \bigcup_{i=1}^{k-1} V'_i$. If

$$p \notin \bigcup_{i=1}^{k-1} V'_i,$$

then it follows from (4) that p is contained in at least k members of $\mathcal{U}$. Since $p \notin U(\alpha_1 \ldots \alpha_k)$ holds by (5), $p \in U_\alpha$ for some $\alpha \in A$ with $\alpha \neq \alpha_i$, $i = 1, \ldots, k$. Then U_α is an open nbd of p which does not intersect $U(\alpha_1 \ldots \alpha_k) - \bigcup_{i=1}^{k-1} V'_i$. Thus in any case

$$p \notin \overline{U(\alpha_1 \ldots \alpha_k) - \bigcup_{i=1}^{k-1} V'_i},$$

which proves our assertion. Observe that for each $x \in X$ either $\bigcup_{i=1}^{k-1} V'_i$ or $W = \bigcap \{U_\alpha \mid x \in U_\alpha \in \mathcal{U}\}$ is a nbd of x which intersects at most one member of the concerned collection. Thus the collection is a discrete closed collection. Therefore by the collectionwise normality of X, there is a discrete open collection $\{V'(\alpha_1 \ldots \alpha_k) \mid \alpha_1, \ldots, \alpha_k \in A\}$ such that

$$U(\alpha_1 \ldots \alpha_k) - \bigcup_{i=1}^{k-1} V'_i \subset V'(\alpha_1 \ldots \alpha_k).$$

Since

$$U(\alpha_1 \ldots \alpha_k) - \bigcup_{i=1}^{k-1} V_i' \subset U_{\alpha_1} \cap \cdots \cap U_{\alpha_k}$$

is obvious, putting

$$V(\alpha_1 \ldots \alpha_k) = V'(\alpha_1 \ldots \alpha_k) \cap U_{\alpha_1} \cap \cdots \cap U_{\alpha_k},$$

we get open sets $V(\alpha_1 \ldots \alpha_k)$ which satisfy (1) and (3). Finally we put

$$V_k' = \bigcup \{V(\alpha_1 \ldots \alpha_k) \mid \alpha_1, \ldots, \alpha_k \in A\}.$$

This completes our induction process to define $V(\alpha_1 \ldots \alpha_k)$ and V_k'.

Now we note that $\{V_k' \mid k = 1, 2, \ldots\}$ is a point-finite open covering of X. Because if $p \in V_k'$, then $p \in V(\alpha_1 \ldots \alpha_k)$ for some $\alpha_1, \ldots, \alpha_k \in A$, and hence by (1) p is contained in at least k members of $\mathcal{U}$. Since $\mathcal{U}$ is point-finite, this means that $\{V_k' \mid k = 1, 2, \ldots\}$ is point-finite. Thus by I), we can construct a locally finite open refinement $\{W_k \mid k = 1, 2, \ldots\}$ of $\{V_k' \mid k = 1, 2, \ldots\}$ such that $W_k \subset V_k'$. Put

$$W(\alpha_1 \ldots \alpha_k) = V(\alpha_1 \ldots \alpha_k) \cap W_k ;$$

then $\{W(\alpha_1 \ldots \alpha_k) \mid \alpha_1, \ldots, \alpha_k \in A, \ k = 1, 2, \ldots\}$ is the desired locally finite open refinement of $\mathcal{U}$ (see (1)–(3)).

K) *Let X be a collectionwise normal space, Y a normed linear space and ψ a l.s.c. carrier of X into $\mathscr{C}(Y)$. Then for a given convex open nbd V of 0, there is a continuous mapping f of X into Y such that*

$$f(p) \in \psi(p) + V \quad \text{for all } p \in X.$$

Proof. Put

$$U = \{p \in X \mid 0 \in \psi(p) + V\}.$$

Then we can easily see that U is an open set of X, because ψ is a l.s.c. carrier. Since Y is paracompact, there is a locally finite open covering $\mathcal{W} = \{W_\beta \mid \beta \in B\}$ satisfying

$$\mathcal{W} < \{y - V \mid y \in Y\}. \tag{1}$$

Now it is obvious that $\{\psi^{-1}\{W_\beta\} \mid \beta \in B\}$ is an open covering of X. We can say moreover that the covering is point-finite on $X - U$. For, if $p \in X - U$, then

$$0 \not\in \psi(p) + V,$$

which implies that $\psi(p) \neq Y$. Therefore $\psi(p)$ is compact. Hence $\psi(p)$ intersects at most finitely many of W_β because $\mathscr{W}$ is locally finite. This implies that p is contained in at most finitely many of $\psi^{-1}\{W_\beta\}$.

Since $X - U$ is a closed set of a collectionwise normal space, it is also collectionwise normal. Therefore, by J), there is a locally finite open covering $\{V'_\beta \mid \beta \in B\}$ of $X - U$ such that

$$V'_\beta \subset \psi^{-1}\{W_\beta\} \cap (X - U).$$

Thus by V.3.C) there is a locally finite open covering $\{V_\beta \mid \beta \in B\}$ of X such that

$$V_\beta \cap (X - U) \subset V'_\beta.$$

Put

$$\mathscr{V} = \{V_\beta \cap \psi^{-1}\{W_\beta\}\} \mid \beta \in B\};$$

then $\mathscr{V}$ is a locally finite open collection of X and covers $X - U$.

Thus $\mathscr{V}' = \mathscr{V} \cup \{U\}$ is a locally finite open covering of X.

Therefore by G), there is a collection $\{f_\alpha \mid \alpha \in A\}$ of non-negative, real-valued, continuous functions over X satisfying (i)–(iii) of G). Especially in view of the condition (iii), to each $\alpha \in A$ we can assign $\beta(\alpha) \in B$ such that

$$f_\alpha(X - V_{\beta(\alpha)} \cap \psi^{-1}\{W_{\beta(\alpha)}\}) = 0$$

unless $f_\alpha(X - U) = 0$. To each $\alpha \in A$ we assign $q(\alpha) \in Y$ as follows. If $f_\alpha(X - U) = 0$, then $q(\alpha) = 0$. Otherwise $q(\alpha)$ is a point of Y satisfying

$$W_{\beta(\alpha)} \subset q(\alpha) - V \quad \text{(see (1))}.$$

Put

$$f(p) = \sum \{f_\alpha(p)q(\alpha) \mid \alpha \in A\}, \quad p \in X.$$

By (ii) of G), f is a continuous mapping of X into Y.

To prove $f(p) \in \psi(p) + V$, let $p \in U$. (In the case that $p \notin U$ the proof proceeds in a similar manner.) Then for each α such that $f_\alpha(p) \neq 0$, $f_\alpha(X - U) \neq 0$, we obtain

$$p \in V_{\beta(\alpha)} \cap \psi^{-1}\{W_{\beta(\alpha)}\} \, .$$

Thus

$$\psi(p) \cap W_{\beta(\alpha)} \neq \emptyset \, ,$$

and hence

$$\psi(p) \cap (q(\alpha) - V) \neq \emptyset \, ,$$

which implies

$$q(\alpha) \in \psi(p) + V \, .$$

For each α such that $f_\alpha(p) \neq 0$, $f_\alpha(X - U) = 0$, we obtain $p \in U$ and accordingly

$$q(\alpha) = 0 \in \psi(p) + V \, .$$

Since $\psi(p)$ and V are both convex, $\psi(p) + V$ is also convex. Therefore, from (i) of G), we get

$$f(p) = \sum f_\alpha(p) q(\alpha) \in \psi(p) + V \, ,$$

proving the assertion.

Theorem VII.12. *A T_1-space X is collectionwise normal if and only if for every Banach space Y and for every l.s.c. carrier φ of X into $\mathscr{C}(Y)$, there is a selection.*

Proof. The 'only if' part is directly derived from K) in a quite similar way as Theorem VII.11 was derived from H). The 'if' part is a direct consequence of the following proposition.

Corollary. *A T_1-space X is collectionwise normal if and only if every continuous mapping of every closed set F of X into a Banach space can be continuously extended over X.*[1]

---

[1] Essentially proved first by C. H. Dowker [4].

Proof. The 'only if' part is a direct consequence of the 'only if' part of the theorem. To prove the 'if' part, we denote by $\{F_\alpha \mid \alpha \in A\}$ a discrete closed collection of X. We denote by $H(A)$ the generalized Hilbert space with the index set A. Define a mapping f of the closed set $\cup \{F_\alpha \mid \alpha \in A\}$ of X into $H(A)$ by

$$f(F_\alpha) = p^\alpha,$$

where $p^\alpha = \{x_{\alpha'} \mid \alpha' \in A\}$, $x_\alpha = 1$, $x_{\alpha'} = 0$ for $\alpha' \neq \alpha$. Then f is easily seen to be continuous over $\cup \{F_\alpha \mid \alpha \in A)$, and hence by the hypothesis there is a continuous extension g of f over X. We consider a spherical nbd $S_{1/2}(p^\alpha)$ for each p^α. Then

$$S_{1/2}(p^\alpha) \cap S_{1/2}(p^{\alpha'}) = \emptyset \quad \text{if } \alpha \neq \alpha'.$$

Therefore the sets

$$U_\alpha = g^{-1}(S_{1/2}(p^\alpha)), \quad \alpha \in A,$$

form a disjoint open collection in X satisfying $U_\alpha \supset F_\alpha$. This proves that X is collectionwise normal.

L) *Let $\{F_i \mid i = 1, 2, \ldots\}$ be a countable, discrete, closed collection in a normal space X, then there is a discrete, open collection $\{V_i \mid i = 1, 2, \ldots\}$ such that $V_i \supset F_i$.*

Proof. Since F_1 and $\cup_{i=2}^\infty F_i$ are disjoint closed sets, there is an open set U_1 which satisfies

$$F_1 \subset U_1 \subset \bar{U}_1 \subset X - \bigcup_{i=2}^\infty F_i.$$

Since F_2 and $\bar{U}_1 \cup (\cup_{i=3}^\infty F_i)$ are disjoint closed sets, there is an open set U_2 which satisfies

$$F_2 \subset U_2 \subset \bar{U}_2 \subset X - \bar{U}_1 \cup \left(\bigcup_{i=3}^\infty F_i \right).$$

Repeating this process, we obtain disjoint open sets $U_1, U_2, \ldots$ such that $U_i \supset F_i$. Since $\cup_{i=1}^\infty F_i$ and $X - \cup_{i=1}^\infty U_i$ are disjoint closed sets, there is an open V which satisfies

$$X - \bigcup_{i=1}^{\infty} U_i \subset V \subset \bar{V} \subset X - \bigcup_{i=1}^{\infty} F_i.$$

Now $V_i = U_i - \bar{V}$, $i = 1, 2, \ldots$, are the desired open sets.

M) *Let F be a closed set of a normal space X and $\mathcal{U} = \{U_i \mid i = 1, 2, \ldots\}$ a locally finite, countable, open covering of F. Then there is a locally finite open covering $\mathcal{V} = \{V_i \mid i = 1, 2, \ldots\}$ of X such that $V_i \cap F \subset U_i$.*

Proof. Using L) we can easily prove this proposition (see V.3.C) for a similar proof).

N) *Let X be a normal space, Y a separable normed linear space, φ a l.s.c. carrier of X into $\mathscr{C}(Y)$. Then, for a given convex open nbd V of 0, there is a continuous mapping f of X into Y such that*

$$f(p) \in \varphi(p) + V \quad \text{for all } p \in X.$$

Proof. The proof is quite analogous to that of K). This time we must take a countable locally finite open covering $\mathcal{W}$ such that

$$\mathcal{W} < \{y - V \mid y \in Y\}.$$

For the rest of the proof, we can proceed in a quite similar way as in the proof of K), but considering only countable coverings and using I) and M). The proof in detail is left to the reader.

Theorem VII.13. *A T_1-space X is normal if and only if one of the following conditions is satisfied:*
　　(i) *for every l.s.c. carrier φ of X into $\mathscr{C}(E^1)$, there is a selection,*
　　(ii) *for every separable Banach space Y and every l.s.c. carrier φ of X into $\mathscr{C}(Y)$, there is a selection.*

Proof. Assuming that X is normal, we can derive the condition (ii) from N) as we derived the necessity of Theorem VII.12 from K). It is obvious that (ii) implies (i) and that (i) implies the normality of X (Tietze's extension theorem).

Example VII.10. Let us denote by C^n and S^{n-1} the n-dimensional open ball with radius 1 and its boundary in E^n, i.e.,

$$C^n = \{(x_1, \ldots, x_n) \mid x_1^2 + \cdots + x_n^2 < 1\},$$

$$\overline{C^n} = \{(x_1, \ldots, x_n) \mid x_1^2 + \cdots + x_n^2 \leq 1\}$$

and

$$S^{n-1} = \{(x_1, \ldots, x_n) \mid x_1^2 + \cdots + x_n^2 = 1\}.$$

Then every topological space which is homeomorphic with C^n is called an *n-cell*. A T_2-space X is called a *CW-complex*[1] if it has a decomposition $\mathscr{D}$ consisting of cells which satisfies the following conditions:

(i) let us denote by K^m the union of the cells $\in \mathscr{D}$ with dimension $\leq m$. If $D \in \mathscr{D}$ is an *n*-cell, then there is a topological mapping f of $\overline{C^n}$ onto $\bar{D}$ such that $f(S^{n-1}) = \bar{D} - D \subset K^{n-1}$;

(ii) for each $D \in \mathscr{D}$, $\bar{D} - D$ intersects at most finitely many members of $\mathscr{D}$;

(iii) a subset F of X is closed provided $F \cap \bar{D}$ is closed for each $D \in \mathscr{D}$.

We can prove that every *CW*-complex is paracompact. In fact we can prove more generally that a topological space X is paracompact if there is a closed covering $\mathscr{K}$ of X such that each member of $\mathscr{K}$ is paracompact, and such that X is dominated by $\mathscr{K}$. To see this we suppose φ is a l.s.c. carrier of X into $\mathscr{F}(Y)$, where Y is a given Banach space. We consider every subcollection $\mathscr{K}'$ of $\mathscr{K}$ such that the restriction of φ to $\cup \{K \mid K \in \mathscr{K}'\}$ has a selection h and denote by $\mathscr{A}$ the collection of all pairs $(\mathscr{K}', h)$. We define that $(\mathscr{K}', h) < (\mathscr{K}'', h')$ if and only if $\mathscr{K}' \subset \mathscr{K}''$ and h' is an extension of h. Then $\mathscr{A}$ is a partially ordered set. Let $\mathscr{A}' = \{(\mathscr{K}_\alpha, h_\alpha) \mid \alpha \in A\}$ be a totally ordered subset of $\mathscr{A}$. Then putting

$$\mathscr{K}' = \cup \{\mathscr{K}_\alpha \mid \alpha \in \mathscr{A}\}; \qquad h(p) = h_\alpha(p), \quad p \in \cup \{K \mid K \in \mathscr{K}_\alpha\},$$

we get a member $(\mathscr{K}', h)$ of $\mathscr{A}$, because the continuity of h follows from the fact that $\mathscr{K}$ dominates X. It is clear that $(\mathscr{K}', h)$ is the least upper bound of $\mathscr{A}'$. Thus we can apply Zorn's lemma to $\mathscr{A}$ to obtain a maximal element $(\mathscr{K}_0, h_0)$ of $\mathscr{A}$. To prove that $\mathscr{K}_0 = \mathscr{K}$, we assume the contrary. Then there is $K_1 \in \mathscr{K}$ such that $K_1 \notin \mathscr{K}_0$. Put

$$K' = K_1 \cap (\cup \{K \mid K \in \mathscr{K}_0\}).$$

Since $\mathscr{K}$ dominates X, K' is a closed subset of K_1. We denote by h' the

---

[1] This terminology is due to J. H. C. Whitehead [1].

restriction of h_0 to K'; then h' is a selection for $\varphi_{K'}$, the restriction of φ to K'. Since K_1 is paracompact, by Theorem VII.11 every l.s.c. carrier of K_1 into $\mathcal{F}(Y)$ has a selection. Hence by C), h' can be extended to a selection h_1 for φ_{K_1}. Put

$$\mathcal{K}_2 = \mathcal{K}_0 \cup \{K_i\},$$

and define a mapping h_2 of $K_2 = \cup\{K \mid K \in \mathcal{K}_2\}$ into Y by

$$h_2(p) = \begin{cases} h_0(p) & \text{if } p \in \cup\{K \mid K \in \mathcal{K}_0\}, \\ h_1(p) & \text{if } p \in K_1. \end{cases}$$

Then h_2 is obviously continuous over K_2, satisfying $h_2(p) \in \varphi(p)$ for all $p \in K_2$, i.e. it is a selection for φ_{K_2}. Thus $(\mathcal{K}_2, h_2)$ is a member of $\mathcal{A}$ satisfying $(\mathcal{K}_2, h_2) > (\mathcal{K}_0, h_0)$, which contradicts that $(\mathcal{K}_0, h_0)$ is maximal. Hence we obtain $\mathcal{K}_0 = \mathcal{K}$ which means that h_0 is a selection for φ. Thus by Theorem VII.11, X is paracompact.

Example VII.11. Let X be a topological space and f a real-valued function defined on X. If for every $p \in X$ and $a > f(p)$ ($a < f(p)$), there is a nbd U of p such that $a > f(p')$ ($a < f(p')$) for all $p' \in U$, then f is called *upper semi-continuous* (*lower semi-continuous*).

A T_1-space X is normal if and only if for every upper semi-continuous function g and lower semi-continuous function h satisfying

$$g(p) \leq h(p) \quad \text{for all } p \in X,$$

there is a continuous function f over X such that

$$g(p) \leq f(p) \leq h(p) \quad \text{for all } p \in X. \tag{1}$$

To prove the sufficiency of the condition, let F and G be disjoint closed sets of X. Define functions g and h over X by

$$g(G) = 1, \qquad g(X - G) = 0;$$
$$h(F) = 0, \qquad h(X - F) = 1.$$

Then g and h are upper and lower semi-continuous functions respectively, satisfying $g \leq h$. Thus there is a continuous function f such that $g \leq f \leq h$. Now, put

$$U = \{x \in X \mid f(x) < \tfrac{1}{2}\}, \qquad V = \{x \in X \mid f(x) > \tfrac{1}{2}\}.$$

Then U and V are open sets such that

$$U \supset F, \qquad V \supset G, \qquad U \cap V = \emptyset.$$

Hence X is normal.

To prove the necessity, let g and h be upper semi-continuous and lower semi-continuous function, respectively, satisfying $g \leq h$. Define a mapping φ from X into $\mathscr{C}(E^1)$ by

$$\varphi(x) = [g(x), h(x)] \quad \text{(a closed interval in } E^1\text{) for every } x \in X.$$

Then it is easy to see that φ is a l.s.c. carrier. Thus by Theorem VII.13 there is a selection f for φ. Then f is a continuous function satisfying

$$g(x) \leq f(x) \leq h(x) \quad \text{for all } x \in X.$$

Thus the proposition is proved.[1]

6. More of extension theory

In the present section we shall discuss a development of Tietze's extension theorem into another direction, where a standard method is given to extend simultaneously all continuous maps defined on a closed set over the whole space. The main theorem (Theorem VII.14) in the following was originally proved by J. Dugundji [1] for metric spaces and extended by C. Borges [1] to M_3-spaces.

A) Let X be an M_3-space with a stratification $F(U, n)$, $U \in \mathcal{O}$, $n \in N$, such that $F(U, n) \supset F(U, m)$ whenever $n \geq m$. Further, let U be an open

---

[1] M. Katětov [3] and Hing Tong [1] gave a similar characterization for countably paracompact spaces. As for other applications and further developments of selection theory, see E. Michael [3]. C. Kuratowski–C. Ryll-Nardzewski [1] also obtained remarkable results in this aspect. See also C. Kuratowski [5].

nbd of $x \in X$. *Then we define* $n(U, x) \in N$ *and an open set* U_x *by*

$$n(U, x) = \min\{n \in N \mid x \in \text{Int } F(U, n)\},$$

$$U_x = \text{Int } F(U, n(U, x)) - F(X - \{x\}, n(U, x)).$$

Then for any open nbds U *and* V *of* x *and* y, *respectively, the following hold*:

   (i) U_x *is an open nbd of* x,
  (ii) $U_x \cap V_y \neq \emptyset$ *and* $n(U, x) \leqslant n(V, y)$ *imply* $y \in U$.

Proof. (i) is obvious. To prove (ii), assume $y \notin U$ and $n(U, x) \leqslant n(V, y)$. Then $F(X - \{y\}, n) \supset F(U, n)$ for each n. Therefore

$$F(X - \{y\}, n(V, y)) \supset F(X - \{y\}, n(U, x)) \supset F(U, n(U, x)) \supset U_x.$$

On the other hand

$$V_y \cap F(X - \{y\}, n(V, y)) = \emptyset,$$

which implies $U_x \cap V_y = \emptyset$. Thus (ii) is proved.

Theorem VII.14. *Let* X *be an* M_3-*space,* G *a closed set of* X, E *a locally convex*[1] *linear topological space and* $C(X, E)$ *the linear space of all continuous maps from* X *into* E. *Then there is a map* φ *from* $C(G, E)$ *into* $C(X, E)$ *such that*:

   (i) $\varphi(f)$ *is an extension of* f *over* X,
  (ii) *the range of* $\varphi(f)$ *is contained in the convex hull*[2] *of the range of* f,
 (iii) φ *is a linear map.*[3]

Proof. Let $W = X - G$. Consider a stratification F of X such that

$$F(X, n) = X, \quad n = 1, 2, \ldots,$$

$$F(X - \{x\}, 1) = \emptyset \quad \text{for all } x \in X.$$

Then for every $x, y \in X$, there is an open nbd U of y such that $x \in U_y$,

---

[1] There is a nbd base consisting of convex sets.
[2] The *convex hull* of a set D is the smallest convex set containing D.
[3] Namely, $\varphi(af + bg) = a\varphi(f) + b\varphi(g)$ for any real numbers a, b and f, $g \in C(G, E)$.

where we use the notation used in A). Because $n(X, y) = 1$, and thus $X_y = X - \emptyset = X \ni x$. Now, we put

$$m(x) = \max\{n(U, y) \mid y \in G \text{ and } x \in U_y\} \quad \text{for each } x \in W. \quad (1)$$

We claim that $m(x) < n(W, x)$. Because, if not, then for some $y \in G$ and an open nbd U of y satisfying $x \in U_y$, we have $n(U, y) \geq n(W, x)$. Since

$$x \in U_y \cap W_x \neq \emptyset,$$

from A) it follows that $y \in W$, which is not true. Hence our claim is proved.

Since W is paracompact, the open cover $\{W_x \mid x \in W\}$ of W has a locally finite open refinement $\mathcal{V}$ (which is an open cover of W). For each $V \in \mathcal{V}$, select $x(V) \in W$ such that $V \subset W_{x(V)}$. Furthermore, select $y(V) \in G$ and an open nbd $Q(V)$ of $y(V)$ such that

$$x(V) \in (Q(V))_{y(V)}, \quad (2)$$

$$n(Q(V), y(V)) = m(x(V)). \quad (3)$$

Now, construct a partition of unity $\{h(V) \mid V \in \mathcal{V}\}$ subordinated to $\mathcal{V}$, where

$$h(V)(W - V) = 0 \quad \text{for all } V \in \mathcal{V}.$$

Then define a map g from X into E by

$$g(x) = \begin{cases} f(x) & \text{if } x \in G, \\ \sum \{h(V)(x)f(y(V)) \mid V \in \mathcal{V}\} & \text{if } x \in W. \end{cases}$$

Then it is obvious that $g(X)$ is contained in the convex hull of $f(G)$, and g is continuous on W. To show that g is continuous at $z \in G$, let P be an open nbd of $f(z)$ in E. Since E is locally convex, there is a convex open set K in E such that

$$f(z) \in K \subset P.$$

Since f is continuous, there is an open nbd M of z such that

$$f(G \cap M) \subset K. \tag{4}$$

We claim that

$$g((M_z)_z) \subset K \subset P.$$

To see it, assume $x \in (M_z)_z \cap G$. Then $x \in M \cap G$, and hence by (4)

$$g(x) = f(x) \in K \subset P.$$

Assume $x \in (M_z)_z - G$; then pick any $V \in \mathcal{V}$ such that $x \in V$. Since $z \not\in W$ and

$$x \in (M_z)_z \cap V \subset M_z \cap W_{x(V)} \neq \emptyset,$$

from A) it follows that

$$x(V) \in M_z.$$

Hence by (1), (3) we obtain

$$n(M, z) \leq m(x(V)) = n(Q(V), y(V)). \tag{5}$$

From (2) it follows that

$$x(V) \in (Q(V))_{y(V)} \cap M_z \neq \emptyset.$$

Hence $y(V) \in M$ follows from (5). Thus by (4) we obtain

$$f(y(V)) \in f(G \cap M) \subset K.$$

Since K is convex, this implies

$$g(x) \in K \subset P.$$

Namely, it is proved that

$$g((M_z)_z) \subset K \subset P.$$

Hence g is continuous at z, i.e. $g \in C(X, E)$. Put

$$\varphi(f) = g \; ;$$

then it is easy to see that φ satisfies the desired conditions.

It is an interesting problem to determine the class of spaces X for which Theorem VII.14 holds, especially in case that E is the real line. It is necessary that X is at least collectionwise normal. In fact R. Heath–D. Lutzer [1] proved the following.

B) *Suppose that Theorem* VII.14 *holds for a* T_1-*space* X *and* $E = E^1$. *Then* X *is collectionwise normal.*

Proof. Let $\mathcal{G} = \{G_\alpha \mid \alpha \in A\}$ be a discrete closed collection in X. Put $G = \bigcup \mathcal{G}$. For each $\alpha \in A$, we define a map $g_\alpha : G \to [0, 1]$ by

$$g_\alpha(x) = \begin{cases} 1 & \text{if } x \in G_\alpha, \\ 0 & \text{otherwise}. \end{cases}$$

Then $g_\alpha \in C(X)$. Let φ be a map from $C(G)$ into $C(X)$ satisfying the conditions of Theorem VII.14. For each $\alpha \in A$, we put

$$V_\alpha = \{x \in X \mid \varphi(g_\alpha)(x) > \tfrac{1}{2}\} .$$

Then V_α is an open set containing G_α. Suppose

$$\alpha_1, \alpha_2 \in A, \qquad \alpha_1 \neq \alpha_2 \quad \text{and} \quad x \in V_{\alpha_1} \cap V_{\alpha_2} .$$

Then

$$1 < \varphi(g_{\alpha_1})(x) + \varphi(g_{\alpha_2})(x) = \varphi(g_{\alpha_1} + g_{\alpha_2})(x)$$
$$\leqslant \sup\{g_{\alpha_1}(x) + g_{\alpha_2}(x) \mid x \in G\} \leqslant 1 ,$$

which is a contradiction. Thus $\{V_\alpha \mid \alpha \in A\}$ is a disjoint open collection proving that X is collectionwise normal.

D. Lutzer–H. Martin [1] proved that for a collectionwise normal space X, a closed metrizable G_δ-set G of X and $E = E^1$, Theorem VII.14 holds.[1]

---

[1] R. Heath–D. Lutzer [2] showed that for the Michael line X and the set G of the rational points of X, the same theorem does not hold. See also R. Arens [2], H. Banilower [1], R. Heath–D. Lutzer–P. Zenor [1], E. v. Douwen–D. Lutzer–T. Przymusiński [1].

Generally, let $\mathscr{P}$ be a class of topological spaces such that whenever $X \in \mathscr{P}$ and F is a closed set of X, then $F \in \mathscr{P}$. A topological space Y is called an *absolute extensor* (*absolute nbd extensor*) for $\mathscr{P}$ or briefly AE (ANE) for $\mathscr{P}$, if whenever $X \in \mathscr{P}$ and F is a closed set of X, then every continuous map from F into Y can be extended to a continuous map from X into Y (from some open nbd of F into Y). Thus every sphere is ANE for the normal spaces. We have seen in the previous section that every Banach space is AE for the collectionwise normal spaces and also in this section that every locally convex linear topological space is AE for the M_3-spaces. O. Hanner [1] and E. Michael [2] studied relations between AE (ANE) for different classes. For example, for a metrizable space Y which is AE (ANE) for the metric spaces, the following facts are known. Y is AE (ANE) for the paracompact, perfectly normal spaces. Y is AE (ANE) for the paracompact T_2-spaces if and only if it is Čech complete. Y is AE (ANE) for the perfectly normal spaces if and only if it is separable. Y is AE (ANE) for the normal spaces if and only if it is separable and Čech complete.

Generally a topological space Y is called an *absolute retract* (*absolute nbd retract*) for a class $\mathscr{P}$ of topological spaces, or briefly AR (ANR) for $\mathscr{P}$, if, whenever Y is a closed set of $X \in \mathscr{P}$, Y is a retract of X. These concepts are due to K. Borsuk [1], [2], who proved that every finite polyhedron is ANR for the compact metric spaces. It is obvious that AE (ANE) for $\mathscr{P}$ implies AR (ANR) for $\mathscr{P}$. Conversely, O. Hanner [2] proved that AR (ANR) for $\mathscr{P}$ implies AE (ANE) for $\mathscr{P}$ if $\mathscr{P}$ is either one of the following classes: the normal, collectionwise normal, paracompact T_2, Lindelöf, compact metric or separable metric spaces.[1]

7. Characterization of topological properties in terms of $C(X)$

In the present section we are going to discuss characterizations of topological properties of X in terms of $C(X)$ or $C^*(X)$ as a ring, lattice or metric space. Our description cannot be very systematic as results of research on this aspect are rather sporadic.

A) *A topological space X is connected if and only if there are no $f, g \in C(X)$ such that neither f^{-1} nor g^{-1} exists, $fg = 0$, and $(f^2 + g^2)^{-1}$ exists, where $C(X)$ is regarded as a ring.*

---

[1] The last two cases were first proved by K. Borsuk [2] and C. Kuratowski [1]. Y. Kodama [1] also proved interesting properties of ANE for the metric spaces.

Proof. The easy proof is left to the reader.

B) *Let X be a compact T_2-space. Then X is second countable if and only if C(X) is second countable, where we regard C(X) as a metric space with respect to the norm.*

Proof. Assume that X is a 2nd countable space with a countable base $\mathcal{U} = \{U_1, U_2, \ldots\}$. Consider a countable set $\mathcal{P} = \{(U_i, U_j) \mid \bar{U}_i \subset U_j\} \subset \mathcal{U} \times \mathcal{U}$. For each $(U_i, U_j) \in \mathcal{P}$ we define $f_{ij} \in C(X)$ such that

$$f_{ij}(\bar{U}_i) = 0, \qquad f_{ij}(X - U_j) = 1 .$$

Denote by D_1 and D_2 the set of all polynomials of $\{f_{ij}\}$ with real coefficients and with rational coefficients, respectively. Then by the corollary to Theorem VII.2 D_1 is dense in $C(X)$. Since D_2 is obviously dense in D_1, $\bar{D}_2 = C(X)$ follows. Thus $C(X)$ is separable and accordingly 2nd countable.

Conversely, assume that $C(X)$ is 2nd countable and $\{f_i \mid i = 1, 2, \ldots\}$ is a dense subset of $C(X)$. Put

$$\mathcal{U} = \{P_i \mid i = 1, 2, \ldots\} ,$$

where $P_i = \{x \in X \mid f_i(x) > 0\}$. Then $\mathcal{U}$ is a base of X. Because, if W is an open nbd of $x \in X$, then there is $\varphi \in C(X)$ such that

$$\varphi(x) = 1, \qquad \varphi(X - W) = -1 .$$

Select f_i such that $\|f_i - \varphi\| < \frac{1}{2}$. Then

$$f_i(x) > \tfrac{1}{2}, \qquad f_i(X - W) < -\tfrac{1}{2} ;$$

hence $x \in P_i \subset W$. Thus $\mathcal{U}$ is a countable base of X, and the 2nd countability follows.

Some important properties like metrizability and 1st countability of a general Tychonoff space X are impossible to characterize by ring properties of $C^*(X)$ and accordingly by topological properties (with respect to the norm topology) either. Because $\beta(X)$ is neither first countable nor metrizable whenever $\beta(X) \neq X$ while $C^*(\beta(X)) \cong C^*(X)$. For a similar reason it is impossible to characterize 1st countability in terms of $C(X)$. However, it could be an interesting problem to characterize, e.g., metrizability of a realcompact space X by a 'nice property' of $C(X)$. It is obvious that every topological property of X can be characterized by a ring property of $C(X)$ in one way or the other, because X can be reconstructed from $C(X)$ by use of its ring structure only. Thus in the above problem the word 'nice' is emphasized. It can be also a problem to characterize metrizability and 2nd countability of a Tychonoff space X in terms of $C(X)$. On the other hand, it is possible to characterize 2nd countability and metrizability of a Tychonoff space X in terms of $C^*(X)$ when infinite operations $\vee_a f_\alpha$ and $\wedge_a f_\alpha$ are taken into consideration, where

$$\left(\bigvee_{\alpha\in A} f\right)(x) = \sup\{f_\alpha(x) \mid \alpha \in A\}, \quad \left(\bigwedge_{\alpha\in A} f_\alpha\right)(x) = \inf\{f_\alpha(x) \mid \alpha \in A\}$$

for $\{f_\alpha \mid \alpha \in A\} \subset C^*(X)$. Note that $\vee_a f_\alpha$ ($\wedge_a f_\alpha$) does not necessarily exist in $C^*(X)$.

C) *A Tychonoff space X is second countable if and only if $C^*(X)$ has a countable subset $\{f_n \mid n \in N\}$ such that for each $f \in C^*(X)$ there is $N' \subset N$ for which $\wedge \{f_n \mid n \in N'\} = f$.*

Proof. Suppose X is 2nd countable; then it is separable and metrizable. Denote by ρ a bounded metric of X and by $\{a_1, a_2, \ldots\}$ a dense countable subset of X. For each $n \in N$, an integer m and a rational q we define $f_{nmq} \in C^*(X)$ by

$$f_{nmq}(x) = m\rho(a_n, x) + q, \quad x \in X.$$

Then the countable set $L = \{f_{nmq}\}$ satisfies the following. For every $f \in C^*(X)$ and $\varepsilon > 0$ there is a subset L' of L such that

$$f(x) \leq \inf\{f'(x) \mid f' \in L'\} < f(x) + \varepsilon \quad \text{for all } x \in X. \tag{1}$$

Suppose $f \leq k \in N$. For each $x \in X$ select a rational $q(x)$ such that

$$f(x) < q(x) < f(x) + \varepsilon$$

and $\delta(x) > 0$ such that

$$f(x') < q(x) \quad \text{whenever } x' \in S_{\delta(x)}(x),$$

and $m(x) \in N$ such that

$$\tfrac{1}{2} m(x)\delta(x) + q(x) \geq k.$$

Select $\eta > 0$ such that

$$\eta < \tfrac{1}{2}\delta(x), \qquad m(x)\eta + q(x) < f(x) + \varepsilon.$$

Then select $n(x) \in N$ for which $a_{n(x)} \in S_\eta(x)$. Now, $f_{n(x)m(x)q(x)}$ is a member of L satisfying

$$f_{n(x)m(x)q(x)} \geq f, \qquad f_{n(x)m(x)q(x)}(x) < f(x) + \varepsilon.$$

Thus $\inf\{f_{n(x)m(x)q(x)} \mid x \in X\}$ satisfies the desired condition (1).
Now, for each $i \in N$ we find $L_i \subset L$ such that

$$f(x) \leq \inf\{f'(x) \mid f' \in L_i\} < f(x) + 1/i \quad \text{for all } x \in X.$$

Then

$$\wedge \left\{ f' \mid f' \in \bigcup_{i=1}^{\infty} L_i \right\} = f.$$

Conversely, suppose that $C^*(X)$ contains a countable subset $\{f_n \mid n \in N\}$ satisfying the condition. Then put

$$U_n = \{x \in X \mid f_n(x) < 0\}, \quad n = 1, 2, \ldots.$$

Now we claim that $\{U_n \mid n = 1, 2, \ldots\}$ is a base of X. To see it, let U be a nbd of $x \in X$. Then there is $f \in C^*(X)$ such that $f(x) = -1$, $f(X - U) = 1$. There is $N' \subset N$ for which $\wedge \{f_n \mid n \in N'\} = f$. Thus for some $n \in N'$ $f_n(x) < 0$ while $f_n(x') \geq f(x') = 1$ for all $x' \in X - U$. Hence $x \in U_n \subset U$, proving our claim. Thus X is 2nd countable.

The proof of the following theorem, due to J. Nagata [13], is based on a similar idea but on a more complicated technique, and thus it is omitted here.[1]

---

[1] See J. Nagata [13] for characterization of paracompact M-space and other spaces.

Theorem VII.15. *A Tychonoff space X is metrizable if and only if $C^*(X)$ contains a sequence L_1, L_2, ... of subsets satisfying:*

(i) *for each n and every $\{f_\alpha \mid \alpha \in A\} \subset L_n$, $\vee_\alpha f_\alpha$ and $\wedge_\alpha f_\alpha$ exist and belong to L_n,*

(ii) *for every $f \in C^*(X)$ there is $\{f_\beta \mid \beta \in B\} \subset \bigcup_{n=1}^{\infty} L_n$ such that $\wedge \{f_\beta \mid \beta \in B\} = f$.*

D) *Let X be a locally compact T_2-space. We endow $C^*(X)$ with the compact open topology. Then $C^*(X)$ is metrizable if and only if X is σ-compact.*[1]

Proof. We observe that $C^*(X)$ is a topological group with respect to the ordinary sum $+$ and the compact open topology and also that $\{U(f, K, \varepsilon) \mid K$ is a compact set of X, $\varepsilon > 0\}$ is a nbd base of each $f \in C^*(X)$, where

$$U(f, K, \varepsilon) = \{g \in C^*(X) \mid |f(x) - g(x)| < \varepsilon \text{ for all } x \in K\}.$$

Suppose $X = \bigcup_{n=1}^{\infty} K_n$ for compact sets K_n, $n = 1, 2, \ldots$. Since X is locally compact, $K_n \subset \bigcup_{i=1}^{h_n} U_i^n$, where U_i^n is an open set whose closure is compact. Put

$$V_n = \bigcup_{i=1}^{h_n} U_i^n.$$

Then V_n is open and $\bar{V}_n$ compact. Put

$$G_n = \bigcup_{m=1}^{n} V_m.$$

Then G_n is open and $\bar{G}_n$ compact satisfying

$$G_n \subset G_{n+1}, \qquad \bigcup_{n=1}^{\infty} G_n = X.$$

Now we claim that $\{U(0, \bar{G}_n, 1/n) \mid n = 1, 2, \ldots\}$ is a nbd base of $0 \in C^*(X)$. To show it, let K be a compact set of X and $\varepsilon > 0$. Then $K \subset G_n$ and $\varepsilon > 1/n$ for some $n \in N$. Thus

$$U(0, \bar{G}_n, 1/n) \subset U(0, K, \varepsilon).$$

[1] Due to M. Rajagopalan.

Hence $C^*(X)$ is 1st countable and accordingly metrizable by the corollary to Theorem VI.16.

Conversely, assume that $C^*(X)$ is metrizable and $0 \in C^*(X)$ has a countable nbd base $\{U(0, K_n, \varepsilon_n) \mid n = 1, 2, \ldots\}$. Then we claim that $X = \bigcup_{n=1}^{\infty} K_n$. Because otherwise we can choose

$$x \in X - \bigcup_{n=1}^{\infty} K_n.$$

Then the nbd $U(0, \{x\}, 1)$ of 0 contains none of $U(0, K_n, \varepsilon_n)$, $n = 1, 2, \ldots$. Because for each n considering $f_n \in C^*(X)$ such that

$$f_n(x) > 1, \qquad f_n(K_n) = 0,$$

we obtain

$$f_n \in U(0, K_n, \varepsilon_n) - U(0, \{x\}, 1).$$

Hence $C^*(X)$ is not 1st countable contradicting our assumption. Therefore $X = \bigcup_{n=1}^{\infty} K_n$, proving that X is σ-compact.[1]

Exercise VII

1. If X is an infinite Tychonoff space, then $C(X)$ is not locally compact with respect to the topology of uniform convergence.

2. The compact open topology is stronger than the weak topology.

3. Let $F(X, Y)$ be a collection of continuous mappings of a topological space X into a regular space Y. Then the mapping space $F(X, Y)$ with the compact open topology is regular.

4. Let X be the space of all maps from a closed interval I into itself with the weak topology. Then there is a continuous map from $\beta(N)$ onto X, where N is the countable discrete space. Thus $|\beta(N)| = 2^c$. (Consider the subspace P of X consisting of all polynomials. B. Pospišil's theorem.)

---

[1] See H. Tamano [5], Z. Frolik [5], J. Guthrie [1], R. A. McCoy [1], J. Gerlits–Zs. Nagy [1], J. Gerlits [1], B. J. Ball–S. Yokura [1] and A. V. Arhangelskii [13] for further results in this and related aspects.

5. For a given set A, the product space $D(A)$ of two point discrete spaces T_α, $\alpha \in A$, is zero-dimensional.

6. Let f be a perfect mapping of a topological space X onto a topological space Y. If X satisfies either one of the following properties, then Y also satisfies the same property: T_2, regular, second axiom of countability.

7. Let f be a perfect mapping of a topological space X onto a topological space Y. If Y is compact (Lindelöf), then X is compact (Lindelöf).

8. Let f be a perfect map from a Čech complete space X onto a Tychonoff space Y. Then Y is Čech complete.

9. Let f be a continuous mapping from X onto Y. Then f is a perfect mapping if and only if the product $f \times i_Z$ of f with the identity mapping on Z is a closed mapping for every space Z. (Hint: To prove the 'if' part, assume $f^{-1}(q)$ is not compact; then it has a non-convergent ultra net $\varphi(\Delta \mid >)$. Let $Z = \Delta \cup \{\infty\}$ to define that $U = \Delta' \cup \{\infty\}$ is a nbd of ∞ if Δ' is a residual subset of Δ.)

10. A regular space X is Lindelöf if and only if for every open cover $\mathcal{U}$ of X, there is a separable metric space Y and a continuous $\mathcal{U}$-map f from X onto Y.

11. X is a paracompact Čech complete space with weight $\leq \alpha$ if and only if it is homeomorphic to a closed set of $H(A) \times I^A$, where $|A| = \alpha$, and I^A denotes the product of α closed intervals.

12. Every paracompact $M\text{-}T_2$-space is homeomorphic to a closed G_δ-set of the product of a metric space and a compact T_2-space.

13. If there is a perfect map from an $M\text{-}T_1$-space X onto Y, then Y is an M^*-*space*, i.e., it has a sequence $\{\mathcal{F}_i \mid i = 1, 2, \ldots\}$ of locally finite closed covers satisfying $w\Delta$-condition. (The converse is also true; see J. Nagata [8].)

14. If there is a compact open map from a metric space X onto a T_1-space Y, then Y has a uniform open basis. (The converse is also true; see A. Arhangelskii [2].)

15. If there is an open π-map f from a metric space X onto Y, then Y is developable. (The converse is also true; see A. Arhangelskii [6].) The map f is called a π-*map* if for any point $y \in Y$ and any nbd V of y,

$$\rho(f^{-1}(y), X - f^{-1}(V)) > 0 \text{ in } X.$$

16. The product space $X \times Y$ is normal for every paracompact M-T_2-space Y if and only if X is paracompact P and T_2.

17. $\mu(X)$ is compact if and only if X is pseudo-compact.

18. Prove Theorem VII.4 for a semi-metric space X.

19. Let f be a perfect map from X onto Y. Then there is a closed subset F of X such that the restriction of f to F is an irreducible map from F onto Y.

20. Give a regular space which has a G_δ-diagonal and a point-countable base without being metrizable.

21. A topological space which is the sum of countably many closed metrizable subsets is itself metrizable if and only if it is collectionwise normal and locally M.

22. A T_2-space X is paracompact and Čech complete if and only if it is the inverse limit of an inverse system of complete metric spaces with perfect bonding mappings.

23. Every realcompact M-space is paracompact.

24. Suppose V is a convex set of a linear space X. Then $2V = V + V$.

25. Let g be an upper semi-continuous function and h a lower semi-continuous function over a topological space X such that $g(p) \leqslant h(p)$, $p \in X$. Then $\varphi(p) = \{y \mid g(p) \leqslant y \leqslant h(p)\}$ is a l.s.c. carrier.

26. If $\{F_\alpha \mid \alpha \in A\}$ is a locally finite closed cover of X, then there is a closed continuous map from the discrete sum of F_α, $\alpha \in A$, onto X.

27. Let $\mathscr{F} = \{F_\alpha \mid \alpha \in A\}$ be a locally finite closed covering of a topological space X. If each F_α is collectionwise normal, then X is also collectionwise normal. Suppose $\mathscr{F}$ is not necessarily locally finite, but X is dominated by $\mathscr{F}$. Then is X still collectionwise normal?

28. Theorem VII.14 holds for any topological space X and its retract G.

29. In case that E is the real line in Theorem VII.14, we can select φ which satisfies, besides (i)–(iii), (iv) $\varphi(f) \leqslant \varphi(g)$ whenever $f \leqslant g$, and (v) φ is a continuous map from $C(G)$ into $C(X)$ (with respect to the topology of uniform convergence).

30. Compactness of a Tychonoff space X cannot be characterized in terms of the ring (or lattice) properties of $C(X)$.

CHAPTER VIII

OTHER ASPECTS

In the present chapter we shall discuss some other aspects of general topology, which were not systematically discussed in the previous chapters.

1. Linearly ordered space

Definition VIII.1. Let X be a linearly ordered set (totally ordered set). We use the symbols (a, b), $[a, b]$, $(a, b]$ and $[a, b)$ in an analogous manner as for the real line, and especially $(a, \infty) = \{x \in X \mid x > a\}$, $(-\infty, a) = \{x \in X \mid x < a\}$, $[a, \infty) = \{x \in X \mid x \geq a\}$ and $(-\infty, a] = \{x \in X \mid x \leq a\}$ will be used frequently.

If X is a topological space with the subbase $\{(a, \infty), (-\infty, a) \mid a \in X\}$, then we call X a *linearly ordered topological space* or *LOTS*. Such a topology of X is called the *order topology* of the linearly ordered set X. On the other hand, a topological space X with a linear order is called a *GO-space* if X is a subspace of a LOTS X', where the order of X is the one induced by the order of X'.

Example VIII.1. The real line E^1, spaces of ordinal numbers and especially the space R_S in Example II.1 are LOTS. The intervals $[a, b]$, $[a, b)$ etc. are not only GO-spaces but also LOTS, and so is the space of all rational numbers in E^1 as well as the one of irrational numbers. On the other hand, the Sorgenfrey line S is a GO-space but no LOTS. Because, let $L = R \times \{0, 1\}$, where R is the set of real numbers, and define a linear order in L as follows:

$$(a, b) < (a', b') \text{ if and only if (i) } a < a' \text{ or (ii) } a = a' \text{ and } b < b'.$$

Then we define the order topology for L. Now it is obvious that S is homeomorphic to the subspace $R \times \{0\}$ of the LOTS L. Hence S is a

GO-space. In a similar way we can see that the Michael line is a GO-space. Generally, let X and Y be linearly ordered sets; then we can define a linear order of the cartesian product $X \times Y$ in the same way as we did for L in the above. Thus defined order of $X \times Y$ is called the *lexicographic order.*

Definition VIII.2. A subset K of a linearly ordered set X is called *convex* if for any a, $b \in K$, $[a, b] \subset K$ holds. Let A be a subset of X. Then for each $a \in A$, $K(a) = \bigcup \{K' \mid K'$ is a convex set of X such that $a \in K' \subset A\}$ is a convex set called a *convex component* of A (in X).

A) *A topological space X with a linear order is a GO-space if and only if* (i) *the topology of X is stronger than its order topology (namely (a, ∞) and $(-\infty, a)$ are open sets), and* (ii) *X has a base consisting of convex sets.*

Proof. Necessity of the condition is obvious. Assume that X satisfies the condition. Then we consider the set $X \times Z$ with the lexicographic order, where Z denotes the linearly ordered set of all integers with the natural order. We denote by X^* a linearly ordered subset of $X \times Z$ defined by

$$X^* = X \times \{0\} \cup \{(x, n) \mid [x, -\infty) \in \mathcal{O} - \mathcal{O}(<), n < 0\}$$

$$\cup \{(x, n) \mid (-\infty, x] \in \mathcal{O} - \mathcal{O}(<), n > 0\},$$

where $\mathcal{O}$ and $\mathcal{O}(<)$ denote the topology of X and the order topology of X, respectively. Now we regard X^* as a LOTS with the order topology. Then it is easy to see that X is homeomorphic to the closed subspace $X \times \{0\}$ of X^*. Thus X is a GO-space.

B) *Every GO-space X is a closed set of a LOTS X^*.*

Proof. See the proof of A).

C) *Every subset A of a GO-space X can be decomposed into the sum of disjoint convex components (in X), which are open in A.*

Proof. The easy proof is left to the reader.

D) *Every convex subspace of a LOTS is a LOTS.*

Proof. Left to the reader.

Theorem VIII.1. *Every GO-space is hereditarily collectionwise normal.*

Proof.[1] By Exercise V.17 it suffices to show that every open set of a LOTS X is collectionwise normal. By C) this is equivalent with that every convex open set of X is collectionwise normal. Finally by D) it suffices to show that the LOTS X itself is collectionwise normal.

Suppose that $\{F_\lambda \mid \lambda \in \Lambda\}$ is a given discrete closed collection in X. For each $\lambda \in \Lambda$ we define

$$F_\lambda^* = \bigcup \{[a, b] \mid a, b \in F_\lambda, [a, b] \cap F_\mu = \emptyset \text{ for } \mu \neq \lambda\}.$$

Then $F_\lambda \subset F_\lambda^*$ is obvious. We can show that $\{F_\lambda^* \mid \lambda \in \Lambda\}$ is discrete. Let $x \in X$ and (c, d) be a nbd of x which intersects F_λ and no F_μ for $\mu \neq \lambda$. Assume $(c, d) \cap F_\mu^* \neq \emptyset$ for some $\mu \neq \lambda$. Then there are $a, b \in F_\mu$ with $[a, b] \cap F_\nu = \emptyset$ for $\nu \neq \mu$, and $[a, b] \cap (c, d) \neq \emptyset$. Since $a, b \notin (c, d)$, $a \leq c < x < d \leq b$ follows. Thus $[a, b] \cap F_\lambda \neq \emptyset$, which is a contradiction. Namely $(c, d) \cap F_\mu^* = \emptyset$ is proved. In the other cases, too, we can easily show that x has a nbd which intersects at most one of $\{F_\lambda^* \mid \lambda \in \Lambda\}$.

Now, decompose each F_λ^* into its convex components as

$$F_\lambda^* = \bigcup \{F_{\lambda\alpha} \mid \alpha \in A_\lambda\}.$$

We also decompose $X - \bigcup\{F_\lambda^* \mid \lambda \in \Lambda\}$ into its convex components as

$$X - \bigcup_\lambda F_\lambda^* = \bigcup \{G_\gamma \mid \gamma \in \Gamma\}.$$

Then the collection $\mathcal{M} = \{F_{\lambda\alpha}, \ G_\gamma \mid \lambda \in \Lambda, \ \alpha \in A_\lambda, \ \gamma \in \Gamma\}$ is linearly ordered in a natural manner because each element of $\mathcal{M}$ is convex. We pick

$$q(G_\gamma) \in G_\gamma \quad \text{for each } \gamma \in \Gamma.$$

For each $F_{\lambda\alpha}$ we define

$$S_{\lambda\alpha} = \{x \in X \mid x > y \text{ for all } y \in F_{\lambda\alpha}\}.$$

[1] This proof is due to L. Steen [1].

Then we claim that $F_{\lambda\alpha}$ has an immediate successor in $\mathcal{M}$ if

$$F_{\lambda\alpha} \cap \bar{S}_{\lambda\alpha} \neq \emptyset . \tag{1}$$

To prove it, let

$$F_{\lambda\alpha} \cap \bar{S}_{\lambda\alpha} = \{p\} .$$

Assume that (a, b) is a nbd of p such that

$$(a, b) \cap F^*_\mu = \emptyset \quad \text{for all } \mu \neq \lambda . \tag{2}$$

Now, it follows that

$$(p, b) \cap F^*_\lambda = \emptyset . \tag{3}$$

Because, if not, then there is $x \in (p, b) \cap F^*_\lambda$, which implies that $[p, x] \subset F^*_\lambda$. Since $p \in F_{\lambda\alpha}$, $[p, x] \subset F_{\lambda\alpha}$ follows contradicting $p \in \bar{S}_{\lambda\alpha}$. Hence (3) is proved, and this combined with (2) implies

$$(p, b) \subset X - \cup \{F^*_\lambda \mid \lambda \in \Lambda\} .$$

Hence $(p, b) \subset G_\lambda$ for some $\gamma \in \Gamma$. Then it is obvious that G_γ is the immediate successor of $F_{\lambda\alpha}$ in the linearly ordered set $\mathcal{M}$. We denote this immediate successor of $F_{\lambda\alpha}$ by $G_{\lambda\alpha}$ $(= G_\gamma)$.

Now, for each $F_{\lambda\alpha}$ satisfying (1) we put

$$I_{\lambda\alpha} = [p, q(G_{\lambda\alpha})) .$$

For each $F_{\lambda\alpha}$ which does not satisfy (1), we put $I_{\lambda\alpha} = \emptyset$. In a similar way we define

$$S'_{\lambda\alpha} = \{x \in X \mid x < y \text{ for all } y \in F_{\lambda\alpha}\}$$

and put

$$\bar{S}'_{\lambda\alpha} \cap F_{\lambda\alpha} = \{p'\}$$

if the intersection is non-empty. Denote by $G'_{\lambda\alpha}$ the immediate predecessor of $F_{\lambda\alpha}$ in $\mathcal{M}$ and put

$$I'_{\lambda\alpha} = (q(G'_{\lambda\alpha}), p'] .$$

If $\bar{S}'_{\lambda\alpha} \cap F_{\lambda\alpha} = \emptyset$, then $I'_{\lambda\alpha} = \emptyset$. Now, define an open set $U_{\lambda\alpha}$ by

$$U_{\lambda\alpha} = I'_{\lambda\alpha} \cup F_{\lambda\alpha} \cup I_{\lambda\alpha}.$$

Put

$$U_{\lambda} = \cup\{U_{\lambda\alpha} \mid \alpha \in A_{\lambda}\}.$$

It is easy to see that $\{U_{\lambda} \mid \lambda \in \Lambda\}$ is a disjoint open collection such that $U_{\lambda} \supset F_{\lambda}$. Hence X is collectionwise normal.

E) *A GO-space with a G_{δ}-diagonal is 1st countable.*

Proof. Left to the reader.

F) *A LOTS X is metrizable if and only if it has a G_{δ}-diagonal.*[1]

Proof. We prove only the 'if' part. Since X is collectionwise normal, by the corollary of Theorem VI.4 it suffices to show that X is developable. Observe that X has a sequence $\{\mathcal{U}_i \mid i = 1, 2, \ldots\}$ of open covers such that each $\mathcal{U}_i$ consists of convex sets and such that $\cap_{i=1}^{\infty} S(x, \mathcal{U}_i) = \{x\}$ for each $x \in X$. Let U be a nbd of $x \in X$ and $a < x$ in X. Then there is n for which $a \notin S(x, \mathcal{U}_n)$. Since $S(x, \mathcal{U}_n)$ is convex, this means that

$$S(x, \mathcal{U}_n) \subset (a, \infty).$$

Similarly we can show that

$$S(x, \mathcal{U}_m) \subset (-\infty, b) \quad \text{for every } b > x \text{ and for some } m.$$

Thus $\{\mathcal{U}_i \mid i = 1, 2, \ldots\}$ is a development of X, proving that X is a developable space.

Example VIII.2. Sorgenfrey line S has a G_{δ}-diagonal but is non-metrizable. Thus F) cannot be extended to GO-spaces.

Definition VIII.3. Let X be a GO-space. Suppose (A, B) is an ordered pair of disjoint open sets of X such that:
  (i) $X = A \cup B$,
  (ii) $a < b$ whenever $a \in A$ and $b \in B$.

---

[1] Due to D. Lutzer [1].

Then (A, B) is called a *gap* if it satisfies (i), (ii) and

(iii) A has no maximal point, and B has no minimal point.

If furthermore $A = \emptyset$ or $B = \emptyset$, then (A, B) is called an *end gap*. (A, B) is called a *pseudo-gap* if it satisfies (i), (ii),

(iv) $A \neq \emptyset$, $B \neq \emptyset$,

and (iv)$_l$ or (iv)$_r$ stated in the following:

(iv)$_l$ A has no maximal point, and B has a minimal point,

(iv)$_r$ A has a maximal point, and B has no minimal point.

Generally, let A be a subset of a linearly ordered set X. A subset A' of A is called *cofinal (coinitial)* in A if for each $a \in A$ there is $a' \in A'$ such that $a' \geqslant a$ ($a' \leqslant a$).

Suppose (A, B) is a (pseudo-) gap of a GO-space X. If there are discrete subsets A' of A which is cofinal in A and a discrete subset B' of B which is coinitial in B, then (A, B) is called a *Q- (pseudo-) gap*.

Theorem VIII.2. *A GO-space X is compact if and only if X has no gaps and no pseudo-gaps.*

Proof. Necessity of the condition is obvious. Suppose that X satisfies the condition. Then X has the minimal point a and the maximal point b. Suppose $\mathcal{U}$ is a given open cover of X by non-empty convex sets. Put $Y = \{x \in X \mid$ there are finitely many elements $U_1, \ldots, U_k$ of $\mathcal{U}$ such that $a \in U_1$, $x \in U_k$, $U_i \cup U_{i+1}$ is convex for $i = 1, \ldots, k - 1\}$. If $Y \neq X$, then $(Y, X - Y)$ is a gap or pseudo-gap of X. Because, Y and $X - Y$ are obviously open sets. It is also easy to see that either Y has no maximal point, or $X - Y$ has no minimal point. This contradicts that X has neither gap nor pseudo-gap. Hence $Y = X$ follows. Thus $b \in Y$, which implies that $X = [a, b]$ is covered by finitely many elements of $\mathcal{U}$. Hence X is compact.

Example VIII.3. Let X be a LOTS. Then X has no pseudo-gap. We denote by X^+ the set of all gaps of X plus X itself. We can define a linear order for X^+ in a natural way. Namely, if $(A, B) = c$ is a gap of X, then we define that

$$a < c \text{ for all } a \in A \quad \text{and} \quad c < b \text{ for all } b \in B.$$

Introduce the order topology into X^+; then X^+ turns out to be a compactification of X. More generally, let X be a GO-space. Then $(X^*)^+$ is a compact space which contains X as a subspace, where X^* denotes the LOTS given in B). The closure of X in $(X^*)^+$ is a compactification of X, which we call *Dedekind compactification.*

Let X be a LOTS and U a subset of X. A gap (A, B) of X is said to be *covered* by U if there is a convex set V such that $V \subset U$, $V \cap A \neq \emptyset$ and $V \cap B \neq \emptyset$. A cover $\mathcal{U}$ of X is said to *cover* the gap (A, B) if $\mathcal{U}$ has an element which covers (A, B).

G) *An open cover $\mathcal{U}$ of a LOTS X has a finite subcover if every gap of X is covered by $\mathcal{U}$.*

Proof. We consider Dedekind compactification X^+ of X. For each $U \in \mathcal{U}$ we define an open set U' of X^+ by

$$U' = U \cup \{\text{all gaps of } X \text{ which are covered by } U\}.$$

Then $\mathcal{U}' = \{U' \mid U \in \mathcal{U}\}$ is an open cover of X. Thus $\mathcal{U}'$ has a finite cover $\mathcal{V}'$. Then $\mathcal{V} = \{V' \cap X \mid V' \in \mathcal{V}'\}$ is a finite subcover of $\mathcal{U}$.

Theorem VIII.3.[1] *Every GO-space X is countably paracompact.*

Proof. By virtue of B) we may assume that X is a LOTS. Let $\mathcal{U} = \{U_1, U_2, \ldots\}$ be a given countable open cover of X such that $U_i \subset U_{i+1}$. It suffices to show that $\mathcal{U}$ has a locally finite open refinement. Decompose each element U_i of $\mathcal{U}$ into its convex components $V_{i\alpha}$, $\alpha \in A_i$, to denote

$$\mathcal{V}_i = \{V_{i\alpha} \mid \alpha \in A_i\}, \qquad \mathcal{V} = \bigcup_{i=1}^{\infty} \mathcal{V}_i.$$

For each $V \in \mathcal{V}_i$, we define

$$W(V) = V_1 \cup V_2 \cup \cdots,$$

where

$$V_1 = V, \qquad V_{j+1} \in \mathcal{V}_{i+j} \quad \text{and} \quad V_j \subset V_{j+1} \quad \text{for } j = 1, 2, \ldots.$$

Then X is decomposed into the sum of the disjoint clopen sets $W(V)$'s. If $\{V_i \mid i = 1, 2, \ldots\}$ has a locally finite open refinement in $W(V)$ for each V, then $\mathcal{V}$ and accordingly $\mathcal{U}$, too, has a locally finite open refinement in X. Thus we may assume without loss of generality that $\mathcal{U}$ consists of convex sets.

---

[1] Due to L. Gillman–M. Henriksen [1].

Pick a fixed point $x_0 \in X$. Assume that $[x_0, \infty) \not\subset U_i$, $i = 1, 2, \ldots$. Then select $x_i \in [x_0, \infty) - U_i$, $i = 1, 2, \ldots$, such that $x_0 < x_1 < x_2 < \cdots$. Then $\{x_i \mid i = 1, 2, \ldots\}$ is cofinal in X. Since $\{x_i\}$ is discrete, there is a discrete open collection $\{P_i \mid i = 0, 1, \ldots\}$ in X such that $x_i \in P_i$ and such that P_i is contained in some element of $\mathcal{U}$. Now, $\{(x_i, x_{i+1}), P_i \mid i = 0, 1, \ldots\}$ is a locally finite open collection in X which covers $[x_0, \infty)$ and refines $\mathcal{U}$. In a similar way $(-\infty, x_0]$ can be covered by a locally finite open collection which refines $\mathcal{U}$. Thus $\mathcal{U}$ has a locally finite open refinement proving that X is countably paracompact.

Theorem VIII.4.[1] *A GO-space X is paracompact if and only if every gap of X is a Q-gap, and every pseudo-gap is a Q-pseudo-gap.*

Proof. Suppose that X is paracompact and (A, B) a gap of X with $A \neq \emptyset$. Then $\mathcal{U} = \{(-\infty, a) \mid a \in A\}$ is an open cover of A. Let $\mathcal{V}$ be a locally finite open refinement of $\mathcal{U}$. Decompose each element of $\mathcal{V}$ into its convex components and denote by $\mathcal{W}$ the open cover of A consisting of the convex components. Obviously $\mathcal{W}$ is point-finite. Let $\mathcal{P}$ be a locally finite open refinement of $\mathcal{W}$. Pick $x(P) \in P$ for each $P \in \mathcal{P}$. Then $\{x(P) \mid P \in \mathcal{P}\}$ is discrete. Let $a \in A$. Select $x_1 \in A$ such that $x_1 > a$. Suppose $x_1 \in P_1 \in \mathcal{P}$, $P_1 \subset W_1 \in \mathcal{W}$. If $P_1 \cap (-\infty, a] \neq \emptyset$, then $a \in W_1$. Suppose $W_1 \subset (-\infty, a_1]$; then select $x_2 \in A$ such that $x_2 > a_1$. Let $x_2 \in P_2 \in \mathcal{P}$, $P_2 \subset W_2 \in \mathcal{W}$. If $P_2 \cap (-\infty, a] \neq \emptyset$, then $a \in W_2$. We cannot continue this process indefinitely because $\mathcal{W}$ is point-finite. Hence

$$P_n \cap (-\infty, a] = \emptyset \quad \text{for some } n,$$

and thus $x(P_n) > a$. Therefore $\{x(P) \mid P \in \mathcal{P}\}$ is cofinal in A. In a similar way we can select a discrete set which is coinitial in B. Hence (A, B) is a Q-gap. Similarly, every pseudo-gap is a Q-pseudo-gap.

Conversely, assume that X satisfies the said condition. Then it is easy to see that every gap of X^* (defined in B)) is a Q-gap. Thus by virtue of B) we may assume without loss of generality that X is a LOTS whose every gap is a Q-gap. Let $\mathcal{U}$ be an open cover of X. Denote by F the set of all gaps of X which are not covered by $\mathcal{U}$. We regard F as a subset of X^+. Then F is obviously a closed set. Now we decompose $X^+ - F$ into its convex components G_γ, $\gamma \in \Gamma$, to put

$$H_\gamma = G_\gamma \cap X.$$

[1] Due to L. Gillman–M. Henriksen [1] in case that X is a LOTS. This general form is due to M. J. Faber [1].

Then it is obvious that $\{H_\gamma \mid \gamma \in \Gamma\}$ is a disjoint open cover of X by convex sets. Regard H_γ as a LOTS covered by the open cover $\mathcal{U}$. Then $\mathcal{U}$ covers every gap of H_γ possibly except its end gaps. Select an inner point a of H_γ. If H_γ has the maximal point, then by G) $H_\gamma' = \{x \in H_\gamma \mid x \geq a\}$ is covered by finitely many elements of $\mathcal{U}$. If $(H_\gamma, \emptyset)$ is an end gap of H_γ, then it determines a Q-gap of X. Thus there is a discrete set cofinal in H_γ. Thus by use of G) and an argument similar to the one in the proof of Theorem VIII.3, we can prove that H_γ' is covered by a locally finite open collection which refines $\mathcal{U}$. We apply the same argument on the left half of H_γ to find a locally finite open cover of H_γ which refines $\mathcal{U}$. Thus X is covered by a locally finite open cover which refines $\mathcal{U}$. Namely X is paracompact.

Corollary. *Let X be a GO-space with a G_δ-diagonal. Then X is hereditarily paracompact.*

Proof. It suffices to show that X is paracompact because every subset of X has a G_δ-diagonal. Suppose not; then by Theorem VIII.4 X has a non-Q-gap or a non-Q-pseudo-gap. Assume, e.g., that (A, B) is a gap such that A has no discrete cofinal subset. Let $\{\mathcal{U}_n \mid n = 1, 2, \ldots\}$ be a sequence of open covers of X by convex sets such that $\mathcal{U}_n > \mathcal{U}_{n+1}$ and such that

$$\bigcap_{n=1}^{\infty} S(x, \mathcal{U}_n) = \{x\} \quad \text{for each } x \in X.$$

We claim that for each n there is $x_n \in A$ such that $S(x_n, \mathcal{U}_n) \cap A$ is cofinal in A. Because, otherwise, by use of a transfinite induction we can select an increasing (transfinite) sequence $\{x^\alpha \mid \alpha < \tau\}$ such that $x^\beta \notin S(x^\alpha, \mathcal{U}_n)$ whenever $\alpha < \beta < \tau$, and $\{x^\alpha \mid \alpha < \tau\}$ is cofinal in A. Then $\{x^\alpha\}$ is obviously discrete contradicting our assumption. Thus our claim is verified.

It follows from our assumption that $\{x_n\}$ is not cofinal in A. Suppose $a \in A$ satisfies $a \geq x_n$, $n = 1, 2, \ldots$. Then $S(a, \mathcal{U}_n)$ is cofinal in A for every n. Since $S(a, \mathcal{U}_n)$ is convex,

$$S(a, \mathcal{U}_n) \supset \{x \in A \mid x \geq a\}, \quad n = 1, 2, \ldots,$$

which contradicts that $\bigcap_{n=1}^{\infty} S(a, \mathcal{U}_n) = \{x\}$. Thus X is paracompact.

H) *Let X be a GO-space with a G_δ-diagonal. If $H(X) = \{x \in X \mid [x, \infty) \in \mathcal{O} - \mathcal{O}(<)$ or $(-\infty, x] \in \mathcal{O} - \mathcal{O}(<)\}$ is σ-discrete, then X is metrizable, where $\mathcal{O}$ and $\mathcal{O}(<)$ denote the topology of X and its order topology, respectively.*

Proof. Let $H(X) = \bigcup_{n=1}^{\infty} H_n$, where each H_n is discrete. To each $x \in H_n$ we assign a countable open nbd basis $\{U_m(x) \mid m = 1, 2, \ldots\}$ such that $U_m(x) \supset U_{m+1}(x)$ and such that $\{U_1(x) \mid x \in H_n\}$ is discrete. This is possible because X is 1st countable and hereditarily collectionwise normal by E) and Theorem VIII.1. Decompose $X - H_n$ into the sum of convex components $\{G_\alpha \mid \alpha \in A_n\}$. Then put

$$\mathcal{V}_{nm} = \{G_\alpha \mid \alpha \in A_n\} \cup \{U_m(x) \mid x \in H_n\}.$$

Suppose $\{\mathcal{W}_i \mid i = 1, 2, \ldots\}$ is a sequence of open covers of X by convex sets such that

$$\mathcal{W}_{i+1} < \mathcal{W}_i, \qquad \bigcap_{i=1}^{\infty} S(x, \mathcal{W}_i) = \{x\} \quad \text{for each } x \in X.$$

Then $\{S(x, \mathcal{W}_i) \mid i = 1, 2, \ldots\}$ is a nbd basis for each $x \in X - H(X)$. Hence $\{\mathcal{V}_{nm}, \mathcal{W}_i \mid n, m, i = 1, 2, \ldots\}$ is a development of X. Thus X is metrizable.

I) *Let X be a perfectly normal GO-space and Y a subset of X. If Y is a discrete space when regarded as a subspace, then Y is a σ-discrete set in X.*

Proof. Since X is hereditarily collectionwise normal, to each $y \in Y$ we can assign a convex open nbd $U(y)$ of y in X such that $\{U(y) \mid y \in Y\}$ is disjoint. Put

$$U = \bigcup \{U(y) \mid y \in Y\}.$$

Then, since X is perfectly normal, $U = \bigcup_{k=1}^{\infty} F_k$ for closed sets F_k, $k = 1, 2, \ldots$. Then $Y_k = F_k \cap Y$ is obviously discrete in X. Hence $Y = \bigcup_{k=1}^{\infty} Y_k$ is a σ-discrete set in X.

Theorem VIII.5.[1] *A GO-space X is metrizable if and only if it is semistratifiable.*

---

[1] Due to D. Lutzer [2].

Proof. It suffices to show the 'if' part. Since X is semi-stratifiable, it has a G_δ-diagonal. Thus by H) it suffices to show that $H(X)$ is σ-discrete. Let $\{U(n, x) \mid n = 1, 2, \ldots\}$ be a sequence of convex open nbds of $x \in X$ satisfying the condition (iii) of Theorem VI.25. Put

$$H'(X) = \{x \in H(X) \mid [x, \infty) \in \mathcal{O} - \mathcal{O}(<)\} .$$

Then it follows that

$$H'(X) = \bigcup_{n=1}^{\infty} H'_n ,$$

where
$$H'_n = \{x \in H'(X) \mid \text{for every } y < x, \ U(n, y) \subset (-\infty, x)\} .$$

Then each H'_n is obviously discrete as a subspace. Hence by I) H'_n is a σ-discrete set in X, and so is $H'(X)$. In a similar way $H''(X) = \{x \in H(X) \mid (-\infty, x] \in \mathcal{O} - \mathcal{O}(<)\}$ is also σ-discrete in X. Hence $H(X)$ is σ-discrete in X. Therefore X is metrizable.

R. Engelking–D. Lutzer [1], M. J. Faber [1], H. Bennett–D. Lutzer [1], and J. v. Wouwe [1] got further interesting results on paracompactness, metrizability and generalized metrizability of GO-spaces. For example, J. v. Wouwe proved:

If a GO-space X is a p-space, then it is an M-space.

A GO-space X is an M-space if and only if it is a $w\Delta$-space. Bennett–Lutzer proved:

A GO-space X is hereditarily M if and only if it is metrizable.

Another interesting aspect of the study of ordered spaces is to find topological conditions in order that a given topological space X be orderable (suborderable), where X is called *orderable* (*suborderable*) if it is homeomorphic to a LOTS (GO-space).

For example, H. Herrlich [2] proved:

A totally disconnected metric space X is orderable if and only if $\dim X = 0$.

By use of the method of J. de Groot–P. Schnare [1], J. v. Dalen–E. Wattel [1] gave necessary and sufficient conditions in order that a T_1-space be orderable (suborderable).[1]

---

[1] See S. Eilenberg [1], M. E. Rudin [6], H. Herrlich [3], [4] and S. Purisch [1] for further results in this aspect.

2. Cardinal functions

There are various topologically invariant cardinal numbers that are assigned to each topological space X, e.g., the cardinality $|X|$, weight $w(X)$, character $\chi(X)$, density $d(X)$ and Lindelöf number $l(X)$, which were defined before.[1] The study of relations between those and other *cardinal functions* is an important part of the modern general topology, especially since A. V. Arhangelskii's theorem (Theorem VIII.6) appeared. The purpose of this section is to discuss just a few of the important results obtained in this aspect. The reader, who is interested in more details of the subject, is referred to A. Arhangelskii's article [8], W. Comfort's article [2] and I. Juhász's book [1].

A) *Let X be a T_1-space. Then $|X| \leqslant 2^{w(X)}$.*

Proof. Left to the reader.

B) *Let X be a T_2-space. Then $|X| \leqslant [d(X)]^{\chi(X)}$.*

Proof. Let $\{U_\alpha(x) \mid \alpha \in A\}$ be a nbd base at $x \in X$ with cardinality $\leqslant \chi(X)$ and Y a dense subset of X with $|Y| = d(X)$. To each $x \in X$ we assign a map y_x from $A \times A$ into Y as follows. Well-order the points of Y; then $y_x(\alpha, \beta)$ is the first point of Y which is in $U_\alpha(x) \cap U_\beta(x)$. Since X is T_2, for distinct points x and x' of X, there are α, β such that

$$U_\alpha(x) \cap U_\beta(x') = \emptyset .$$

Then

$$(U_\alpha(x) \cap U_\beta(x)) \cap (U_\alpha(x') \cap U_\beta(x')) = \emptyset ,$$

and hence

$$y_x(\alpha, \beta) \neq y_{x'}(\alpha, \beta) .$$

This means that $y_x \neq y_{x'}$. Thus the map : $x \to y_x$ is one-to-one. Hence

$$|X| \leqslant [d(X)]^{\chi(X)\chi(X)} = [d(X)]^{\chi(X)} .$$

----

[1] We may use the symbol $\chi_x(X)$ to mean the minimum (infinite) cardinality of a nbd base of $x \in X$.

Theorem VIII.6 (Arhangelskii's theorem). *Let* X *be a* T_2-*space. Then* $|X| \leq 2^{\chi(X)l(X)}$.

Proof.[1] For each $x \in X$ we choose a nbd base $\mathscr{U}_x$ such that $|\mathscr{U}_x| \leq \chi(X)$. Let τ be the initial ordinal number with cardinality $l(X) + 1$.

We define closed sets F_β, $0 \leq \beta < \tau$ such that:

(1) $F_\alpha \subset F_\beta$ if $0 \leq \alpha < \beta < \tau$,

(2) $|F_\beta| \leq 2^{\chi(X)l(X)}$ for each β,

(3) if $\mathscr{U} \subset \cup \{\mathscr{U}_x \mid x \in \cup_{\alpha<\beta} F_\alpha\}$, $|\mathscr{U}| \leq l(X)$ and $X - \cup \mathscr{U} \neq \emptyset$, then $F_\beta - \cup \mathscr{U} \neq \emptyset$.

The construction is done by use of induction on β. Let $F_0 = \{p\}$, where p is an arbitrary point of X. Assume $0 < \beta < \tau$ and F_α have been defined for all $\alpha < \beta$. Put

$$\mathcal{O} = \cup \left\{ \mathscr{U}_x \mid x \in \bigcup_{\alpha<\beta} F_\alpha \right\},$$

$$\mathscr{U}' = \{X - \cup \mathscr{U} \mid \mathscr{U} \subset \mathcal{O}, |\mathscr{U}| \leq l(X), X - \cup \mathscr{U} \neq \emptyset\}.$$

Note that $|\mathcal{O}| \leq 2^{\chi(X)l(X)}$ follows from (2), and hence

$$|\mathscr{U}'| \leq (2^{\chi(X)l(X)})^{l(X)} = 2^{\chi(X)l(X)}.$$

Select a point $p(V)$ from each element V of $\mathscr{U}'$ to construct a set

$$E = \{p(V) \mid V \in \mathscr{U}'\}.$$

Then $|E| \leq 2^{\chi(X)l(X)}$, and hence

$$\left| E \cup \left(\bigcup_{\alpha<\beta} F_\alpha \right) \right| \leq 2^{\chi(X)l(X)}.$$

Thus it follows from (3) that

$$\overline{E \cup \left(\bigcup_{\alpha<\beta} F_\alpha \right)} \leq (2^{\chi(X)l(X)})^{\chi(X)} = 2^{\chi(X)l(X)}.$$

Now, define that

---

[1] This proof is due to R. Pol [1].

$$F_\beta = \overline{E \cup \left(\bigcup_{\alpha < \beta} F_\alpha \right)}.$$

Then F_β obviously satisfies (1) and (2). To prove (3), suppose

$$\mathcal{U} \subset \bigcup \left\{ \mathcal{U}_x \mid x \in \bigcup_{\alpha < \beta} F_\alpha \right\}, \qquad |\mathcal{U}| \leq l(X) \quad \text{and} \quad X - \bigcup \mathcal{U} \neq \emptyset.$$

Then $X - \bigcup \mathcal{U} \in \mathcal{U}'$, and thus

$$p(X - \bigcup \mathcal{U}) \in E \cap (X - \bigcup \mathcal{U}) = E - \bigcup \mathcal{U} \neq \emptyset.$$

Hence

$$F_\beta - \bigcup \mathcal{U} \neq \emptyset,$$

proving (3).

Now, we claim that

$$X = \bigcup_{0 \leq \alpha < \tau} F_\alpha.$$

To this end we first prove that $F = \bigcup_{0 \leq \alpha < \tau} F_\alpha$ is closed. Suppose $x \in \bar{F}$. Let

$$\mathcal{U}_x = \{ U_\gamma \mid \gamma \in \Gamma \}, \qquad |\Gamma| \leq \chi(X);$$

then $U_\gamma \cap F \neq \emptyset$ for all $\gamma \in \Gamma$. To each $\gamma \in \Gamma$ we assign $\alpha(\gamma) < \tau$ such that $U_\gamma \cap F_{\alpha(\gamma)} \neq \emptyset$. Put

$$\sup \{ \alpha(\gamma) \mid \gamma \in \Gamma \} = \beta < \tau.$$

Then by (1) $F_\beta \cap U_\gamma \neq \emptyset$ for all $\gamma \in \Gamma$. This implies $x \in F_\beta \subset F$ since F_β is closed. Hence F is closed.

Assume $y \in X - F \neq \emptyset$. Then for each $x \in F$ we select $U(x) \in \mathcal{U}_x$ such that $y \notin U(x)$. Then F has a subset G such that

$$\mathcal{U} = \{ U(x) \mid x \in G \}$$

covers F, and $|\mathcal{U}| \leq l(X)$. Since $G \subset \bigcup_{0 \leq \alpha < \tau} F_\alpha$ and $|G| \leq l(X)$, by (1) $G \subset F_\alpha$ for some $\alpha < \tau$. Let $\beta = \alpha + 1$; then

$$\mathcal{U} \subset \cup \{\mathcal{U}_x \mid x \in F_\alpha\} = \cup \left\{ \mathcal{U}_x \mid x \in \bigcup_{\gamma < \beta} F_\gamma \right\},$$

$$|\mathcal{U}| \leq l(X) \quad \text{and} \quad y \in X - \cup \mathcal{U} \neq \emptyset.$$

On the other hand, $F_\beta - \cup \mathcal{U} = \emptyset$ follows from

$$F_\beta \subset F \subset \cup \mathcal{U}.$$

This contradicts (3), and hence

$$X = F = \bigcup_{0 \leq \alpha < \tau} F_\alpha.$$

Since by (2) $|F_\alpha| \leq 2^{\chi(X)l(X)}$ for all $\alpha < \tau$, we have

$$|X| \leq (l(X) + 1)2^{\chi(X)l(X)} = 2^{\chi(X)l(X)}.$$

Corollary. *Let X be a first countable Lindelöf T_2-space. Then $|X| \leq 2^{\aleph_0}$.*

Alexandroff's problem: Is every first countable compact T_2-space of cardinality $\leq \mathfrak{c}$? had been one of the most famous problems of general topology during some forty years until it was affirmatively answered by A. V. Arhangelskii [7] in 1969 as we saw in the above, and the work of Arhangelskii greatly enhanced this aspect of research. The converse of the above corollary is not true. In fact a compact T_2-space X with $|X| \leq 2^{\aleph_0}$ is not necessarily first countable. But the following theorem due to S. Mrowka [3] implies that such a space has many first countable points if CH is assumed.

Theorem VIII.7. *Let X be a non-empty compact T_2-space. If X has no first countable point, then $|X| \geq 2^{\aleph_1}$.*

Proof. First observe that if a zero set of X contains at least two points, then it contains two disjoint non-empty zero sets. Also note that each zero set F of X indeed contains at least two points. Because if $F = \{x\}$, then $\{x\}$ is a G_δ-set, and accordingly x has a countable nbd base, which is impossible.

Thus we assign to each zero set F of X two disjoint non-empty zero sets contained in F and denote them by F_0 and F_1. Let $A = \{a_\xi \mid 0 \leq \xi < \omega_1\}$ be a transfinite sequence consisting of 0 and 1. To A we assign a decreasing sequence $\{F_\xi^{(A)} \mid 0 \le \xi < \omega_1\}$ of non-empty zero sets in X as follows. First we put

$$F_0^{(A)} = X_{a_0}.$$

Suppose $0 < \xi < \omega_1$, and $F_\eta^{(A)}$ have been defined for all $\eta < \xi$ in such a way that

$$F_\eta^{(A)} \supset F_{\eta'}^{(A)} \quad \text{whenever } \eta < \eta' < \xi.$$

Then, let

$$F = \bigcap \{F_\eta^{(A)} \mid \eta < \xi\}.$$

Since X is compact, F is a non-empty zero set. Define that

$$F_\xi^{(A)} = F_{a_\xi}.$$

Now, assume that $A = \{a_\xi \mid 0 \le \xi < \omega_1\}$ and $B = \{b_\xi \mid 0 \le \xi < \omega_1\}$ are distinct sequences such that $a_\xi \ne b_\xi$, and $a_\eta = b_\eta$ for all $\eta < \xi$. Then $F_\xi^{(A)} \cap F_\xi^{(B)} = \emptyset$ follows from the above construction. Thus $F^{(A)} = \bigcap \{F_\xi^{(A)} \mid 0 \le \xi < \omega_1\}$ and $F^{(B)} = \bigcap \{F_\xi^{(B)} \mid 0 \le \xi < \omega_1\}$ are disjoint non-empty closed sets. Thus X contains at least $2^{\aleph_1}$ distinct points.

Corollary 1. *If Z is a non-empty zero set in a compact T_2-space X and if Z contains no first countable point, then $|Z| \ge 2^{\aleph_1}$.*

Proof. Left to the reader.

Corollary 2. *Let X be a compact T_2-space such that $\aleph_0 < |X| < 2^{\aleph_1}$. Then X has uncountably many first countable points.*

Proof. Assume the contrary that $p_1, p_2, p_3, \ldots$ are the only first countable points of X. Then we claim that there is a non-empty zero set of X which is disjoint from $X_0 = \{p_1, p_2, \ldots\}$. To this end, let p_0 be an arbitrary point such that $p_0 \notin X_0$. For each n construct a continuous function $f_n : X \to [0, 1]$ such that

$$f_n(p_0) = 0, \qquad f_n(p_n) = 1.$$

Put

$$f = \sum_{n=1}^{\infty} \frac{1}{2^n} f_n.$$

Then $f(p_0) = 0$, and $f(p_n) \neq 0$ for all $n \geq 1$. Thus $Z = f^{-1}(0)$ satisfies the desired condition. Since $|Z| \leq |X| < 2^{\aleph_1}$, this contradicts Corollary 1.

Example VIII.4. Let X be the product of $\aleph_1$ copies of the two point discrete space. Then X has no first countable point, and $|X| = 2^{\aleph_1}$. Assume CH. Then by Corollary 2 every compact T_2-space with cardinality $\mathfrak{c}$ has uncountably many first countable points.

We define another cardinal function which is often used in this aspect of general topology.

Definition VIII.4. The *Souslin number* or *cellularity* $c(X)$ of X is the smallest infinite cardinal m such that every pairwise disjoint collection of non-empty open sets of X has cardinality $\leq m$.

Theorem VIII.8.[1] *Let X be a T_2-space. Then $|X| \leq 2^{\chi(X)c(X)}$.*

Proof. For each $x \in X$ we select an open nbd base $\mathcal{U}_x$ such that $|\mathcal{U}_x| \leq \chi(X)$. Let τ be the initial ordinal number with cardinality $\chi(X)c(X) + 1$ and Y a fixed set with $|Y| = \chi(X)c(X)$. We define subsets X_β, $0 \leq \beta < \tau$, of X such that:
  (1) $X_\alpha \subset X_\beta$ if $0 \leq \alpha < \beta < \tau$,
  (2) $|X_\beta| \leq 2^{\chi(X)c(X)}$ for each β,
  (3) if $\mathcal{V}_y$, $y \in Y$, are subcollections of $\cup \{\mathcal{U}_x \mid x \in X\}$ such that

$$\cup\{\mathcal{V}_y \mid y \in Y\} \subset \cup\left\{\mathcal{U}_x \mid x \in \bigcup_{\alpha < \beta} X_\alpha\right\},$$

$$|\mathcal{V}_y| \leq \chi(X)c(X) \quad \text{for all } y \in Y,$$

and

$$X - \cup\{\bar{V}_y \mid y \in Y\} \neq \emptyset,$$

then

$$X_\beta - \cup\{\bar{V}_y \mid y \in Y\} \neq \emptyset,$$

where we put

$$V_y = \cup\{V \mid V \in \mathcal{V}_y\}.$$

---

[1] We owe this theorem to A. Hajnal–I. Juhász [1] and the proof to R. Pol [1].

The construction is done by use of induction on β. Let $X_0 = \{p\}$, where p is an arbitrary point of X. Assume $0 < \beta < \tau$ and X_α have been defined for all $\alpha < \beta$. Put

$$\mathcal{O} = \cup \left\{ \mathcal{U}_x \mid x \in \bigcup_{\alpha < \beta} X_\alpha \right\},$$

$$\mathcal{V} = \left\{ X - \cup\{\bar{V}_y \mid y \in Y\} \mid \cup\{\mathcal{V}_y \mid y \in Y\} \subset \mathcal{O}, \ |\mathcal{V}_y| \leq \chi(X)c(X) \right.$$

$$\left. \text{for all } y \in Y, \ X - \cup\{\bar{V}_y \mid y \in Y\} \neq \emptyset \right\}.$$

It follows from (2) that

$$|\mathcal{O}| \leq 2^{\chi(X)c(X)},$$

and thus

$$|\mathcal{V}| \leq 2^{\chi(X)c(X)}.$$

Select a point $p(V)$ from each element V of $\mathcal{V}$ to define

$$E = \{p(V) \mid V \in \mathcal{V}\}.$$

Put

$$X_\beta = E \cup \left(\bigcup_{\alpha < \beta} X_\alpha \right).$$

Then $|X_\beta| \leq 2^{\chi(X)c(X)}$. Namely X_β satisfies (2). It is easy to see that X_β also satisfies (1) and (3). Thus $\{X_\alpha \mid 0 \leq \alpha < \tau\}$ is defined.

Now we claim that $X = \bigcup_{0 \leq \alpha < \xi} X_\alpha$. To show it, put

$$X' = \cup\{X_\alpha \mid 0 \leq \alpha < \tau\} \tag{4}$$

and assume that

$$x_0 \in X - X' \neq \emptyset.$$

We can put

$$\mathcal{U}_{x_0} = \{U_y \mid y \in Y\},$$

where $U_y = U_{y'}$ may happen for distinct y and y'. For each $y \in Y$ we put

$$\mathcal{W}_y = \{U \mid U \in \bigcup\{\mathcal{U}_x \mid x \in X'\}, \ U \cap U_y = \emptyset\}. \tag{5}$$

Denote by $\mathcal{V}_y$ a maximal disjoint subcollection of $\mathcal{W}_y$. Then we can put

$$\mathcal{V}_y = \{U'_x \mid x \in X'_y\},$$

where $U'_x \in \mathcal{U}_x$, $X'_y \subset X'$. Further note that $|\mathcal{V}_y| \leq c(X)$, and thus we may assume $|X'_y| \leq c(X)$. By (1) and (4), $X'_y \subset X_\alpha$ for some $\alpha < \tau$. Hence

$$\mathcal{V}_y \subset \bigcup\{\mathcal{U}_x \mid x \in X_\alpha\} \quad \text{for each } y \in Y \text{ and for some } \alpha < \tau.$$

Since $|Y| = \chi(X)c(X)$, we have

$$\bigcup\{\mathcal{V}_y \mid y \in Y\} \subset \bigcup\left\{\mathcal{U}_x \mid x \in \bigcup_{\alpha < \beta} X_\alpha\right\} \quad \text{for some } \beta < \tau.$$

Now observe that $U_y \cap V_y = \emptyset$ follows from (5). Since U_y is a nbd of x_0, this implies $x_0 \notin \bar{V}_y$, and hence

$$x_0 \in X - \bigcup\{\bar{V}_y \mid y \in Y\} \neq \emptyset.$$

Thus by (3) there is $x' \in X_\beta - \bigcup\{\bar{V}_y \mid y \in Y\}$. Select $U_y \in \mathcal{U}_{x_0}$ and $U \in \mathcal{U}_{x'}$ such that

$$U \cap U_y = \emptyset \quad \text{and} \quad U \cap V_y = \emptyset.$$

Then $\{U\} \cup \mathcal{V}_y$ is a disjoint open subcollection of $\mathcal{W}_y$, which contradicts the maximality of $\mathcal{V}_y$. Hence $X = X'$ is concluded. From (2) it follows that $|X| \leq 2^{\chi(X)c(X)}$.

The following theorem on density of product spaces is extremely interesting and important.

Theorem VIII.9.[1] *Let $\{X_\alpha \mid \alpha \in A\}$ be a collection of topological spaces such that $d(X_\alpha) \leq m$ for all $\alpha \in A$ and $|A| \leq 2^m$. Then $d(\prod_\alpha X_\alpha) \leq m$.*

Proof. To begin with, we prove the theorem in case that each X_α is discrete, $|X_\alpha| = m$ and $|A| = 2^m$. Then $\prod_\alpha X_\alpha$ is regarded as the space of

---

[1] Due to E. S. Pondiczery [1].

all maps from D^m (the product of m copies of the two point discrete space D) into X (the discrete space of m points) with the weak topology. Denote by $\mathcal{U}$ a base of D^m with $|\mathcal{U}| = m$ and by $\mathcal{U}'$ the set of all finite disjoint subcollections of $\mathcal{U}$. Then $|\mathcal{U}'| = m$. Denote by x_0 a fixed point of X. We denote by P the set of all maps $f : D^m \to X$ such that for some $\{U_1, \ldots, U_n\} \in \mathcal{U}'$, f is constant on each of $U_1, \ldots, U_n$, and $f(z) = x_0$ for all $z \in D^m - U_1 \cup \cdots \cup U_n$. Then $|P| = m$ is obvious.

Now, let us prove that P is dense in $\prod_\alpha X_\alpha$. Suppose that $\{x_\alpha \mid \alpha \in A\}$ is a given point of $\prod_\alpha X_\alpha$ and $\alpha_1 \ldots \alpha_n \in A$, where $\alpha_1, \ldots, \alpha_n$ are all distinct from each other. Regard $\{\alpha_1, \ldots, \alpha_n\}$ as a finite subset of D^m. Then there is $\{U_1, \ldots, U_n\} \in \mathcal{U}'$ such that $\alpha_i \in U_i$, $i = 1, \ldots, n$. Thus there is $f \in P$ such that

$$f(U_i) = x_{\alpha_i}, \quad i = 1, \ldots, n.$$

Hence

$$f \in P \cap \left\{ \{x'_\alpha\} \in \prod_\alpha X_\alpha \mid x'_{\alpha_i} = x_{\alpha_i}, \quad i = 1, \ldots, n \right\} \neq \emptyset,$$

and hence

$$\{x_\alpha\} \in \bar{P} \quad \text{holds in } \prod_\alpha X_\alpha.$$

Namely P is dense in $\prod_\alpha X_\alpha$. This proves that $d(\prod_\alpha X_\alpha) \leq m$.

Generally, if $d(X_\alpha) = m$ and $|A| = 2^m$, then we select a dense set X'_α of X_α with $|X'_\alpha| = m$ for each α. Denote by X''_α the discrete space with m points. Then it follows from the above result that $d(\prod_\alpha X''_\alpha) \leq m$. Since $\prod_\alpha X'_\alpha$ has a topology weaker than $\prod_\alpha X''_\alpha$, this implies $d(\prod_\alpha X'_\alpha) \leq m$ and eventually $d(\prod_\alpha X_\alpha) \leq m$. The validity of theorem in the most general case also easily follows.

Corollary 1. *Let X_α, $\alpha \in A$, be separable spaces and $|A| \leq \mathfrak{c}$. Then $\prod_\alpha X_\alpha$ is separable.*

Proof. Obvious.

Corollary 2. *Any product of separable spaces has CCC.*

Proof. Left to the reader.

3. Dyadic space

In this section we study dyadic spaces which form an important subclass of the compact T_2-spaces. Recall VII.2.D) before reading the following definition due to P. S. Alexandroff.

Definition VIII.5. A topological space X is called a *dyadic space* if it is the image of D^m for some cardinal number m by a continuous mapping.

A) *Every dyadic space is a compact T_2-space with CCC.*

Proof. The last property follows from Corollary 2 to Theorem VIII.9.

Now, let us prove that every compact metric space is a dyadic space.

B) *Let F and G be totally disconnected compact subsets of the real line E^1, both with no isolated point. Then F and G are homeomorphic.*

Proof. Let

$$a = \inf F, \qquad b = \sup F, \qquad c = \inf G, \qquad d = \sup G.$$

Then $[a, b] - F$ is decomposed into countably many disjoint open intervals (convex components), say P_i, $i = 1, 2, \ldots$. Then $\mathcal{P} = \{P_i \mid i = 1, 2, \ldots\}$ is an ordered set with the natural order: $P_i < P_j$ if and only if P_i is on the left of P_j. Similarly $[c, d] - G$ is decomposed into countably many disjoint open intervals Q_i, $i = 1, 2, \ldots$. We regard $\mathcal{Q} = \{Q_i \mid i = 1, 2, \ldots\}$ as an ordered set. We define an order-preserving one-to-one map f from $\mathcal{P}$ onto $\mathcal{Q}$ as follows.

First define that $f(P_1) = Q_1$. Assume that we have defined $f(P_i) = Q_{n_i}$ for $i = 1, \ldots, k - 1$. Then we define $f(P_k)$ as follows. Put

$$N_k = \{n \in N \mid \text{(i) } n \neq n_1, \ldots, n_{k-1}, \text{ (ii) if } i \leq k - 1 \text{ and } P_i < P_k,$$
$$\text{then } Q_{n_i} < Q_n, \text{ (iii) if } i \leq k - 1 \text{ and } P_i > P_k, \text{ then}$$
$$Q_{n_i} > Q_n\}.$$

It is easy to see that $N_k \neq \emptyset$ follows from the property of G. Then define that

$$f(P_k) = Q_{n_k}, \qquad n_k = \min N_k.$$

Now f is an order-preserving map. It follows from the property of X that for each pair (P_i, P_j) of distinct elements of $\mathscr{P}$, there is a third element $P_k \in \mathscr{P}$ between P_i and P_j. Thus f is proved to be onto. Since f is obviously one-to-one, it satisfies the desired conditions. We define a map φ from F onto G as follows. Let $x \in F$. Then put

$$\mathscr{U}_x = \{P_i \in \mathscr{P} \mid P_i \text{ is on the left of } x\}.$$

Now it is obvious that $x = \sup (\bigcup \mathscr{U}_x)$. Then define that

$$\varphi(x) = \sup [\bigcup \{f(P) \mid P \in \mathscr{U}_x\}].$$

It is easy to see that φ is a topological map from F onto G, but the details are left to the reader.

C) *Let K be a totally disconnected compact subset of E^1. Then there is a totally disconnected compact subset F of E^1 with no isolated point and a continuous map from F onto K.*

Proof. Let $a = \inf K$, $b = \sup K$. Decompose $[a, b] - K$ into the convex components as

$$[a, b] - K = \bigcup_{k=1}^{\infty} I_k,$$

where each I_k is an open interval. We construct a Cantor discontinuum C_k in each $\bar{I}_k$ to define that

$$F = K \cup \left(\bigcup_{k=1}^{\infty} C_k \right).$$

Then F is obviously a totally disconnected compact set with no isolated point because K is totally disconnected and compact. We can define a continuous map f from F onto K as follows. For each k, select

$$p_k \in I_k - C_k.$$

Then

$$f(x) = \begin{cases} x & \text{if } x \in K, \\ \inf I_k & \text{if } x \in C_k \text{ and } x < p_k, \\ \sup I_k & \text{if } x \in C_k \text{ and } x > p_k. \end{cases}$$

It is easy to show that f satisfies the desired condition.

Theorem VIII.10.[1] *Every compact metric space X is a dyadic space.*

Proof. By VII.2.D) and Exercise II.41 there is a closed subset K of the Cantor discontinuum and a continuous map f from K onto X. K is obviously totally disconnected and compact. Thus by C) there is a totally disconnected compact subset F of E^1 with no isolated point and a continuous map g from F onto K. By B) there is a homeomorphism h from the Cantor discontinuum $D^{\aleph_0}$ onto F. Then $f \circ g \circ h$ is a continuous map from $D^{\aleph_0}$ onto X, proving the theorem.

D) *Every zero set Z of a dyadic space X is dyadic.*[2]

Proof. Let f be a continuous map from $D^\tau = \prod \{D_t \mid t \in T\}$ onto X, where each D_t is a two point discrete space. Then $f^{-1}(Z)$ is a zero set of D^τ. So suppose that

$$f^{-1}(Z) = \bigcap_{n=1}^{\infty} U_n \quad \text{for open sets } U_n, \ n = 1, 2, \ldots, \text{ of } D^\tau.$$

To each $x \in f^{-1}(Z)$ and $n \in N$ we assign a basic open nbd

$$U_n(x) = \prod \{P_t \mid t \in T(x)\} \times \prod \{D_t \mid t \in T - T(x)\} \subset U_n,$$

where $T(x)$ is a finite subset of T, and P_t an open set of D_t for each $t \in T(x)$. For a fixed n we select $x_1, \ldots, x_k \in f^{-1}(Z)$ such that

$$f^{-1}(Z) \subset \bigcup_{i=1}^{k} U_n(x_i) \subset U_n. \tag{1}$$

Observe that

$$\bigcup_{i=1}^{k} U_n(x_i) = V_n \times \prod \{D_t \mid t \in T - T'\},$$

where T' is a finite subset of T, and V_n an open set of $\prod \{D_t \mid t \in T'\}$. Put

$$W_n = \bigcup_{i=1}^{k} U_n(x_i).$$

---

[1] Due to P. S. Alexandroff [6].
[2] E. Efimov [1] proved that every closed subset of a dyadic space X is dyadic if and only if X is metrizable.

Then $f^{-1}(Z) = \bigcap_{n=1}^{\infty} W_n$ follows from (1), and hence

$$f^{-1}(Z) = Y \times \prod \{D_t \mid t \in T - T''\}$$

for a countable subset T'' of T and a closed set Y of $\prod \{D_t \mid t \in T''\}$. Since Y is a compact metric space, it is dyadic by Theorem VIII.10. Thus $f^{-1}(Z)$ is also dyadic, and so is Z.

E) *Let f be a continuous map from a compact T_2-space X onto Y. Then $w(Y) \leq w(X)$.*

Proof. Left to the reader.

Theorem VIII.11.[1] *If X is a dyadic space, then $w(X) = \chi(X)$ holds.*

Proof. Let f be a continuous map from $D^T = \prod \{D_t \mid t \in T\}$ onto X. Suppose $\chi(X) = m$. Then for each $x \in X$

$$f^{-1}(x) = \bigcap \{U_\lambda \mid \lambda \in M(x)\},$$

where $M(x)$ satisfies

$$|M(x)| \leq m, \tag{1}$$

and U_λ is an open subset of D^T. For each $\lambda \in M(x)$ we can find an open set

$$W_\lambda = V_\lambda \times \prod \{D_t \mid t \in T - T'\} \subset D^T,$$

where T' is a finite subset of T, and V_λ is an open set of $\prod \{D_t \mid t \in T'\}$ such that

$$f^{-1}(x) \subset W_\lambda \subset U_\lambda.$$

Then it follows that

$$f^{-1}(x) = \bigcap \{W_\lambda \mid \lambda \in M(x)\}.$$

---

[1] Due to A. Esenin–Volpin [1].

Hence

$$f^{-1}(x) = Y(x) \times \prod \{D_t \mid t \in T - T(x)\},$$

where $T(x)$ is a subset of T with $|T(x)| \leq m$, and $Y(x)$ is a subset of $\prod \{D_t \mid t \in T(x)\}$. Observe that $|X| \leq 2^m$ follows from Arhangelskii's theorem. Thus $T' = \bigcup \{T(x) \mid x \in X\}$ satisfies

$$|T'| \leq 2^m.$$

Now there are subsets $Y'(x)$, $x \in X$, of $\prod \{D_t \mid t \in T'\}$ such that

$$f^{-1}(x) = Y'(x) \times \prod \{D_t \mid t \in T - T'\} \quad \text{for all } x \in X.$$

Then the restriction of f to $D^T = \prod \{D_t \mid t \in T'\} \times \{0\}$ is a continuous map from $D^{T'}$ onto X, where 0 denotes the point of $\prod \{D_t \mid t \in T - T'\}$ all of whose coordinates are zero. By Theorem VIII.9 it holds that $d(D^{T'}) \leq m$. Therefore X has a dense subset X_0 with $|X_0| \leq m$.

Let

$$T'' = \bigcup \{T(x) \mid x \in X_0\}.$$

Then $|T''| \leq m$. Now, the restriction f_0 of f to $D^{T''} = \prod \{D_t \mid t \in T''\} \times \{0\}$ maps $D^{T''}$ onto X, where 0 denotes the point of $\prod \{D_t \mid t \in T - T''\}$ all of whose coordinates are zero. Because

$$f_0(D^{T''}) \supset X_0,$$

and hence

$$X = \bar{X}_0 \subset \overline{f_0(D^{T''})} = f_0(D^{T''}).$$

Thus by E) we have

$$w(X) \leq w(D^{T''}) \leq m,$$

which proves the theorem.[1]

---

[1] More generally B. A. Efimov [1] proved: If X is a dyadic space, then $w(X) = \sup \{\chi_x(X) \mid x \in M\}$, where M is any dense subset of X which is not discrete as a subspace. If M is dense and discrete as a subspace of X, then $w(X) \leq \aleph_0$.

Corollary. *Every first countable dyadic space is metrizable.*

Proof. Obvious.

Example VIII.5. Let $[0, 1] \times \{0, 1\}$ be the cartesian product of the closed segment and the two elements set. We introduce a topology into $X = [0, 1] \times \{0, 1\} - \{(0, 0), (1, 1)\}$ by use of a nbd base as follows.
If $p = (x, 0) \in X$ and $\varepsilon > 0$, then

$$U_\varepsilon(p) = \{(x', 0) \in X \mid x - \varepsilon < x' \leqslant x\} \cup \{(x', 1) \mid x - \varepsilon < x' < x\}.$$

If $p = (x, 1) \in X$ and $\varepsilon > 0$, then

$$U_\varepsilon(p) = \{(x', 0) \in X \mid x < x' < x + \varepsilon\} \cup \{(x', 1) \in X \mid x \leqslant x' < x + \varepsilon\}.$$

Now $\{U_\varepsilon(p) \mid \varepsilon > 0\}$ is a nbd base of each $p \in X$. It is easy to see that X is a compact T_2-space with CCC and the first countability. This space is called *Alexandroff–Urysohn's two arrows space*. On the other hand, X is not metrizable because its subspace $(0, 1) \times \{0\}$ is homeomorphic to Sorgenfrey line, which is not metrizable. Thus by the corollary of Theorem VIII.11 X is no dyadic space.

Example VIII.6. Let X be a separable first countable non-compact Tychonoff space. Then we can prove that $\beta(X)$ is not dyadic. Assume the contrary that there is a continuous map f from $D^T = \prod \{D_t \mid t \in T\}$ onto $\beta(X)$. Then, since $\beta(X)$ has a countable dense subset each of whose points is first countable, by use of the method of the proof of Theorem VIII.11, we can find a countable subset T' of T such that

$$f(\prod \{D_t \mid t \in T'\} \times \{0\}) = \beta(X).$$

Thus $\beta(X)$ is a perfect image of a metrizable space, and hence $\beta(X)$ itself is metrizable, which is impossible. Therefore $\beta(X)$ is not dyadic. Now, we can prove that if X is a non-pseudo-compact Tychonoff space, then $\beta(X)$ is not dyadic.[1] Because, there is an unbounded real-valued continuous function f defined on X. Thus $f(X) = Y$ is non-compact, first countable and separable. We can extend f to a continuous map βf from

---

[1] Due to R. Engelking–A. Pełczyński [1].

$\beta(X)$ onto $\beta(Y)$. Since $\beta(Y)$ is not dyadic as observed before, $\beta(X)$ cannot be dyadic either.

Let us state some other results on dyadic spaces. N. A. Shanin [4] proved that a dyadic space cannot be represented as the sum of a well-ordered increasing sequence of nowhere dense subsets, which is a reinforcement of Baire's theorem. L. Ivanovskiĭ–V. Kuz'minov [1] proved that every compact topological group is dyadic. R. Engelking–A. Peĺc-zyński [1] proved that every extremely disconnected dyadic space is finite. B. A. Efimov [2] made an extensive study of dyadic spaces and especially proved that a dyadic space is metrizable if and only if it is hereditarily normal if and only if it is Fréchet–Urysohn. B. Efimov [2] and B. Efimov–R. Engelking [1] studied *irreducible dyadic spaces*, i.e., the images of D^T by irreducible continuous maps. R. Engelking [2] and A. Arhan-gelskii–V. Ponomarev [1] also obtained interesting results on dyadic spaces.

Another interesting class of compact spaces was introduced by J. de Groot [6]. A topological space X is called *supercompact* if it has a subbase $\mathcal{U}$ such that every open cover of X by elements of $\mathcal{U}$ has a binary subcover (i.e. a binary subbase). It is easy to see that every compact LOTS is supercompact. It is proved by M. Strok–A. Szymánski [1] that every compact metric space is supercompact.[1] C. F. Mills [2] proved that every compact topological group is supercompact. Every supercompact space is of course compact. On the other hand, it is known that $\beta(N)$ is not supercompact. The two arrows space (Example VIII.5) is a supercompact space which is not dyadic. It is unknown if every dyadic space is supercompact. It is known that closed subsets of a supercompact space are not necessarily supercompact[2] even if they are G_δ. Continuous images of supercompact spaces are not necessarily supercompact even if they are T_2.[3] See A. Verbeek [1], J. v. Mill [1] and E. v. Douwen–J. v. Mill [1] for further results in this aspect.

4. Measure and topological space

The purpose of this section is to characterize some topological proper-ties in terms of measure and show some relations between the two

[1] See also C. F. Mills [1] and E. K. v. Douwen [1].
[2] See M. G. Bell [1].
[3] See C. F. Mills–J. v. Mill [1].

concepts, measure and topology. Throughout this section all spaces are at least Tychonoff. We use some terminologies concerning measure in a manner slightly different from the customary use.

Definition VIII.6. Let $\mathcal{A}(X)$ be the smallest collection of subsets of a space X satisfying (i) $B_1 \cap B_2 \in \mathcal{A}(X)$ whenever B_1, $B_2 \in \mathcal{A}(X)$, (ii) $X - B \in \mathcal{A}(X)$ whenever $B \in \mathcal{A}(X)$, (iii) $\mathcal{A}(X) \supset \mathcal{Z}(X)$, where $\mathcal{Z}(X)$ denotes the collection of all zero sets of X. A non-negative real-valued function μ defined on $\mathcal{A}(X)$ is called a *measure* if μ is *finitely additive*, i.e.,

$$\mu\left(\bigcup_{i=1}^{k} B_i\right) = \sum_{i=1}^{k} \mu(B_i)$$

whenever $\{B_i \mid i = 1, \ldots, k\}$ is a disjoint finite subcollection of $\mathcal{A}(X)$.

A measure μ is called *regular* if $\mu(B) = \inf\{\mu(U) \mid B \subset U \in \mathcal{P}(X)\}$ for each $B \in \mathcal{A}(X)$, where $\mathcal{P}(X)$ denotes the collection of all cozero sets of X.

A measure μ is called *σ-additive* if

$$\mu\left(\bigcup_{i=1}^{\infty} B_i\right) = \sum_{i=1}^{\infty} \mu(b_i)$$

whenever $\{B_i \mid i = 1, 2, \ldots\}$ is a disjoint countable subcollection of $\mathcal{A}(X)$.

A measure μ is called *τ-additive* if for every open cover $\mathcal{U}$ of X by cozero sets and for every $\varepsilon > 0$ there is a finite subcollection $\mathcal{V}$ of $\mathcal{U}$ such that $\mu(\bigcup \mathcal{V}) > \mu(X) - \varepsilon$. A measure μ is called *tight* if for every $\varepsilon > 0$ there is a compact subset C of X such that $\mu^*(C) > \mu(X) - \varepsilon$, where μ^* is defined for every subset C of X by

$$\mu^*(C) = \inf\{\mu(U) \mid C \subset U \in \mathcal{P}(X)\}.$$

Remark. Let $\mathcal{B}_a(X)$ be the smallest Borel field which contains $\mathcal{Z}(X)$. Then every element of $\mathcal{B}_a(X)$ is called a *Baire set*. A non-negative real-valued function λ defined on $\mathcal{B}_a(X)$ is called a (finite) *Baire measure* on X if $\lambda(\bigcup_{i=1}^{\infty} B_i) = \sum_{i=1}^{\infty} \lambda(B_i)$ for every disjoint countable subcollection $\{B_i \mid i = 1, 2, \ldots\}$ of $\mathcal{B}_a(X)$. If μ is a σ-additive measure, the function μ^* defined in the above is a so-called outer measure, and its restriction to $\mathcal{B}_a(X)$ is a Baire measure.[1]

---

[1] See e.g. H. L. Royden [1].

A) *The following conditions are equivalent for a measure μ on X:*

(i) μ *is σ-additive,*

(ii) μ *is regular, and for every increasing sequence* $\{U_i \mid i = 1, 2, \ldots\}$ *of cozero sets of X such that* $\bigcup_{i=1}^{\infty} U_i = X$, $\lim_{i \to \infty} \mu(U_i) = \mu(X)$,

(iii) μ *is regular, and for every decreasing sequence* $\{Z_i \mid i = 1, 2, \ldots\}$ *of zero sets of X such that* $\bigcap_{i=1}^{\infty} Z_i = \emptyset$, $\lim_{i \to \infty} \mu(Z_i) = 0$.

Proof. Note that every measure μ satisfies $\mu(\emptyset) = 0$, $\mu(B) \leqslant \mu(B')$ whenever $B \subset B'$, and $\mu(\bigcup_{i=1}^{k} B_i) \leqslant \sum_{i=1}^{k} \mu(B_i)$ whenever $B_i \in \mathcal{A}(X)$, $i = 1, 2, \ldots, k$. It is obvious that (ii) and (iii) are equivalent. It is also easy to see that (ii) implies (i). Assume (i); then for every decreasing sequence $\{V_i \mid i = 1, 2, \ldots\}$ of cozero sets with $\bigcap_{i=1}^{\infty} V_i \in \mathcal{A}(X)$, we have

$$\mu(V_1) = \mu(V_1 - V_2) + \mu(V_2 - V_3) + \cdots + \mu\left(\bigcap_{i=1}^{\infty} V_i \right).$$

Thus $\mu(\bigcap_{i=1}^{\infty} V_i) = \lim_{i \to \infty} \mu(V_i)$. Observe that every element B of $\mathcal{A}(X)$ is expressed in the form $B = \bigcup_{i=1}^{n} (Z_i \cap U_i)$ for some $n \in N$, $Z_i \in \mathcal{Z}(X)$ and $U_i \in \mathcal{P}(X)$. Since each Z_i is a countable intersection of cozero sets, so is B. Hence $B = \bigcap_{i=1}^{\infty} V_i$ for some decreasing sequence $\{V_i\}$ of cozero sets. Thus $\lim_{i \to \infty} \mu(V_i) = \mu(B)$, proving that μ is regular. The rest of (ii) is easy to verify.

Definition VIII.7. From now on we mean by a measure a regular measure and denote by $\mathcal{M}(X)$, $\mathcal{M}_\sigma(X)$, $\mathcal{M}_\tau(X)$ and $\mathcal{M}_t(X)$ the sets of all regular measures, σ-additive measures, τ-additive measures, and tight measures on X, respectively.

B) $\mathcal{M}_t(X) \subset \mathcal{M}_\tau(X) \subset \mathcal{M}_\sigma(X) \subset \mathcal{M}(X)$ *holds for any space X.*

Proof. Obvious.

Definition VIII.8. A measure μ on X is called a *two-valued measure* if $\mu(\mathcal{A}(X)) = \{0, 1\}$. Let x be a fixed point of X. Then a *Dirac measure* δ_x is defined by

$$\delta_x(B) = \begin{cases} 1 & \text{if } x \in B \in \mathcal{A}(X), \\ 0 & \text{if } x \notin B \in \mathcal{A}(X). \end{cases}$$

We denote by $\mathcal{T}(X)$, $\mathcal{T}_\sigma(X)$, $\mathcal{T}_\tau(X)$ and $\mathcal{D}(X)$ the set of all two-valued measures, two-valued σ-additive measures, two-valued τ-additive measures and Dirac measures on X, respectively.

C) $\mathscr{T}_\tau(X) \subset \mathscr{D}(X) \subset \mathscr{T}_\sigma(X) \subset \mathscr{T}(X)$ holds for any space X.

Proof. We shall prove only the first inclusion relation. Suppose $\mu \in \mathscr{T}_\tau(X)$. Put

$$\mathscr{L}_1 = \{ Z \in \mathscr{L}(X) \mid \mu(Z) = 1 \} .$$

Observe that $\mu(\cap \mathscr{L}) = 1$ for every finite subcollection $\mathscr{L}$ of $\mathscr{L}_1$, because μ is finitely additive. Hence by the τ-additivity of μ, we have $\mu(\cap \mathscr{L}_1) = 1$. Thus there is $x \in \cap \mathscr{L}_1$. Now, let $B \in \mathscr{A}(X)$ satisfy $\mu(B) = 1$. Then, since μ is regular, there is $Z \in \mathscr{L}(X)$ such that $Z \subset B$, $\mu(Z) = 1$. Hence $Z \in \mathscr{L}_1$, and thus $x \in Z \subset B$. Assume $B \in \mathscr{A}(X)$ and $\mu(B) \neq 1$, i.e. $\mu(B) = 0$. Then $\mu(X - B) = 1$, and hence $x \in X - B$. This proves that μ is a Dirac measure.

D) *Let $\mathscr{F}$ be a maximal zero filter of a space X. Then define a map $\mu : \mathscr{A}(X) \to \{0, 1\}$ by $\mu(B) = 1$ if $Z \in \mathscr{F}$ for some Z with $\mathscr{L}(X) \ni Z \subset B$, and $\mu(B) = 0$ if $Z \notin \mathscr{F}$ for every Z with $\mathscr{L}(X) \ni Z \subset B$. Then $\mu \in \mathscr{T}(X)$.*

Proof. To prove that μ is regular, suppose $B = Z \cap U$, where $Z \in \mathscr{L}(X)$ and $U \in \mathscr{P}(X)$. Further assume what $\mu(B) = 0$. If $Z \notin \mathscr{F}$, then $Z \cap Z_1 = \emptyset$ for some $Z_1 \in \mathscr{F}$. Hence $X - Z_1 \in \mathscr{P}(X)$ satisfies

$$B \subset X - Z_1 \quad \text{and} \quad \mu(X - Z_1) = 0 .$$

If $Z \in \mathscr{F}$, then for any $Z' \in \mathscr{L}(X)$ with $Z' \subset U$, we have $Z' \notin \mathscr{F}$. Because otherwise $B \supset Z \cap Z' \in \mathscr{F}$, contradicting $\mu(B) = 0$. Thus $\mu(U) = 0$ and $B \subset U \in \mathscr{P}(X)$. This proves that $\mu(B) = \inf \{\mu(U) \mid B \subset U \in \mathscr{P}(X)\}$. Since every member of $\mathscr{A}(X)$ is in the form of $\cup_{i=1}^n (Z_i \cap U_i)$ for $Z_i \in \mathscr{L}(X)$ and $U_i \in \mathscr{P}(X)$, the same holds for every $B \in \mathscr{A}(X)$.

Suppose $B_1, B_2 \in \mathscr{A}(X)$ and $B_1 \cap B_2 = \emptyset$. If $\mu(B_1) = 1$ or $\mu(B_2) = 1$, then $\mu(B_1 \cup B_2) = 1$ is obvious. If $\mu(B_1) = \mu(B_2) = 0$, then we select U_1, $U_2 \in \mathscr{P}(X)$ such that

$$U_i \supset B_i , \quad \mu(U_i) = 0 , \quad i = 1, 2 .$$

Suppose $Z \in \mathscr{L}(X)$, and $Z \subset U_1 \cap U_2$. Then $Z \cap (X - U_1)$ and $Z \cap$

$(X - U_2)$ are disjoint zero sets. Thus there is a continuous function $f : X \to [0, 1]$ such that

$$f(Z \cap (X - U_1)) = 0 , \qquad f(Z \cap (X - U_2)) = 1 .$$

Put

$$Z_1' = \{x \in Z \mid f(x) \geqslant \tfrac{1}{2}\}, \qquad Z_2' = \{x \in Z \mid f(x) \leqslant \tfrac{1}{2}\}.$$

Then Z_1' and Z_2' are zero sets such that

$$Z_1' \subset U_1 , \qquad Z_2' \subset U_2 , \qquad Z_1' \cup Z_2' = Z .$$

If $Z \in \mathscr{F}$, then either Z_1' or Z_2' belongs to $\mathscr{F}$, i.e. $\mu(U_1) = 1$ or $\mu(U_2) = 1$, which is impossible. Hence $Z \notin \mathscr{F}$, which implies $\mu(U_1 \cup U_2) = 0$ and accordingly $\mu(B_1 \cup B_2) = 0$. Thus μ is a regular measure.

Theorem VIII.12.[1] *The following conditions are equivalent for a space X*:
   (i) *X is pseudo-compact*,
   (ii) *$\mathcal{M}(X) = \mathcal{M}_\sigma(X)$*,
   (iii) *$\mathcal{T}(X) = \mathcal{T}_\sigma(X)$*.

Proof. (i) $\Rightarrow$ (ii). Let $\mu \in \mathcal{M}(X)$. Suppose $\{Z_i \mid i = 1, 2, \ldots\}$ is a decreasing sequence of zero sets in X such that $\bigcap_{i=1}^{\infty} Z_i = \emptyset$. For each i there is a continuous function $f_i : X \to [0, 1/2^i]$ satisfying $Z_i = f_i^{-1}(0)$. Then $f = \sum_{i=1}^{\infty} f_i$ is a positive-valued continuous function on X such that $f(x) \leqslant \sum_{i=k+1}^{\infty} 1/2^i$ whenever $x \in Z_k$. Since X is pseudo-compact, $Z_k = \emptyset$ for some k, because otherwise $1/f$ would be unbounded. Thus $\mu(Z_k) = 0$, proving that $\mu \in \mathcal{M}_\sigma(X)$.

   (ii) $\Rightarrow$ (iii) is obvious.

   (iii) $\Rightarrow$ (i). Assume that X is not pseudo-compact and f an unbounded continuous function from X into $[0, \infty)$. Select a point sequence $\{x_i\} \subset X$ such that

$$f(x_{i+1}) > f(x_i) + 1 , \quad i = 1, 2, \ldots .$$

Put $Z_i = f^{-1}(f(x_i))$; then $\{Z_i \mid i = 1, 2, \ldots\}$ is a collection of zero sets in X. Obviously there is a discrete collection $\{U_i \mid i = 1, 2, \ldots\}$ of cozero sets satisfying $Z_i \subset U_i$. Hence $Z_k' = \bigcup_{i=k}^{\infty} Z_i$ is a zero set for each k. Thus $\mathscr{F}_0 = \{Z_k' \mid k = 1, 2, \ldots\}$ is a decreasing sequence of non-empty zero sets

---

[1] Due to A. D. Alexandroff [1] and I. Glicksberg [1].

with an empty intersection. Denote by $\mathscr{F}$ a maximal filter such that $\mathscr{F} \supset \mathscr{F}_0$. Define $\mu \in \mathscr{T}(X)$ by use of D) and $\mathscr{F}$. Then $\mu(Z_k') = 1$ for $k = 1, 2, \ldots$, while $\bigcap_{k=1}^{\infty} Z_k' = \emptyset$, which means that μ is not σ-additive. Hence $\mathscr{T}(X) \neq \mathscr{T}_\sigma(X)$.

Theorem VIII.13.[1] *The following conditions are equivalent for a space X:*
(i) *X is compact,*
(ii) *$\mathscr{M}(X) = \mathscr{M}_\tau(X)$,*
(iii) *$\mathscr{T}(X) = \mathscr{T}_\tau(X)$,*
(iv) *$\mathscr{T}(X) = \mathscr{D}(X)$.*

Proof. (i) $\Rightarrow$ (ii) $\Rightarrow$ (iii) $\Rightarrow$ (iv) is obvious.

To prove (iv) $\Rightarrow$ (i), suppose that $\mathscr{F}$ is a zero filter of X, and X satisfies (iv). Define $\mu \in \mathscr{T}(X)$ by use of D) and $\mathscr{F}$. Then $\mu \in \mathscr{D}(X)$ follows from (iv). Thus there is $x \in X$ such that $x \in Z$ for every $Z \in \mathscr{Z}(X)$ with $\mu(Z) = 1$. Hence $x \in \{Z \mid Z \in \mathscr{F}\}$, which means that $\mathscr{F} \to x$. Therefore X is compact.

Theorem VIII.14.[2] *X is realcompact if and only if $\mathscr{T}_\sigma(X) = \mathscr{D}(X)$.*

Proof. Let X be realcompact and $\mu \in \mathscr{T}_\sigma(X)$. Then $\mathscr{F} = \{Z \in \mathscr{Z}(X) \mid \mu(Z) = 1\}$ is a maximal zero filter with c.i.p., because μ is regular and σ-additive. Hence $\mathscr{F}$ converges to $x \in X$. Then $\mu = \delta_x \in \mathscr{D}(X)$ is easy to see, as follows. Suppose $x \in B \in \mathscr{A}(X)$. If $\mu(B) = 0$, then there is $U \in \mathscr{P}(X)$ with $U \supset B$, $\mu(U) = 0$. Thus $X - U \in \mathscr{Z}(X)$ satisfies $\mu(X - U) = 1$, while $\mathscr{F} \to x \in U$ implies $X - U \notin \mathscr{F}$, which is a contradiction. Hence $\mu(B) = 1$. Suppose $x \notin B \in \mathscr{A}(X)$; then $x \in X - B \in \mathscr{A}(X)$, and hence $\mu(X - B) = 1$. This implies $\mu(B) = 0$. Namely $\mu = \delta_x \in \mathscr{D}(X)$. Therefore $\mathscr{T}_\sigma(X) = \mathscr{D}(X)$ is proved.

Conversely, assume $\mathscr{T}_\sigma(X) = \mathscr{D}(X)$ and that $\mathscr{F}$ is a maximal zero filter with c.i.p. Define $\mu \in \mathscr{T}(X)$ by use of D) and $\mathscr{F}$. Then $\mu \in \mathscr{T}_\sigma(X)$ can be proved as follows.

Suppose $\{Z_i \mid i = 1, 2, \ldots\}$ is a decreasing sequence of zero sets with $\bigcap_{i=1}^{\infty} Z_i = \emptyset$. If $\mu(Z_i) = 1$, $i = 1, 2, \ldots$, then $Z_i \in \mathscr{F}$, $i = 1, 2, \ldots$, which contradicts c.i.p. of $\mathscr{F}$. Thus $\mu(Z_i) = 0$ for some i, and hence by A) μ is σ-additive. Thus $\mu \in \mathscr{D}(X)$. Suppose $\mu = \delta_x$. Then $\mathscr{F}$ converges to x. Therefore X is realcompact.

Corollary 1. *Every Baire set B of a realcompact space X is realcompact.*

---

[1] Due to A. D. Alexandroff [1].
[2] Due to E. Hewitt [4].

Proof. It suffices to show $\mathcal{T}_\sigma(B) \subset \mathcal{D}(B)$. Suppose $\mu \in \mathcal{T}_\sigma(B)$. Then we define a map $\bar{\mu} : \mathcal{A}(X) \to \{0, 1\}$ by

$$\bar{\mu}(B') = \mu(B' \cap B) \quad \text{for } B' \in \mathcal{A}(X).$$

It is obvious that $\bar{\mu} \in \mathcal{T}_\sigma(X)$. Since X is realcompact, $\bar{\mu} \in \mathcal{D}(X)$ follows from the theorem. Suppose $\bar{\mu} = \delta_x$, and $x \in X$. It suffices to show that every Baire set of X is an intersection of members of $\mathcal{A}(X)$. Because, assume the claim is true, and $x \notin B$. Then $B \subset B' \not\ni x$ for some $B' \in \mathcal{A}(X)$. Thus $\bar{\mu}(B') = \mu(B) = 1$, which contradicts that $\bar{\mu} = \delta_x$. Hence $x \in B$, i.e. $\mu \in \mathcal{D}(B)$. Hence the corollary is proved. Now, to prove our claim, we define a transfinite sequence $\{B_\alpha \mid 0 \leq \alpha < \omega_1\}$ of subsets of X by induction on α as follows. $\mathcal{B}_0 = \mathcal{A}(X)$. For α with $0 < \alpha < \omega_1$,

$$\mathcal{B}_\alpha = \left\{ \bigcap_{i=1}^{\infty} B_i, \, X - B_0 \mid B_i, B_0 \in \bigcup_{0 \leq \beta < \alpha} \mathcal{B}_\beta \right\}.$$

Then it is obvious that $\mathcal{B}_a(X) = \bigcup \{\mathcal{B}_\alpha \mid 0 \leq \alpha < \omega_1\}$.

Now we prove the following assertion by use of induction on α. Let $x \notin B_0 \in \mathcal{B}_\alpha$ or $x \in X - B_0 \in \mathcal{B}_\alpha$; then there is $B' \in \mathcal{A}(X)$ such that $B_0 \subset B' \not\ni x$. The assertion is obviously true if $\alpha = 0$. Assume that the assertion is true for all $\beta < \alpha$. Suppose $x \notin B_0 \in \mathcal{B}_\alpha$. If $B_0 = \bigcap_{i=1}^{\infty} B_i$ for $B_i \in \bigcup_{\beta < \alpha} \mathcal{B}_\beta$, $i = 1, 2, \ldots$, then $x \notin B_i$ for some i. Thus by the induction hypothesis there is $B' \in \mathcal{A}(X)$ satisfying $B_0 \subset B_i \subset B' \not\ni x$. If $X - B_0 \in \bigcup_{\beta < \alpha} \mathcal{B}_\beta$, then again by the induction hypothesis there is $B' \in \mathcal{A}(X)$ such that $B_0 \subset B' \not\ni x$. Suppose $x \in X - B_0 \in \mathcal{B}_\alpha$. If $X - (X - B_0) = B_0 \in \bigcup_{\beta < \alpha} \mathcal{B}_\beta$, then by the induction hypothesis $B_0 \subset B' \not\ni x$ for some $B' \in \mathcal{A}(X)$. If $X - B_0 = \bigcap_{i=1}^{\infty} B_i$ for $B_i \in \bigcup_{\beta < \alpha} \mathcal{B}_\beta$, $i = 1, 2, \ldots$, then

$$B_0 = \bigcup_{i=1}^{\infty} (X - B_i), \qquad x \notin X - B_i, \quad i = 1, 2, \ldots.$$

Thus by the induction hypothesis there are $B'_i \in \mathcal{A}(X)$ such that $X - B_i \subset B'_i \not\ni x$, $i = 1, 2, \ldots$. Since each B'_i is a finite sum of intersections of a zero set and a cozero set, we can put

$$\bigcup_{i=1}^{\infty} B'_i = \bigcup_{n=1}^{\infty} (Z_n \cap U_n) \quad \text{for } Z_n \in \mathcal{Z}(X), \, U_n \in \mathcal{U}(X), \, n = 1, 2, \ldots.$$

Thus

$$B_0 \subset \bigcup_{n=1}^{\infty} (Z_n \cap U_n) \not\ni x.$$

Since each Z_n is a countable intersection of cozero sets, and a countable sum of cozero sets is a cozero set, we can find $V \in \mathcal{U}(X)$ such that $B_0 \subset V \not\ni x$. Thus the assertion and accordingly the corollary is proved.

Theorem VIII.15.[1] *A subparacompact normal space X is realcompact if and only if every discrete closed subset of X has a non-measurable cardinality.*

Proof. Suppose that X satisfies the said condition and denote by m_0 the minimum measurable cardinal number. Then every discrete closed subset of X has a cardinality $< m_0$. Let p_0 be a fixed point of $\beta(X) - X$. For each $x \in X$ we select an open nbd $U(x)$ in $\beta(X)$ such that $\overline{U(x)} \not\ni p_0$. Then there is a σ-discrete closed cover $\mathcal{F} = \bigcup_{n=1}^{\infty} \mathcal{F}_n$ of X such that

$$\mathcal{F} < \{U(x) \cap X \mid x \in X\},$$

where each $\mathcal{F}_n$ is a discrete collection in X, because X is subparacompact. Put $F_n = \bigcup \mathcal{F}_n$, $n = 1, 2, \ldots$.

If $p_0 \notin \bar{F}_n$ for each n, then there is a continuous function $f_n : \beta(X) \to [0, 1/2^n]$ satisfying

$$f_n(p_0) = 0, \qquad f_n(F_n) = 1/2^n.$$

Put

$$f = \sum_{n=1}^{\infty} f_n;$$

then f is a continuous function on $\beta(X)$ such that

$$f(p_0) = 0 \quad \text{and} \quad f > 0 \quad \text{on } X.$$

Thus the restriction of $1/f$ to X cannot be continuously extended to p_0 unless being allowed to take on the value ∞. Namely $p_0 \notin \gamma(X)$, where $\gamma(X)$ denotes the realcompactification of X (see IV.6.E)).

If $p_0 \in \bar{F}_n$ for some n, then we consider the subspace $Y = \{p_0\} \cup F_n$ of $\beta(X)$. Then $\mathcal{D} = \{p_0\} \cup \mathcal{F}_n$ is a decomposition of Y into closed sets.

---

[1] Due to M. Katětov [2]. The proof is due to S. Mrówka [6].

Denote by $\tilde{Y}$ the quotient space of Y, i.e.,

$$\tilde{Y} = Y/\mathcal{D} .$$

Also denote by φ the quotient map : $Y \to \tilde{Y}$. Note that $\varphi(p_0)$ is the only non-isolated point of $\tilde{Y}$, because each member of $\mathcal{F}_n$ is clopen in Y. Thus Y is a Tychonoff space. Further note that $\varphi(F_n)$ is realcompact, because it is a discrete space with a non-measurable cardinality $< m_0$. (Observe that every discrete collection of subsets of X has a cardinality $< m_0$.) Let us show that $\varphi(F_n)$ is C^*-embedded in $\tilde{Y} = \varphi(F_n) \cup \{\varphi(p_0)\}$. Suppose that g is a bounded real-valued continuous function on $\varphi(F_n)$. Then $g \circ \varphi \in C^*(F_n)$. Since F_n is a closed set of the normal space X, $g \circ \varphi$ can be extended to $h \in C^*(Y)$. Define a real-valued function $\tilde{g}$ on $\tilde{Y}$ by

$$\tilde{g} = g \quad \text{on } \varphi(F_n), \qquad \tilde{g}(\varphi(p_0)) = h(p_0) .$$

It is easy to see that $\tilde{g}$ is a continuous extension of g. Thus our claim is proved, and hence (by Exercise IV.4(i)) Y is homeomorphic to a subset of $\beta(\varphi(F_n))$, i.e.,

$$\varphi(p_0) \in \beta(\varphi(F_n)) - \varphi(F_n) .$$

Since $\varphi(F_n)$ is realcompact,

$$\varphi(p_0) \notin \gamma(\varphi(F_n)) = \varphi(F_n) .$$

Thus there is a continuous function $\psi : \tilde{Y} \to S^1$ such that

$$\psi(\varphi(p_0)) = \infty \quad \text{and } \psi(z) \neq \infty \quad \text{for all } z \in \varphi(F_n)$$

(see IV.6.E)). The restriction of $\psi \circ \varphi$ to F_n can be extended to $f \in C(X)$. Now regard f as a continuous map into S^1 and let $f^* : \beta(X) \to S^1$ be the continuous extension of f. Then $f^*(p_0) = \infty$ is easy to verify, and hence $p_0 \notin \gamma(X)$. Thus $\gamma(X) = X$ is proved. Therefore X is realcompact.

Definition VIII.9. A space X is called *measure-compact* if $\mathcal{M}_\sigma(X) = \mathcal{M}_\tau(x)$.

E) *Every Lindelöf space is measure-compact. Every measure-compact space is realcompact.*

Proof. Use C) and Theorem VIII.14 to prove the second statement.

Definition VIII.10. Let μ be a measure on X. Then we mean by the *support* of μ the set

$$\cap\{Z \in \mathcal{Z}(X) \mid \mu(Z) = \mu(X)\} = X - \cup\{V \in \mathcal{U}(X) \mid \mu(V) = 0\}.$$

F) *X is measure-compact if and only if the trivial measure 0 is the only σ-additive measure on X with an empty support.*

Proof. The 'only if' part is obvious. To prove the 'if' part, let $\mu \in \mathcal{M}_\sigma(X)$, where X satisfies the said condition. Let $\mathcal{U}$ be a cover of X by cozero sets. Put

$$a = \sup\{\mu(\cup \mathcal{U}') \mid \mathcal{U}' \text{ is a finite subcollection of } \mathcal{U}\}.$$

We claim that $a = \mu(X)$. For each $n \in N$, select a finite subcollection $\mathcal{U}_n$ of $\mathcal{U}$ such that $a - 1/n < \mu(\cup \mathcal{U}_n)$. We may assume $\mathcal{U}_n \subset \mathcal{U}_{n+1}$ for each n. Put

$$\mathcal{U}_\infty = \bigcup_{n=1}^\infty \mathcal{U}_n, \qquad U_\infty = \{U_1, U_2, \ldots\}, \qquad U_\infty = \bigcup_{n=1}^\infty U_n.$$

Then $\mu(U_\infty) = a$. Define $\mu' \in \mathcal{M}_\sigma(X)$ by

$$\mu'(B) = \mu(B - U_\infty), \quad B \in \mathcal{A}(X).$$

Let us prove that the support of μ' is empty. Let U be an arbitrary member of $\mathcal{U}$. Then

$$\mu(U \cup U_\infty) = \sup\left\{\mu\left(U \cup \left(\bigcup_{i=1}^n U_i\right)\right) \mid n = 1, 2, \ldots\right\} \leq a = \mu(U_\infty).$$

This implies

$$\mu(U - U_\infty) = \mu(U_\infty).$$

Thus

$$\mu'(U) = \mu(U - U_\infty) = \mu(U \cup U_\infty) - \mu(U_\infty) = 0.$$

Since $\mathcal{V}$ covers X, the support of μ' is empty. Hence $\mu' \equiv 0$ follows from the condition of X, and hence

$$\mu'(X) = \mu(X - U_\infty) = 0 .$$

Hence

$$\mu(X) = \mu(U_\infty) = a ,$$

proving that $\mu \in M_\tau(X)$. Therefore X is measure-compact.

Example VIII.7. Sorgenfrey line S is measure-compact, because it is Lindelöf. We can prove that $S \times S$ is not measure-compact. First observe, as follows, that every cozero set of $S \times S$ is an F_σ-set with respect to the Euclidean topology of E^2. To do so, suppose U is a cozero set of $S \times S$. Then there is a continuous function $f : S \times S \to [0, 1]$ such that $U = f^{-1}((0, 1])$. Put

$$F_n = \left\{ (x, y) \in E^2 \mid f\left(\left(x - \frac{1}{n}, x \right) \times \left(y - \frac{1}{n}, y \right) \right) \subset \left[\frac{1}{n}, 1 \right] \right\}, \quad n = 1, 2, \ldots .$$

Then it is obvious that $U = \bigcup_{n=1}^\infty F_n$. To prove that each F_n is closed in E^2, let $(x, y) \in E^2 - F_n$. Then we can select $(p, q) \in (x - 1/n, x) \times (y - 1/n, y)$ such that

$$f(p, q) \subset [0, 1/n) . \tag{1}$$

Then

$$V = (p, p + 1/n) \times (q, q + 1/n)$$

is an open nbd of (x, y) in E^2. Let (x', y') be an arbitrary point of V. Then $(p, q) \in (x' - 1/n, x') \times (y' - 1/n, y')$, which combined with (1) implies that

$$f((x' - 1/n, x') \times (y' - 1/n, y')) \not\subset [1/n, 1] .$$

Thus $(x', y') \notin F_n$, proving that $V \cap F_n = \emptyset$. Namely $(x, y) \notin \bar{F}_n$. Therefore F_n is closed in E^2, and U is an F_σ-set there. Hence $\mathcal{B}_a(E^2) = \mathcal{B}_a(S \times S)$ follows. Let

$$Y = \{ (x, y) \in E^2 \mid x \geq 0, y \geq 0, x + y = 1 \} ,$$

and λ the Lebesgue measure on Y. Define a Baire measure μ on E^2 by

$$\mu(B) = \lambda(B \cap Y), \quad B \in \mathcal{B}_a(E^2) .$$

Then μ is a Baire measure on $S \times S$ by the previous observation. Hence we may regard $\mu \in \mathcal{M}_\sigma(S \times S)$. We claim that μ has an empty support. Because $\mu(S \times S - Y) = 0$, and for each $(x, y) \in Y$, $U = (x - 1, x] \times (y - 1, y) \in \mathcal{U}(S \times S)$ satisfies $\mu(U) = 0$. Since μ is not trivial, it follows from F) that $S \times S$ is not measure-compact. Thus $S \times S$ also gives an example of a realcompact, non-measure-compact space.

It is known that a measure-compact space is not necessarily paracompact. A space X is called *strongly measure-compact* if $\mathcal{M}_\sigma(X) = \mathcal{M}_t(X)$ and *measure-complete* if $\mathcal{M}_\tau(X) = \mathcal{M}_t(X)$. It is known that the countable product of strongly measure-compact (measure-complete) spaces is strongly measure-compact (measure-complete). Every Baire set of a strongly measure-compact (measure-complete) space is strongly measure-compact (measure-complete). Obviously X is strongly measure-compact if and only if it is measure-complete and measure-compact. It would be an interesting problem to give purely topological characterizations to those concepts defined in terms of measure. See V. S. Varadarajan [1], W. Moran [1], [2], [3], Z. Frolik [1] and J. J. Dijkstra [1] for further discussions in this aspect.

Exercise VIII

1. Every separable GO-space is first countable.

2. Let X be a GO-space. If $X = A \cup B$, where $A \neq \emptyset$, $B \neq \emptyset$, A has a maximum point, B has a minimum point, and $a < b$ whenever $a \in A$ and $b \in B$, then (A, B) is called a *jump* of X. A GO-space X is compact if and only if it has neither jump, nor pseudo-gap, nor gap except possible endgaps.

3. A gap (pseudo-gap) (A, B) is called *countable* if there is a strictly increasing countable sequence cofinal in A or a strictly decreasing countable sequence cofinal in B. A GO-space X is countably compact if and only if X has neither countable gap nor countable pseudo-gap.

4. If X is a metrizable GO-space, then the set $H(X)$ defined in 1.H) is σ-discrete.

5. Every perfectly normal GO-space is paracompact (D. Lutzer).

6. Prove the converse of 1.I). (Hint: Decompose a given open set U into its convex components, each of which, as easily seen, is F_σ.)

7. If Z is a non-empty zero set in $\beta(X)$ such that $Z \subset \beta(X) - X$, where X is a Tychonoff space, then $|Z| \geqslant 2^{\aleph_1}$.

8. If $\{X_\alpha \mid \alpha \in A\}$ is a collection of topological spaces such that $d(X_\alpha) \leqslant m$ for all $\alpha \in A$, then $c(\prod_\alpha X_\alpha) \leqslant m$.

9. If X is a T_2-space, then $|X| \leqslant 2^{2^{d(X)}}$.

10. The *tightness* $t(X)$ of a topological space X is defined by $t(X) = \min\{m \mid$ for every $x \in X$ and $Y \subset X$ with $x \in \bar{Y}$, there is $Z \subset Y$ such that $x \in \bar{Z}$, $|Z| \leqslant m\} + \aleph_0$. If Y is a closed continuous image of X, then $t(Y) \leqslant t(X)$.

11. $t(X) \leqslant \chi(X) \leqslant w(X) \leqslant |X|\chi(X)$, $c(X) \leqslant d(X) \leqslant w(X)$, $l(X) \leqslant w(X)$, $|X| \leqslant 2^{w(X)}$, where X is a T_1-space in the last relation and a general topological space in the others.

12. If U is an open set of a dyadic space X, then $\bar{U}$ is a dyadic space.

13. If X is a dyadic space, then there is a continuous map from D^T onto X, where $|T| = w(X)$.

14. Every dyadic LOTS is metrizable (N. Shanin).

15. If X is a non-compact metric space, then $\beta(X)$ is not dyadic (B. Efimov).

16. Give an example of a Tychonoff space which is no dense subset of any dyadic space.

17. The two arrows space given in Example VIII.5 is perfectly normal.

18. The two arrows space is a compactification of Sorgenfrey line S, and thus S is not Čech complete.

19. Every closed set of a measure-compact space is measure-compact.

20. Give an example of a non-Lindelöf, measure-compact space.

21. Every σ-compact regular space is strongly measure-compact. Every locally compact T_2-space is measure-complete.

EPILOGUE

Naturally it is impossible to cover all topics of general topology by a single book of moderate size. It is even truer because the area defined by the terminology 'general topology' is not so clear. Thus various factors were taken into consideration to choose the topics for this book. One of the factors was 'custom'. It is a funny phenomenon that an algebraic aspect of topology related with homology group and homotopy group is treated as 'algebraic topology' while another algebraic aspect related with rings of continuous functions is counted in 'general topology'. However, the author just followed today's custom to adopt rings of continuous functions and reject general homological topology as a topic though the latter was treated as a part of general topology by P. S. Alexandroff [5]. Some topics (e.g., box product, theory of absolute, fuzzy topological space, etc.) were omitted because they were, according to the author's opinion, special or isolated topics no matter how interesting they might be. Further, some important aspects were omitted because the author believed that they should be treated in other books. For examples, dimension theory, categorical topology, infinite-dimensional topology and theory of spaces with a transformation group are very interesting areas deeply related with general topology, but they are left to specialized books and articles like R. Engelking [4], J. Nagata [5], H. Herrlich [1], [5], H. Herrlich–G. Strecker [2], C. Bessaga–A. Pełczyński [2], Yu. Smirnov [8] and J. de Vries [1]. The author hopes more books will be written especially for the last three subjects.

BIBLIOGRAPHY

((R) indicates that the paper is written in Russian)

J. M. Aarts
[1] Every metric compactification is a Wallman-type compactification, International Symposium on Topology and its Applications (Herceg-Novi) 1968, 29–34.

O. T. Alas
[1] On a characterization of collectionwise normality, Canad. Math. Bull. 14 (1971) 13–15.

A. D. Alexandroff
[1] Additive set-functions in abstract spaces, Mat. Sb. 8(50) (1940) 307–345, 9(51) (1941) 563–625, 13(55) (1943) 169–238.

P. S. Alexandroff
[1] On bicompact extension of topological spaces, Sbornik N.S. 5 (1939) 403–423 (R),
[2] On the metrization of topological spaces, Bull. Acad. Polon. Math. Ser. 8 (1960) 135–140 (R),
[3] On some results in the theory of topological spaces obtained during the last twenty five years, Uspehi Mat. Nauk SSSR 15 (1960) 25–95 (R), Russian Math. Surveys 15 (1960) 23–83,
[4] On some results concerning topological spaces and their continuous mappings, Proc. Prague Symposium (1962) 41–54,
[5] On some basic directions in general topology, Russian Math. Surveys 19 (1964) No. 6, 1–39,
[6] Über stetige Abbildungen kompakter Räume, Math. Ann. 96 (1927) 555–571.

P. S. Alexandroff–V. I. Ponomarev
[1] On bicompact extensions of topological spaces, Vestnik Moskov Univ. Ser. Mat. (1959) No. 5, 91–108 (R),
[2] Projection spectra, Proc. 2nd Prague Symposium, 1966, 25–30.

E. M. Alfsen–J. E. Fenstad
[1] On the equivalence between proximity structures and totally bounded uniform structures, Math. Scand. 17 (1959) 353–360.

E. M. Alfsen–O. Njåstad
[1] Proximity and generalized uniformity, Fund. Math. 52 (1963) 235–252.

R. A. Alo–H. L. Shapiro
[1] Wallman compact and realcompact spaces, Proc. Berlin Symposium on Extensions of Topological Structures, 1967, 9–14,
[2] Extensions of totally bounded continuous pseudometrics, Proc. Amer. Math. Soc. 19 (1968) 877–884.

K. Alster–R. Engelking
[1] Subparacompactness and product spaces, Bull. Polon. Acad. Math. Ser. 20 (1972) 763–767.

R. D. Anderson
[1] Hilbert space is homeomorphic to the countable infinite product of lines, Bull. Amer. Math. 72 (1966) 515–519,
[2] Some open questions in infinite-dimensional topology, Proc. 3rd Prague Symposium, 1971, 29–35.

M. Antonovskii
[1] Metric spaces over semifields, Proc. Prague Symposium (1962) 64–68.

M. Antonovskii–V. Boltjanskii–T. Sarymsakov
[1] Metric spaces over semifields, Trudy Tashkent Univ. 191 (1960).

C. Aquaro
[1] Completions of uniform spaces, Proc. Prague Symposium (1962) 69–71.

R. F. Arens
[1] A topology for spaces of transformations, Ann. Math. 47 (1946) 480–495,
[2] Extensions of functions on fully normal spaces, Pacific J. Math. 2 (1952) 11–22.

R. F. Arens–J. Eells, Jr.
[1] On imbedding uniform and topological spaces, Pacific J. Math. 6 (1956) 397–403.

A. V. Arhangelskii
[1] On the metrization of topological spaces, Bull. Acad. Polon. Math. Ser. 8 (1960) 589–595 (R),
[2] On mappings of metric spaces, Doklady SSSR 145 (1962) 245–247 (R), Soviet Math. 3 (1962) 953–956,
[3] Bicompact sets and the topology of spaces, Soviet Math. 4 (1963) 561–564,
[4] On a class of spaces containing all metric spaces and all locally bicompact spaces, Soviet Math. 4 (1963) 751–754,
[5] Some types of factor mappings and the relations between classes of topological spaces, Soviet Math. 4 (1963) 1335–1338,
[6] Mappings and spaces, Russian Math. Surveys 21 (1966) 115–162,
[7] On the cardinality of bicompacta satisfying the first axiom of countability, Soviet Math. 10 (1969) 951–955,
[8] On cardinal invariants, Proc. 3rd Prague Symposium, 1971, 37–46,
[9] Some metrization theorems, Uspehi Mat. Nauk 18 (1963) No. 5 (113) 139–145 (R),
[10] A characterization of very k-spaces, Czech. Math. J. 18 (1968) 392–395,
[11] On left-separated subspaces, Vestnik Moskov Univ. Ser. Math. Meh. (1977) No. 5, 30–36,
[12] On hereditary properties, Gen. Topol. Appl. 3 (1973) 39–46,
[13] On relationships between topological properties of X and $C_p(X)$, Proc. 5th Prague Symposium, 1981, 24–36.

A. V. Arhangelskii–V. I. Ponomarev
[1] On dyadic bicompacta, Soviet Math. 9 (1968) 1221–1224.

M. Atsuji
[1] Uniform continuity of continuous functions on uniform spaces, Canad. J. Math. 13 (1961) 654–663,
[2] Uniform extension of uniformly continuous functions, Proc. Japan Acad. 37 (1961) 10–13,
[3] Uniform continuity of continuous functions of metric spaces, Pacific J. Math. 8 (1958) 11–16,
[4] On normality of the product of two spaces, Proc. 4th Prague Symposium, 1976, Part B, 25–27.

C. E. Aull
[1] A note on countably paracompact spaces and metrization, Proc. Amer. Math. 16 (1965) 1316–1317.

R. W. Bagley–E. H. Connell–J. D. McKnight, Jr.
[1] On properties characterizing pseudo-compact spaces, Proc. Amer. Math. 9 (1958) 500–506.

B. J. Ball–S. Yokura
[1] Functional bases for subsets of $C^*(X)$, to appear.

Z. Balogh
[1] On the structure of spaces which are paracompact p-spaces hereditarily, Acta Math. Acad. Sci. Hungar 33 (1979) 361–368,
[2] On the metrizability of F_{pp}-spaces and its relationship to the normal Moore space conjecture, Fund. Math. 113 (1981) 45–58.

S. Banach
[1] Théorie des opérations linéaires, Warsaw, 1932.

B. Banaschewski
[1] Local connectedness of extension spaces, Canad. J. Math. 8 (1956) 359–398.

C. Bandt
[1] On Wallman–Shanin compactification, Math. Nachr. 77 (1977) 283–351.

H. Banilower
[1] Simultaneous extensions and projections in $C(S)$, to appear.

M. G. Bell
[1] Not all compact spaces are supercompact, Gen. Topol. Appl. 8 (1978) 151–155.

H. R. Bennett–D. J. Lutzer
[1] Certain hereditary properties and metrizability in generalized ordered spaces, Fund. Math. 107 (1980) 71–84.

E. S. Berney
[1] A regular Lindelöf semimetric space which has no countable network, Proc. Amer. Math. 26 (1970) 361–370.

C. Bessaga–A. Pełczyński
[1] Some remarks on homeomorphism of Banach spaces, Bull. Acad. Polon. Math. Ser. 8 (1960) 757–761,
[2] Selected topics in infinite-dimensional topology, Warsaw, 1975.

R. H. Bing
[1] Metrization of topological spaces, Canad. J. Math. 3 (1951) 175–186.

J. R. Boone
[1] A characterization of metacompactness in the class of θ-refinable spaces, Gen. Topol. Appl. 3 (1973) 253–264.

C. J. R. Borges
[1] On stratifiable spaces, Pacific J. 17 (1966) 1–16,
[2] A study of multivalued functions, Pacific J. 23 (1967) 451–461.

C. J. R. Borges–D. J. Lutzer
[1] Characterizations and mappings of M_i-spaces, Proc. Topology Conference, VPI and State University 1973 (1974) 34–40.

K. Borsuk
[1] Sur les rétracts, Fund. Math. 17 (1931) 152–170,
[2] Über eine Klasse von lokal zusammenhängenden Räume, Fund. Math. 19 (1932) 220–242,
[3] On some metrization of the hyperspace of compact sets, Fund. Math. 41 (1955) 168–202.

H. Brandenburg
[1] On spaces with a G_δ-basis, to appear,
[2] Some characterizations of developable spaces (1979) preprint.

J. B. Brown
[1] Stochastic metrics, Z. Wahrscheinlichkeitstheorie Verw. Geb. 24 (1972) 49–62.

G. Bruns–J. Schmidt
[1] Die punktalen Typen topologischer Räume, Math. Japonica 4 (1957) 133–177.

D. K. Burke
[1] On p-spaces and $w\Delta$-spaces, Pacific J. 35 (1970) 285–296,
[2] On subparacompact spaces, Proc. Amer. Math. 23 (1969) 655–663.

D. K. Burke–E. K. v. Douwen
[1] On countably compact extensions of normal locally compact M-spaces, Set-theoretic Topology (1977) 81–89.

D. K. Burke–D. J. Lutzer
[1] Recent advances in the theory of generalized metric spaces, Proc. Memphis State Univ. Topology Conference (1975) 1–70.

D. K. Burke–R. A. Stoltenberg
[1] A note on p-spaces and Moore spaces, Pacific J. 30 (1968) 601–608.

E. Čech
[1] On bicompact spaces, Ann. Math. 38 (1937) 823–844.

J. Ceder
[1] Some generalizations of metric spaces, Pacific J. 11 (1961) 105–125.

J. Chaber
[1] Conditions which imply compactness in countably compact spaces, Bull. Acad. Polon. Sci. Ser. Math. 24 (1976) 993–998.

M. M. Čoban
[1] On σ-paracompact spaces, Vestnik Moskov Univ. Ser. I. Mat. Meh. 24 (1969) No. 1, 20–27.

D. E. Cohen
[1] Spaces with weak topology, Quart. J. Math. Oxford 5 (1954) 77–80.

J. Colmez
[1] Sur les espaces précompacts, C. R. Acad. Sci. Paris 233 (1951) 1552–1553.

W. W. Comfort
[1] A short proof of Marczewski's separability theorem, Amer. Math. Monthly 76 (1969) 1041–1042,
[2] A survey of cardinal invariants, Gen. Topol. Appl. 1 (1971) 163–199,
[3] On the Hewitt realcompactification of a product space, Trans. Amer. Math. 131 (1968) 107–118.

H. H. Corson–J. R. Isbell
[1] Some properties of strong uniformities, Quart. J. Math. Oxford, Ser. 1 (1960) 17–31.

H. H. Corson–E. Michael
[1] Metrizability of certain countable unions, Ill. J. Math. 8 (1964) 351–360.

G. D. Creede
[1] Semi-stratifiable spaces, Proc. Topology Conference, Arizona State Univ., 1967, 313–323.

A. Császár
[1] Foundation of general topology, New York, 1963.

J. v. Dalen–E. Wattel
[1] Characterization of ordered spaces, Gen. Topol. Appl. 3 (1973) 347–354.

J. Dieudonné
[1] Une généralisation des espaces compacts, J. Math. Pures Appl. 23 (1944) 65–76.

J. J. Dijkstra
[1] Measures in topology, Master thesis, Univ. of Amsterdam, 1977.

D. Doičinov
[1] A unified theory of topological spaces, proximity spaces and uniform spaces, Soviet Math. 5 (1964) 595–598.

R. Doss
[1] Uniform spaces with a unique structure, Amer. J. Math. 71 (1949) 19–23.

E. K. v. Douwen
[1] Special bases for compact metrizable spaces, Fund. Math. 111 (1981) 201–209.

E. K. v. Douwen–D. Lutzer–T. Przymusiński
[1] Some extensions of the Tietze–Urysohn theorem, preprint.

E. K. v. Douwen–J. v. Mill
[1] Supercompact spaces, Topol. Appl. 13 (1982) 21–32.

C. H. Dowker
[1] On imbedding theorem for paracompact metric spaces, Duke Math. J. 14 (1947) 639–645,
[2] An extension of Alexandroff's mapping theorem, Bull. Amer. Math. 54 (1948) 336–391,
[3] On countably paracompact spaces, Canad. J. Math. 1 (1951) 219–224,
[4] On a theorem of Hanner, Archiv für Math. 2 (1952) 307–313,
[5] Homotopy extension theorems, Proc. London Math. Soc. 6 (1956) 100–116.

E. Duda
[1] One to one mappings and applications, Gen. Topol. Appl. 1 (1971) 135–142.

R. Duda
[1] On biconnected sets with dispersion point, Rozprawy Mat. 37 (1964).

J. Dugundji
[1] An extension of Tietze's theorem, Pacific J. Math. 1 (1951) 353–367.

M. Edelstein
[1] On the representation of contractive homeomorphisms as transformations in Hilbert spaces, Proc. London Math. Soc. 21 (1970) 462–474.

B. A. Efimov
[1] On dyadic bicompacta, Soviet Math. 4 (1963) 496–500,
[2] Dyadic bicompacta, Trans. Moscow Math. Soc. 14 (1965) 229–267.

B. A. Efimov–R. Engelking
[1] Remarks on dyadic spaces II, Colloq. Math. 13 (1965) 181–197.

V. A. Efiremovič
[1] The geometry of proximity I, Mat. Sbornik N.S. 31 (1952) 189–200 (R).

V. A. Efiremovič–A. S. Švarc
[1] A new definition of uniform spaces, Metrization of proximity spaces, Dokl. SSSR 89 (1953) 393–396 (R).

S. Eilenberg
[1] Ordered topological spaces, Amer. J. Math. 63 (1941) 39–45.

R. Engelking
[1] On the Freudenthal compactification, Bull. Acad. Polon. Math. Ser. 9 (1961) 379–383,
[2] Cartesian products and dyadic spaces, Fund. Math. 57 (1965) 287–304,
[3] General topology, Warsaw, 1977,
[4] Dimension theory, Amsterdam, 1978.

R. Engelking–D. Lutzer
[1] Paracompactness in ordered spaces, Fund. Math. 94 (1978) 49–58.

R. Engelking–S. Mrówka
[1] On E-compact spaces, Bull. Acad. Polon. Math. Ser. 6 (1958) 429–436.

R. Engelking–A. Pełczyński
[1] Remarks on dyadic spaces, Colloq. Math. 11 (1963) 55–63.

A. Esenin–Volpin
[1] On the relation between the local and integral weight in dyadic bicompact spaces, Dokl. SSSR 68 (1949) 441–444 (R).

M. J. Faber
[1] Metrizability in generalized ordered space, Math. Centrum Tract 53 (1974) Amsterdam.

V. V. Filippov
[1] On feathered paracompacta, Soviet Math. 9 (1968) 161–164,
[2] On the perfect images of a paracompact p-space, Soviet Math. 8 (1967) 1151–1153.

J. Flachsmeyer
[1] Zur Spektralentwicklung topologischer Räume, Math. Ann. 144 (1961) 253–274,
[2] Topologische Projektivräume, Math. Nachr. 26 (1963) 37–66,
[3] On the Busemann metrization of the hyperspaces, Math. Nachr. 80 (1979) 51–56.

W. G. Fleissner
[1] Normal Moore spaces in the constructible universe, Proc. Amer. Math. 46 (1974) 294–298,
[2] If all normal Moore spaces are metrizable, then there is an inner model with a measurable cardinal, Trans. Amer. Math. 273 (1982) 365–373.

S. Fomin
[1] Extensions of topological spaces, Ann. Math. 44 (1943) 471–480.

R. H. Fox
[1] On topologies for function spaces, Bull. Amer. Math. 51 (1943) 429–432.

A. Fraenkel
[1] Zehn Vorlesungen über die Grundlagen der Mengenlehre, Leipzig–Berlin, 1927.

502   BIBLIOGRAPHY

S. Franklin
[1] Spaces in which sequences suffice, Fund. Math. 57 (1965) 107–115.

S. Franklin–B. Thomas
[1] On the metrizability of k_ω-spaces, Pacific J. Math. 72 (1977) 399–402.

H. Freudenthal
[1] Entwicklungen von Räumen und ihren Gruppen, Compositio Math. 4 (1937) 145–234,
[2] Neuaufbau der Endentheorie, Ann. Math. 43 (1942) 261–279,
[3] Enden und Primenden, Fund. Math. 39 (1953) 189–210.

A. H. Frink
[1] Distance functions and the metrization problem, Bull. Amer. Math. 43 (1937) 133–142.

O. Frink
[1] Compactifications and semi-normal spaces, Amer. J. Math. 86 (1964) 602–607.

Z. Frolik
[1] The topological product of countably compact spaces, Czech. Math. J. 10 (1960) 329–338,
[2] The topological product of two pseudocompact spaces, Czech. Math. J. 10 (1960) 339–349,
[3] On the topological product of paracompact spaces, Bull. Acad. Polon. Math. Ser. 8 (1960) 747–750,
[4] Applications of complete families of continuous functions to the theory of Q-spaces, Czech. Math. J. 11 (1961) 115–133,
[5] A characterization of topologically complete spaces in the sense of E. Čech in terms of convergence of functions, Czech. Math. J. 13 (1963) 148–151,
[6] Topological methods in measure theory and the theory of measurable spaces, Proc. 3rd Prague Symposium (1971) 127–139.

I. Gelfand–A. Kolmogoroff
[1] On rings of continuous functions on topological spaces, Dokl. SSSR 22 (1939) 11–15.

J. Gerlits
[1] Some properties of $C(X)$ II, Topol. Appl. 15 (1983) 482–489.

J. Gerlits–Zs. Nagy
[1] Some properties of $C(X)$ I, Topol. Appl. 14 (1982) 151–161.

L. Gillman–M. Jerison
[1] Rings of continuous functions, Princeton–Toronto–London–N.Y., 1960.

S. Ginsburg–J. R. Isbell
[1] Some operators on uniform spaces, Trans. Amer. Math. 93 (1959) 145–168.

A. M. Gleason
[1] Projective topological spaces, Ill. J. Math. 2 (1958) 482–489.

I. Glicksberg
[1] The representation of functionals by integrals, Duke Math. J. 19 (1952) 253–261,
[2] Stone–Čech compactifications of products, Trans. Amer. Math. 90 (1959) 369–382.

K. Gödel
[1] The consistency of the axiom of choice and of the generalized continuum hypothesis with the axioms of set theory, Princeton, 1940.

J. T. Goodykoontz, Jr.
[1] Connectedness im kleinen and local connectedness in 2^X and $C(X)$, Pacific J. Math. 53 (1974) 387–397.

G. Grimeisen
[1] Gefilterte Summation von Filtern und iterierte Grenzprozesse, I, II, Math. Ann. 141 (1961) 318–342, 144 (1961) 386–417.

J. de Groot
[1] Realisations under continuous mappings, Indag. Math. 12 (1950) 483–492,
[2] On some problems of Borsuk concerning a hyperspace of compact sets, Indag. Math. 18 (1956) 95–103,
[3] Subcompactness and the Baire category theorem, Indag. Math. 25 (1963) 761–767,
[4] On the topological characterization of manifold, Proc. 3rd Prague Symposium, 1971, 155–158,
[5] Topological characterization of metrizable cubes, Theory of Sets and Topology (A collection of papers in honour of F. Hausdorff) (1972) 473–494,
[6] Supercompactness and superextension, Contributions to extension theory, Proc. Symposium in Berlin 1967 (1969) 89–90.

J. de Groot–R. H. McDowell
[1] Extension of mappings on metric spaces, Fund. Math. 48 (1959/60) 251–263,
[2] Locally connected spaces and their compactifications, Ill. J. Math. 11 (1967) 353–364.

J. de Groot–P. S. Schnare
[1] A topological characterization of products of compact totally ordered spaces, Gen. Topol. Appl. 2 (1972) 67–74.

G. Gruenhage
[1] Stratifiable spaces are M_2, Topology Proc. 1 (1976) 221–226,
[2] Stratifiable σ-discrete spaces are M_1, Proc. Amer. Math. 72 (1978) 189–190,
[3] On the $M_3 \to M_1$ question, Topology Proc. 5 (1980) 77–104.

J. A. Guthrie
[1] Metrization and paracompactness in terms of real functions, Bull. Amer. Math. 80 (1974) 720–721.

A. W. Hager
[1] On inverse-closed subalgebras of $C(X)$, Proc. London Math. Soc. (3) 19 (1969) 253–257.

A. Hajnal–I. Juhász
[1] Discrete subspaces of topological spaces II, Indag. Math. 31 (1969) 18–30.

S. Hanai
[1] On open mappings II, Proc. Japan Acad. 37 (1961) 233–238,
[2] Inverse images of closed mappings I, Proc. Japan Acad. 37 (1961) 298–301,
[3] Open mappings and metrization theories, Proc. Japan Acad. 39 (1963) 450–454.

F. Hausdorff
[1] Mengenlehre, Berlin–Leipzig, 1935.

O. Hanner
[1] Solid spaces and absolute retract, Arkiv för Mat. 1 (1951) 375–382.
[2] Retraction and extension of mappings of metric and non-metric spaces, Arkiv för Mat. 2 (1952) 315–364.

R. W. Heath
[1] Arcwise connectedness in semi-metric spaces, Pacific J. Math. 12 (1962) 1301–1319,
[2] On open mappings and certain spaces satisfying the first countability axiom, Fund. Math. 62 (1965) 91–96,
[3] Stratifiable spaces are σ-spaces, Notices Amer. Math. Soc. 17 (1969) 761,

[4] An easier proof that certain countable space is not stratifiable, Proc. Washington State Univ. Topology Conference, 1970, 56–59,
[5] On spaces with point-countable bases, Bull. Acad. Polon. Math. Ser. 13 (1965) 393–395.

R. Heath–R. Hodel
[1] Characterizations of σ-spaces, Fund. Math. 77 (1973) 271–275.

R. Heath–H. Junnila
[1] Stratifiable spaces are subspaces and closed continuous images of M_1-spaces, Proc. Amer. Math. 83 (1981) 146–148.

R. Heath–D. Lutzer
[1] The Dugundji extension property and collectionwise normality, Bull. Acad. Polon. Math. Ser. 22 (1974) 827–830,
[2] Dugundji extension theorem for linearly ordered spaces, Pacific J. Math., to appear.

R. Heath–D. Lutzer–P. Zenor
[1] Monotonically normal spaces, Trans. Amer. Math. 178 (1973) 481–493.

G. Helmberg
[1] Uniform convergence of nets of functions, Indag. Math. 21 (1959) 419–427.

M. Henriksen–L. Gillman
[1] Concerning rings of continuous functions, Trans. Amer. Math. 77 (1954) 340–362.

M. Henriksen–J. Isbell
[1] Local connectedness in the Stone–Čech compactification, Ill. J. Math. 4 (1957) 574–552.

H. Herrlich
[1] Categorical topology, Gen. Topol. Appl. 1 (1971) 1–15,
[2] Ordnungsfähigkeit total-diskontinuierlicher Räume, Math. Ann. 159 (1965) 77–80,
[3] Ordnungsfähigkeit topologischer Räume, Dissertation (1962) Berlin,
[4] Ordnungsfähigkeit zuzammenhänger Räume, Fund. Math. 57 (1965) 305–311,
[5] Categorical topology 1971–1981, Proc. 5th Prague Symposium 1981 (1982) 279–383.

H. Herrlich–G. E. Strecker
[1] Algebra $\cap$ topology = compactness, Gen. Topol. Appl. 1 (1971) 283–287,
[2] Category theory, Berlin, 1982.

E. Hewitt
[1] On two problems of Urysohn, Ann. Math. 47 (1946) 503–509,
[2] Certain generalization of the Weierstrass approximation theorem, Duke Math. J. 14 (1947) 419–427,
[3] Rings of real-valued continuous functions I, Trans. Amer. Math. 64 (1948) 54–99,
[4] Linear functionals on spaces of continuous functions, Fund. Math. 37 (1950) 161–189.

R. E. Hodel
[1] Moore spaces and $w\Delta$-spaces, Pacific J. Math. 38 (1971) 641–652,
[2] Spaces defined by sequences of open covers which guarantee that certain sequences have cluster points, Duke Math. J. 39 (1972) 253–263,
[3] Some results in metrization theory, 1950–1972, Proc. Topology Conference VPI and State Univ. 1973 (Springer Lecture Notes 375) (1974) 120–136,
[4] A note on subparacompact spaces, Proc. Amer. Math. 251 (1970) 842–845,
[5] Metrizability of topological spaces, Pacific J. Math. 55 (1974) 446–459,
[6] Extensions of metrization theorems to higher cardinality, Fund. Math. 87 (1975) 219–229.

A. Hohti
[1] On uniform paracompactness, Dissertationes, Helsinki, 1981,
[2] On the Ginsburg–Isbell derivatives and ranks of metric spaces, to appear.

H. Hung
[1] A contribution to the theory of metrization, Canad. J. Math. 29 (1977) 1145–1151,
[2] A general metrization theorem, Topol. Appl. 11 (1980) 275–279.

M. Hušek
[1] The Hewitt realcompactification of a product, Comment. Math. Univ. Carolin. (1970) 393–395.

D. M. Hyman
[1] A category slightly larger than the metric and CW-categories, Mich. Math. J. 15 (1968) 193–214,
[2] A note on closed maps and metrizability, Proc. Amer. Math. 21 (1969) 109–112.

S. Iliadis
[1] Absolutes of Hausdorff spaces, Soviet Math. 4 (1963) 295–298.

T. Inagaki
[1] Contribution à la topologie I, Math. J. Okayama Univ. 1 (1952) 158–166.

J. R. Isbell
[1] Uniform spaces, Providence, 1964.

K. Iseki
[1] A characterization of pseudo-compact spaces, Proc. Japan Acad. 33 (1957) 320–322.

T. Ishii
[1] On product spaces and product mappings, J. Math. Soc. Japan 18 (1966) 166–181,
[2] Paracompactness of topological completions, Fund. Math. 92 (1976) 65–77,
[3] On closed mappings and M-spaces I, II, Proc. Japan Acad. 43 (1967) 752–756, 757–761,
[4] On wM-spaces I, II, Proc. Japan Acad. 46 (1970) 16–21.

F. Ishikawa
[1] On countably paracompact spaces, Proc. Japan Acad. 31 (1955) 686–687.

T. Isiwata
[1] Mappings and spaces, Pacific J. Math. 20 (1967) 255–480,
[2] The product of M-spaces need not be an M-space, Proc. Japan Acad. 45 (1968) 154–156.

M. Ito
[1] The closed image of a hereditary M_1-space is M_1, to appear,
[2] M_3-spaces whose every point has a closure preserving outer base are M_1, to appear.

L. Ivanovskiĭ–V. Kuz'minov
[1] On a conjecture of P. S. Alexandroff, Dokl. SSSR 123 (1958) 785–786 (R).

F. B. Jones
[1] Concerning normal and completely normal spaces, Bull. Amer. Math. 43 (1937) 671–679.

I. Juhász
[1] Cardinal functions in topology – ten years later, Math. Cent. Tracts 123, Amsterdam, 1980.

H. Junnila
[1] Neighbornets, Pacific J. Math. 76 (1978) 83–108.

G. I. Kac
[1] Topological spaces in which one may introduce a complete uniform structure, Dokl. SSSR 99 (1954) 899–900 (R).

S. Kakutani
[1] Über die Metrisation der topologischen Gruppen, Proc. Japan Acad. 12 (1936) 82–84.

I. Kaplansky
[1] Lattices of continuous functions, Bull. Amer. Math. 53 (1947) 617–623.

M. Katětov
[1] On H-closed extension of topological spaces, Časopis Pěst Mat. Fys. 72 (1947) 17–32,
[2] Measures in fully normal spaces, Fund. Math. 38 (1951) 73–84,
[3] On real-valued functions on topological spaces, Fund. Math. 38 (1951) 85–91,
[4] On the dimension of non-separable spaces I, Czech. Math. J. 2 (1952) 333–368,
[5] Extension of locally finite coverings, Colloq. Math. 6 (1958) 145–151.

A. Kato
[1] Solution of Morita's problems concerning countably-compactifications, Gen. Topol. Appl. 7 (1977) 77–87,
[2] Various countably-compactifications and their applications, Gen. Topol. Appl. 8 (1978) 27–46.

Y. Katuta
[1] A theorem on paracompactness of product spaces, Proc. Japan Acad. 43 (1967) 615–618,
[2] On the normality of product of a normal space with a paracompact space, Gen. Topol. Appl. 1 (1971) 259–319,
[3] Characterizations of paracompactness by increasing covers and normality of product spaces, Tsukuba J. Math. 1 (1977) 27–43.

J. E. Keesling
[1] On the equivalence of normality and compactness in hyperspaces, Pacific J. Math. 33 (1970) 657–667.

J. L. Kelley
[1] General topology, New York, 1955.

V. Kljušin
[1] Perfect mappings of paracompact spaces, Soviet Math. 5 (1964) 1583–1586.

B. Knaster–A. Lelek–J. Mycièlski
[1] Sur les décompositions des ensembles connexes, Colloq. Math. 6 (1958) 227–246.

Y. Kodama
[1] Note on an absolute nbd extensor for metric spaces, J. Math. Soc. Japan 8 (1956) 206–215.

A. P. Kombarov
[1] On hereditarily disconnected spaces, Moscow Univ. Bull. 26 (1971) No. 4, 21–25.

H.J. Kowalsky
[1] Einbettung metrischer Räume, Arch. für Math. 8 (1957) 336–339.

M. Krein–S. Krein
[1] On an inner characterization of the set of all continuous functions defined on a bicompact Hausdorff space, Dokl. SSSR 27 (1940) 427–430.

K. Kunugi
[1] Sur les espaces completes en régulièrement completes I, II, Proc. Japan Acad. 30 (1954) 553–556, 912–916.

C. Kuratowski
[1] Sur les espaces localement connexes et péaniens en dimension n, Fund. Math. 24 (1935) 269–287,
[2] Quelques problèmes concernant les espaces métriques non-séparables, Fund. Math. 25 (1935) 534–545,
[3] Topologie I, Warsaw, 1966,
[4] Topologie II, Warsaw, 1961,
[5] A general approach to the theory of set-valued mapping, Proc. 3rd Prague Symposium, 1971, 271–280.

C. Kuratowski–C. Ryll-Nardzewski
[1] A general theorem on selections, Bull. Acad. Polon. Sci. Math. Ser. 13 (1965) 397–403.

G. Kurepa
[1] Sur les classes (E) et (D), Bull. Math. Belgrade 5 (1936) 124–136,
[2] Le problème de Souslin et les espaces abstraits, Comptes Rendus, Paris 203 (1936) 1049–1052,
[3] Sur l'écart abstrait, Glasnik Mat. Fiz. Astr. 11 (1956) 105–132.

N. S. Lašnev
[1] Continuous decompositions and closed mappings of metric spaces, Soviet Math. 6 (1965) 1504–1506,
[2] Closed mappings of metric spaces, Soviet Math. 7 (1966) 1219–1221.

S. Leader
[1] On completion of proximity spaces by local clusters, Fund. Math. 48 (1960) 201–216.

A. Lelek
[1] Ensembles σ-connexes et la théorie de Gehman, Fund. Math. 47 (1954) 265–276.

E. R. Lorch
[1] Compactifications, Baire functions and Daniel integrations, Acta Sci. Math. (Széged) 24 (1965) 204–218.

D. Lutzer
[1] A metrization theorem for linearly ordered spaces, Proc. Amer. Math. 22 (1969) 357–358,
[2] On generalized ordered spaces, Dissert. Math. 89 (1971) 1–36.

D. Lutzer–H. Martin
[1] A note on the Dugundji extension theorem, Proc. Amer. Math. 45 (1979) 388–390.

J. Mack
[1] On a class of countably paracompact spaces, Proc. Amer. Math. 16 (1965) 467–472,
[2] Directed covers and paracompact spaces, Canad. J. Math. 19 (1967) 649–654.

Z. Mamuzić
[1] Abstract distance and neighborhood spaces, Proc. Prague Symposium (1962) 261–266,
[2] Introduction to general topology, Groningen, 1963.

E. Marczewski (Szpilrajn)
[1] Remarque sur les produits cartésiens d'espaces topologiques, Dokl. SSSR 31 (1941) 525–528.

S. Mardešić
[1] On covering dimension and inverse limits of compact spaces, Ill. J. Math. 4 (1960) 278–291.

S. Mardešić–P. Papic
[1] Sur les espaces dont toute transformation réelle continue et bornée, Glasnik Mat. Fiz. Astr. 10 (1955) 225–232.

H. W. Martin
[1] Weak bases and metrization, Trans. Amer. Math. 222 (1976) 337–344.

W. G. McArthur
[1] Hewitt realcompactifications of products, Canad. J. Math. 22 (1970) 646–656.

L. F. McAuley
[1] A relation between perfect separability completeness and normality in semi-metric spaces, Pacific J. Math. 6 (1956) 315–326,
[2] A note on complete collectionwise normality and paracompactness, Proc. Amer. Math. 9 (1958) 796–799.

B. H. McCandless
[1] On order paracompact spaces, Canad. J. Math. 21 (1969) 400–405.

R. A. McCoy
[1] K-spaces function spaces, Int. J. Math. Sci. 3 (1980) 701–711.

K. Menger
[1] Statistical metrics, Proc. Nat. Acad. Sci. USA 25 (1942) 535–537.

E. A. Michael
[1] A note on paracompact spaces, Proc. Amer. Math. 4 (1953) 831–838,
[2] Some extension theorems for continuous functions, Pacific J. Math. 3 (1953) 784–806,
[3] Continuous selections I, II, III, Ann. Math. 63 (1956) 375–390, 64 (1956) 562–580, 65 (1956) 375–390,
[4] Another note on paracompact spaces, Proc. Amer. Math. 8 (1957) 822–828,
[5] Yet another note on paracompact spaces, Proc. Amer. Math. 10 (1959) 309–314,
[6] The product of a normal space and a metric space need not be normal, Bull. Amer. Math. 69 (1963) 375–376,
[7] $\aleph_0$-spaces, J. of Math. and Mechanics 15 (1966) 983–1002,
[8] Bi-quotient maps and cartesian products of quotient maps, Ann. Inst. Fourier 18, Fasc. 2 (1968) 287–302,
[9] A quintriple quotient quest, Gen. Topol. Appl. 2 (1972) 91–138,
[10] Topologies on spaces of subsets, Trans. Amer. Math. 71 (1951) 152–182,
[11] On Nagami's Σ-spaces and some related matters, Proc. Wash. State Univ. Topol. Conf., 1970, 13–17.

J. v. Mill
[1] Supercompactness and Wallman spaces, Thesis, Math. Cent. Amsterdam, 1977,
[2] Superextensions of metrizable continua are Hilbert cubes, Fund. Math. 107 (1980) 201–224.

J. v. Mill–H. Vermeer
[1] Wallman compactifications and the continuum hypothesis, Vrije Univ. Rapport No. 88 (1978).

C. F. Mills
[1] A simpler proof that compact metric spaces are supercompact, Proc. Amer. Math. 73 (1979) 388–390,
[2] Compact topological groups are supercompact, to appear in Fund. Math.

C. F. Mills–J. v. Mill
[1] A non supercompact continuous image of a supercompact space, Houston J. 5 (1979) 241–247.

A. Miščenko
[1] Spaces with a point-countable base, Soviet Math. 3 (1962) 855–858.

T. Mizokami
[1] On the closed images of paracomplexes, to appear,
[2] On free MD-spaces, to appear.

D. Montgomery
[1] Non-separable metric spaces, Fund. Math. 25 (1935) 527–533.

W. Moran
[1] The additivity of measures on completely regular spaces, J. London Math. Soc. 43 (1968) 633–639,
[2] Measures and mappings on topological spaces, Proc. London Math. Soc. 19 (1969) 493–508,
[3] Measures on metacompact spaces, Proc. London Math. Soc. 20 (1970) 507–524.

K. Morita
[1] Star-finite coverings and the star-finite property, Math. Japonica 1 (1948) 60–68,
[2] On bicompactifications of semicompact spaces, Sci. Rep. Tokyo Bunrika Diagaku, Sec. A, 4 (1952) 200–207,
[3] A condition for metrizability of topological spaces and for n-dimensionality, Sci. Rep. Tokyo Kyoiku Daigaku, Sec. A, 5 (1955) 33–36,
[4] Paracompactness and product spaces, Fund. Math. 50 (1961) 223–236,
[5] Products of normal spaces with metric spaces, Math. Ann. 154 (1964) 365–382,
[6] Topological completions and M-spaces, Sci. Rep. Tokyo Kyoiku Daigaku, Sec. A, 10 (1970) 271–288,
[7] A survey of the theory of M-spaces, Gen. Topol. Appl. 1 (1971) 47–55,
[8] Some properties of M-spaces, Proc. Japan Acad. 43 (1967) 865–872.

K. Morita–S. Hanai
[1] Closed mappings and metric spaces, Proc. Japan Acad. 32 (1956) 10–14.

K. Morita–T. Rishel
[1] Results related to closed images of M-spaces I, II, Proc. Japan Acad. Supplements to 47 (1971) 1004–1007, 1008–1011.

B. Morrel–J. Nagata
[1] Statistical metric spaces as related to topological spaces, Gen. Topol. Appl. 9 (1978) 233–237.

S. Mrówka
[1] Some properties of Q-spaces, Bull. Acad. Polon. Math. Ser. 4 (1954) 947–950,
[2] Compactness and product spaces, Colloq. Math. 7 (1959) 19–22,
[3] On the potency of compact spaces and the 1st axiom of countability, Bull. Acad. Polon. Math. Ser. 6 (1958) 7–9,
[4] Recent results on E-compact spaces and structures of continuous functions, Proc. Univ. Oklahoma Topol. Conf. (1972) 168–221,
[5] Recent results on E-compact spaces, TOPO 72 – General Topology and its Applications (Springer, Berlin, 1974) 298–301,
[6] An elementary proof of Katětov's theorem concerning Q-spaces, Michigan Math. J. 11 (1964) 61–63.

A. Mysior
[1] A union of realcompact spaces, Bull. Acad. Polon. Math. Ser. 29 (1981) 169–172,
[2] A regular space which is not completely regular, Proc. Amer. Math. 81 (1981) 652–653.

A. D. Myškis
[1] On the equivalence of certain methods of definitions of boundary, Mat. Sbornik, N.S.
    26 (1950) 228–236 (R), Amer. Math. Soc. Tr. 51 (1951).

K. Nagami
[1] A note on Hausdorff spaces with the star-finite property II, Proc. Japan Acad. 37 (1961)
    189–192,
[2] Σ-spaces, Fund. Math. 65 (1969) 169–192,
[3] Minimal class generated by open compact and perfect mappings, Fund. Math. 78 (1973)
    227–264.

J. Nagata
[1] On the uniform topology of bicompactification, J. Inst. Polytech. Osaka City Univ. 1
    (1950) 28–39,
[2] On a necessary and sufficient condition of metrizability, J. Inst. Polytech. Osaka City
    Univ. 1 (1950) 93–100,
[3] A contribution to the theory of metrization, J. Inst. Polytech. Osaka City Univ. 8 (1957)
    185–192,
[4] Note on dimension theory for metric spaces, Fund. Math. 45 (1958) 143–181,
[5] Modern dimension theory, revised and extended edition, Berlin, 1983,
[6] Mappings and M-spaces, Proc. Japan Acad. 45 (1969) 140–144,
[7] A survey of the theory of generalized metric spaces, Proc. 3rd Prague Symposium, 1971,
    321–331,
[8] Some theorems on generalized metric spaces, Theory of Sets and Topology (A collection
    of papers in honour of F. Hausdorff, 1972) 377–390,
[9] On G_δ-sets in the product of a metric space and a compact space I, II, Proc. Japan Acad.
    49 (1973) 179–186,
[10] On generalized metric spaces and q-closed mappings, Symposia Mathematicae 16
    (Proc. Roma Topology Conference in 1973) (1975) 55–66,
[11] A note on M-spaces and topologically complete spaces, Proc. Japan Acad. 45 (1969)
    541–543,
[12] A note on Filippov's theorem, Proc. Japan Acad. 45 (1969) 30–33,
[13] On rings on continuous functions, Proc. 4th Prague Symposium, 1976, Part A,
    136–153.

G. Nöbeling–H. Bauer
[1] Über die Erweiterung topologischer Räume, Math. Ann. 130 (1955) 20–45.

J. Novák
[1] Regular space, on which every continuous function is constant, Časopis Pěst Mat. Fys.
    73(1948) 58–68 (in Czech, with an English summary),
[2] On the certain product of two compact spaces, Fund. Math. 40 (1953) 106–112.

P. Nyikos
[1] Prabir Roy's space Δ is not N-compact, Gen. Topol. Appl. 3 (1973) 197–210,
[2] A provisional solution to normal Moore space problem, Proc. Amer. Math. 75 (1980)
    429–435.

P. Nyikos–H.-C. Reichel
[1] Some results on cardinal functions in metrization theory, Glas. Mat. Ser. III 15 (35)
    183–202.

H. Ohta
[1] Local compactness and Hewitt realcompactifications of products, Proc. Amer. Math. 69 (1978) 339–343,
[2] Local compactness and Hewitt realcompactifications II, Topol. Appl. 13 (1982) 155–165.

A. Okuyama
[1] On multi-valued monotone closed mappings, Proc. Japan Acad. 36 (1960) 106–110,
[2] On metrizability of M-spaces, Proc. Japan Acad. 40 (1964) 176–179,
[3] Some generalizations of metric spaces, their metrization theorems and product spaces, Sci. Rep. Tokyo Kyoiku Daigaku, Sec. A, 9 (1967) 236–254,
[4] A survey of the theory of σ-spaces, Gen. Topol. Appl. 1 (1971) 57–63,
[5] σ-spaces and closed mappings I, Proc. Japan Acad. 44 (1968) 472–477.

B. Pasynkov
[1] ω-mappings and inverse spectra, Soviet Math. 4 (1963) 706–709,
[2] On universal bicompacta of given weight and dimension, Soviet Math. 5 (1964) 245–246,
[3] On the spectral decomposition of topological spaces, Math. Sb. 66 (1965) 35–79.

E. Pol–R. Pol
[1] An open perfect mapping of a hereditarily disconnected space onto a connected space, Fund. Math. 86 (1974) 271–278.

R. Pol
[1] Short proofs of two theorems on cardinality of topological spaces, Bull. Acad. Polon. Math. Ser. 22 (1974) 1245–1249.

E. S. Pondiczery
[1] Power problems in abstract spaces, Duke Math. J. 11 (1944) 835–837.

V. I. Ponomarev
[1] Axioms of countability and continuous mappings, Bull. Acad. Polon. Math. Ser. 8 (1960) 127–134 (R),
[2] Normal spaces as images of zero dimensional spaces, Soviet Math. 1 (1960) 774–777,
[3] Properties of topological spaces preserved under many-valued continuous mappings, Mat. Sb. 51 (1960) 515–536 (R),
[4] On paracompact and finally compact spaces, Soviet Math. 2 (1961) 1510–1512,
[5] The absolute of a topological space, Soviet Math. 4 (1963) 299–302,
[6] Spaces co-absolute with metric spaces, Uspehi Mat. Nauk 21, No. 4 (1966) 101–132,
[7] Open mappings of normal spaces, Dokl. SSSR 126 (1959) 716–718 (R).

L. S. Pontrjagin
[1] Topological groups, Princeton, 1939.

V. V. Projzvolov
[1] Decomposition maps of locally connected spaces, Soviet Math. 5 (1964) 1540–1542.

T. Przymusiński
[1] Topological properties of product spaces and the notion of n-cardinality, Proc. Lousiana State Univ. Topol. Conf., 1977, 1–11.

T. Przymusiński–F. Tall
[1] The undecidability of the existence of a non-separable normal Moore space satisfying the countable chain condition, Fund. Math. 83 (1974) 291–297.

S. Purisch
[1] The orderability and suborderability of metrizable spaces, Trans. Amer. Math. 226 (1977) 59–76.

J. Rainwater
[1] Spaces whose finest uniformity is metric, Pacific J. Math. 9 (1959) 567–570.

H.-C. Reichel
[1] Some results on distance functions, Proc. 4th Prague Symposium, 1976, 371–380,
[2] Basic properties of topologies compatible with (not necessarily symmetric) distance functions, Gen. Topol. Appl. 8 (1978) 283–289,
[3] Toward a unified theory of (m-) metric and (m-) semimetric spaces, to appear.

H.-C. Reichel–W. Ruppert
[1] Über Metriserbarkeit durch Distanzfunktionen mit Werten in angeordneten Halbgruppen, Monatshefte für Math. 83 (1977) 223–251.

H. L. Royden
[1] Real analysis, London, 1968.

M. E. Rudin
[1] A new proof that metric spaces are paracompact, Proc. Amer. Math. 20 (1969) 603,
[2] Souslin's conjecture, Amer. Math. Monthly 76 (1969) 1113–1119,
[3] A normal space X for which $X \times I$ is not normal, Bull. Amer. Math. 77 (1971) 246,
[4] The box product of countably many compact metric spaces, Gen. Topol. Appl. 2 (1972) 293–298,
[5] The normality of products with one compact factor, Gen. Topol. Appl. 5 (1975) 45–59,
[6] Interval topology in subsets of totally orderable spaces, Trans. Amer. Math. 118 (1965) 376–389,
[7] The metrizability of normal Moore spaces, Studies in Topology (1975) 507–516.

M. E. Rudin–M. Starbird
[1] Products with a metric factor, Gen. Topol. Appl. 5 (1975) 235–248.

J. Schmidt
[1] Beiträge zum Filtertheorie I, II, Math. Nachr. 7 (1952) 359–378, 10 (1953) 197–231,
[2] Eine Studie zum Begriff der Teilfolge, Deutsche Mathematiker Vereiningung 63 (1960) 28–50.

B. Schweizer–A. Sklar
[1] Statistical metric spaces, Pacific J. Math. 10 (1960) 313–334.

B. Schweizer–A. Sklar–E. Thorp
[1] The metrization of statistical metric spaces, Pacific J. Math. 10 (1960) 673–675.

N. A. Shanin
[1] On special extensions of topological spaces, Dokl. SSSR 38 (1943) 6–9,
[2] On separation in topological spaces, Dokl. SSSR 38 (1943) 110–113,
[3] On the theory of bicompact extensions of topological spaces, Dokl. SSSR 38 (1943) 154–156,
[4] On the product of topological spaces, Trudy Mat. Inst. Steklov 24 (1948) 3–111.

T. Shiraki
[1] M-spaces, their generalizations and metrization theorems, Sci. Rep. Tokyo Kyoiku Daigaku Sec. A, 11 (1971) 57–67.

T. Shirota
[1] A class of topological spaces, Osaka Math. J. 4 (1952) 23–40.

W. Sierpiński
[1] Un théoreme sur les continus, Tohoku Math. J. 13 (1918) 300–303,
[2] Sur les ensembles connexes et non connexes, Fund. Math. 2 (1921) 1–95.

G. Silov
[1] Ideals and subrings of the ring of continuous functions, Dokl. SSSR 22 (1939) 7–10.

F. E. Siwiec
[1] A study of $\mathscr{A}$-spaces and mappings, including M-spaces, σ-spaces, quotient mappings and $\mathscr{A}$-covering mappings, Thesis, Pittsburgh, 1972.

F. E. Siwiec–J. Nagata
[1] A note on nets and metrization, Proc. Japan Acad. 44 (1968) 623–627.

E. G. Sklyarenko
[1] Bicompact extension of semi bicompact spaces, Dokl. SSSR 120 (1958) 1200–1203 (R),
[2] On the imbedding of normal spaces in bicompact spaces of the same weight and of the same dimension, Dokl. SSSR 123 (1958) 36–39 (R).

F. G. Slaughter, Jr.
[1] The closed images of a metrizable space are M_1, Proc. Amer. Math. 37 (1973) 309–314.

F. G. Slaughter–J. M. Atkins
[1] Pull-back theorems for closed mappings, to appear in Trans. Amer. Math.

J. van der Slot.
[1] A survey of realcompactness, Theory of Sets and Topology (A collection of papers in honour of F. Hausdorff, 1972) 473–494.

Yu. M. Smirnov
[1] On normally disposed sets of normal spaces, Mat. Sb. N.S. 27 (1951) 173–176 (R),
[2] A necessary and sufficient condition for metrizability of topological spaces, Dokl. SSSR 77 (1951) 197–200 (R),
[3] On proximity spaces in the sense of V. A. Efremovič, Dokl. SSSR 84 (1952) 895–898 (R),
[4] On the completeness of proximity spaces, Dokl. SSSR 88 (1953) 761–764 (R),
[5] On the completeness of proximity spaces, Trudy Moskov. Mat. Obsc. 3 (1954) 271–306 (R),
[6] On strongly paracompact spaces, Izv. Akad. Nauk-SSSR, Math. Ser. 20 (1956) 253–274 (R),
[7] On the metrizability of bicompacta decomposable into a sum of sets with countable basis, Fund. Math. 43 (1956) 387–393 (R),
[8] Some topological aspects of the theory of topological transformation groups, Proc. 4th Prague Symposium, 1976, Part A (1977) 196–204.

R. H. Sorgenfrey
[1] On the topological product of paracompact spaces, Bull. Amer. Math. 53 (1947) 631–632.

L. A. Steen
[1] A direct proof that a linearly ordered space is hereditarily collectionwise normal, Proc. Amer. Math. 24 (1970) 727–728.

A. K. Steiner–E. F. Steiner
[1] Products of compact metric spaces are regular Wallman, Indag. Math. 30 (1968) 428–430.

F. W. Stevenson–W. J. Thron
[1] Results on ω_μ-metric spaces, Fund. Math. 65 (1969) 317–324.

R. A. Stoltenberg
[1] A note on stratifiable spaces, Proc. Amer. Math. 23 (1969) 294–297.

514   BIBLIOGRAPHY

A. H. Stone
[1] Paracompactness and product spaces, Bull. Amer. Math. 54 (1948) 977–982,
[2] Metrizability of decomposition spaces, Proc. Amer. Math. 7 (1956) 690–700,
[3] Metrizability of unions of spaces, Proc. Amer. Math. 10 (1959) 361–366,
[4] A note on paracompactness and normality of mapping spaces, Proc. Amer. Math. 14 (1963) 81–83,
[5] Absolute F_σ-spaces, Proc. Amer. Math. 13 (1962) 495–499,
[6] Non-separable Borel sets, Rozprawy Math. 28 (1962),
[7] On σ-discreteness and Borel isomorphism, Amer. J. Math. 85 (1963) 655–666.

M. H. Stone
[1] Applications of the theory of Boolean rings to general topology, Trans. Amer. Math. 41 (1937) 375–481.

M. Strok–A. Szymański
[1] Compact metric spaces have binary bases, Fund. Math. 89 (1975) 81–89.

J. Suzuki
[1] Paracompactness of product spaces, Proc. Japan Acad. 45 (1969) 457–460.

F. D. Tall
[1] Some set-theoretic consistency results, Proc. 3rd Prague Symposium, 1971, 419–425,
[2] Set-theoretic consistency results and topological theorems concerning the Moore space conjecture and related problems, Thesis Univ. of Wisc. 1968, Dissert. Math. 148 (1977),
[3] The normal Moore space problem, Math. Cent. Tracts 116 (1979) 243–261.

H. Tamano
[1] On paracompactness, Pacific J. Math. 10 (1960) 1043–1047,
[2] On compactification, J. Math. Kyoto Univ. 1 (1962) 161–193,
[3] On some characterizations of paracompactness, Proc. Topol. Conf. Arizona State Univ., 1967, 277–285,
[4] A characterization of paracompactness, Fund. Math. 72 (1971) 189–201,
[5] On rings of realvalued continuous functions, Proc. Japan Acad. 34 (1958) 361–366.

H. Tamano–J. E. Vaughan
[1] Paracompactness and elastic spaces, Proc. Amer. Math. 28 (1971) 299–301.

T. Tanaka
[1] On the family of connected subsets and the topology of spaces, J. Math. Soc. Japan 7 (1955) 389–393.

Y. Tanaka
[1] Metrizability of certain quotient spaces, to appear in Fund. Math.

Y. Tanaka–Zhou Hao-xuan
[1] Spaces dominated by metric subsets, to appear.

R. Telgársky
[1] C-scattered and paracompact spaces, Fund. Math. 73 (1971) 59–74,
[2] Covering properties and product spaces, Proc. 3rd Prague Symposium, 1971, 437–439.

H. Terasaka
[1] On the cartesian product of compact spaces, Osaka Math. J. 4 (1952) 11–15.

J. Terasawa
[1] On the zero-dimensionality of some non-normal product spaces, Sci. Rep. Tokyo Kyoiku Daigaku Sec. A, 11 (1972) 95–102.

E. Thorp
[1] Generalized topologies for statistical metric spaces, Fund. Math. 51 (1962) 9–21.

H. Tong
[1] Some characterization of normal and perfectly normal spaces, Duke Math. J. 19 (1952) 289–292.

D. R. Traylor
[1] A note on metrization of Moore spaces, Proc. Amer. Math. 14 (1963) 804–805.

V. Trnkova
[1] Unions of strongly paracompact spaces, Soviet Math. 3 (1962) 1248–1250.

J. W. Tukey
[1] Convergence and uniformity in topology, Ann. of Math. Studies, 1940.

A. Tychonoff
[1] Über die topologische Erweiterung von Räumen, Math. Ann. 102 (1929) 544–561.

V. M. Ul'janov
[1] Bicompact extensions with first axiom of countability, Vestnik Moscow Univ. Math. 28 (1973) No. 2, 13–18,
[2] Solution of a basic problem on compactifications of Wallman type, Soviet Math. 18 (1977) 567–571.

V. S. Varadarajan
[1] Measures on topological spaces, Amer. Math. Soc. Tr. (2) 48 (1965) 161–228.

J. E. Vaughan
[1] Some recent results in the theory of $[a, b]$-compactness, TOPO 72, Proc. 2nd Pittsburgh Conf. 1972 (Springer Lecture Notes 278) (1974) 534–550.

A. Verbeek
[1] Superextensions of topological spaces, Math. Cent. Tracts 41, Amsterdam, 1972.

H. de Vries
[1] Compact spaces and compactifications, Thesis, Amsterdam, 1962.

J. de Vries
[1] Glicksberg's theorem for G-spaces, Proc. 5th Prague Symposium 1981 (1982) 663–673.

F. J. Wagener
[1] Normal base compactifications, Indag. Math. 26 (1964) 78–83.

K. Wagner
[1] Zur Metrizierbarkeit topologischer Räume, Abhandl. Math. Hamburg 19 (1954) 14–22.

A. D. Wallace
[1] Extensional invariance, Trans. Amer. Math. 70 (1951) 97–102.

H. Wallman
[1] Lattices and topological spaces, Ann. Math. 39 (1938) 112–126.

A. Weil
[1] Sur les espaces à structure uniforme et sur la topologie générale, Paris, 1937.

J. H. C. Whitehead
[1] Combinatorial topology I, Bull. Amer. Math. 55 (1949) 213–245.

G. T. Whyburn
[1] Topological analysis, Princeton, 1958,

[2] On quasi-compact mappings, Duke Math. J. 19 (1952) 445–496.

H. H. Wicke
[1] On the Hausdorff open continuous images of Hausdorff paracompact p-spaces, Proc. Amer. Math. 22 (1969) 136–140,
[2] The regular open continuous images of complete metric spaces, Pacific J. Math. 23 (1967) 621–625.

H. H. Wicke–J. M. Worrell
[1] Topological completeness of first countable Hausdorff spaces I, Fund. Math. 75 (1972) 209–222.

S. Willard
[1] Metric spaces all of whose decompositions are metric, Proc. Amer. Math. 21 (1965) 126–128.

J. M. Worrell
[1] Upper semi-continuous decompositions of developable spaces, Proc. Amer. Math. 16 (1965) 485–490,
[2] The closed continuous images of metacompact topological spaces, Portugal Math. 25 (1966) 175–177.

J. M. Worrell–H. H. Wicke
[1] Characterizations of developable spaces, Canad. J. Math. 17 (1965) 820–830.

J. M. v. Wouwe
[1] GO-spaces and generalizations of metrizability, Thesis, Amsterdam, 1978.

O. Wyler
[1] Top categories and categorical topology, Gen. Topol. Appl. 1 (1971) 17–28.

Y. Yasui
[1] Unions of strongly paracompact spaces, Proc. Japan Acad. 43 (1967) 263–268,
[2] On ω_μ-metrizable spaces, Math. Japonica 20 (1975) 159–180,
[3] On the characterization of the $\mathcal{B}$-property by the normality of product spaces, Topol. Appl. 15 (1983) 323–326.

A. Zarelua
[1] A universal bicompactum of given weight and dimension, Soviet Math. 5 (1964) 214–218.

P. Zenor
[1] A class of countably paracompact spaces, Proc. Amer. Math. 42(1970) 258–262,
[2] A metrization theory, Colloq. Math. 27 (1973) 241–243,
[3] On continuously perfectly normal spaces, Proc. Univ. Oklahoma Topology Conf. 1972 (1972) 334–336.

INDEX